6th International Conference on Mechatronics 2017 (ICOM'17)

IOP Conference Series: Materials Science and Engineering
Volume 260

Kuala Lumpur, Malaysia
8 - 9 August 2017

Editors:

Rini Akmeliawati
Muhammad Mahbubur
Syamsul Bahrin Abdul Hamid

ISBN: 978-1-5108-5200-6
ISSN: 1757-8981

Printed from e-media with permission by:

Curran Associates, Inc.
57 Morehouse Lane
Red Hook, NY 12571

Some format issues inherent in the e-media version may also appear in this print version.

Printed by Curran Associates, Inc. (2017)

For permission requests, please contact the Institute of Physics at the address below.

Institute of Physics
Dirac House, Temple Back
Bristol BS1 6BE UK

Phone: 44 1 17 929 7481
Fax: 44 1 17 920 0979

techtracking@iop.org

Additional copies of this publication are available from:

Curran Associates, Inc.
57 Morehouse Lane
Red Hook, NY 12571 USA
Phone: 845-758-0400
Fax: 845-758-2633
Email: curran@proceedings.com
Web: www.proceedings.com

Table of contents

Volume 260

6th International Conference on Mechatronics - ICOM'17
8–9 August 2017, Kuala Lumpur, Malaysia

Accepted papers received: 19 October 2017
Published online: 7 November 2017

Preface

6th International Conference on Mechatronics - ICOM'17

Peer review statement

Papers

Effects of Coil Diameter in Thickness Measurement Using Pulsed Eddy Current Non-destructive Testing

Khairun Nisa Azaman, Ali Sophian and Faris Nafiah.....1

Realization and Addressing Analysis In Blockchain Bitcoin

Raja Sakti Arief Daulay, Surya Michrandi Nasution and Marisa W. Paryasto.....7

Implementation and Analysis of the use of the Blockchain Transactions on the Workings of the Bitcoin

Muhammad Reza Rizky Fauzi, Surya Michrandi Nasution and Marisa W. Paryasto.....16

Classification of the Correct Quranic Letters Pronunciation of Male and Female Reciters

Safiah Khairuddin, Salmiah Ahmad, Abdul Halim Embong, Nik Nur Wahidah Nik Hashim, Tareq M.K. Altamas, Syarifah Nuratikah Syd Badaruddin and Surul Shahbudin Hassan.....25

Flip-Chip Bonding Fabrication Technique

Tengku Muhammad Afif bin Tengku Azmi and Nadzril bin Sulaiman.....37

Mechatronics technology in predictive maintenance method

Nurul Afiqah A. Majid and Asan G.A. Muthalif.....45

Human location estimation using thermopile array sensor

S Parnin and M M Rahman.....51

Speech-based Class Attendance

Umar Faizel Amri, Nik Nur Wahidah Nik Hashim and Noor Hazrin Hany Mohamad Hanif.....61

Traffic intensity monitoring using multiple object detection with traffic surveillance cameras

H G Muhammad Hamdan and O O Khalifah.....67

Control of cooperative manipulators in holding deformable objects

A A Alkathiri and N Z Azlan.....75

Dragon Stream Cipher for Secure Blackbox Cockpit Voice Recorder

Fadira Akmal, Surya Michrandi Nasution and Fairuz Azmi.....82

Enhancement web proxy cache performance using Wrapper Feature Selection methods with NB and J48

Dua'a Mahmoud Al-Qudah, Rashidah Funke Olanrewaju and Amelia Wong Azman.....89

UAV State Estimation Modeling Techniques in AHRS

Shikin Razali and Amzari Zhahir.....101

Modelling and Simulation Analysis of Rolling Motion of Spherical Robot

N N Kamis, A H Embong and S Ahmad.....107

Development of Fuzzy Logic Controller for Quanser Bench-Top Helicopter

M. H. Jafri, H. Mansor and T.S. Gunawan.....115

Design and analysis of planar spiral resonator bandstop filter for microwave frequency

S. M. A. Motakabber and Muhammad Shaifudin Suharsono.....125

Real-time Human Activity Recognition

N. Albukhary and Y.M. Mustafah.....131

Optimized Vibration Chamber for Landslide Sensory and Alarm System

Eliza Sabira Binti Ismail, Mohamed Hadi Habaebi, Jamal I. Daoud and Md Rafiqul Islam.....136

Miniaturized Water Flow and Level Monitoring System for Flood Disaster Early Warning

Salami Ifedapo Abdullahi, Mohamed Hadi Habaebi, Teddy Surya Gunawan and MD Rafiqul Islam.....146

The Effect of Electrical Stimulation in Improving Muscle Tone (Clinical)

M F Azman and A W Azman.....156

An Experimental Design of Bypass Magneto-Rheological (MR) damper

MM Rashid, Mohammad Abdul Aziz and Md. Raisuddin Khan.....169

Prediction of Beck Depression Inventory (BDI-II) Score Using Acoustic Measurements in a Sample of Iium Engineering Students

Muhamad Fikri Zanil, Nik Nur Wahidah Nik Hashim and Huda Azam.....178

Prediction of dry ice mass for firefighting robot actuation

M T Ajala, Md R Khan, A A Shafie, MJE Salami and M I Mohamad Nor.....185

Modified Ashworth Scale (MAS) Model based on Clinical Data Measurement towards Quantitative Evaluation of Upper Limb Spasticity

A Ahmad Puzi, S N Sidek, H Mat Rosly, N Daud and H Md Yusof.....193

Network-Based Real-time Integrated Fire Detection and Alarm (FDA) System with Building Automation

F Anwar, R I Boby, M M Rashid, M M Alam and Z Shaikh.....200

Trajectory tracking in quadrotor platform by using PD controller and LQR control approach

Maidul Islam, Mohamed Okasha and Moumen Mohammad Idres.....215

Assistive-as-Needed Strategy for Upper-Limb Robotic Systems: *An Initial Survey*

I. M. Khairuddin, S. N. Sidek, H. Md Yusof, K. Baarath and A. P. P. A. Majeed.....223

Nanosatellites constellation as an IoT communication platform for near equatorial countries

A Narayanasamy, Y A Ahmad and M Othman.....229

Wireless Battery Management System of Electric Transport

Ataur Rahman, Mizanur Rahman and Mahbubur Rashid.....242

Time response for sensor sensed to actuator response for mobile robotic system

N S Amir and A A Shafie.....253

Improvement of Dimensional Accuracy of 3-D Printed Parts using an Additive/Subtractive Based Hybrid Prototyping Approach

A N M Amanullah Tomal, Tanveer Saleh and Md. Raisuddin Khan.....261

The effect of electrohydrodynamic force on the lift coefficient of a NACA 0015 airfoil

Y Yusof, A Hossain, A H Abdullah, Rizal M E Nasir, A Hamid, N Muthmainnah.....269

Network Function Virtualization (NFV) based architecture to address connectivity, interoperability and manageability challenges in Internet of Things (IoT)

Shariq Haseeb, Aisha Hassan A. Hashim, Othman O. Khalifa and Ahmad Faris Ismail.....279

Determining the Time of Flight and Speed of Sound on Different types of Edible Oil

N. A. Azman and S. B. Abd Hamid.....290

Closed-loop model identification of cooperative manipulators holding deformable objects

A A Alkathiri, R Akmeliawati and N Z Azlan.....296

Effect of Temperature on Ultrasonic Signal Propagation for Extra Virgin Olive Oil Adulteration

N.A. Alias, S.B. Abdul Hamid and A. Sophian.....302

Design of Dual Band Microstrip Patch Antenna using Metamaterial

Md Rafiqul Islam, A.A Alsaleh Adel, Aminah W. N. Mimi, M. Sarah Yasmin and Farihah A.M. Norun.....309

Optimum concentric circular array antenna with high gain and side lobe reduction at 5.8 GHz

Mohammed Zaid, Md Rafiqul Islam, Mohamed H Habaebi, AHM Zahirul Alam and Khaizuran Abdullah.....316

Applying Mechatronics to Improve the Safety of Children in Vehicles - What Can Be Done?

Khairul Hazziq Zufar and Ahmad Jazlan.....324

Object Detection Technique for Small Unmanned Aerial Vehicle

M. Faiz Bin Ramli, Ari Legowo and Syariful Syafiq Shamsudin.....334

Performance analysis of rain attenuation on earth-to-satellite microwave links design in Libya

Md Rafiqul Islam, Asma Ali Hussein Budalal, Mohamed H Habaebi, Khairayu Badron and Ahmad Fadzil Ismail.....342

Speech to Text Translation for Malay Language

Rami Ali Al-khulaidi and Rini Akmeliawati.....348

Rapid lard identification with portable electronic nose

Marsad Latief, Aida Khorsidtalab, Irwan Saputra, Rini Akmeliawati, Anis Nurashikin,
Irwandi Jaswir and Gunawan Witjaksono.....357

IOP Conf. Series: Materials Science and Engineering **260** (2017) 011001 doi:10.1088/1757-899X/260/1/011001

PREFACE

6th International Conference on Mechatronics 2017 (ICOM'17), was held from 8th-9th August 2017 at International Islamic University Malaysia (IIUM) Gombak Campus, Kuala Lumpur, Malaysia. The present volume contains papers from a selection from the invited and oral presentations.

ICOM is the regular international conference series in Malaysia. It was initially held at Kuala-lumpur bi-annual from 2008. ICOM is designed to bring together scientists, engineers researchers from a variety of disciplines engaged in applying platform for exchange of ideas on advances of mechatronics and their applications as well as to foster research network and worldwide collaboration. The scope of this conference ranges from basic research to Advanced instrumentation and control, and Robotics and automation studies. It is recognized as one of the comprehensive international conferences in the field of mechatronics engineering. An increasing number of young researchers present and exchange their innovative ideas on this friendly and professional platform, which has made ICOM a unique meeting in Kuala-Lumpur.

This year ICOM attracted distinguished scholars in the field of mechatronics from all over the world. The major topics covered at the conference and presented in this volume include: Intelligent systems, advanced instrumentation and control, signal and image processing, Machine vision, Human-machine interface, Robotics, Automation, Manufacturing mechatronics, Smart materials and structures, Sensors and actuators, Active vibration control, Computer and information technology, MEMS and NEMS, Biomechatronics and rehabilitation engineering, Mechatronics education, Autonomous systems, Energy and sustainability, Transportation systems, Fault detection and Machine learning. The conference voted for the three best student papers; awards were presented to the participant students whose posters were recognized as excellent and who took part in the oral presentation.

The Conference Secretariat and Local Organizing Committee deserve recognition for planning a smoothly run and productive conference with comprehensive, instructive lectures and innovative work. The faculties and students from International Islamic University Malaysia were dedicated to their work in reception and service during the conference. It is a pleasure to thank all of them for their efficient and

IOP Conf. Series: Materials Science and Engineering **260** (2017) 011001 doi:10.1088/1757-899X/260/1/011001

hard work. We are also grateful for the financial support from Kulliyyah of Engineering, Research Management Centre (RMC) of International Islamic University Malaysia.

Finally, we would like to thank all the authors for their contributions to COM'17 and all the members of the Committees for their cooperation and time spent reviewing submissions.

Editors

Rini Akmeliawati, Muhammad Mahbubur Rashid and Syamsul Bahrin Abdul Hamid

Conference Chairs

Syamsul Bahrin Abdul Hamid

6th International Conference on Mechatronics (ICOM'17)

8th-9th August 2017

Sponsored and Organized by

Kulliyyah of Engineering, International Islamic University Malaysia.

CONFERENCE COMMITTEES

Chairman	Syamsul Bahrin Abdul Hamid
Co-chairman	Tanveer Saleh
Secretary	Md. Mozasser Rahman Nadzril Sulaiman
Technical Committee	Md. Raisuddin Khan Amir Akramin Shafie Momoh -Jimoh Eyiomika Salami Shahrul Na'im Sidek Abd Halim Embong Noor Hazrin Hany Mohamad Hanif
Publication/Programme	Rini Akmeliawati Muhammad Mahbubur Rashid Nik Nur Wahidah Nik Hashim Salmiah Ahmad Siti Fauziah Toha@Tohara

	Asan Gani Abdul Muthalif Yasir Mohd Mustafah
Website and Publicity	Zaw Zaw Htike Ali Sophian Ahmad Imran Ibrahim Ahmad Jazlan Haja Mohideen
International Advisory Board and Keynote Speakers	Tanveer Saleh Iskandar Al-Thani Mahmood Zukifli Zainal Abidin Rini Akmeliawati Amir Akramin Shafie Md. Raisuddin Khan Momoh -Jimoh Eyiomika Salami Nadzril Sulaiman
Finance, Treasurer and Sponsorship	Amir Akramin Shafie Zukifli Zainal Abidin Sharifah Junita Syed Idrus
Souvenir	Norsinnira Zainul Azlan Ahmad Imran Ibrahim
Registration/Payment	Hazlina Md. Yusof Wahju Sediono
Local arrangement and accommodation	Sharifah Junita Syed Idrus

International Advisory Board

Aysegul Ucar (Firat University, Turkey)

Bambang Riyanto Trilaksono (Institut Teknologi Bandung, Indonesia)

Dina S. Laila (Coventry University, UK)

Endra Joelianto (Institut Teknologi Bandung, Indonesia)

Essameddin Badreddin (Universitat Heidelberg, Germany)

Ismaila Tijani (Higher Colleges of Technology, UAE)

Khansa Bdirina (University of Djelfa, Algeria)

Lafifa Jamal (University of Dhaka, Bangladesh)

Mahmoud Shafik (University of Derby, UK)

IOP Conf. Series: Materials Science and Engineering **260** (2017) 011001 doi:10.1088/1757-899X/260/1/011001

Man Zhi Hong (Swinburne University of Technology, Australia)

Mohamed Trabia (University of Nevada, Las Vegas, USA)

Mohammad Ali (Bangladesh University of Engineering and Technology, Bangladesh)

Mohammad Habibur Rahman (University of Wisconsin-Milwaukee, USA)

Mounir Bouhedda (Dr. Yahia Fares University of Médéa, Algeria)

Prashanta Dutta (Washington State University, USA)

Qudrat Khan (COMSATS Institute of Information Technology, Pakistan)

Toshio Fukuda (Nagoya University, Japan)

Winda Astuti (Bina Nusantara University, Indonesia) Baoyong Zheng, Hua Wei Technologies Corporation, Inc (China)

Keynote Speakers

Prof. Manukid Parnichkun (AIT Thailand)

Prof. Dr.sc.techn. Essam Badreddin (University of Heidelberg, Germany)

Prof. Prashanta Dutta (University of South Carolina and Texas A&M University USA)

IOP Conf. Series: Materials Science and Engineering **260** (2017) 011002 doi:10.1088/1757-899X/260/1/011002

Peer review statement

All papers published in this volume of *IOP Conference Series: Materials Science and Engineering* have been peer reviewed through processes administered by the proceedings Editors. Reviews were conducted by expert referees to the professional and scientific standards expected of a proceedings journal published by IOP Publishing.

IOP Conf. Series: Materials Science and Engineering **260** (2017) 012001 doi:10.1088/1757-899X/260/1/012001

Effects of Coil Diameter in Thickness Measurement Using Pulsed Eddy Current Non-destructive Testing

Khairun Nisa Azaman, Ali Sophian, Faris Nafiah

Department of Mechatronics Engineering, Faculty of Engineering, International Islamic University Malaysia, Malaysia

ali_sophian@iium.edu.my

Abstract. Non-destructive testing (NDT) techniques are used in industry to evaluate the properties of a material, component or structure without causing any permanent damage. Among the techniques, pulsed eddy current (PEC) NDT is regarded as a new technique where a broadband pulse excitation is used, as opposed to single frequencies employed in conventional eddy current NDT. In this study, a 2D axisymmetric electromagnetic model of a PEC probe has been developed and it has been used to study the effects of the excitation coil diameter on the performance of PEC probes in sample thickness measurement. A PEC system has also been built to validate the model. Aluminium plates are used as the sample and they can be stacked up to replicate thickness from 1 mm to 10 mm. The results show that there is a very good correlation between the simulation and experimental results, with an average error of less than 10%. The results also suggest that the larger the diameter of the excitation coil, the deeper the penetration and therefore the larger the thickness measurement range. It has also been shown that although the larger diameters have deeper penetration, the smallest diameter has the highest sensitivity if normalization is not used. These conclusions indicate that coil diameter is an important parameter in a PEC probe design for thickness measurement applications.

1. Introduction

Non-destructive testing (NDT) technologies are widely used in various types of industries such as aerospace, manufacturing, automotive, construction, military, oil and gas to characterize material properties, detect and characterize the flaws in structures or materials. NDT applications assure that the tested materials are not damaged during the testing so that their future usage is not affected. Safety, quality and operational costs benefit greatly from the use of NDT.

Pulsed eddy current (PEC) is in general still considered as a new emerging technique in non-destructive testing and evaluation (NDT&E). Being a member of eddy current testing (ECT) family, PEC shares the advantages of ECT, such as reliability and minimal preparation of samples. PEC has been employed in detection and characterization of defects, measurement of stress and measurement of thickness, among others [1]. The advantage of the PEC technique comes with its wide bandwidth of excitation frequencies that are excited simultaneously.

An example of the work in PEC for thickness measurement is [2], where Q235 steel plates with a thickness of up to 30 mm was measured. Shin et al. also investigated thickness measurement for different materials, including copper and titanium [3]. Röntgen Technische Dienst's PEC system is capable of measuring the wall thickness made of low alloy carbon steel of up to 65 mm [4]. Other works done on thickness measurement for wall thickness of pipes include [5], [6], [7], and [8].

IOP Conf. Series: Materials Science and Engineering **260** (2017) 012001 doi:10.1088/1757-899X/260/1/012001

The success of thickness measurement by using PEC depends on the depth of penetration of the induced eddy currents, which is affected by many factors, such as the height of the coil, the number of turns and the lift-off. The understanding on the effects of these parameters is paramount for designing an optimum PEC probe. The depth of penetration is increased when the height of the coil is increased although the magnetic field density is decreased at the surface of the material. The number of turns of the coil is very important in determining the strength of the magnetic field generated by the coil. However, it has been reported that the number of turns does not affect the depth of penetration [9]. Only the strength and sensitivity will be affected by varying the number of turns of the coil. The effect of another parameter, which is the lift-off, on depth of penetration has been studied by Janoušek [10], who found that the larger the lift-off, the higher the depth of penetration but results in lower sensitivity.

The aim of this work is to investigate the relation between PEC's excitation coil diameter and the depth of penetration and the sensitivity in thickness measurement of non-ferromagnetic metallic samples. The study will be carried out through both finite element modelling and experimental tests. In the following sections, the modelling, experimental setup and results will be described. Discussion and conclusions are presented subsequently.

2. Modelling

In this study, a 2D axisymmetric finite element model has been built by using Comsol, as illustrated in figure 1. The model includes an excitation coil and an aluminium plate as the sample that has an electrical conductivity of 26.33 x 10^6 S/m. The number of turns is fixed at 144 turns while the coil current is 3A. The sample's thickness is varied from 1 mm up to 10 mm with an increment of 1 mm. The dimension of the plate is 130 mm x 130 mm. The magnetic flux density is measured at the coordinate (0, 1) mm.

Initially, the simulation was started with 20 mm diameter of coil with sample thickness of 1 mm. Then, the thickness of the sample was increased by 1 mm until it reached 10 mm. Subsequently, the same steps were repeated for coil diameters of 30 mm and 40 mm.

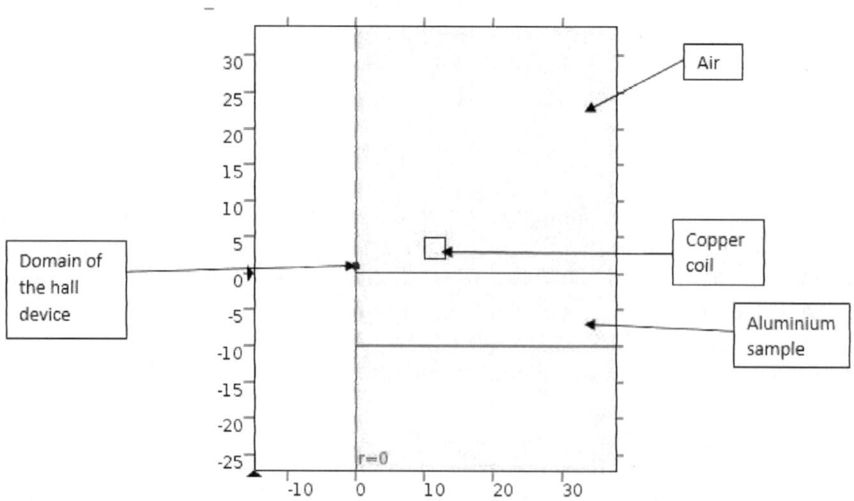

Figure 1. The Geometry of the 2D axisymmetric Model

3. Experimental Setup

A PEC system has been implemented for the experimental setup that will be used to gather data for validating the built model. The steady-state excitation current is 3 A. Probes of different diameters have been built, where each of them contains an excitation coil and a sensing device. The coil bobbin was made by using a 3D printer. The excitation coil windings use copper wire and have 144 turns. A Hall-device (A1324) with a sensitivity of 5mV/G has been used in each probe. Figure 2 shows the illustration of the probe and the sample. For the specimens, aluminium plates have been used whose dimensions are 130 mm x 130 mm x 1 mm. Different sample thicknesses were achieved by stacking the plates up accordingly.

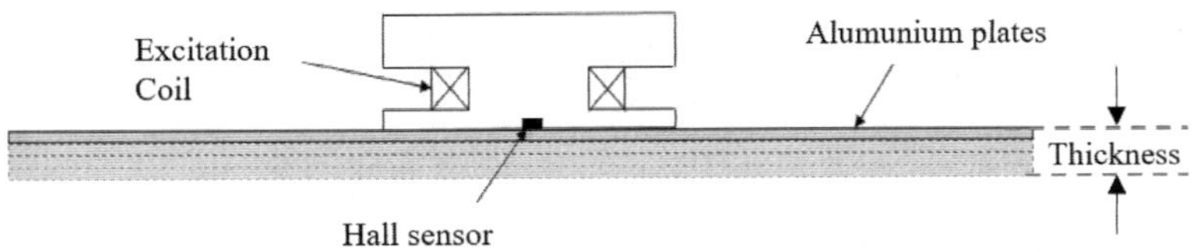

Figure 2. Set-up of the Experiment

For acquisition of the transient magnetic field signals, a data acquisition (DAQ) system was used with a sampling frequency of 50 kHz. The data of the magnetic flux density has been recorded for each variation of thickness by using a LabVIEW-based application. Figure 3 shows the block diagram of the overall system. The DAQ card was also used to generate the excitation signal with 1 Hz frequency and 0.5% duty cycle. The low duty cycle was chosen to minimize the heating effects of the coil and the components within the excitation circuit.

Figure 3. System Block Diagram

For thickness measurement in this study, the parameter to be used is the peak value of the differential PEC signal [1], which is illustrated in Figure 4.

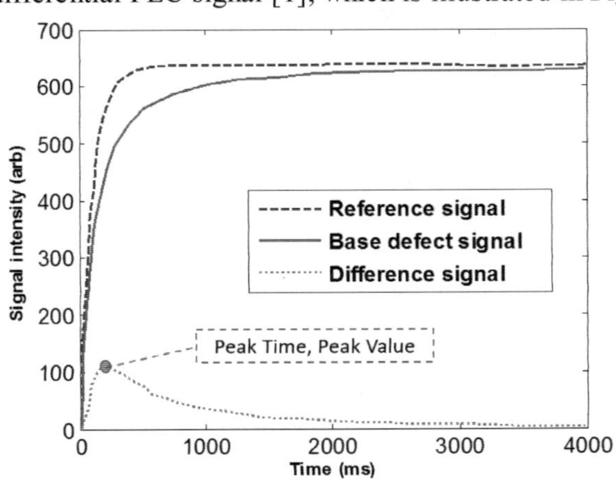

Figure 4. Typical PEC Signals: Base, Reference and Differential

4. Results and Discussion

In order to compare the trend of the peak values for different thicknesses, normalization of the measured magnetic field signals has been used to reduce the effects of the excitation current variations during the tests. The normalized PEC signals were calculated as

$$B_{norm}(t) = \frac{B(t)}{B_{ss}} \quad (1)$$

, where B_{norm} is the normalized magnetic field density, B is the measured magnetic field density and B_{ss} is its steady state value. Signal normalization is commonly used in PEC.

IOP Conf. Series: Materials Science and Engineering **260** (2017) 012001 doi:10.1088/1757-899X/260/1/012001

Figures 5 and 6 show typical normalization results obtained by using both simulation and experiment, which show that both sets of results show similar trends in the transient magnetic field density. The rate of change of the magnetic flux density decreases when the thickness of the plate is increased. However, the rate of change increases when the coil diameter is increased.

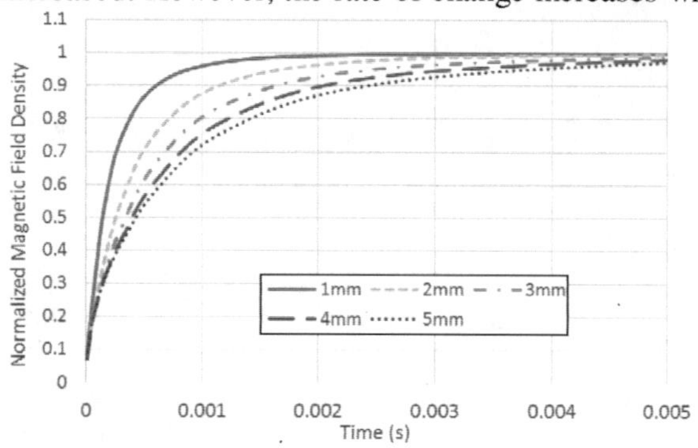

Figure 5. Normalized Signals from the 30-mm-Diameter Probe (Simulation)

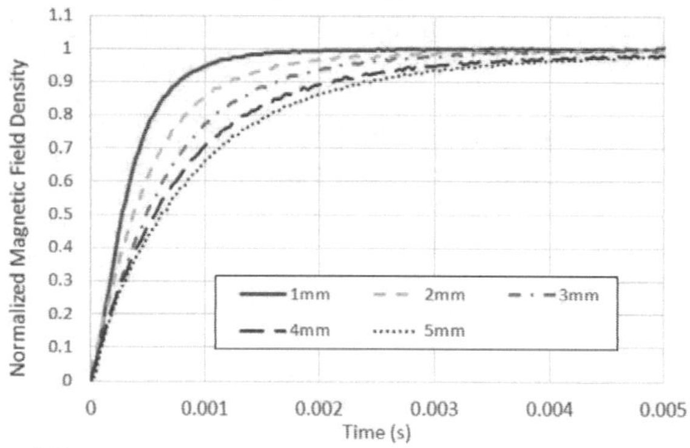

Figure 6. Normalized Signals from the 30-mm-Diameter Probe (Experiment)

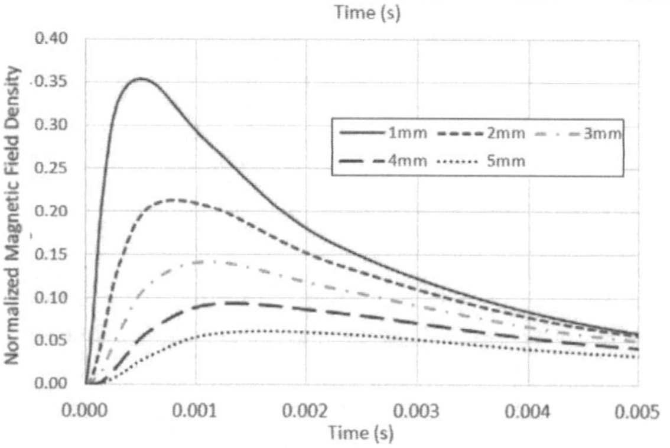

Figure 7. Examples of the Resulting Differential Signals (Simulation)

Figure 7 shows typical differential signals, which were derived by subtracting the normalized reference signal from the normalized response signal. The reference signal used was the signal obtained from the sample thickness of 10 mm. It can be clearly seen that the intensity of the differential signal is higher for lower thicknesses.

4.1. Model Validation

The peak values were then extracted from the differential signal obtained through both experiment and simulation, which are plotted in Figure 8 and tabulated in Table 1, which also highlights the good correlation between the simulation and the experiment with most of the errors are less than 10%. This result indicates that the developed model is reliable to be used in further analysis and prediction of thickness measurement and depth of penetration.

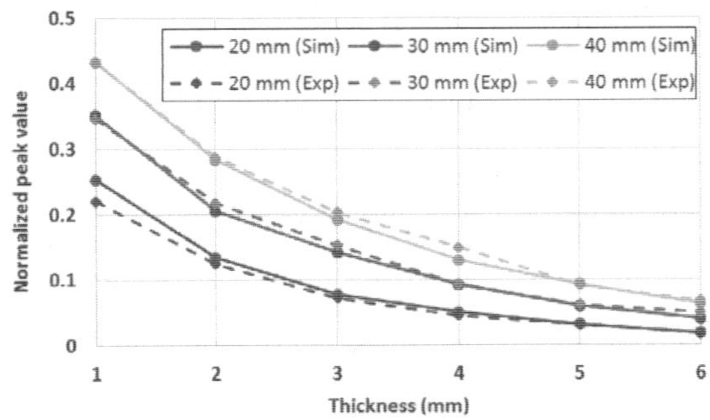

Figure 8. Normalized Peak Values vs. Sample Thickness for Different Coil Diameters (Simulation and Experiment)

Table 1. Peak Values from Both Simulation and Experiment

		Normalized Peak Value (no unit)								
Probe diameter →		20mm			30mm			40mm		
Method →		Simulation	Experiment	Error (%)	Simulation	Experiment	Error (%)	Simulation	Experiment	Error (%)
Sample Thickness (mm)	1	0.254	0.220	-13.0	0.351	0.346	-1.4	0.432	0.432	0.0
	2	0.134	0.124	-7.6	0.206	0.218	5.8	0.283	0.288	1.6
	3	0.078	0.073	-6.7	0.142	0.153	7.9	0.192	0.203	5.8
	4	0.051	0.046	-9.6	0.093	0.093	0.3	0.130	0.149	14.8
	5	0.032	0.031	-3.9	0.060	0.062	4.4	0.093	0.092	-1.7
	6	0.019	0.018	-6.9	0.041	0.050	23.8	0.064	0.068	6.8

4.2. Range of Thickness Measurement

By using the model, the relationship between the normalized peak value and the thickness was derived for each coil diameter, as can be seen in figure 9.

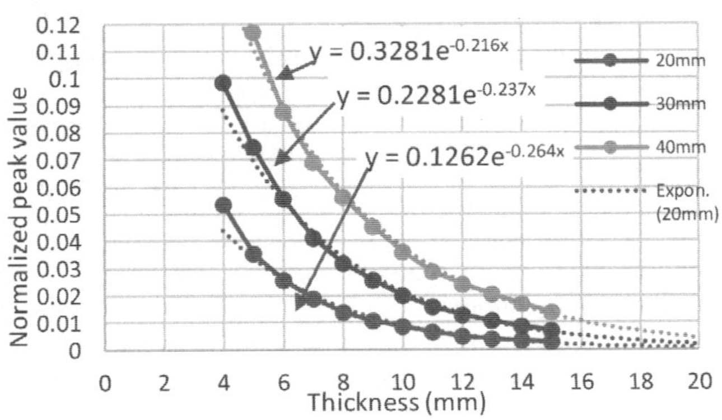

Figure 9. Plots of the Peak Values vs Thicknesses with their Exponential Curve-fittings (Based on Simulation Results)

In order to determine the range of measurement, a signal-to-noise ratio of 3 was assumed. Due to different field strengths of the coils, the minimum normalized for each coil diameter is different and shown in table 2. The table shows that the range of measurement gets higher as the diameter is increased. It shows that if the area of induced eddy currents is larger on the surface of the sample, then the eddy currents penetrate deeper into the sample.

Table 2. Prediction of the Maximum Measurable Thickness

Coil Diameter	20 mm	30 mm	40 mm
Steady State Value (T)	0.0219	0.0155	0.0123
Min Peak Value (V), SNR = 3		0.0106	
Min Peak Field Density (T), SNR = 3		0.00021	
Min Normalized Peak Value, SNR = 3	0.0097	0.0136	0.0171
Predicted Max Thickness (mm)	9.7	11.9	13.6

IOP Conf. Series: Materials Science and Engineering **260** (2017) 012001 doi:10.1088/1757-899X/260/1/012001

4.3. Sensitivity
Although the largest diameter has the largest measurement range, the smallest diameter looks to have the highest sensitivity for thinner sample measurement when normalization is not used. This can be seen from the change in the measured field density when the thickness is changed from 1 to 2 mm, which shows the 20-mm-diameter coil has the highest change, as can be seen in Table 3. This is in contrast to the thickness change from 6 to 7 mm, where the 40-mm-diameter coil has the highest output change. Therefore, it can be concluded that the sensitivity of the probe does not only depend on the diameter, but only on the measured thickness range.

Table 3. Sensitivity in Different Thickness Ranges

		ΔB (T)	
Thickness change →		1 mm - 2mm	6 mm - 7 mm
Coil diameter (mm)	20	0.00276	0.00012
	30	0.00235	0.00022
	40	0.00185	0.00027

5. Conclusion
A few conclusions can be drawn from the results obtained in this work. Firstly, the 2D axisymmetric model has been validated by the experimental results with an average error of less than 10%, indicating that it is a useful and reliable tool for predicting the performance of a PEC probe design in sample thickness measurement. Secondly, the larger the diameter of the excitation coil, the deeper the penetration and therefore the larger the thickness measurement range. Thirdly, although the larger diameters have deeper penetration, the smallest diameter has the highest sensitivity if normalization is not used. Overall, the work shows that coil diameter is an important parameter the optimization of a PEC probe design.

References

[1] Sophian A, Tian G and Fan M 2017 Pulsed Eddy Current Non-destructive Testing and Evaluation: A Review *Chinese J. Mech. Eng.* **30** 500–14

[2] Huang C and Wu X 2015 An improved ferromagnetic material pulsed eddy current testing signal processing method based on numerical cumulative integration *NDT E Int.* **69** 35–9

[3] Shin Y-K, Choi D-M, Kim Y-J and Lee S-S 2009 Signal characteristics of differential-pulsed eddy current sensors in the evaluation of plate thickness *NDT E Int.* **42** 215–21

[4] Robers M A and Scottini R 2002 Pulsed Eddy Currents in Corrosion Detection *NDT.net* **7**

[5] Chen X and Lei Y 2014 Time-domain analytical solutions to pulsed eddy current field excited by a probe coil outside a conducting ferromagnetic pipe *NDT E Int.* **68** 22–7

[6] Xu Z, Wu X, Li J and Kang Y 2012 Assessment of wall thinning in insulated ferromagnetic pipes using the time-to-peak of differential pulsed eddy-current testing signals *NDT E Int.* **51** 24–9

[7] Xie S, Chen Z, Takagi T and Uchimoto T 2015 Quantitative non-destructive evaluation of wall thinning defect in double-layer pipe of nuclear power plants using pulsed ECT method *NDT E Int.* **75** 87–95

[8] Park D G, Angani C S, Kim G D, Kim C G and Cheong Y M 2009 Evaluation of Pulsed Eddy Current Response and Detection of the Thickness Variation in the Stainless Steel *IEEE Trans. Magn.* **45** 3893–6

[9] Majidnia S, Nilavalan R and Rudlin J 2016 A numerical study of the depth of penetration of eddy currents *Insight - Non-Destructive Test. Cond. Monit.* **58** 129–34

[10] Janoušek L 2012 Impact of selected parameters on eddy current attenuation in conductive materials *2012 ELEKTRO* 419–22

Realization and Addressing Analysis In Blockchain Bitcoin

Raja Sakti Arief Daulay, Surya Michrandi Nasution, Marisa W. Paryasto
Faculty of Electrical Engineering, Telkom University, Bandung, Indonesia

rajasaktidaulay@gmail.com, michrandi@telkomuniversity.ac.id,
marisa.paryasto@telkomuniversity.ac.id

Abstract. The implementation research and analyze address blockchain on this bitcoin will have the results that refers to making address bitcoin a safe and boost security of address the bitcoin. The working mechanism of blockchain in making address bitcoin which is already in the blockchain system.

Keywords : Address, Bitcoin, SHA-256, Base58Check, Blockchain, Simulation bitcoin sandbox.

1. Introduction

The development of computers on this day is growing rapidly and also affect the financial system in the world. Example is bitcoin, which is an electronic money made by Satoshi Nakamoto in 2009. But bitcoin also developed a system called blockchain. Blockchain is a publicly accessible financial bookkeeping and blockchain can show all transaction data that has occurred in a bitcoin network.

This journal is implementation of blockchain in the addressing sector of bitcoin. Where the bitcoin address on the blockchain is used to be able to receive payments from others. And we can tell your bitcoin address to others. This bitcoin address consists of different byte versions and your public key hash byte combines them which are then encoded into Base58Check.

In this case the study analyzes how to use of bitcoin addresses in securing the sender and recipient address for transactions on bitcoin.

2. Related work

Reinhard [1], In this paper, the advantages of a bitcoin are defined as a digital signature chain. The privacy or confidentiality of user data is done in a unique way on bitcoin, in example each address in the form of hash value. By default bitcoins also replaces the user address with a new hash value that is different from the previous address. That is, transactions occur by way of digital signature on the hash of the previous transactions and the public key of the receiver of the coin and the receiver further verifies the received digital signature to ensure the ownership of the coins. However the problem with the system is that the coin recipient can not be sure that the sender is not sending the same coin more than once (double-spend). Bitcoin stores all transaction data that ever happened. This then raise the question of how space on a user's computer is enough to hold all the bitcoin transaction data that has occurred. Bitcoin solves this problem by using a data structure called Merkle Tree. The basic logic of using this data structure is longer transaction data can be deleted and stored its hash only.

Danielle Drainville [2], In this research explains bitcoin using some type of cryptography for validation and system security. Namely using the hash function used for transaction and block generation is SHA-

IOP Conf. Series: Materials Science and Engineering **260** (2017) 012002 doi:10.1088/1757-899X/260/1/012002

256. As well as signature algorithm such as elliptic curve digital signature algorithms such (ECDSA). The algorithm is used to prevent malicious users from destroying the system. The function of this hash is to prevent malicious users from stealing and creating their own bitcoins. The conclusion in this paper is that cryptography is very useful for bitcoin security as well as few cases of theft or cases caused by cracker to break down this system, as for cases that occur because in the resulting by users who are negligent in closing his account.

Nicolas Dorier 2015 [3], From this book explains the bitcoin address is used to be able to receive payments from others. That private key is used to be able to spend the bitcoin owned. These two keys are not stored into the network, they can also be generated without having to connect to the internet. To get the bitcoin address, you can get it easily with your public key in the network you use for that address. This bitcoin address consists of different byte versions (on both networks) and your public key hash byte combines them, then encoded into Base58Check. The Base58Check encoding has some neat features like checksums to prevent typos and also reduce ambiguous characters. Base58Check encoding of an address to ensure that bitcoin wallet users do not send bitcoins to an address that should be used on different networks.

Dody Dharma [6], This paper describes the hash function SHA-256 uses 6 logical functions, each of which operates on a 32-bit word represented as x, y, and z. Where on the functional will be a new 32 bit word. The hash function is most commonly used to speed up the process of reading data in a table or data comparison process such as data search in a database, detecting the duplication of records on a large file, and searching for sequence similarities. The hash algorithm recommended by Adobe to be utilized is SHA-256, as it is considered quite safe against cryptographic attacks.

3. System design and implementation

3.1. System design

The concept of making the design us to implement and analyze in blockchain bitcoin, but discussed in this concept only in addressing the bitcoin and analyzing the system is safe or not. What is needed in the research is a program in bitcoin using python programming language, as well as software to test and analyze how the system running and tested to the system is safe or not.

The author analyzes how blockchain creates addresses on bitcoins, the concept being made this time is to see how the public key creates the address, and how to transfer the bitcoin from sender and the transacted receiver bitcoin. Bitcoin is not also used in terms of sale and purchase transactions but is also used as a mining, so research aims to determine the security of bitcoin[1].

The addresses on the bitcoin is like a key for the bitcoin to be known and usable, bitcoin addressing is designed so that bitcoin can not be known by others[3]. The bitcoin address is unique, the address is created using cryptography type Base58Check and SHA-256[3].

A bitcoin address is the same as an email address that serves to receive and store bitcoins. If an email user has only one email addresses while bitcoin users have multiple addresses. Bitcoin addresses usually have 26-35 characters and it's made very unique.

1MgErLiH1DuGMrd58fuL4CLQHc4VSboqKn

Figure 1. Example of bitcoin address

The address contains several numbers and letters. The bitcoin address has an error-checking code called checksum. The computations on the checksum will detect addresses that have one faulty nature. This process helps to prevent errors when sharing our address. Usually wallets do validation and reject if address is wrong.

IOP Conf. Series: Materials Science and Engineering **260** (2017) 012002 doi:10.1088/1757-899X/260/1/012002

- Valid bitcoin address:
1MgErLiH1DuGMrd58fuL4CLQHc4VsboqKn
- Invalid bitcoin address:
1MgErLiH1DuGMrd58fuL4CLOHc4VSboqKN

Figure 2. Example of address validation

From both instances the address resembles a valid address but the second address is invalid at checksum. Both have the same resemblance except for the letter N that is at the end of the second address[5].

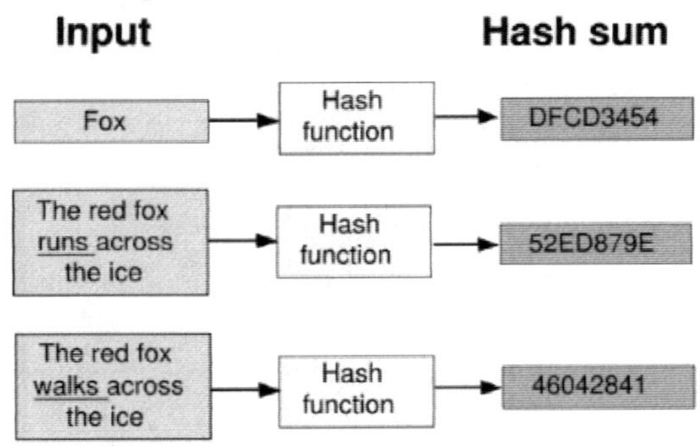

Figure 3. Illustration of hash function[8]

Base58 is a binary-encoded text format developed for use in bitcoins and is widely used in cryptocurrencies. Base58 also offers a balance of compact, readable, and error detection and prevention. Base58 is part of base64, which uses large and small types of letters and numbers, but can eliminate some of the characters that are often mistaken for each other. To add extra security to typos or transcription errors, Base58Check is a Base58 encoding format, often used in bitcoins that have built-in error checking codes. The checksum is an additional four bytes added to the end of the data being encoded. Checksum is derived from data encoded hashes therefore can be used to detect and prevent transcription in typing errors. When presented with the Base58Check code, the decoding software will calculate the checksum of the data and compare it with the checksum already included in the code. If both do not match, it will indicate that the error is known and the Base58Check data is invalid[7].

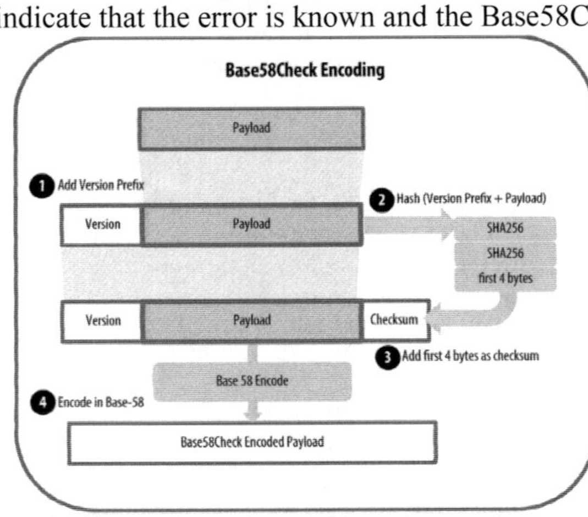

Figure 4. Illustration of Base58Check encoding process[7]

After learning about related topics, the next process is to design a system that includes the research flow, special flow, collecting the device used. The design of this system is done by creating a flowchart and use case diagram of how the system runs. Implementation in this research is how a blockchain system can create address on bitcoin and how to combine address bitcoin from public key sender, and whether

IOP Conf. Series: Materials Science and Engineering **260** (2017) 012002 doi:10.1088/1757-899X/260/1/012002

the address is including safe or not[4]. Testing aims to test whether the bitcoin blockchain system is running well or not. Is the bitcoin blockchain address system secure.
This flowchart explains how the research is done as follows.

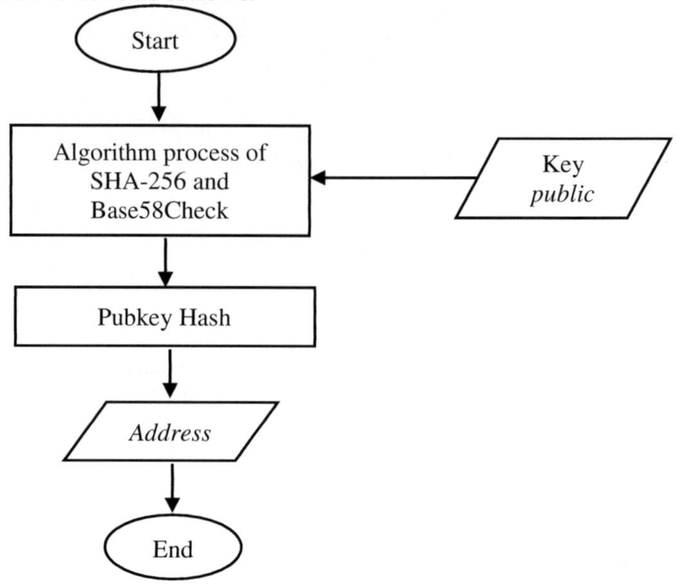

Figure 5. Flowchart

In the flowchart above the first step is to process the SHA-256 and Base58Check algorithm after which the processing that occurs that the address will be combined with the public key will be in hash with SHA-256 algorithm, then after the address has been in the hash with SHA-256 will checksum with Base58 encoding. Then after the new Base58 encoding will be formed onto the final format of the address that is Base58Check.

3.2. Implementation
Implementation of address at this stage is that transactions can be done and address serves as a tool to deliver transaction results. Address is made by using public key and then in hashing to be the correct address format.

Private key : 7b24e0a419659ee37a73765ddb2b05cf13210f614d6cc165aec6aa9722fb9422

Public key :
04a0d6f4c01fe3026e66cc477d54109e1db3d9f7b9cfcfa554e5dcd6af789689404e670dca6936ba10abdd08471
4d9b40f0d2d5415b9480419dbe45cd30fac55da

Bitcoin address : 13wv5ixTLYwFG9scfCpyndEeZjCg2apdZh

Figure 6. Generate address results

Figure 6. shows that the bitcoin address has been created, basically the address has several stages to be formed into the address with the correct format. Format for the correct address is Base58Check, whereas to make the address required public key that has hexadecimal format, it needs hashing for the address can be converted to SHA-256 code. From the result of hashing that becomes SHA-256 code will be checksum 32 bytes, after the result obtained then will be checksum again with 4 bytes. From the results in checksum 4 bytes, then will be generated by using the library of Base58Check format. Then the address will be generated which can be seen in Figure 6.

Verification Key :
a0d6f4c01fe3026e66cc477d54109e1db3d9f7b9cfcfa554e5dcd6af789689404e670dca6936ba10abdd084714
d9b40f0d2d5415b9480419dbe45cd30fac55da

Hashed public key : 002053060e600097fead4f753b5ca845e371ac5284

this is your full checksum 32 bytes :
dcc9bfc03bd004808ca615ff059722e632baeec44c69536b5d016130853f6599

this is your checksum 4 bytes : dcc9bfc0

this is my bin_addr : 002053060e600097fead4f753b5ca845e371ac5284dcc9bfc0

Figure 7. Address creation process

In the application design result created for the transaction, then the address is inserted into a wallet application to know the identify of bitcoin user.

IOP Conf. Series: Materials Science and Engineering **260** (2017) 012002 doi:10.1088/1757-899X/260/1/012002

Figure 8. Info address on wallet

If the user wants to make a transaction then the user must fill the account box with the recipient address so that the bitcoin sent is successful. Here the sender must also know the address of the recipient.

Figure 9. How to transaction using wallet

If the transaction is successful it will be displayed as shown in Figure. 10. and the balance of the bitcoin of the user will be reduced. If the recipient's address is invalid, then the transaction status column will show "Transaction Failed".

IOP Conf. Series: Materials Science and Engineering **260** (2017) 012002 doi:10.1088/1757-899X/260/1/012002

Figure 10. Transaction delivery succeeded

After the transaction is successful then the data will be sent to the database in the form of record results from transactions containing information in the form of bitcoin sent, byte on the public key, and public key hash that contains the address of the recipient.

```
======== Blockchain Transaction: Output =======
Number of outputs          : 01
Satoshis (BTC)             : 3.0
Bytes in pubkey script     : 19
OP_DUP                     : 76
OP_HASH160                 : a9
Push 20 bytes as data      : 4
PubKey Hash                : receiver address
OP_EQUALVERIFY             : 88
OP_CHECKSIG                : ac
```

Figure 11. Data record

Once the record is sent to the database, the database will contain information such as the table containing the delivery record that has been done before. In the database is also known in the form of a note about the address used during the transaction process. Such a process is called blockchain where the delivery process will be related to the previous one, the security gained in the blockchain is also very useful because all the shipping process will be recorded in full.

Table 1. Blockchain Database

Kode_Transaction	Sender	Receiver	Bitcoin value	Time_stamp
53bf2502	16ByDscY2dMnCfwMdJvi V2CBcbauHqaZJ3	1PoDhtBLLHbx5rhZXTc1 DapsqA3dkAwCjK	5	2017-06-03 14:09:59.27481
6d71fa0a	16ByDscY2dMnCfwMdJvi V2CBcbauHqaZJ3	19sacEBHM2HxC2iB9uV FXjzMe4Yjm1Smih	3	2017-06-04 11:21:40.40417

IOP Conf. Series: Materials Science and Engineering **260** (2017) 012002 doi:10.1088/1757-899X/260/1/012002

4. Analysis

In this section the test results that have been done is to find out whether the address bitcoin is safe for delivery bitcoin. In bitcoin security system is located on private and public key as well as the address. To generate the address required public key, while the path to send bitcoin is required also address so that bitcoin will be sent right to the receiver.

Address is made using public key for the security of user data is maintained, the format of the address has also been encrypted. Address encryption begins with the public key format which then in hashing first, after in hashing then the format will change to SHA-256. Then SHA-256 will be checksum 2 times, the first with a 32 byte checksum and then followed by a 4 byte checksum. Then the result of the 4 byte checksum will be encrypted into Base58Check code.

The security of the address is also aimed at the ability of Base58 to use large and small fonts and numbers, to add extra security to typos or transcription errors. The Base58 encoding also has a built-in error checking code, when presented with Base58Check code, the decoding software will calculate the checksum of the data and compare it with the checksum already included in the code. So if both do not match, will indicate that the error is known and the Base58Check data is invalid.

So with this security address on bitcoin is located on the public key and checksum process that can make into Base58Check code. By encrypting the public key that is converted into the address then the transaction will be safe and run well.

5. Conclusion

The conclusion that can be in this research is that way to get address that is by using public key and then used hashing method to make public key to SHA-256 code, after in hashing become SHA-256 then checksum will be 32 byte and 4 byte . Once encoded into the checksum the result becomes a valid Base58Check code used for an address in bitcoin.

In performing address transactions are also required for successful transactions. The process in the blockchain output also show the Pubkey Hash in the contents of the address of the recipient and after arriving at the database all the delivery results that occur also indicate the address of the sender and receiver are also recorded well without any changes made, which where the address is safe and encrypted with good.

The security of the address is also indicated by the encoding of Base58 which has a built-in error code, that is, when presented with Base58Check code, the decoding software will calculate the checksum of the data and compare it with the checksum already included in the code.

References

[1] Reinhard Denis Najogie, mei 2013, "Analisa Keamanan Bitcoin". Program Studi Teknik Informatika. Sekolah Teknik Elektro dan Informatika Institut Teknologi Bandung.

[2] Danielle Drainville, 21 Desember 2012, "*An Analysis of the Bitcoin Electronic Cash System*". University of Waterloo

[3] Nicolas Dorier, Will Bill Strait, 10 Januari 2015, "*Blockchain Programming in C#*"

[4] Satoshi Nakamoto. **Bitcoin: A Peer-to-Peer Electronic Cash System**.

[5] Richard Caetano, 2015, "*Learning Bitcoin: Embrace the new world of finance by leveraging the power of crypto-currencies using Bitcoin and the Blockchain*".

[6] Dody Dharma, "Studi Perbandingan Penggunaan Algoritma Hash SHA-256 dengan Simetrik dan Asimetrik Chipers dalam Perancangan Secure SWF Rich Internet Application (RIA)

[7] Andreas M. Antonopoulos, 2015, "*Mastering Bitcoin*"

[8] Rama Febriyan, "Perbandingan Digital Signature Algorithm dan Elliptic Curve Digital Signature Algorithm".Institut Teknologi Bandung.

[9] Andreas M. Antonopoulos, 2015 "**Bitcoin & The Blockchain**". O'Reilly E-Book Advantage.

Implementation and Analysis of the use of the Blockchain Transactions on the Workings of the Bitcoin

Muhammad Reza Rizky Fauzi, Surya Michrandi Nasution, Marisa W. Paryasto

Faculty of Electrical Engineering, Telkom University, Bandung, Indonesia

muhammadreza1801@gmail.com, surya.michrandi@gmail.com, marisa.paryasto@gmail.com

Abstract. The present study discusses the workings of blockchain bitcoin in the transactions sector, on the development of today's emerging computers in the financial sector of blockchain bitcoin traction. In this case the author analyzes how the transaction is running, as well as how blockchain bitcoin is doing work in transactions on the system to make transactions. Transaction is a data structure that encodes the transfer of values between users with in a bitcoin system. Every transaction is a public entry in this bitcoin blockchain. And become a large transaction bookkeeping global.

Keywords : Bitcoin, Blockchain, Transaction Bitcoin, Simulation, Blockchain Bitcoin Sandbox.

1. Introduction

The development of computers today is very fast and continues to grow rapidly. One of the current computer developments is in the world financial sector, The financial sector is called the digital currency in the presence of this digital currency makes it easier for users to make financial investments in the form of digital currency called bitcoin and security in investment in the financial sector . It is now very difficult for the user to invest and secure his finances because there are so many crimes and irresponsible people taking and financing, Then with the development of this computer in the financial sector is very important and in needed the present so that our finances are not taken and tapped by the irresponsible parties and finances that we have saved is under control. The development of computers in the financial sector is also in use to facilitate the user perform transactions send and receive digital currency bitcoin in a large scope is of the world in the form of digital currency is bitcoin. With the development of the computer helps users to send or receive bitcoin inside the country and abroad, And with the development of the computer today users can make financial investments in the form of digital currency called bitcoin. in the presence of this user do not have to worry about the crime because with the development of computers in the field of finance is the user feel safe with the user save money in the form of bitcoin investment.

In this study researchers discussed the workings of blockchain bitcoin in the transactions sector, on the development of the computer that is developing at this time in the financial sector in the blockchain transaction bitcoin. In this case the researchers make the implementation and analyze how the

transaction is running, and how blockchain bitcoin stores the transaction data that has been done and do work in transactions in bitcoin system to make transactions.
Transaction is the most important part of a bitcoin system. Everything else in bitcoin is designed to be done, spread into the network, validated, and then added to a ledger (blockchain) called a ledger (bookkeeping). Transaction is a data structure that encodes the transfer of values between users within a bitcoin system. Each transaction is a public entry in blockchain bitcoin. And become a large transaction bookkeeping global.

2. Related work

Bitcoin is one of the first implementations of cyptocurrency or digital currency. Bitcoin is a unique currency in which there is no central body governing transactions or publications. Bitcoin is open source and peer-to-peer. Open source means the currency system is developed jointly by anyone who wants to contribute. Peer-to-peer means that each transaction is recorded by a network of computers directly connected like a torrent system, Not through a mediating party such as a bank or merchant as is the case with most existing online payment systems (paypal, credit card, etc.). In this paper we will discuss the security aspects underlying the operation of bitcoin. Starting from account security and privacy, how digital signature systems are on bitcoins, how bitcoin overcomes double spending issues, How bitcoin mining works, as well as how bitcoin overcomes fraud from users who want to cheat. Keyword bitcoin, cryptocurrency, bitcoin mining, double spending .[1].
Bitcoin is widely used as a digital currency and is widely used by hundreds or even thousands of merchants worldwide online, as a recognized currency. The bitcoin system itself is based on cryptocurrency, regardless of government regulation and decentralized the technology used is peer-to-peer networking and cryptography to maintain data integrity. In this research will be studied on how to get bitcoin through one way that is mining. Keywords bitcoin, cryptography, merchant, mining, peer-to-peer networking. [2].
Recently many reported incidents of infringement, surveillance and security that compromised privacy on users by form or model Calls to questions where the third party controls and collects some personal data. With Bitcoin providing trustworthy financial credentials for administrative computing to be under control and the system uses a peer-to-peer network with a decentralized general ledger. In this paper, we describe a decentralized personal data management system that ensures users control their own data. We implemented a protocol that converts blockchain into automatic access control management that does not require a third party. Unlike bitcoins, transactions in our systems are not really financial, they are used to carry instructions, such as storing queries and data. [3].
In a technology-dependent world, privacy is now much sought after. The privacy required among others is desirable especially when wanting to make online payments. With this online payment it motivates users to switch to electronic money and not using normal money normally. The shape of electronic money is not like that of banknotes used every day. The most successful use of electronic money is a decentralized bitcoin with a peer-to-peer system. This paper provides a broad introduction to bitcoin, While analyzing the construction and investigating some benefits and shortcomings. In this case it can be seen that when compared to paper money and money electron called bitcoin, bitcoin has its own class that has its own system. [4].

3. System design and implementation

3.1. System design
In this case the concept of making the design we made is implementing and analyzing how blockchain bitcoin works, and analyzing what is inside the blockchain bitcoin. but discussion in this concept is addresing on bitcoin, key on bitcoin, and transaction on bitcoin, and analyzing whether the system is safe or not for use in bitcoin transactions. The things needed for research to be made is a program on

IOP Conf. Series: Materials Science and Engineering **260** (2017) 012003 doi:10.1088/1757-899X/260/1/012003

bitcoin by using python programming language and java programming, as well as a simple software application to test and analyze the validity of how the system is running and to test whether the system is safe to use or not.
Below is the work of blockchain bitcoin transactions on an existing system:

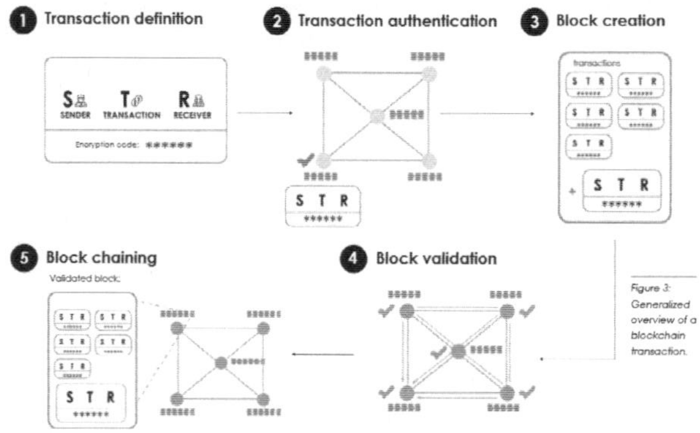

Figure 1. How Transaction Works [5]

In the picture above transaction definition sender as a tool to make a transaction and transaction result into a network. The transaction message consists of the recipient's public address details containing the value of the transaction and an exchange, and the address itself serves as a tool for conducting sales transactions between the sender and the recipient authenticating the transaction.
Transaction authentication nodes (computers or users) receive network messages and verify message validation by decrypting key exchanges between sender and receiver, as well as transactions confirmed and placed on databases that provide information about transaction delivery.
Block creation of pending transactions is put together in the latest version of the ledger or called the data base (blockchain) by one of the nodes in the localhost network. at certain time intervals the node broadcasts the block into the localhost network to be validated.
Block validation node receives localhost network of proposals from blocks and works to validate it by going through repeated processes that require awareness of most networks. Different blockchain localhost networks use different techniques to validate them. Bitcoin in a simple localhost network block for storing transaction data. Various techniques have different pros and cons. The common deminator ensures that any transactions are in effect, and cannot create fake transactions.
Block chaining if all transactions have been validated, the new block is chained in blockchain, and the data base is broadcast to the localhost network, and the whole process can be completed in a few seconds.

Example illustration on key to be transactions on ECDSA algorithm :

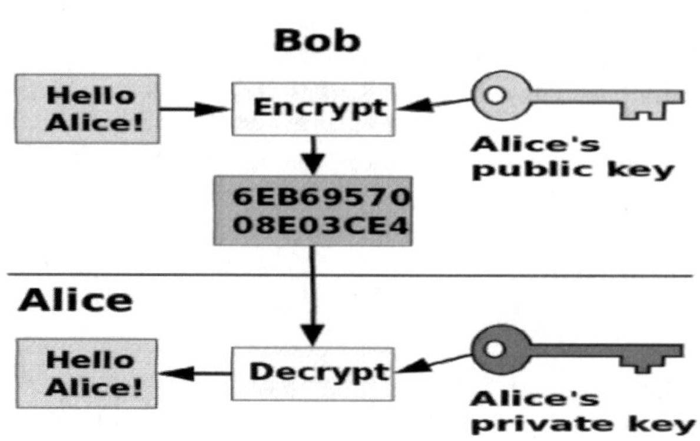

Figure 2. An illustrated example of how key works to transact [8]

IOP Conf. Series: Materials Science and Engineering **260** (2017) 012003 doi:10.1088/1757-899X/260/1/012003

The picture above explains ECDSA is the result algorithm of DSA development. This algorithm uses elliptic curve over the prime finite field to specify the key. Stage at ECDSA is the stage of determining key, signing stage, and verification stage. In ECDSA, the party that will perform the digital signature, has an elliptic curve domain parameter of D = {q, FR, a, b, G, n, h} and the secret key pair dA and the QA public key. Then the party who will verify the signature has a copy of the authentic D document and the QA public key.

- Key Generation
  1. Selecting a random integer dA, whose value is between [1, n-1].
  2. Calculating QA = dA • G = (x1, y1)
  3. Secret key = dA, and public key = QA [8]

Example illustration on address to be transactions Base58 on algorith :

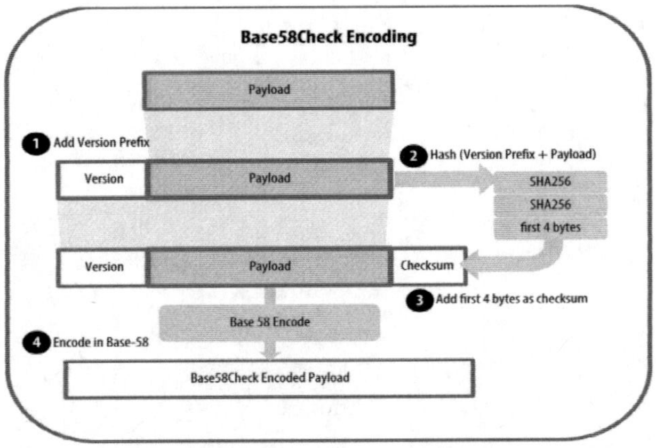

Figure 3. Illustration of Base58 process Check encoding [9]

Figure above explains that Base58 is a binary-encoded text format developed for use in bitcoins and is widely used in cryptocurrencies. Base58 also offers a balance on a compact, readable, and detect errors and prevention. Base58 is part of base64, which uses large and small types of letters and numbers, but can eliminate some of the characters that are often mistaken for each other. To add extra security to typos or transcription errors, Base58 Check is a Base58 encoding format, often used in bitcoins that has error-checking codes Built-in. The checksum is an additional four bytes added to the end of the data being encoded. Checksum is derived from data encoded hashes therefore can be used to detect and prevent transcription in typing errors. When presented with the Base58Check code, the decoding software will calculate the checksum of the data and compare it with the checksum already included in the code. If both do not match, it will indicate that the error is known and the Base58Check data is invalid. [9]

After learning about related topics, the next process is to design a system that includes the research flow, special flow, collecting the device used. The design of this system is done by creating a flowchart and use case diagram of how the system runs. Implementation in this research is how a blockchain system can create address on bitcoin and how Combining the bitcoin address of the sender and receiver, and whether the address is safe or not. Testing aims to test whether the bitcoin blockchain system is running well or not. Is the bitcoin blockchain address system secure.

In the picture below Specific Design above explains that after the user has obtained the key and address bitcoin, it will enter into the transaction phase, the transaction phase here is a public exchange Key and address of the sender so that the sender can send the bitcoin which is then validated by the sender and the receiver so that the transaction is valid or invalid key and address. Once the transaction is valid then the transaction will be sent into the blockchain ledger, the blockchain is a database that stores transaction data,after transaction is completed then there is a transaction confirmation that tells the sender that the

IOP Conf. Series: Materials Science and Engineering **260** (2017) 012003 doi:10.1088/1757-899X/260/1/012003

bitcoin that has been sent to the recipient has arrived at the recipient and the contents of the transaction confirmation itself in the form of transaction code, data sender, receiver data, the number of bitcoin, and time stamp.
This flowchart explains how the research is done as follows :

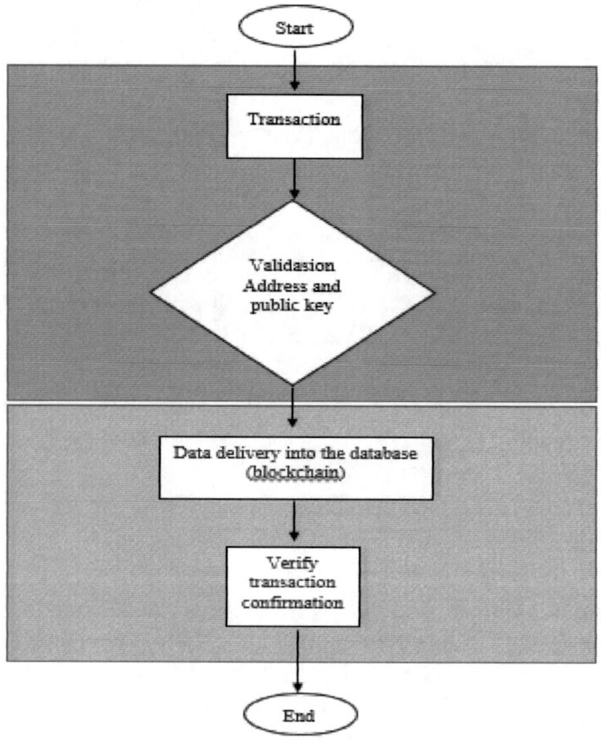

Figure 4. Spesific Design

3.2. Implementation

This stage discusses about the design of system implementation, the program that has been made is in the form of an application system to implement transaction experiments on bitcoin with a simple method in the form of GUI applications.
After that stage of the application is used to determine the identity data of bitcoin users in the form of bitcoin wallet. The definition of bitcoin wallet itself is a place (wallet) as a storage bitcoin, Users will not be able to have bitcoins without having bitcoin storage. This wallet also serves to do bitcoin transactions on blockchain bitcoin. And this wallet application also serves as a tool for storing bitcoin user data such as balance, address, private key, public key of a user, as well as a transaction tool on blockchain bitcoin you want to do.
This application is designed in a simple form by using GUI method in java, because this application is just an example and a replica of the wallet application on the real bitcoin.
The design of this application system is used to perform a transaction on bitcoin. The application of this wallet makes it easier for users to transact both send and receive bitcoin with other users.
In this design we use several applications to do the making. Things needed to do Making this program is a program on bitcoin by using python programming language, as well as a simple software application that is GUI application to do testing. As well as analyzing the validity of how the system works with what it wants, Such as the transaction program is running and the application for testing was running

IOP Conf. Series: Materials Science and Engineering **260** (2017) 012003 doi:10.1088/1757-899X/260/1/012003

well in order to make the transaction process to be done. And to test whether the system is safe to use or not in transacting on bitcoin.

3.3. Program Structure

Stage explains how the parameter system in a simple application software that has been created that is with programming algorithms that use python language that raised from a transaction work on bitcoin, then after generating transaction data on the bitcoin, the next experiment is to combine an address and key on the bitcoin into the transaction to be performed.

What is required at this stage is to work on raising an address and key on bitcoin to perform transactions on bitcoins. Because if you do not do that stage to bring up an address and user key then the transaction will not be done.

Design of the transactions that liaises between the user address and user key to do the transaction work on bitcoin. After getting the address and key users of the next stage is to unite the program into the transaction program for the transaction can run, After the program has been integrated and running well then input the program into a simple GUI application that has been created, the application is a wallet of bitcoin is a bitcoin user data storage. With the wallet users can do bitcoin transactions, without the wallet then the user cannot use bitcoin.

The result of the planning that has been achieved is the program can transact with what has been desired, this is the result of testing of how transactions work on blockchain bitcoin:

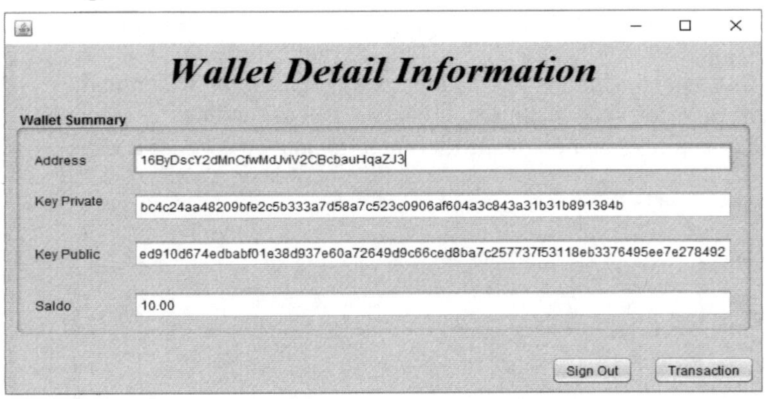

Figure 5. Bitcoin Wallet App

Bitcoin Wallet application that shows the contents of the bitcoin wallet, the wallet contents of the wallet contains several pieces of data are: Address or user id, private key as a tool to verify transaction results and as a bitcoin address backup tool, public key as a tool for selling transactions, and balances on bitcoins from coin ID users owned by bitcoin users.

Figure 6. Sending Transaction Process

IOP Conf. Series: Materials Science and Engineering **260** (2017) 012003 doi:10.1088/1757-899X/260/1/012003

In Sending Transaction Process transaction process to be performed, in this process the sender sells its bitcoin to the recipient. The sender needs the address of the recipient to do delivery,If the recipient does not provide the address then the sender will not be able to send the bitcoin to the recipient because what is needed to transact on the bitcoin is the recipient and the sender must do that step first ie the exchange of address and the public key to conduct the transaction and verify it.

Figure 7. Transaction Status Process

Process of transaction status, with the transaction status on a bitcoin wallet makes it easy for the sender to see if the bitcoin is already sent or not sent. If successful in sending the balance of the sender bitcoin will decrease and if unsuccessful then the sender bitcoin balance will not decrease.

Blockchain Transaction : Output

Number of outputs	: 01
Satoshis (BTC)	: 5.0
Bytes in pubkey script	: 19
OP_DUP	: 76
OP_HASH160	: a9
Push 20 bytes as data	: 14
PubKey Hash	: [receiver address]
OP_EQUALVERIFY	: 88
OP_CHECKSIG	: ac

locktime: 0 (a block height) : 000000000
StatusTransaction:Success

Figure 8. Transaction Record Process

The figure above explain Transaction Record Process, once the transaction is received the data sent to the recipient and to the database Changed data contained in it contains public key hash, the amount Bitcoin sent and transaction status that has been sent to the recipient.

Table 1. Process From Database

Kode_ Transa ction	Sender	Receiver	Bitcoin value	Time_stamp
53bf25 02	16ByDscY2dMnCfwM dJviV2CBcbauHqaZJ3	1PoDhtBLLHbx5rhZXTc1 DapsqA3dkAwCjK	5	2017-06-03 14:09:59.27481

IOP Conf. Series: Materials Science and Engineering **260** (2017) 012003 doi:10.1088/1757-899X/260/1/012003

6d71fa0a	16ByDscY2dMnCfwMdJviV2CBcbauHqaZJ3	19sacEBHM2HxC2iB9uVFXjzMe4Yjm1Smih	3	2017-06-04 11:21:40.40417
8a3a7204	16ByDscY2dMnCfwMdJviV2CBcbauHqaZJ3	1JLqrCyEQAwiy5iszQKehTNLvuzWB9mo1m	3	2017-06-04 11:32:04.60249

Above describes the table process of the database, the database itself here is as a replica tool of a data blockchain (ledger). This database serves as a notice to the seller and the buyer that the transaction process is done and as a mean of data transaction, transaction that has been going on or have done the transaction process before. The database also serves as a large storage tool to store some data that transacts from the number of users who do bitcoin transactions.

2017-06-04
12:48:56.802733;3;3;1CbuckSG2GtUfzXLhZGm2
yHo4646FuQgCQ2 2017-06-04

Figure 9. Transaction Confirmation Process

Figure this explain the Transaction confirmation process. This process is a confirmation tool to notify the sender and the recipient that the bitcoin has been already sent by the sender has been sent and received by the receiver. With the confirmation of the transaction itself provides convenience and security to the sender that the bitcoin delivery transaction is already in progress and has been sent to the recipient.

4. Analysis

At this stage, the test results that have been done are to find out whether the bitcoin system is safe when it wants to transact on bitcoin. The bitcoin system is the transaction security system, in the bitcoin itself the security lies in the key (public key and private key) and address. When you want to trade on bitcoin itself, you requires a key and an address to check the validity and validity of a transaction on bitcoin to know how transaction works on bitcoin, if the transaction does not have or have key and address then the transaction will not run and can not be used To conduct a transaction, Because key and address is needed by transaction to do bitcoin sale, and with key and address make sense of security and sense of comfort of bitcoin user to do bitcoin trading transaction, key exchange and address itself serves as a tool to conduct sales transaction Between sender and receiver.

The bitcoin transaction system itself lies in the exchange of the user's key and address, when transacting the sender and the recipient must first perform the key and address exchange in order to make the transaction process, if there is no exchange first then the transaction will not run. With the key and address exchange system on the transaction itself provides a sense of security to the user when they want to do bitcoin transactions.

So with this security the transaction on the bitcoin lies in key and address on bitcoin. By exchanging legitimate keys and validity that has been run between the sender and the recipient then the transaction goes well.

5. Conclusion

The result of the hypothesis that the researcher wants is a safe transaction to use when transacting bitcoin between sender and receiver after transaction, transaction data is stored on blockchain bitcoin so that user feel secure in transaction bitcoin blockchain own also as tool of storage of transaction data which

IOP Conf. Series: Materials Science and Engineering **260** (2017) 012003 doi:10.1088/1757-899X/260/1/012003

have been going on and know that The transaction is done. The test expected by the researcher is a test for safety in the transaction on blockchain bitcoin. the transaction security system on the bitcoin itself lies in the key and address then in the test this time is to analyse whether the system is safe to transactions on bitcoin as well as to implement how the transaction is running, By performing a valid key exchange and validity that has been running between the sender and The recipient then the transaction goes well.

References

[1] Reinhard Denis Najogie, *mei 2013,* **"Analisis Keamanan Bitcoin".**

[2] Ferry Mulyanto, M Tirta Mulia. **"Analisis Mining System pada Bitcoin".**

[3] Guy Zyskind, Oz Nathan, dan Alex 'Sandy' Pentland, 2015 IEEE. **" Decentralizing Privacy : Using Blockchain to Protect Personal Data".** MIT Media Lab, Tel-Aniv University, MIT Media Lab. Cambridge,Massachusetts. Tel-Aniv, Israel. Cambridge,Massachusetts.

[4] Danielle Drainville, Desember 2012." **An Analysis of The Bitcoin Electronic Cash System".** University of Waterloo.

[5] Peter Froystad, Jarle Holm, 2015 **"Whitepaper Blockchain : Powering the internet of Value".** EVRY Financial Service

[6] Michael Crosby, Nachippan, Pradhan Pattanayak, Sanjeev Verma, Vignesh Kalyanaraman, October 2016 **"Sutardja Center for Entrepreneurship and Technology (Blockchain Technology)".** Barkeley Enggneering.

[7] Richard Caetano, 2015 **"Learning Bitcoin Embrace teh new word of finance by leveraging the power of crypto-currencies using Bitcoin and the Blockchain".** Packt publishing open source, www. It-ebooks.info

[8] Rama Febriyan, 2014/2015 **"Perbandingan Digital Signature Algorithm dan Elliptic Curve Digital Signature Algorithm".**

[9] Andreas M. Antonopoulos, 2015 **"Bitcoin & the Blockchain"**. O'Reilly Ebook Advantage.

Classification of the Correct Quranic Letters Pronunciation of Male and Female Reciters

Safiah Khairuddin, Salmiah Ahmad, Abdul Halim Embong, Nik Nur Wahidah Nik Hashim, Tareq M.K. Altamas, Syarifah Nuratikah Syd Badaruddin, Surul Shahbudin Hassan*

Department of Mechatronics Engineering, International Islamic University Malaysia, Kuala Lumpur, Malaysia

* Centre for Languages and Pre-University Academic Development, International Islamic University Malaysia, Kuala Lumpur, Malaysia

Abstract – Recitation of the Holy Quran with the correct Tajweed is essential for every Muslim. Islam has encouraged Quranic education since early age as the recitation of the Quran correctly will represent the correct meaning of the words of Allah. It is important to recite the Quranic verses according to its characteristics (*sifaat*) and from its point of articulations (*makhraj*). This paper presents the identification and classification analysis of Quranic letters pronunciation for both male and female reciters, to obtain the unique representation of each letter by male as compared to female expert reciters. Linear Discriminant Analysis (LDA) was used as the classifier to classify the data with Formants and Power Spectral Density (PSD) as the acoustic features. The result shows that linear classifier of PSD with band 1 and band 2 power spectral combinations gives a high percentage of classification accuracy for most of the Quranic letters. It is also shown that the pronunciation by male reciters gives better result in the classification of the Quranic letters.

Keywords *– Quranic Letters, Formants, Power Spectral Density, Linear Discriminant Analysis*

1. Introduction

Nowadays, speech processing applications has been widely used in various sectors such as wireless communication, education, medicine and many more. Speech is used as a communication platform for human to connect with each other. Arabic language is one of the languages that is widely used in speech all over the world, especially for Muslims, as it is the language of the Quran, which is the holy book of Islam. Pronunciation of Quranic utterances is considered as a part of general speech production process. It is compulsory for all Muslims to read the Quran regularly throughout their life. Therefore, it is important to start the education process in learning Quranic recitation from early age. People usually learn to recite the Quran through a teacher that will teach them the correct ways of

IOP Conf. Series: Materials Science and Engineering **260** (2017) 012004 doi:10.1088/1757-899X/260/1/012004

pronunciation and recitation of Quranic verses. The student will also read the Quran in front of the teacher and the teacher will listen and corrects the wrong pronunciation or recitation [1].

This conventional way of learning Quran has been practiced for hundreds of years, ever since the Quran was sent down to Prophet Muhammad s.a.w.. The process of learning will usually take a couple of years before the student becomes expert and can recite the Quran by themselves [2]. Technically, the Quran should be recited using the correct Tajweed, or the rules of pronunciation during recitation of Quran. Tajweed literally means to make better or to improve. It can also be defined as articulating every letter from its articulation point (Makhraj) and giving them its right and dues of characteristics (Sifaat) [3]. Point of articulation of a specific letter is the place that is used to emit the letter. It is necessary to use the right articulation points while producing the utterances. The study on makhraj using speech processing has been done in [4] [5] and [6].

To date, there are no published study on identifying Makhraj in the pronunciation of all 28 Quranic letters that compares male and female pronunciation. Thus, the motivation of this paper is to identify the possible features that can represent each of the Quranic letters uniquely and to see whether there are significant differences between male and female Quranic letters pronunciation.

2. Database collection and formulation

2.1 Data Collection

Our sample consisted of 22 sets of audio data which were obtained via recording session with 11 males and 11 females who were selected based on their expertise in Quranic recitation. All data were verified by a certified expert which specialized in Quranic recitation, Ust. Dr. Surur Shahbudin. The audio data was recorded using a portable high-quality field recorder, TASCAM DR-05, with frequency response of 40Hz to 20 kHz. The recordings were done in a normal room environment such as classrooms and labs in engineering faculty of the university, which will later may represent a normal learning environment. During the data recording, the participant will need to recite 28 combinations of Quranic letters, which are based on *Rasm 'Uthmani* narrated by Hafs bin Sulaiman for *Qira'at Asim* [7]. Table 1 shows the *sokoon* (ْ) combination of *Quranic* letters and its pronunciation that is recited for the recording. The participants are required to recite each *sukoon* letters only once without repeating them. This *sokoon* (ْ) combination of *Quranic* letters dataset can be used to describe the *Makhraj* (point of articulations) and *Sifaat* (characteristics) of each letter.

Table 1. The *sokoon* (ْ) combination of Quranic letters and its pronunciation.

Combination of Quranic letters	Pronunciation
أء	aa
أب	ab
أت	at
أث	ath
أج	aj
أح	ah
أخ	akh
أد	ad
أذ	az
أر	ar
أز	azz
أس	as
أش	ash
أص	asd
أض	adh
أط	athd
أظ	azd
أع	a'
أغ	agh
أف	af
أق	aq
أك	ak
أل	al
أم	am
أن	an
أه	ahh
أو	aww
أي	aii

2.2 Pre-processing

The audios were digitized with a sampling rate of 99.6 kHz and edited using *Audacity,* open source software for audio signal. The processing of data was started with de-identification of personal information as well as removal of interviewer's voice and undesirable sound. Unwanted background noise was also removed using *Audacity*. For the sake of simplicity, all stereo data have been converted to mono channel prior to the noise reduction step. The audio signal has been normalized with the amplitude in between -1 to 1.

IOP Conf. Series: Materials Science and Engineering **260** (2017) 012004 doi:10.1088/1757-899X/260/1/012004

3. Feature extraction

Features extraction was done in speech recognition system to identify the components of the audio signal that are good for representing the acoustic content.

3.1 Formant

A formant is a concentration of acoustic energy around a frequency in the speech wave [8]. Each formant corresponds to a resonance in the vocal tract. They represent the frequencies that pass most of the acoustic energy from the source to the output. The formant frequency representation shows an efficient and compact representation of time-varying characteristics of speech [9]. Usually the first two formants are enough to distinguish between the vowels. Formant frequencies of the audio signal were first obtained using spectrogram features in *PRAAT* software.

The first and second formant frequency (F1 and F2) are extracted from 22 samples, which contains 11 male data and 11 female data. The average formant frequency was calculated. Table 2 shows the average F1 and F2 value of 11 male experts and 11 female experts for all Quranic letters *sukoon* combinations.

Table 2. Average F1 and F2 value (Hz) for 11 males and 11 female samples.

Quranic letters	**Male average**		**Female average**	
	F1 (Hz)	**F2 (Hz)**	**F1 (Hz)**	**F2 (Hz)**
أءْ	770.99	1718.26	796.48	1692.11
أبْ	674.04	1752.69	658.21	1706.89
أتْ	773.82	1885.74	775.96	1888.05
أثْ	851.46	1973.26	807.26	1918.27
أجْ	708.54	2002.06	632.34	1908.15
أحْ	990.37	1965.40	946.95	1985.06
أخْ	898.11	1992.48	839.76	1652.43
أدْ	632.93	1825.73	657.17	1892.83
أذْ	702.64	1878.27	679.53	1821.29
أرْ	742.87	1669.10	723.77	1463.68
أزْ	734.17	1953.30	668.46	1850.18
أسْ	986.86	2171.90	845.37	2002.96
أشْ	1263.66	2442.06	995.29	2277.63
أصْ	947.03	2138.36	840.56	1924.64
أضْ	676.95	1638.52	671.06	1483.90
أطْ	755.98	1664.30	714.73	1427.16
أظْ	754.98	1789.21	607.38	1496.56

IOP Conf. Series: Materials Science and Engineering **260** (2017) 012004 doi:10.1088/1757-899X/260/1/012004

أعْ	928.86	1741.28	795.37	1607.24
أغْ	754.89	1749.50	662.78	1517.99
أفْ	987.61	2065.16	855.68	1911.81
أقْ	775.82	1626.50	733.76	1450.35
أكْ	799.22	1967.71	794.13	1754.01
ألْ	620.81	1860.35	616.13	1855.05
أمْ	661.91	1820.03	624.39	1677.02
أنْ	667.69	1906.23	538.91	1692.22
أهْ	850.59	1901.06	837.53	1788.08
أوْ	716.52	1679.28	675.86	1527.66
أيْ	656.27	2063.66	624.45	1846.38

3.2 Power Spectral Density (PSD)

Power Spectral Density is the distribution of power over a frequency in signals [10]. Three spectral energy bands were extracted using Matlab from all 28 letters of *sukoon* combinations of 22 sets of male and female data. The energy bands named PSD_1, PSD_2, and PSD_3 will be paired by two and undergoes data classification with the different band combinations.

4. Classification based on selected features

Linear Discrimant Analysis (LDA) or linear classifier used in this study as classifier. The analysis is done based on pairwise analysis classification between male and female sample. Resampling method was applied because of the small number of data sets. Jackknife (Leave-One-Out) was used as the resampling method in this project as it can give an optimum result [11]. The Jackknife method leaves one sample from the data set for testing and constructs the classification function with the remaining N-1 samples as the training data [10].

5. Results and discussion

All the features that have been extracted are classified using LDA with Jackknife method as the resampling technique. In this study, we evaluated the features that have been extracted and see what is the feature combinations that will give the best result. All data are classified according to male and female group, and the data is further being classified by each quranic letters *sukoon* combination to group them into male or female class. The percentage indicates the accuracy of the classifier, which represent the correct classification of the data into its class. There would be 28 categories of Quranic letters after the classification process which is represented by male and female recitation.

5.1 Formant Frequency

F1 and F2 features extracted from the samples undergo classification process to classify them into male or female experts group. Table 3 shows the percentage of correct classification of all data according to its class.

Table 3. Male and Female experts' data classification with all data (Feature: F1 and F2).

Classifier	**Percentage (%)**		
LDA (resampled with Jackknife method)	All	Male	Female
	62.0	60.7	63.3

Using LDA, classification of all data using the first two formants gives 62.0% accuracy as it gives the result of 60.7% of correct classification for male group while female group exhibit 63.3% of the correct classification. The percentage of correct classification was also calculated using the same method to each of the Quranic letters *sukoon* combination. The results were tabulated in Table 4.

Table 4. Percentage of correct classification for each letter's *sukoon* combination using LDA (Feature: F1 and F2).

Letter Combination	أءْ	أبْ	أتْ	أثْ	أجْ	أحْ	أخْ	أدْ	أذْ	
All (%)	50.0	31.8	0.0	40.9	59.1	50.0	72.7	59.1	45.5	
Male (%)	36.4	36.4	0.0	45.5	54.5	36.4	63.6	54.5	45.5	
Female (%)	63.6	27.3	0.0	36.4	63.6	63.6	81.8	63.6	45.5	
Letter Combination	أرْ	أزْ	أسْ	أشْ	أصْ	أضْ	أطْ	أظْ	أعْ	
All (%)	63.6	59.1	72.7	72.7	63.6	50.0	63.6	68.2	77.3	
Male (%)	63.6	63.6	63.6	72.7	63.6	45.5	63.6	63.6	90.9	
Female (%)	63.6	54.5	81.8	72.7	63.6	54.5	63.6	72.7	63.6	
Letter Combination	أغْ	أفْ	أقْ	أكْ	ألْ	أمْ	أنْ	أهْ	أوْ	أيْ
All (%)	81.8	72.7	54.5	59.1	0.0	59.1	72.7	54.5	68.2	68.2
Male (%)	81.8	72.7	54.5	63.6	0.0	63.6	81.8	54.5	63.6	72.7
Female (%)	81.8	72.7	54.5	54.5	0.0	54.5	63.6	54.5	72.7	63.6

The result shows that the combination of F1 and F2 give the highest percentage of correct classification for the letter combination 'أغْ' for male experts, which is 90.9%. For female experts group, the letter combination of 'أخْ' 'أسْ' and 'أغْ' can be represented correctly using F1 and F2 features with 81.8% of accuracy respectively. Meanwhile, the result indicates that the combination of F1 and F2 features cannot represent the letter 'أثْ' and 'ألْ' correctly at all, as the percentage of classification accuracy is at zero percent. In average, this feature combination is not a good representation to specifically classify the Quranic letters into groups of male and female as the result shows low percentage accuracy for most of the letters. There are also not many differences in male and female percentage result using formants as acoustic features. Therefore, it can be concluded that formants are not a good feature representation in differentiating the gender in Quranic letters pronunciation.

5.2 Power Spectral Density (PSD)

Three PSD energy bands that have been extracted (PSD_1, PSD_2 and PSD_3) undergoes data classification process with three different combination of bands as the features. The combination of energy bands is in Table 5.

Table 5. Combination of PSD energy band.

Combination	PSD band
1	PSD_1 and PSD_2
2	PSD_1 and PSD_3
3	PSD_2 and PSD_3

The result of data classification that was done using LDA with Jackknife as resampling method was recorded in Table 6.

Table 6. Percentage of correct classification of PSD bands combination using LDA.

Letters		Combination	All (%)	Male (%)	Female (%)
1	أءْ	1	77.3	72.7	81.8
		2	77.3	72.7	81.8
		3	77.3	63.6	90.9
2	أبْ	1	90.9	81.8	100
		2	90.9	81.8	100
		3	81.8	72.7	90.9
3	أتْ	1	77.3	81.8	72.7
		2	77.3	81.8	72.7
		3	77.3	81.8	72.7
4	أثْ	1	90.9	81.8	100
		2	90.9	81.8	100
		3	95.5	90.9	100
5	أجْ	1	86.4	90.9	81.8
		2	77.3	90.9	63.6
		3	81.8	90.9	72.7
6	أحْ	1	63.6	54.5	72.7
		2	54.5	54.5	54.5
		3	68.2	63.6	72.7
7	أخْ	1	77.3	72.7	81.8
		2	68.2	81.8	54.5
		3	59.1	63.6	54.5
8	أدْ	1	40.9	45.5	36.4
		2	45.5	45.5	45.5
		3	50.0	36.4	63.6
9	أذْ	1	50.0	54.5	45.5
		2	54.5	63.6	45.5
		3	54.5	63.6	45.5
10	أرْ	1	59.1	45.5	72.7
		2	54.5	36.4	72.7
		3	54.5	36.4	72.7
11	أزْ	1	68.2	72.7	63.6
		2	50.0	27.3	72.7
		3	45.5	27.3	63.6
12	أسْ	1	95.5	100	90.9
		2	72.7	81.8	63.6
		3	86.4	90.9	81.8

13	أشْ	1	81.8	81.8	81.8
		2	68.2	63.6	72.7
		3	59.1	63.6	54.5
14	أصْ	1	63.6	72.7	54.5
		2	72.7	63.6	81.8
		3	63.6	63.6	63.6
15	أضْ	1	72.7	90.9	54.5
		2	72.7	90.9	54.5
		3	72.7	81.8	63.6
16	أطْ	1	68.2	72.7	63.6
		2	54.5	54.5	54.5
		3	50.0	45.5	54.5
17	أظْ	1	68.2	72.7	63.6
		2	63.6	81.8	45.5
		3	68.2	72.7	63.6
18	أعْ	1	72.7	90.9	54.5
		2	72.7	90.9	54.5
		3	72.7	90.9	54.5
19	أغْ	1	45.5	45.5	45.5
		2	50.0	54.5	45.5
		3	50.0	54.5	45.5
20	أفْ	1	59.1	63.6	54.5
		2	59.1	63.6	54.5
		3	59.1	63.6	54.5
21	أقْ	1	54.5	54.5	54.5
		2	54.5	54.5	54.5
		3	50.0	72.7	27.3
22	أكْ	1	100	100	100
		2	100	100	100
		3	100	100	100
23	ألْ	1	77.3	72.7	81.8
		2	81.8	81.8	81.8
		3	77.3	72.7	81.8
24	أمْ	1	77.3	81.8	72.7
		2	50.0	45.5	54.5
		3	72.7	72.7	72.7
25	أنْ	1	68.2	81.8	54.5
		2	68.2	72.7	63.6
		3	54.5	45.5	63.6

26	أهْ	1	54.5	45.5	63.6
		2	36.4	45.5	27.3
		3	22.7	36.4	9.1
27	أوْ	1	100	100	100
		2	59.1	54.5	63.6
		3	95.5	100	90.9
28	أيْ	1	100	100	100
		2	81.8	72.7	90.9
		3	95.5	90.9	100

Based on the observation of the percentage result, it was found that combination 1 which is the combination of band 1:2 (PSD_1 & PSD_2) gave highest classification accuracy for all, as well as male and female data for each of the Quranic letters *sukoon* pronunciation using LDA methods. This method also shows to be better in classifying male speaker in the letters pronunciation as it demonstrates a higher percentage in correctly classifying the letters into male group. With LDA, the combination of band 1:2 indicates the highest percentage accuracy in classifying the letter أبْ, and أثْ in female group, the letter أسْ in male group, and أكْ , أوْ , and أيْ in all of the groups, which shows 100% accuracy.

However, this combination may not be suitable in classifying the letters أدْ , أذْ , أغْ , أقْ , and أهْ as they displays a low percentage classification accuracy. This may be because the features are not enough to distinguish the unique characteristic of the letters to be classified into its own group. On the other hand, the other letters *sukoon* combination shows a good result in classification percentage with band 1:2 combination.

5.3 Power Spectral Density and Formant

We've combined the features of PSD and Formant to see if the result would give a better percentage accuracy. Table 7 shows the percentage accuracy of the classification of each letter into groups of all, male and female.

Table 7. Percentage classification of combined features of each letter using LDA.

Letter Combination	أءْ	أبْ	أتْ	أثْ	أجْ	أحْ	أخْ	أدْ	أذْ	
All (%)	77.3	90.9	77.3	90.9	86.4	63.6	77.3	40.9	50.0	
Male (%)	72.7	81.8	81.8	81.8	90.9	54.5	72.7	45.5	54.5	
Female (%)	81.8	100	72.7	100	81.8	72.7	81.8	36.4	45.5	
Letter Combination	أرْ	أزْ	أسْ	أشْ	أصْ	أضْ	أطْ	أظْ	أعْ	
All (%)	59.1	68.2	95.5	81.8	63.6	72.7	68.2	68.2	72.7	
Male (%)	45.5	72.7	100	81.8	72.7	90.9	72.7	72.7	90.9	
Female (%)	72.7	63.6	90.9	81.8	54.5	54.5	63.6	63.6	54.5	
Letter Combination	أغْ	أفْ	أقْ	أكْ	ألْ	أمْ	أنْ	أهْ	أوْ	أيْ
All (%)	45.5	59.1	54.5	100	77.3	77.3	68.2	54.5	100	100
Male (%)	45.5	63.6	54.5	100	72.7	81.8	81.8	45.5	100	100
Female (%)	45.5	54.5	54.5	100	81.8	72.7	54.5	63.6	100	100

The result exhibit a slight increase in percentage accuracy of classification of letters when all the features are combined. However, the changes are not significant as we could see that the percentage results are almost the same with classification of PSD band 1:2. Only the letter أكْ , أوْ , and أيْ can be represented using the features correctly with 100% accuracy. The other letter shows the same behaviour as in classification using PSD band 1:2 only. This shows that the combination of features has no effect in increasing the accuracy of the classifier to classify each of the letter.

6. Conclusion

We have done classification of Quranic letters to Male and Female groups using Linear Discriminant Analysis with Formant and Power Spectral Density as the features. Among the two features, power spectral density with combination of band 1:2 exhibit a better result to classify the letters into male and female. Overall, the classification percentage was higher for male reciters identification. For future study, it is recommended to explore other acoustical features and classifiers that will give better results in classifying the Quranic letters into its group accurately.

Acknowledgement

The author would like to express sincere appreciations to Ustaz Dr. Surul Shahbudin Hassan from the Centre For Languages And Pre-University Academic Development, International Islamic University Malaysia, Kuala Lumpur, Malaysia for his efforts in making the experiment successfully conducted by providing the experts, also thanks to all Ustaz and Ustazah who have kindly agreed and spent time in recording the pronunciation and recitation. Thank you also to the undergraduate students under my supervision who were committed in performing their final year project successfully.

References

[1] Ibrahim, N. J., Yusoff, Z. M., & Razak, Z. (2008). Quranic verse Recitation Recognition Module for Educational Programme.

[2] Rashid, N. R. R., Venkat, I., Damanhoori, F., Mustaffa, N., Husain, W., & Khader, A. T. (2013, December). Towards Automating the Evaluation of Holy Quran Recitations: A Pattern Recognition Perspective. In *Advances in Information Technology for the Holy Quran and Its Sciences (32519), 2013 Taibah University International Conference on* (pp. 424-428). IEEE.

[3] Wardhani, A., & Nurhayati, T. Makhraj Al-Jauf Dalam Kajian Ilmu Tajwid.

[4] Wahidah, A., Suriazalmi, M., Niza, M., Rosyati, H., Faradila, N., Hasan, A., Farizan, Z. (2012). Makhraj recognition using speech processing. Paper presented at the Computing and Convergence Technology (ICCCT), 2012 7th International Conference.

[5] Altalmas, T., Ahmad, S., Sediono, W., & Hassan, S. S. (2015). Quranic Letter Pronunciation Analysis based on Spectrogram technique: Case Study on Qalqalah Letters.

[6] N.W.Arshad, S.N.Abdul Aziz, R. Hamid, R. Abdul Karim, F. Naim, N. F. Zakaria (2011), Speech Processing for Makhraj Recognition: The Design of Adaptive Filter for Noise Removal, International Conference on Electrical, Control and Computer Engineering, Pahang, Malaysia, June 21-22, 2011

[7] Arshad, N. W., Sukri, S. M., Muhammad, L. N., Ahmad, H., Hamid, R., Naim, R., & Naharuddin, N. Z. (2013). Makhraj Recognition for Al-Quran Recitation using MFCC. *International Journal of Intelligent Information Processing (IJIIP)*, 45-53.

[8] Rabiner, L., & Juang, B. H. (1993). Fundamentals of speech recognition. (pp. 26-28).

[9] Jorgensen, C., Lee, D. D., & Agabont, S. (2003, July). Sub auditory speech recognition based on EMG signals. In Neural Networks, 2003. Proceedings of the International Joint Conference on (Vol. 4, pp. 3128-3133). IEEE.

[10] Azam, H., Hashim, N. N. W. N., Sediono, W., Mukhtar, F., Ibrahim, N., Mokhtar, S. S. S., & Aziz, S. A. (2016, December). Classifications of clinical depression detection using acoustic measures in Malay speakers. In *Biomedical Engineering and Sciences (IECBES), 2016 IEEE EMBS Conference on* (pp. 606-610). IEEE.

[11] Hashim, N. N. W. N. (2011). Analysis of power spectrum density of male speech as indicators for high risk and depressed decision (Doctoral dissertation, Vanderbilt University).

IOP Conf. Series: Materials Science and Engineering **260** (2017) 012005 doi:10.1088/1757-899X/260/1/012005

Flip-Chip Bonding Fabrication Technique

Tengku Muhammad Afif bin Tengku Azmi, Nadzril bin Sulaiman

International Islamic University Malaysia, Faculty of Engineering, Gombak P.O. Box 10, 50728 Kuala Lumpur, Malaysia

afif5785@gmail.com,nadzril@iium.edu.my

Military systems, outer space exploration and even in medical diagnosis and treatment used magnetic field detection. Low magnetic field detection is particularly important in tracking of magnetic. Traditional magnetometer tends to be bulky that hinders its inclusion into micro-scaled environment. This concern has brought the magnetometer into the trend of device miniaturization. Miniaturized magnetometer is usually fabricated using conventional microfabrication method particularly surface micromachining in which micro structures are built level by level starting from the surface of substrates upwards until completion of final structure. Although the miniaturization of magnetometer has been widely researched and studied, the process however is not. Thus, the process governing the fabrication technique is studied in this paper. Conventional method of fabrication is known as surface micromachining. Besides time consuming, this method requires many consecutive steps in fabrication process and careful alignment of patterns on every layer which increase the complexity. Hence, studies are done to improve time consuming and reliability of the microfabrication process. The objective of this research includes designing micro scale magnetometer and complete device fabrication processes. A micro-scale search coil magnetometer of 15 windings with 600µm thickness of wire and 300µm distance between each wire has been designed.

1. INTRODUCTION

Magnetometer is a device used to detect external magnetic field. Basic magnetometer contains only a thin layer of metal that will detect the change in voltage when placed in magnetic field [1]. Common magnetometers are bulky. MEMS have brought an advancement to the development of magnetometer which are small, light, low power consumption, high sensitivity and high resolution [2].

Basic types of magnetometer are scalar and vector magnetometer. Scalar measured the magnitude of the vector magnetic field while the vector magnetometer measures the vector components of a magnetic field [3]. A vector is a mathematical entity with both magnitude and direction. The Earth's magnetic field at a given point is a vector. A magnetic compass is designed to give a horizontal bearing direction, whereas a vector magnetometer measures both the magnitude and direction of the total magnetic field. Three orthogonal sensors are required to measure the components of the magnetic field in all three dimensions.

The advancement in technology lead to the miniaturization of magnetometer. Some common and actively research micro magnetometer are SQUID, ferromagnetic and magnetorisistor. The early develop micro SQUID was introduced by Mark Ketchen at IBM [4]. The need to miniaturized high sensitivity low power magnetometer has led to the development of ferromagnetic MEMS magnetometer [5]. This new micro magnetometer has minimum the power intake and maintain its sensitivity although being scaled-down. This device consists of dual 4.2×2×200 µm3 polysilicon torsion bars and a 100×100×13.4 µm3 ferromagnetic plate.

2. SEARCH COIL

Search coil can be miniaturized by means of MEMS technology. Its components include, coil and core (air). When exposed to an external magnetic field, the micro coil will generate a voltage different and it represent the magnitude of magnetic flux density. When measuring the magnetic flux density, which is in Tesla (T), a voltmeter (v) is used as a representation.

A planar design of search coil magnetometer is used in this work. The main reason is to simplify the fabrication process. Figure 1 shows micro coil magnetometer while figure 2 shows the side view. The design for both driving and sensing coil are made identical to each other with different orientation. It will then apply the flip chip bonding technique.

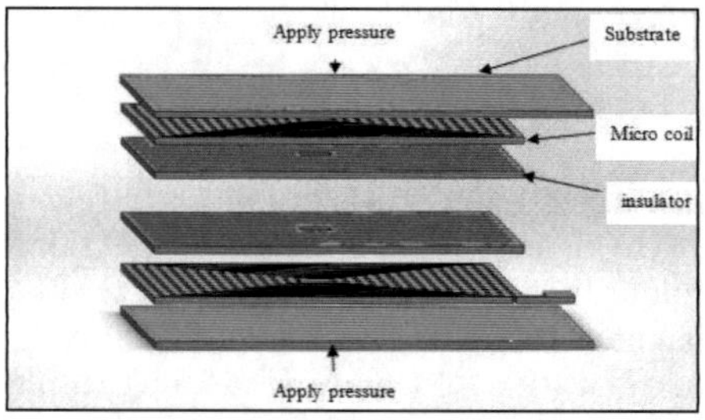

Figure 1 Micro coil magnetometer

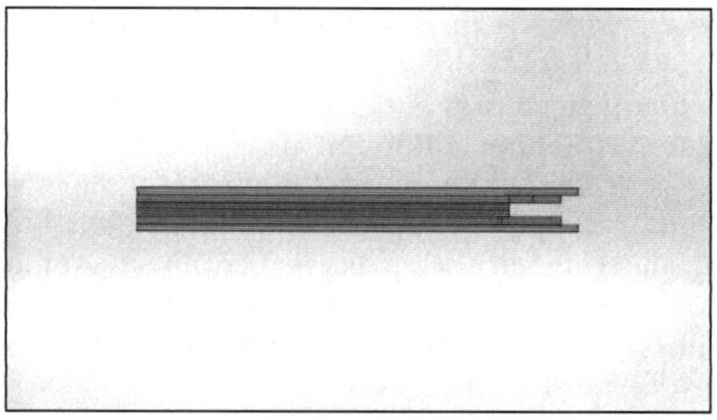

Figure 2 Side view

Search coil magnetometer can detect a magnetic field as low 20fT and no upper limit [6]. It works based on the principal of faraday's law of induction. It is suitable to use in detecting low and high magnetic field. The advantage over fluxgate magnetometer meter is that it consumes less power. Moreover, the design of search coil is simpler compare to fluxgate. Hence, it will make it more easy to focus on the fabrication technique part which is the main idea in this paper.

Search coil is a type of magnetometer that operates based on law of voltage induction. This law was introduced by Faraday which states a change in magnetic fields could produce an electromotive force or voltage. This voltage will in turn produce electric currents in a closed circuit. It is capable to produce magnetic fields in the range of 20 femtoTesla up to unlimited upper range. It is a simple and low cost magnetic field sensor. The downside is that it is unable to measure static or slow varying magnetic fields. Figure 3 shows the comparison of magnetometer in term of sensitivity and frequency. It shows that the coil-based (search coil) can detect the low magnetic field at high frequency.

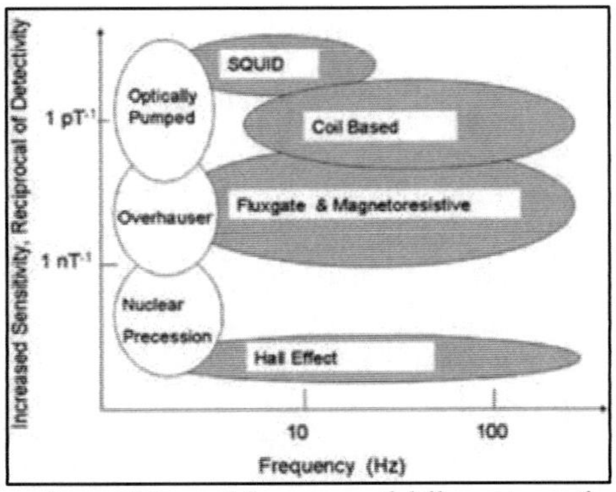

Figure 3 Comparison of sensitivity and frequency of different type of magnetometer [7]

3. RESULTS

The idea of this part is to determine the fastest possible technique that can be achieved to fabricate a micro magnetometer. A conventional way of fabrication depends on surface micro machining technique only. Here, surface micro machining is combined with the flip-chip based technique to shorten the time taken. Besides, it can have increased the possibility of refabricating and minimized the used of substances

3.1. Search Coil Fabrication

This section will show the comparison between the surface microfabrication and surface microfabrication + flip-chip bonding technique. Here, the steps taken for complete fabrication of search coil are shown.

3.1.1. Surface Micromachining + Flip Chip Bonding

From figure 4 it shows that it takes nine steps to complete the fabrication process provided both coils are fabricated in line with each other. Not only that the steps are lesser than the process which depends on surface micromachining only, it is easier to redo the process if there are mistakes along way.

IOP Conf. Series: Materials Science and Engineering **260** (2017) 012005 doi:10.1088/1757-899X/260/1/012005

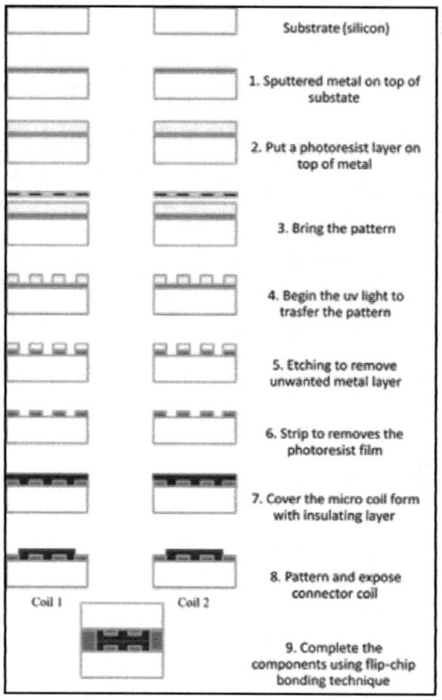

Figure 4 Surface Micromachining + Flip-Chip Bonding

3.1.2. Surface Micromachining

As shown in figure 5 and figure 6, it takes fourteen steps to complete the fabrication which is five steps more than the previously technique. The process cannot proceed unless the previous process is done. This will prolong the duration and steps compared to previous technique which enable the process to be done simultaneously. Besides, it is much more complicated to correct mistakes done during the process since the steps are depending on each other.

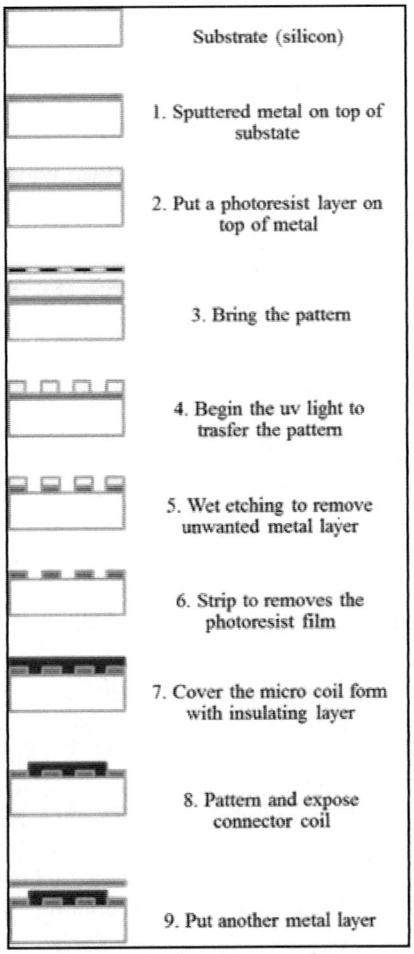

Figure 5 Surface Micromachining

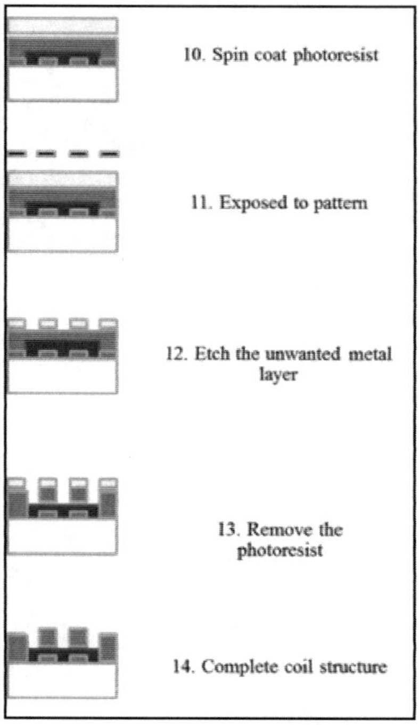

Figure 6 Surface Micromachining (cont)

Note that Figure 7 and Figure 8 shows the image taken during the process.

1) Prepare the substrate (silicon).
2) Insert the substrate into Physical Vapor Deposition (PVD) machine to get a surface of metal (aluminum) on the substrate.
3) Clean the aluminum surface with DI water to remove unwanted object.
4) Start putting positive photoresist on top of aluminum and spin coat to get evenly distributed surface.
5) Soft bake to harden a bit the photoresist.
6) Expose pattern from transparency mask.
7) Develop the pattern in developer solution and rinse with Deionized (DI) water once the pattern is visible with naked eye. Spin coat to dry.
8) Hard baked to make sure the patterned photoresist is well preserved before putting in aluminum etch solvent.
9) Put in aluminum etch solvent and rinsed with DI water once the unwanted aluminum has been etched out. Spin coat to dry.
10) Used acetone to remove the remaining photoresist on aluminum pattern. Spin coat.
11) Insulate and combined the two coil (flip-chip).

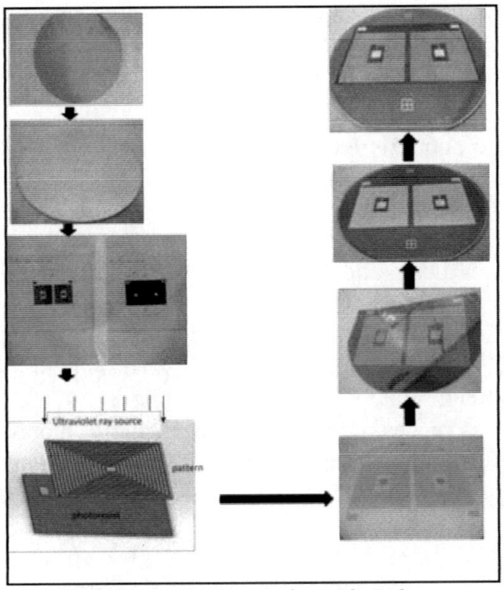

Figure 7 Fabrication of Search Coil

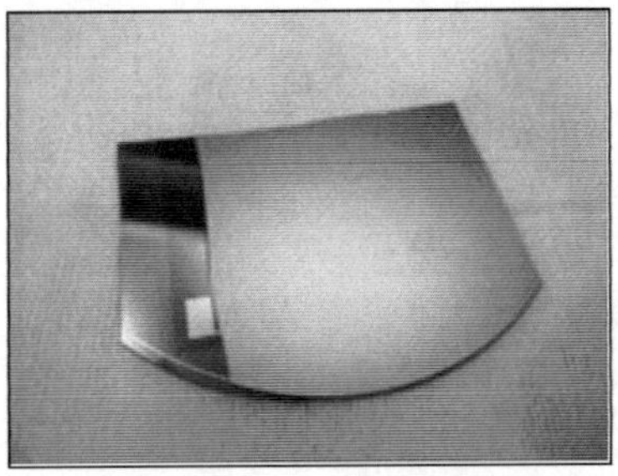

Figure 8 Complete Prototype

4. CONCLUSIONS

It is important to take notes about the steps taken during fabrication process. Types of fabrication techniques used should be chosen per the product that is intended to produce. Such reasons are to ensure the process to be reliable and efficient. The purposed of this paper have been fulfilled through the results obtained. Simultaneous simulations need to be done to get a fine and accurate results. The results are then compared and are used in designing the prototype. It will help to produce a workable prototype as the efficiency has been confirm beforehand. Although, the main idea is to produce a micro-scaled prototype, the sensitivity should be considered too. Thus, when choosing the specifications of the micro coil, it is important to compare both simulations from thickness and spacing factors. Finally, the fabrication of micro-scaled devices can be improved in many ways. Important aspect for the micro coil should be simulated repeatedly to ensure the outcome is usable. Though past researcher also did not see the flip-chip bonding approach proposed in this paper. Flip-chip bonding is not the optimum process yet. In future, through thorough investigation and studies, new processes of fabricating micro-scaled devices will be developed. In fabricating the devices, cleanroom should be utilized. Cleanrooms were initially a solution to particle contamination reduction (were invented for microelectronics dedicated mechanical assembly). The main features are; Overpressure (50 Pa) for keeping particles outside, filtered air (99.9999% at 0.15 μm particle size), heating/ cooling/ humidification/ drying of incoming air, laminar (unidirectional) air flow in the working areas, materials compatibility and mechanical and electrical interference minimization. Initially, search coil magnetometer is chosen due to its durable, room temperature operation and having non-movable components. After completing the fabrication

IOP Conf. Series: Materials Science and Engineering **260** (2017) 012005 doi:10.1088/1757-899X/260/1/012005

process, the performance testing and evaluation should be done. It is hope that researcher will continue to adapt new method to improve the current founding

Acknowledgments
The authors acknowledge the supports from the Ministry of Higher Education and the International Islamic University Malaysia under grant no. RAGS14-033-0096.

References
[1] B. Y. Y. Cai, Y. Zhao, X. Ding, and J. Fennelly February, 2012, Magnetometer basics for mobile phone applications.
[2] A. L. Herrera-may, L. A. Aguilera-cortés, P. J. García-ramírez, N. B. Mota-carrillo, W. Y. Padrón-hernández, and E. Figueras 2011 Development of Resonant Magnetic Field Microsensors: Challenges and Future Applications.
[3] Edelstein, Alan 2007. Advances in magnetometry. J. Phys.: Condens. Matter 19: 165217 (28pp).
[4] Wikswo, J. P. 1995. Squid magnetometers for biomagnetism and non-destructive testing: important questions and initial answers. IEEE Transactions on Applied Superconductivity 5(2): 74-120.
[5] Yang, H. H., Myung, N.V., Yee, J., Park, D-Y., Yoo, B-Y., Schwartz, M., Nobe, K. & Judy, J.W. 2002 ferromagnetic micromechanical magnetometer. Sensors and actuator A 97-98: 88-97.
[6] James Lenz, J., & Edelstein, A. s. 2006 magnetic sensors and their applications. IEEE Sensors Journal 6(3): 631-649.
[7] A. S. Edelstein, J Burnette, G. A. Fischer, S. F. Cheng, W. F. Egelhoff, Jr., P. W. T. Pong, R. D. Mcmichael, E. R. nowak 2008 advance in magnetometry through miniaturization.

IOP Conf. Series: Materials Science and Engineering **260** (2017) 012006 doi:10.1088/1757-899X/260/1/012006

Mechatronics technology in predictive maintenance method

Nurul Afiqah A. Majid[1], Asan G.A. Muthalif[2]

Smart Structures, Systems and Control Research Lab (S[3]CRL), Department of Mechatronics Engineering, International Islamic University Malaysia, Kuala Lumpur

[1]afiqahmajid93@gmail.com, [2]asan@iium.edu.my

Abstract. This paper presents recent mechatronics technology that can help to implement predictive maintenance by combining intelligent and predictive maintenance instrument. Vibration Fault Simulation System (VFSS) is an example of mechatronics system. The focus of this study is the prediction on the use of critical machines to detect vibration. Vibration measurement is often used as the key indicator of the state of the machine. This paper shows the choice of the appropriate strategy in the vibration of diagnostic process of the mechanical system, especially rotating machines, in recognition of the failure during the working process. In this paper, the vibration signature analysis is implemented to detect faults in rotary machining that includes imbalance, mechanical looseness, bent shaft, misalignment, missing blade bearing fault, balancing mass and critical speed. In order to perform vibration signature analysis for rotating machinery faults, studies have been made on how mechatronics technology is used as predictive maintenance methods. Vibration Faults Simulation Rig (VFSR) is designed to simulate and understand faults signatures. These techniques are based on the processing of vibrational data in frequency-domain. The LabVIEW-based spectrum analyzer software is developed to acquire and extract frequency contents of faults signals. This system is successfully tested based on the unique vibration fault signatures that always occur in a rotating machinery.

Keywords. Condition monitoring, lab-VIEW based spectrum, signal analysis

1. Introduction

Mechatronics engineering involves multidisciplinary branches of engineering studies that is capable of providing the necessary skills to install, calibrate, modify, troubleshoot, repair and maintain automated systems [1]. In industrial applications, most of the equipment or machinery are exposed to external vibration, especially in rotating machinery. Machine condition monitoring is the process of monitoring the condition of a machine with the intention to predict mechanical failures. Vibration is a key indicator of the state of the machine and widely used for predictive maintenance. Hence, a simulation rig, incorporating mechatronics technologies such as sensors, actuators, signal processing and control, is vital to understand and implementation predictive maintenance[2].

The basic maintenance strategies can be divided into three parts which are breakdown maintenance, preventive maintenance and predictive maintenance [3]. No actions or maintenance will take place until the design life is reached in breakdown maintenance. It is a reactive process and thus, it is unable to prevent downtime from happening. On the other hand, preventive maintenance is performed on a schedule basis. However, predictive maintenance focus on periodic or continuous equipment

IOP Conf. Series: Materials Science and Engineering **260** (2017) 012006 doi:10.1088/1757-899X/260/1/012006

condition monitoring which can eventually help in reducing the cost. It aims at knowing the earliest possible occurrence of damage to a unit in order to prevent equipment breakdown that could lead to stagnation of a production process [4].

Vibration signals produced by rotating machinery contains information on the nature of the fault [5]. The parameters needed for vibration indicator are listed in Table 1 [6].

Table 1 Parameter needed for machine health assessments

Fault	**Parameters**				
	Temperature	Pressure	Flow	Oil Analysis	Vibration
Out Of Balance					x
Misalignment Bent Shaft	x				x
Damaged Rolling Element Bearings	x			x	x
Damaged Journal Bearings	x	x	x	x	x
Damaged Or Worn Gears				x	x
Mechanical Looseness					x
Noise					x
Cracking					x

Effective vibration analysis starts with obtaining an accurate signal from standard vibration transducer with the help of an accelerometer. The analog signal is then converted into digital signal using analog to digital converter. The digital signals can be processed directly or using formula defined by the user.

In this paper, Vibration Fault Simulation System (VFSS) is introduced to further improve the understanding of vibration fault signals. Here, Vibration Fault Simulation Rig (VFSR) is developed to study the different maintenance techniques as well as for further development of vibration faults system. LabVIEW software is used to acquire the experimental data. This paper is organized as below: Section 1 provides the introduction to mechanical health whereas Section 2 discusses on the Vibration Fault Simulation System (VFSS). Sections 3 and 4 presents the experimental setup and results.

2. Vibration Fault Simulation System (VFSS)

The design proposed in this paper aims at detecting the unique vibration faults signatures of a rotating machine. To facilitate the detection of mechanical defects in digital format, the mechanical vibration of the motor is converted to electrical analog signals by a sensor. This model of vibration faults can be used for the purposes to study vibration pattern in various faults and as a training equipment for technical staff and engineering students in vibration monitoring and analysis of rotating machinery [6].

The fault simulation that occur in a rotating machinery are imbalance, misalignment, mechanical looseness, bearing fault, bent shaft, missing blade, balancing and critical speed. The system is divided into four subsystems, which are:

i. Main shaft system - to simulate imbalance, angular misalignment, parallel misalignment, mechanical looseness, bearing fault and missing blade.
ii. Bent shaft system - to simulate bent shaft faults.
iii. Balancing mass system - can simulate balancing faults.
iv. Critical speed system - to simulate critical speed in rotating machinery.

2.1. Main shaft system setup procedures

The setup of the main shaft system can be describe in the following steps:

1. Insert rotor disc through the shaft (10 mm diameter) and slot the shaft through the bearing 1 and bearing 2 as shown in Figure 1.
2. Before tightening its setscrew, slot flexible coupling to the shaft.
3. Tighten the setscrews of rotor disc after placing it in the middle of shaft.
4. The propeller was slotted at the end of the shaft and tighten its setscrews.

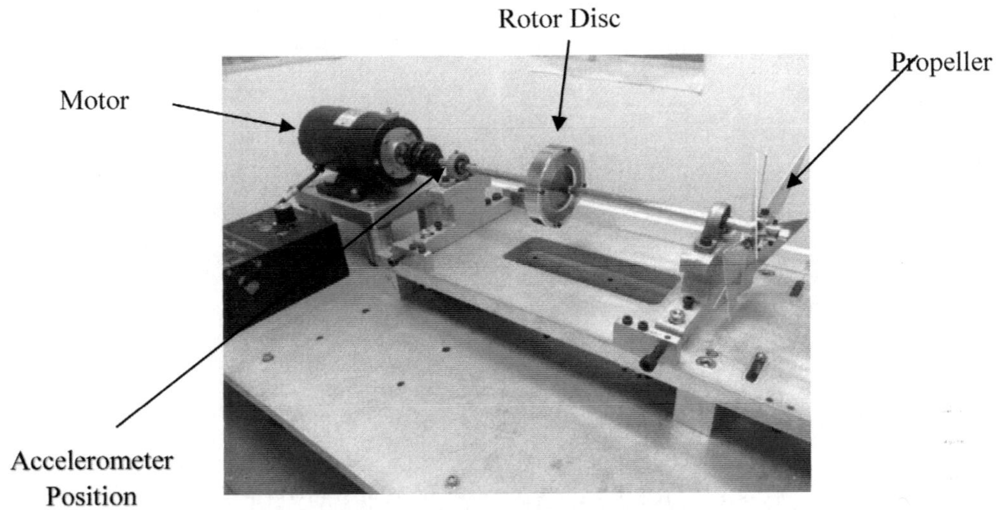

Figure 1 Main shaft system.

3. Methods of simulation

Vibration fault signals will be sent into Data Acquisition (DAQ) device where it will be used to interact or to read the accelerometer signal before being analyzed and displayed by the LabVIEW software. Every fault in the rotating machine has its own unique signature that can be detected and analyzed. For every fault, the result recorded is in the form of frequency analysis. All the simulations were done using the procedures explained in the Table 2.

Table 2 Methods of simulation.

Machinery Faults	Procedures
Imbalance	1. The accelerometer will be mounted to the bearing 1. 2. Then, run the motor and view the spectrum display. 3. While the motor is running, drop the ball into rotor disc to create the imbalance faults. 4. Lastly, view the new spectrum display and record the new data.
Mechanical Looseness	1. Run the motor at some speed and view the spectrum display. 2. To create the mechanical looseness, slowly loose the screws at the bearing base stand. 3. Lastly, view the new spectrum display and record the new data.

4. Results and analysis

1. Normal Spectrum:

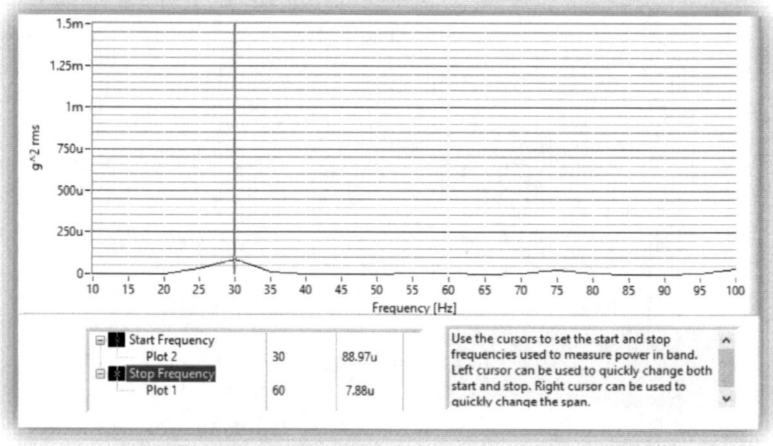

Figure 2 Spectrum analysis for normal spectrum.

Figure 2 shows the result for normal spectrum of the VFSR when there is no faults occurs on the VFSR. The motor is run at 52% of its maximum speed (3450 rpm) which is at 1794 rpm. In frequency domain, 1794 rpm is equivalent to 30 Hz. It can be observed from the graph that the first peak exists at 23 Hz (which means 1 time running speed), while the second peak exists near 60 Hz (2 times running speed). The peaks are found not accurately at the 30 Hz and 60 Hz due to noise disturbance from the surrounding during experiment.

2. Imbalance:

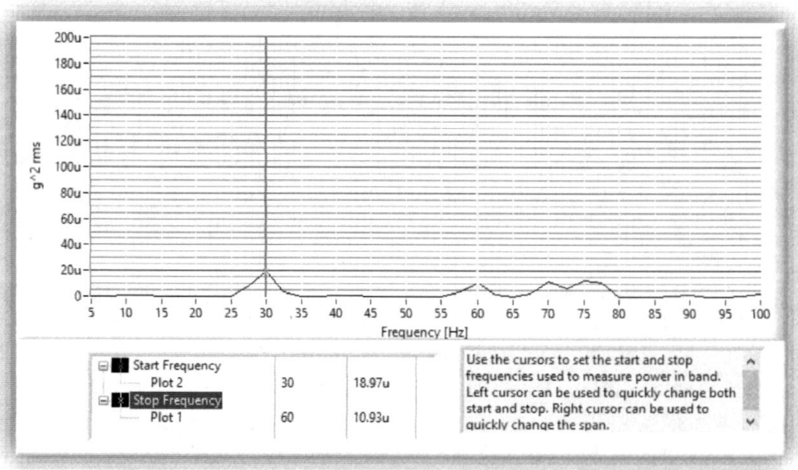

Figure 3 Spectrum analysis for imbalance (rotor disc).

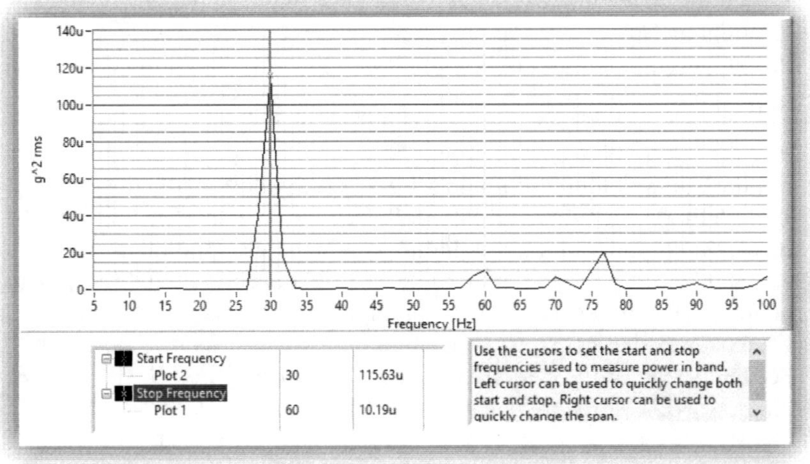

Figure 4 Spectrum analysis for imbalance (rotor disc and ball).

Figure 3 shows the spectrum for imbalance with rotor disc only. It can be observed that the peaks are found at 30 Hz (1 time running speed) and 60 Hz (2 times running speed). When the ball is drop into the rotor disc, the ball allocated itself at the hole inside the rotor disc during rotation. This causes unbalance force and the spectrum is shown in Figure 3. The amplitude due to unbalance compared to Figure 4. Theoretically, the peak at 1 time running speed always present and dominates the spectrum.

Mechanical Looseness:

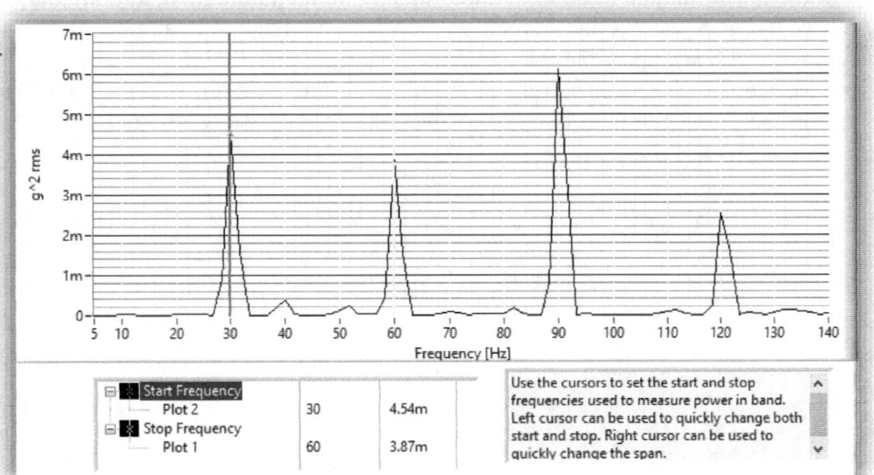

Figure 5: Spectrum analysis for mechanical looseness.

Figure 5 shows the spectrum of mechanical looseness. From the spectrum analysis above, the signal produced harmonically when there is some mechanical looseness occurs at VFSR. The signature occurred due to the improper fit between the bearing base and the plate base cause due to non-linear response of loose parts to the dynamic forces from the motor. From the figure above also, the harmonics

IOP Conf. Series: Materials Science and Engineering **260** (2017) 012006 doi:10.1088/1757-899X/260/1/012006

occur starting from 30 Hz (1 time running speed), 60 Hz (2times running speed), 90 Hz (3 times running speed), 120 Hz (4 times running speed) and etc.

5. Conclusion

Condition monitoring is the best way to monitor and observed the condition of a machine, especially for rotating machine. Vibration of the rotating machine is the best parameter to be monitored as each fault has its own unique signature. The early stage of a failure can be detected and the real failure of certain part of rotating machine can be predicted. Vibration Fault Simulation System (VFSS) is an application of mechatronics technology, which can be used in predictive maintenance. This paper presents the unique vibration fault signatures that occur in a machine by using a sensor to detect the spectrum of rotating machine. Also, the design of the Vibration Faults Simulation Rig (VFSR) has been proven to work well. In addition, t LabVIEW software is used to simulate common vibration faults.

References

[1] C. W. De Silva, *Mechatronics: an integrated approach*: CRC press, 2004.
[2] C. W. De Silva, *Vibration: fundamentals and practice*: CRC press, 2006.
[3] T. M. Hunt, *Condition monitoring of mechanical and hydraulic plant: a concise introduction and guide*: Chapman & Hall, 1996.
[4] S. Saponara, L. Fanucci, F. Bernardo, and A. Falciani, "A network of vibration measuring nodes with integrated signal processing for predictive maintenance of high power transformers," in *Intelligent Signal Processing (WISP), 2015 IEEE 9th International Symposium on*, 2015, pp. 1-4.
[5] N. Nikolaou and I. Antoniadis, "Demodulation of vibration signals generated by defects in rolling element bearings using complex shifted Morlet wavelets," *Mechanical systems and signal processing,* vol. 16, pp. 677-694, 2002.
[6] A. Gani and M. Salami, "Vibration faults simulation system (VFSS): A system for teaching and training on fault detection and diagnosis," in *Research and Development, 2002. SCOReD 2002. Student Conference on*, 2002, pp. 15-18.

IOP Conf. Series: Materials Science and Engineering **260** (2017) 012007 doi:10.1088/1757-899X/260/1/012007

Human location estimation using thermopile array sensor

S Parnin[1] and M M Rahman[2]

Department of Mechatronics Engineering, Kulliyyah of Engineering, International Islamic University Malaysia, 53100 Kuala Lumpur, Malaysia.

[1]sullehaparnin@gmail.com, [2]mozasser@iium.edu.my

Abstract. Utilization of Thermopile sensor at an early stage of human detection is challenging as there are many things that produce thermal heat other than human such as electrical appliances and animals. Therefrom, an algorithm for early presence detection has been developed through the study of human body temperature behaviour with respect to the room temperature. The change in non-contact detected temperature of human varied according to body parts. In an indoor room, upper parts of human body change up to 3°C whereas lower part ranging from 0.58^0 C to 1.71^0 C. The average changes in temperature of human is used as a conditional set-point value in the program algorithm to detect human presence. The current position of human and its respective angle is gained when human is presence at certain pixels of Thermopile's sensor array. Human position is estimated successfully as the developed sensory system is tested to the actuator of a stand fan.

1. Introduction

Localization estimations of a subject prior to the location of the sensor can be interpreted using distance, angle, and coordinates. Depending on the parameters measured by sensor, the output is translated into distance, angle, coordinates, and etc. Our early work in [1], only tested a single Thermopile sensor to distinguish the sensor behaviour with respect to number of human in the sensor's field of view (FOV). Result support that the location of occupants can be determine by the highest and distinguishable temperature value of the pixels. However, we could not solely depend on one, but want to strengthen the finding by adding another stage of human sensing detection precedence the localization stages which is presence detection stage. Thus, the main direction of this work is to develop algorithm for human location detection and to rotate an actuator to the detected location.

2. Thermopile sensor

Thermal sensor or temperature sensor is a device that detects heat, a distinguishable property that human has. In general, operation types of Thermal sensor can be either contact or non-contact based. Examples of contact types of temperature measurement sensor are Thermocouple, RTD and Thermistor. Meanwhile, for non-contact sensors for thermal radiation measurement are Pyroelectric, Thermopile, and Bolometer. To this date, research for applying non-contact Thermal sensor for human sensing is not as common as the other sensors such as Pyroelectric and Ultrasonic. The possible application of this human detection heat-based sensory system are home automation, security, and other indoor types of application.

Many of the Thermal sensor used MEMS technology, a process to create a tiny microscopic size of mechanical devices (i.e. MEMS Thermopile). Thermopile array types of Thermal sensor is one of the

IOP Conf. Series: Materials Science and Engineering **260** (2017) 012007 doi:10.1088/1757-899X/260/1/012007

famous passive sensor among the researcher for human detection. It comprises of a stack of junction (i.e. serially connected Thermocouple) [2] of two diverse materials that when the voltage is applied at the junction, voltage difference and temperature difference will be produced. Passive characteristics does not allow it to radiate energy to the surrounding. Therefore, Thermopile sensor will not be affected by the reflectivity of the materials or object [3]. As Thermopile is a non-contact Temperature sensor, energy radiated by the object is collected by the absorber attached to the Thermocouple junction make it possible for the temperature to be measured from a distance location.

Method proposed in [3] for human detection in application for pedestrian counting at doorway used TPA81 Thermal sensor manufactured by Devantech Ltd. A person is said to be detected when the temperature exceeds the normalized or threshold temperature. Furthermore, different result of the correct detection rate for different walking speed are recorded. It was found that correct detection rate decreases as the speed is increasing from 98% to 90% for 0.5 $\frac{m}{s}$ to 2 $\frac{m}{s}$ walking speed. For the high pedestrian traffic entrance experiments, 95.7% correct detection was documented. It has a total of 18 times false detection in the duration of 30 minutes.

Kuki et al. in [4] proposed a multiple human location estimation system using Thermopile array sensors. Sensors installed at the ceiling detect the temperature of an empty room space. They employ a background removal and fuzzy inference on the temperature distributions. The system does not give a good result in human presence detection but it is excellent in human counting. It has 52.4±12.4% total correct rate for a dynamic or moving human test. A higher correct rate is obtained for a stationary position as thermal distribution of stationary person is clear than a moving person.

There is also a research attempt to detect human temperature wirelessly using 40X60 Thermal array sensor for smart camera application [5]. When the sensor reading is recorded, it can produce a thermal image similar from the output of thermal camera. That is the reason it can produce an image if the sensor FOV is wide. As such in his research, many set of Thermal sensors are needed for having a bigger picture frame. The thermal images are then filtered and isolated using MATLAB to the areas of high heat. Their result shows the physical shape of human body can be clearly seen at distance of 0.5 meter from the sensor.

In [6] using Daubechies Wavelets (i.e. a nonlinear and non-parametric method) to eliminate the high frequency noise from the thermopile array sensor readings. The process of eliminate the noise that coming from the unstable data using Daubechies Wavelets give a better result without altering useful information such as signal rupture produces by human presence. The time series graph human presence denotes that once human is exiting, temperature drops to ambient temperature.

In [7], they design a sensing units that state the status of room occupancy and counts the number of human using two OMRON Thermopile sensors. They found that at a long range of between 6 m to 9 m, temperature reading recorded an average rise of 3^0 C upon the presence of an object known to be 10^0 C warmer than the ambient temperature. Furthermore, overlapping two sensor FOV allows the identification of the human location with respect to the sensor position. Both sensors will give an intersection point for the detected human since both FOV are overlapping.

3. Algorithm development

3.1 Changes in non-contact human body temperature using single Thermopile sensor

Experiment is conducted to analyse the changes of human body temperature with respect to the room temperature using a single Thermopile sensor D6T-8L-06. The sensor is tilt to 90° angle therefore the adjustment can cover the whole length of human body. The measured distance of 2 m from the sensor gives the total width and a width of every pixel to be about 0.2 m each respectively. At this total wideness, it is enough to fit in the height of a man up to 1.7 m. Meanwhile, Thermopile sensor is attached at the height of 70 cm to the fixed position aluminium rod. The pixels of Thermopile sensor are positioned to target different human body parts such as head, neck, shoulder, midriff, waist, thigh, leg, and lower leg.

As controlled parameters, experiment is conducted in an indoor room as outdoor environment have high sunlight penetration exposure that can affect temperature readings. A 5 mX3 m room is used throughout the experiment. Before experiment subjects are rested in an air-conditioned room for a period of time (i.e. 10 to 15 minutes), then is brought to the experimental room with no air conditioning appliance. The purpose is to study the minimum net changes since the body already regulate to a balance thermal comfort after sitting long in the air condition room. Therefore, the changes in temperature can be used as the setting point or condition in the algorithm to detect the human presence.

An image of the sensor FOV is captured right before the subject entered the experimental area. Different configurations of human body (i.e. front, side and back) in front of the sensor is shown in figure 1. The significant of this configuration to illustrates the freedom of the human behaviour in front of the sensor as human may face the presence detection device from front, side or blocking the fan from back.

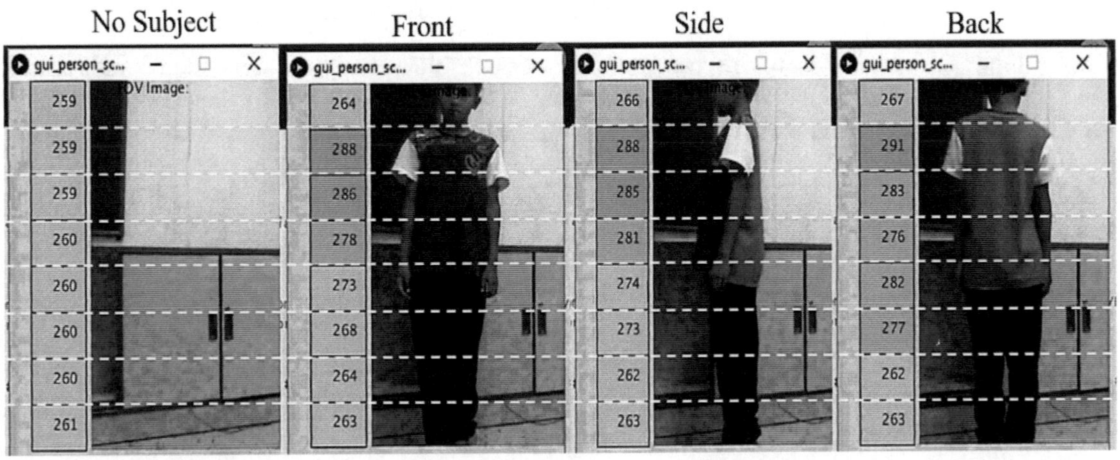

Figure 1. Sample of GUI windows for the experiments for Front, Side, and Back position. The program is developed using Processing user interface. Temperature value is in multiple of 10 and unit is in ^{0}C.

Changes in the temperature is used as the setting point or condition in the algorithm to detect the human presence. This research denoted each body part temperature as T_p. Where the subscript p denotes for the parts of human body (i.e. head, shoulder, chest, midriff, waist, thigh, leg and lower leg). Equation (1) and equation (2) are used to estimate the changes of section of human body temperature, ΔT_p to the room temperature, T_R.

$$\Delta T_p = T_p - T_{R,AVG} \tag{1}$$

$$T_{R,AVG} = \frac{\sum_{0}^{N} T_X}{N} \tag{2}$$

$T_{R,AVG}$ is notation for average temperature for sensor pixels in the room. While, notation T_X is used to describe the temperature data of each pixel and the formulae for derivation of T_X is explain in our previous work in [1]. Table 1 tabulated the result of the experiment for ten subjects.

IOP Conf. Series: Materials Science and Engineering **260** (2017) 012007 doi:10.1088/1757-899X/260/1/012007

Table 1. Average Changes in temperature of body parts for front, side, and back configuration for 10 subjects with respect to average temperature, ΔT_p

Body Parts (p)	*Front* (^{0}C)	*Side* (^{0}C)	*Back* (^{0}C)
Head	1.6042	1.6960	1.4875
Shoulder	2.8990	2.5163	2.6263
Chest	2.5563	2.7563	2.9563
Midriff	2.4763	2.0613	2.1363
Waist	2.2363	1.7412	2.3163
Tight	1.6563	1.5563	1.7088
Leg	1.0153	1.1486	1.2156
Lower leg	0.6833	0.5875	0.6833

Since human wear a variety of types and thickness of clothes, the non-contact temperature will not give the output as high as the normal skin temperature (i.e. 36-37^0 C). Rather the expected output will be less due to the thermal insulation of the clothes. The average changes in temperature resulted that human takes a change of more than 2^0 C for upper parts of the human body except for the head. While, the lower part of a body changes at average of more than 1°C except for the lower leg. This result is used as the conditional set points in the program to detect human presence in the next section.

3.2 Multiple sensor algorithm

The algorithm to combine two Thermopile sensors can be written as in the form of equations. Equation (3) and equation (4) lists the temperature that can be collected by each of two sensors at instant time, t. $TPAT$ is the temperature at the surface of the sensors itself. Both Thermopile sensors will give one TPAT and 1 × 8 array of FOV pixels.

$$T\ Distributions\ sensor1 = \{TPAT(t), T0(t), T1(t), T2(t), \ldots\ldots, T7(t)\} \quad (3)$$

$$T\ Distributions\ sensor2 = \{TPAT(t), T0(t), T1(t), T2(t), \ldots\ldots, T7(t)\} \quad (4)$$

Combination of both sensors into 1×16 array excluding the sensor's surface temperature (i.e. $TPAT$(°C)) give new temperature distribution of T_X as in equation (5).

$$While\ T_X = \{\ T0(t), T1(t), T2(t), \ldots\ldots, T15(t)\}\ °C \quad (5)$$

Recall that T_X is the temperature outputted from the Thermopile sensor which depends on the objective and background temperature. Because of the human skin temperature can be distinguishable and always be higher than any other temperature we could compare the temperature from Equation 5 to find the maximum pixel in the array, T_{max} using equation (6).

$$T_{max} = \max_{0 \leq i \leq 15} T_X \quad °C \quad (6)$$

$$T_{max} = \max_{0 \leq i \leq 15}\{\ T0(t), T1(t), T2(t), \ldots\ldots, T15(t)\}\ °C \quad (7)$$

Equation 6 can also be written as in equation (7). To distinguish the types of object in front of sensor FOV (i.e. human and non-human), T_{max} is further compared with the setting point value from its average temperature data from the previous experiment.

IOP Conf. Series: Materials Science and Engineering **260** (2017) 012007 doi:10.1088/1757-899X/260/1/012007

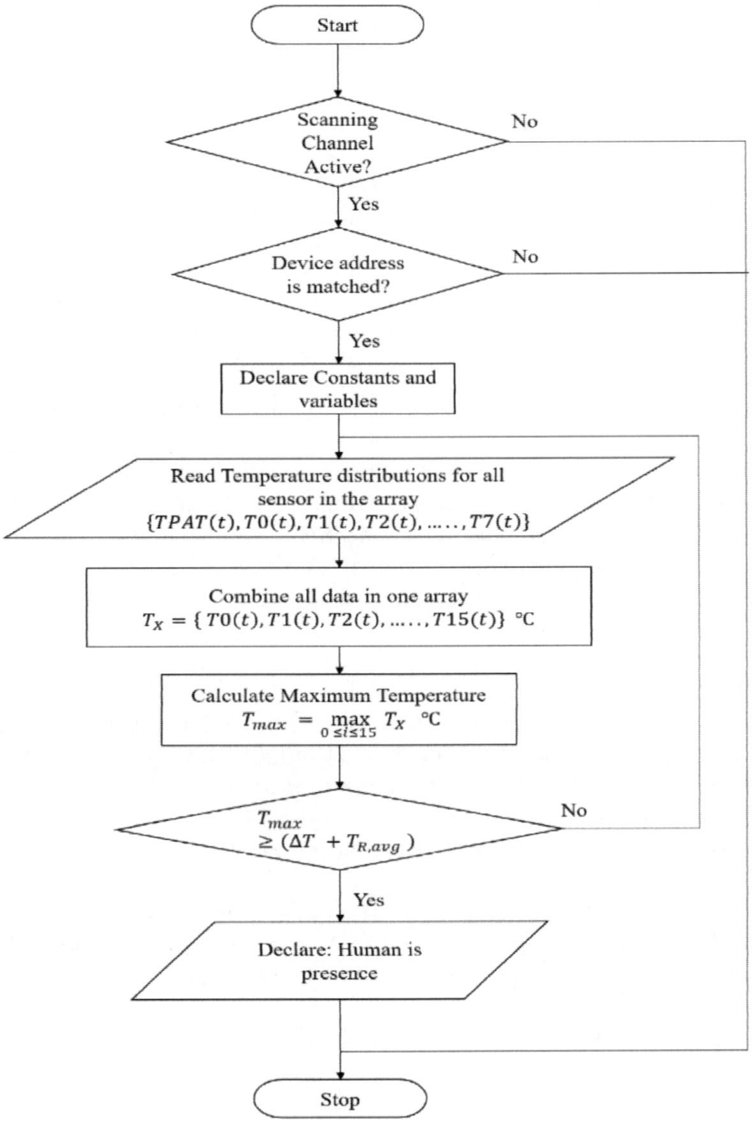

Figure 2. Flowchart for human presence detection algorithm using Thermopile sensor. Maximum number of Thermopile sensor can be used in the algorithm is 4.

$$Human\,presence = \begin{cases} 1, & T_{max} \geq (\Delta T\ + T_{R,avg}\ ) \\ 0, & T_{max}\ < (\Delta T\ + T_{R,avg}\ ) \end{cases} \quad (8)$$

From equation (8), human is detected and returned value of 1 when the condition is true and returned zero if otherwise. The presence of human in the room as estimated in the equation is only depend on the temperature regardless of the state of motion. In this research, change of temperature due to the presence of human, ΔT generally is chosen as 2^0 C as the height of sensor aim to sense at the upper part of the body. Figure 2 shows the algorithm flowchart for human presence detection. Once the human is detected, the direction in terms of angle is estimated as in section 5.

4. System block diagram and schematics

Block diagram of Thermopile sensory system is shown in figure 3. Sensor array is considered as 'Device' and is connected to the 'Master' (i.e. Controller) through the 'Slave' (i.e. multiplexer). When the controller sends the address of $0X70$ to multiplexer, all the four channel is enable. As the channels are enable, controller read the temperature distributions, T by commanding read and write to the device address of $0X0A$.

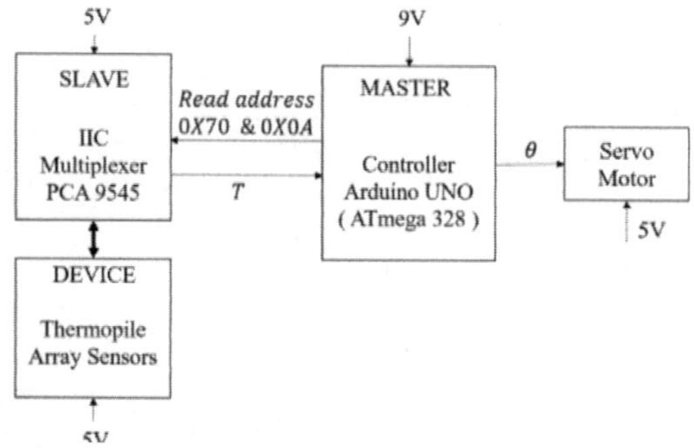

Figure 3. Block Diagram of Thermopile array sensor.

After the human location (angle) is estimated, controller use the value to rotate the actuator toward the current human location.

Thermopile array sensor used Inter-Integrated Circuit (IIC) communication protocol. Therefore, data can be transmitted and received only using two serial lines, which are, Serial Data (SDA) and Serial Clock Line (SCL). Arduino Microcontroller only have one SDA and SCL pin. Therefore, multiplexer switch is needed to connect multiple Thermopile together. PCA9545 is a 4-channel switch multiplexer. PCA 9545 comes in a very small size of surface mounting chips.

5. Sensor array arrangement

5.1 Sensor Array Positioning

Since one D6T-8L-06 can have a wide width angle of 62.8^0, another sensor is added to widen the sensor FOV. The position of one sensor to another matter as much as to reduce the uncovered spot (i.e. blind spot). The following restricts the arrangement of the Thermopile sensors: -

i. Two sensors FOV must not be overlapped
ii. Two sensors must close enough to reduce blind spot between each FOV.

A sensors holder is designed to satisfy the design constraint. Therefore, this combination of two sensors has the maximum of 125.6^0 wide width-angle of sensors FOV and a total of 16 independent pixels. Figure 4 shows the illustration of subject movement in 125.6^0 wide width-angle.

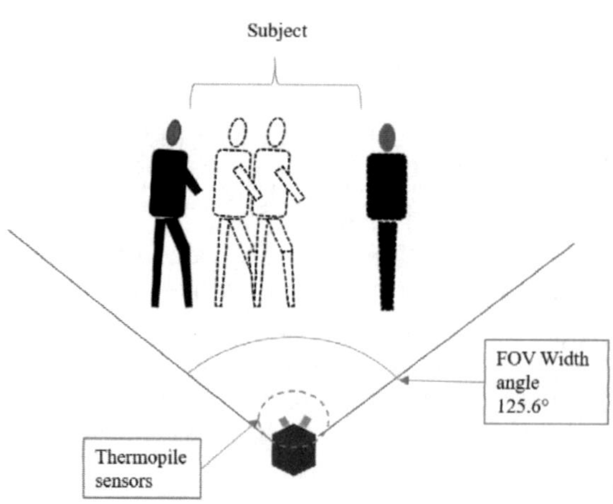

Figure 4. Illustration of the FOV of two Thermopile D6T-8L-06 sensors.

5.2 Actuator angle direction estimation Method

The next step after sensing the human presence is to direct the servo motor to the location that the human is presence. Taking the advantage that the eight pixels have independent temperature value and

IOP Conf. Series: Materials Science and Engineering **260** (2017) 012007 doi:10.1088/1757-899X/260/1/012007

has a horizontal FOV of 62.8^0 for single sensor, we develop an active FOV of 125.6^0. Shown in the figure 5(a), is the position of the eight pixels of sensor 1 and 2. To minimize the blind spot and maximize the total sensor's FOV, sensor 1 is flipped to 180^0. This configuration has interchange the pixel position of the sensor 1 causing both P0 of the sensors to meet in the middle section.
As the horizontal angle of the sensors are 62.8^0 each, and each pixel have wide angle, $\emptyset$ of 7.85^0. Therefore θ_1 and θ_2 are both 27.2^0.

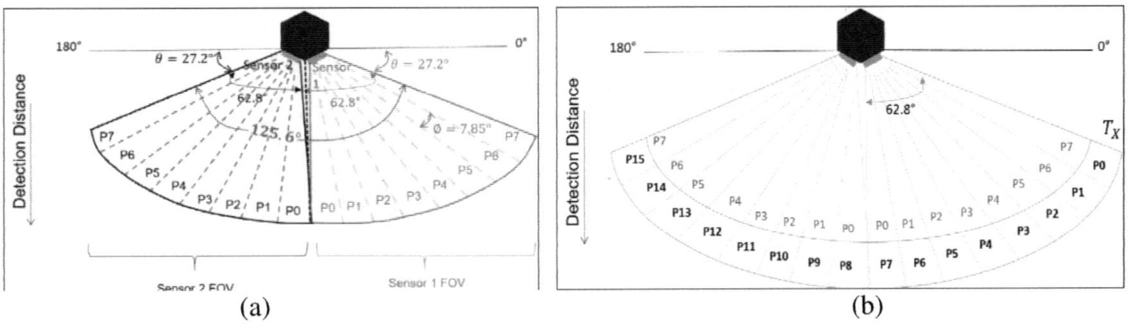

Figure 5. a) Total FOV angles. b) T_X arrangement.

The detection power will decrease over the increase in detection distance. As explained in the algorithm, a set of new pixel number is introduced to store the combination of both sensors pixels containing 16 elements in the array of T_X (equation (5) and illustrated in figure 5(b).

Upon detecting the presence of human in any pixels of T_X, the fan-head servo will move to the centre of the pixel. Let say, human is detected at P0, then servo will rotate at an angle of $\theta + (\emptyset/2)$. The most right of sensor 1 is set to be the 0^0 angle.
The presence of human in the next pixel will increase the angle by $\emptyset$. The last swinging angle is therefore $\theta + \frac{31\emptyset}{2}$ which is equivalent to $(125.6° + 27.2° - \frac{7.85°}{2})$ or $(180° - 27.2° + \frac{7.85°}{2})$. Figure 6 depict the assignation of angle of the servo is directed to go if the certain pixel is occupied with human presence.

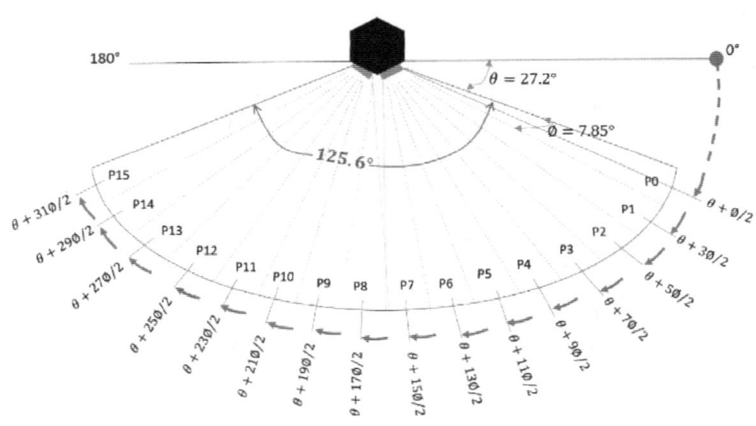

Figure 6. Angle assignation of each pixel location.

6. Result

Similar to [1], we tested the developed algorithm to a Standing Fan which swing according to the presence of human in the room.

IOP Conf. Series: Materials Science and Engineering **260** (2017) 012007 doi:10.1088/1757-899X/260/1/012007

6.1 Single Human Detection (Following Mode)

The fan under test was kept in a 5X3 m room that has one window, one door lead to outside, and two doors lead to another room. All doors have at least a 10 mm allowable distance between the floor to allow the door to be open, therefore the temperature from the other two rooms may affect the test room through the space under it. The window was not exposed to the direct sunlight as the hallway very wide, but still heat transfer from the window could not be neglected. The air-condition from the two room was turn on, and the door is open and closed at any time. The first mode of the test is to see the fan response corresponding to the subject facing to the fan. This means that the Thermopile sensory system will only detect the Front body temperature. Initially, subjects are asked to step on all the grid point marked on the floor and stay stationary. Figure 7 illustrate the motion of the swing fan upon the presence of a subject. It is observed that the fan head follows the subject by estimating the correct angle successfully.

As the subject is move to another point, fan head also rotate to the new position. When the subject is at stationary, fan head also stays at its current position. Furthermore, sensors also could detect subject successfully up to 4 m. The test for the distance of 3 m is shown in figure 7 (a). Despite the successful response, there are cases of slower response of the system feedback. This happen only at the middle pixel location (i.e. pixel 7 and pixel 8).

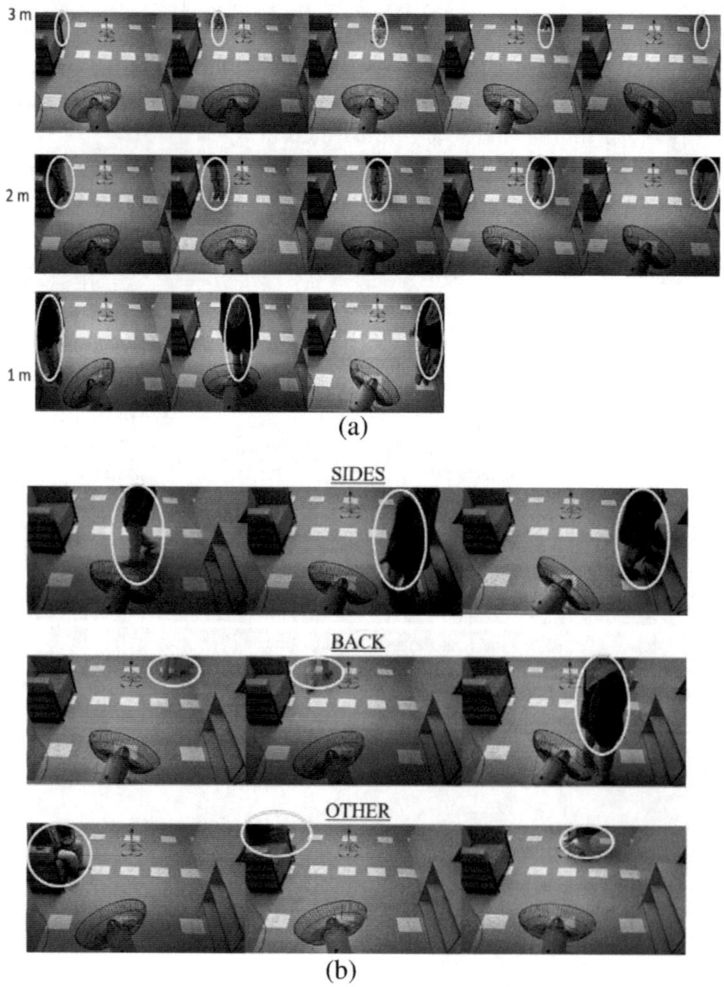

Figure 7. (a) Fan response to a subject facing FRONT of the sensor. (b) Fan response for single subject in free-style motion

In Figure 7 (b), the swing fan is tested during the subject move at free path. This test the sensor ability to detect human from either Front, Side, or Back of the body temperatures. Again, the fan is able to estimate the correct angle of the location of subject efficiently.

6.2 Multiple Human Detection (Ranging Mode)

In the real utilization of a regular Stand fan, the default swinging angle can serve multiple humans. Therefore, to make the fan to become more interactive corresponding to multiple human, ranging mode is developed. In this mode, the fan can estimate the swinging range between two extreme temperature at the pixel location occupied by humans. The larger the distance between the humans, the bigger the angle range. For example, if subject A and subject B move from far to closer, the swing angle also become smaller. Figure 8 illustrate the fan response to multiple human.

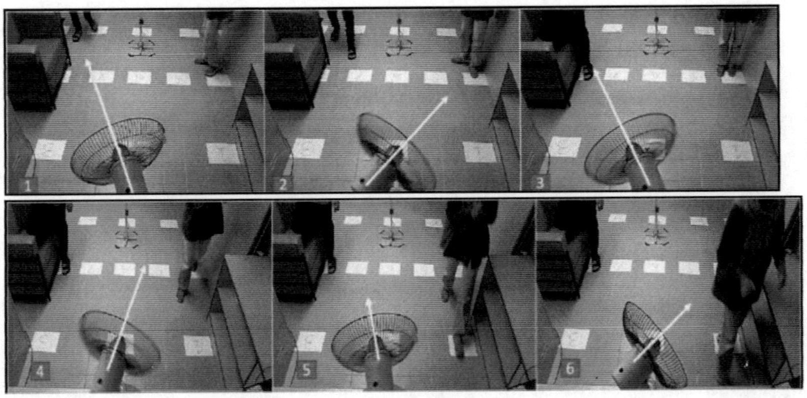

Figure 8. Fan response for multiple subject.

Thermopile sensory system developed have the advantages in detecting a single and multiple human presence and its respective location. However, the controller need to be program separately. It is possible to combined both mode into one program but it will require much deeper programming knowledge and time.

7. Conclusion

The number of Thermopile sensors used and their arrangements or positioning with respect to one another would return a different FOV. Vertical sensor arrangement was used to study the non-contact temperature change of human. While, in general application, horizontally array is used. The experiments result shows significant changes in the non-contact body temperature. Using the algorithm developed, ΔT_p is generally chosen as 2^0 as the height of the swing appliance (i.e. stand fan) is limited to detect the upper part of human body. Condition as in equation 8 is applicable to verify the human presence. However, it is needed for the algorithm to be tested on more subjects and at different other indoor room. Other heat producing device or things such as computer CPU, iron, pets, or even a cup of hot water should be added within FOV to test the reliable of algorithm in near future.

Thermal sensory system developed may even be applied in the security system or even a military purpose for aiming the direction of a target. For instance, in the surveillance system, sensory system can be attached with camera to record and widen the recorded area by following the motion.

To summarize, our approach is only applicable to detect up to 2 persons between FOV. Different programs are used for each situation (i.e. following mode is for a single person is presence and ranging mode is for more than one person is detected). Swing type of electrical appliances such as stand fan that used to test the Thermopile sensory system developed is chosen due to its wide household application. Therefore, producing new types of smarter stand fan that equipped with human detection system is far more important rather than to produce a regular type of fan. In future, it is hope that this system can be upgraded to detect multiple number of human in a single program mode.

IOP Conf. Series: Materials Science and Engineering **260** (2017) 012007 doi:10.1088/1757-899X/260/1/012007

8. References

[1] S. Parnin and M. M. Rahman, "Human Location Detection System Using Micro-Electromechanical Sensor for Intelligent Fan," IOP Conference Series: Materials Science and Engineering, Volume 184, conference 1, pp. 1–8, 2017.

[2] OMRON Corporation, "Infrared MEMS Thermal Sensor D6T Product Series : Infrared MEMS Thermal Sensor," 2013.

[3] L. Honorato, I. Spiniak, and M. Torres-torriti, "Human Detection Using Thermopiles," in IEEE Latin American Robotic Symposium, 2008, pp. 151–157.

[4] M. Kuki, H. Nakajima, N. Tsuchiya, K. Kuramoto, S. Kobashi, and Y. Hata, "Mining multi human locations using thermopile array sensors," Proc. Int. Symp. Mult. Log., pp. 59–64, 2013.

[5] T. Medjeldi and S. Guillemette, "Wireless System for Detecting Human Temperature," pp. 156–159, 2013.

[6] T. Guettari, J. Boudy, B. Benkelfat, G. Chollet, J. Baldinger, P. Dore, and D. Istrate, "thermal signal analysis in smart home environment for detecting a human presence," in 1st International Conference on Advanced Technologies, 2014, pp. 334–339.

[7] Y. Guan, G. Linkowski, and T. Zhao, "Home Automation Localization Sensor (HAL sensor)" 2015.

Speech-based Class Attendance

'Umar Faizel Amri[1], Nik Nur Wahidah Nik Hashim[2] and Noor Hazrin Hany Mohamad Hanif [3]

[1-3] Department of Mechatronics Engineering, International Islamic University Malaysia

Abstract. In the department of engineering, students are required to fulfil at least 80 percent of class attendance. Conventional method requires student to sign his/her initial on the attendance sheet. However, this method is prone to cheating by having another student signing for their fellow classmate that is absent. We develop our hypothesis according to a verse in the Holy Qur'an (95:4), "We have created men in the best of mould". Based on the verse, we believe each psychological characteristic of human being is unique and thus, their speech characteristic should be unique. In this paper we present the development of speech biometric-based attendance system. The system requires user's voice to be installed in the system as trained data and it is saved in the system for registration of the user. The following voice of the user will be the test data in order to verify with the trained data stored in the system. The system uses PSD (Power Spectral Density) and Transition Parameter as the method for feature extraction of the voices. Euclidean and Mahalanobis distances are used in order to verified the user's voice. For this research, ten subjects of five females and five males were chosen to be tested for the performance of the system. The system performance in term of recognition rate is found to be 60% correct identification of individuals.

1. Introduction

The attendance system is considered important and significant in every area. This is to make sure the participants attend the programmes related in order to achieve the KPI and objectives. In this study, the attendance system is projected in the area of class of lecture. As students, they need to attend the class because it is compulsory. Years ago, Kulliyyah of Engineering International Islamic University Malaysia (KOE IIUM) doesn't enforce the obligation for engineering students to attend classes, provided obeying the following constraints; as long as the students scored the subjects of lectures, it matters. Paradoxically, not only the students chose to missed classes, but also don't get the full quality of understandings regarding topics of subjects. Moreover, it is unethical for Muslim students to seek knowledge without having the way to meet the lecturers as the educators; the source of knowledge. This scenario showed how important the attendance system in the field of class of lectures.

The objectives of the research are to investigate the ability of speech features to recognize speaker as well as to analyse the effect of different types of speech in identifying unique individuals and to determine the factors affecting the distortion or increasing errors in speaker recognition. Based on these objectives, it then leads to the following questions, what is the best feature used in speaker recognition application? Is that possible to combine different features in a single set of data? These questions were expected to be answered throughout this research.

2. Previous works

IOP Conf. Series: Materials Science and Engineering **260** (2017) 012008 doi:10.1088/1757-899X/260/1/012008

In this section, other researcher's papers were examined especially on the features that they used to extract the audio signal features and their results. According to [1], they used two types of features i.e. power spectral density (PSD) and also Mel-frequency Cepstral Coefficient (MFCC). For PSD, the results for male interview is 74.25% while female is 84.25% success rate. For the results for reading session, for male the score is 75.5% and female is 87%. For MFCC, they used 2-types which is 4-MFCC and 8-MFCC. For 4-MFCC, the success rate for interview for male is 70.3% and for female is 69.23%. For 8-MFCC, the success rate for male is 77.84% and female is 72.53%. For reading session, for 4-MFCC's male, the score is 71.28% while female scored 67.82%. For 8-MFCC, the male scored 80.62% while female scored 73.3%. Here, the higher the MFCC used, the better the result.

As for the research conducted by [2], the methodology used is text-dependent or fixed-text category. This research is specifically use spectrograms and row mean. Features used to identify the speaker is depended on optimal spectrogram segmentations and Euclidean distance. Spectrogram is used to obtain the feature vectors while row mean is applied on the spectrograms to get feature vector 8, 16 or 32. Then, Euclidean distance is measured to identify the speaker. The result from this research is astonishing that the minimum success rate is 50% while the maximum hits 100% perfectly. Unfortunately, not all researchers get good results. For [3], using MFCC as method for speaker recognition is not the best in noisy environment. Not only MFCC, Cepstral Mean and Variance Normalization (CMVN), Relative Spectral (RS) and Feature Warping also being used. Voice Activity Detection (VAD) is also used to improve MFCC method. However, they managed to improves the correctness rate by increasing to 25.2%, 19.1% and 9.7% for every method respectively. Research by [4] also used MFCC for text-independent speaker recognition based on GMM-BUM. Accuracy of the speaker recognition in increased by 3.10%. The score-level fusion Teager Energy Operator (TEO) phase together with MFCC performs better compare to MFCC alone.

3. Methodology

In this section, several procedures are being made in order to achieved the objective of this study. The model used for this methods is the same as what Desai and Pujara [5]. Figure 3.1 shows the basic model for speaker recognition.

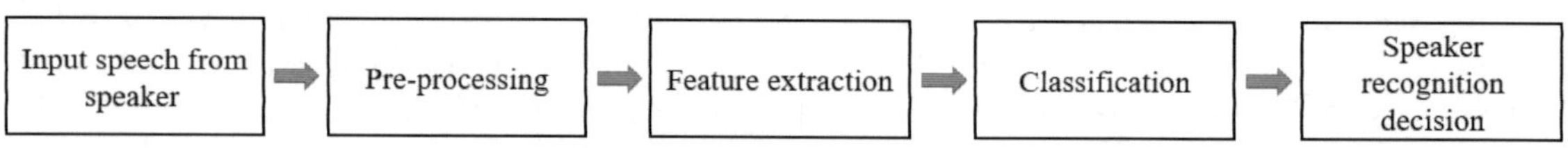

Figure 3.1. Basic model for speaker recognition.

In this research for the early stage, five females and five males' students were chosen to test the system. The database that were collected are from students saying their names for five times. Four data per individual will be used as training data and one left for testing data. The pre-processing method was done using Audacity software. The feature extraction used for this research are PSD and Transition Parameter. The classification method used are the Euclidean and Mahalanobis distance equations.

3.1. Feature extraction

There are two method used for feature extraction i.e. PSD and Transition Parameter. Both of the features use are including in the system. The display of the system using MATLAB GUI function in order for users to choose which method of feature they prefer for.

3.1.1. Power spectral Density (PSD). Power Spectral Density (PSD) is also known as energy in frequency bands (2). PSD is used to characterized a stationary random process in frequency domain. The PSD used is to be limited from 0 Hz to 2 KHz because most of energy in speech signals is mostly present in this range. To minimize the number of feature extraction, the frequency of which 0 Hz to 2

IOP Conf. Series: Materials Science and Engineering **260** (2017) 012008 doi:10.1088/1757-899X/260/1/012008

KHz is being divided into four band equally. In band one, there are only 0 Hz to 500 Hz frequency. For band two, the frequency is from 500 Hz to 1000 Hz. For band three and band four the frequencies are 1000 Hz to 1500 Hz and 1500 Hz to 2000 Hz respectively. Even though there are now four PSD (which regarding to the four band), only the first three band is used in the feature extraction process (band one, band two and band three) for the fourth band can be predicted by the first three band. Figure 3.2 shows the PSD procedures.

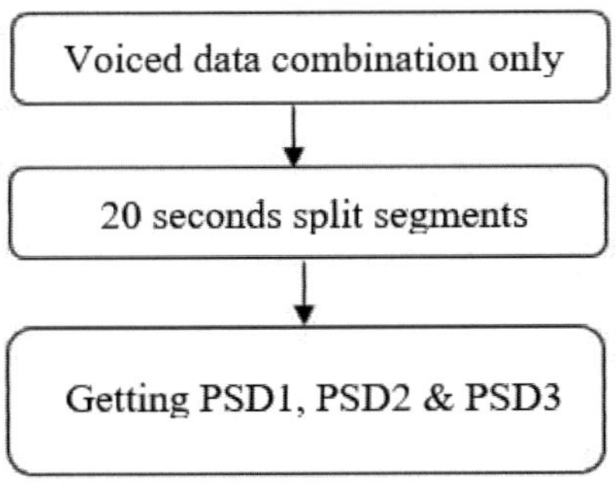

Figure 3.2. The PSD feature procedure. The voiced data only is to be collected because it produced the highest energy compare to silence and unvoiced data. 20 seconds split segments is limited only to the speech data more than 20 seconds. PSD1 is the energy band for 0 Hz to 500 Hz, PSD2 is the energy band for 50 Hz to 1 kHz and PSD3 is the energy band for 1001 Hz to 1.5 kHz.

3.1.2. Transition Parameter. Audio signal that consists voiced, unvoiced and silence was marked as 1, 2 and 3. The variation happens during the timing pattern of the speech. Here, the variations are captured in in the form of transition from one state to another state. Interestingly, these states are interchangeable and may also not change; depending on the set of probabilities concerning the states. The probability is simply estimated using discrete-time Markov process [6]. According to Nik Hashim [6], "One of the output parameters given is the estimated transition matrix (T) which in this case is a three-by-three matrix where t_{ij} = Pr (X_{k+1} = j | X_k = i) for I = 1,2,3 and j = 1,2,3", p. 48. Here, for example, element t_{23} is denoted as the conditional probability of unvoiced-to-silent. The transition matrix, T elements were sequenced into one row vector {t_{11}, t_{12}, t_{13}, t_{21}, t_{22}, t_{23}, t_{31}, t_{32}, t_{33}}.

3.2. Speaker recognition

This is the final part of the speaker recognition method. Which is analysis of the features that were extracted. This project used one method in order to get the authentication of the speaker called Euclidean distance. Euclidean distance is also known as Euclidean metric is a popular method used to get the distance between two connecting points in a straight line in Euclidean space. The computation method used there is depending on the dimensions. Euclidean distance, $\boldsymbol{d_\epsilon}$ is given in the equation,

$$\boldsymbol{d_\epsilon} = \|\boldsymbol{x} - \boldsymbol{\mu_i}\| \quad (1)$$

where $\boldsymbol{x}$ and $\boldsymbol{\mu_i}$ are both present in the form of matrix (that denoted in bold forms). Here, $\boldsymbol{x}$ is the test trained data and $\boldsymbol{\mu_i}$ is the tested data. From (1), we know that feature vectors are assigned to classes which respected to their Euclidean distance from the mean points [8].

Then, Mahalanobis distance, $\boldsymbol{d_m}$ equation is used as the continuation of this Euclidean distance given,

$$\boldsymbol{d_m} = \sqrt{(\boldsymbol{x} - \boldsymbol{\mu_i})^T * (\boldsymbol{x} - \boldsymbol{\mu_i})} \quad (2)$$

where (2) is defined as the square root of the transpose, T of $(\boldsymbol{x} - \boldsymbol{\mu_i})$ multiply by the same matrix $(\boldsymbol{x} - \boldsymbol{\mu_i})$.

By using (2), we can now calculate the Euclidean distance of the tested and trained data. The idea is that, the least the distance between test and train data, the accurate it is; which indicates the authentication the speaker recognition.

To do this, train data from each subject must be calculated together using (2) with the test data from each subject. Here, we know that there are ten subjects. Which means, there are 10 x 10 possible combinations in order to analyse the data. Because this research used three PSDs, all possible combinations of three PSD will be used for testing and training data.

However, for transition parameter feature extraction, the possible combinations are Voiced-Voiced (t_{11}) Voiced-Unvoiced (t_{12}), Voiced-Silent (t_{13}), Unvoiced-Voiced (t_{21}), Unvoiced-Unvoiced (t_{22}), Unvoiced-Silent (t_{23}), Silent-Voiced (t_{31}), Silent-Unvoiced (t_{32}) and Silent-Silent (t_{33}).

3.3. Development of GUI-based Speech Attendance System

We develop a graphical user interface (GUI) based system for the extraction and analysis of features for speaker identification using MATLAB. User will be able to speak into the system and his/her name will be registered as present for that particular day. The flowchart shown in Figure 3.3 illustrates the system operated of this speech-based class attendance and Figure 3.4 shows displayed the GUI for this system.

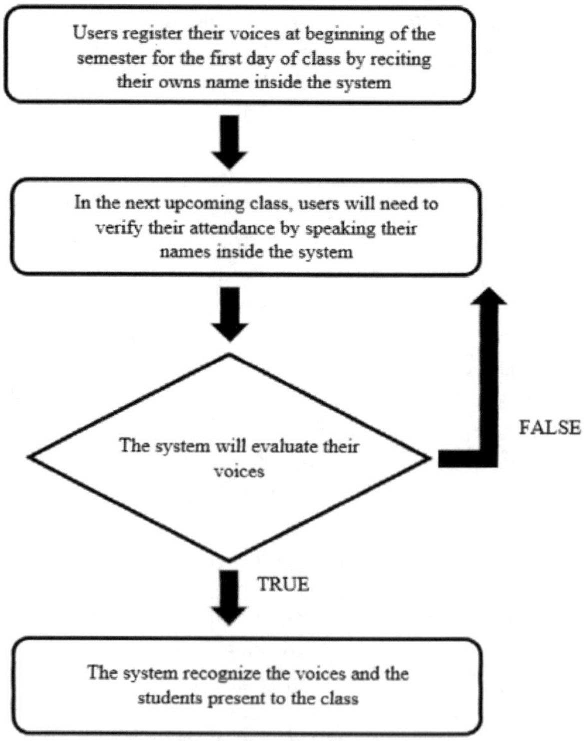

Figure 3.3. General flowchart of speech-based class attendance.

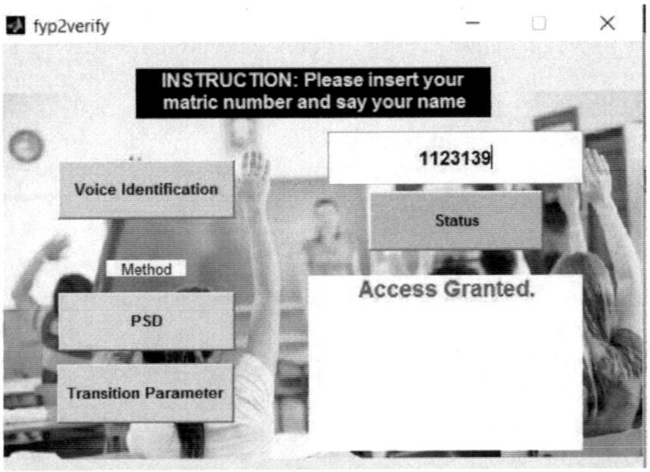

Figure 3.4. The display of GUI-based Speech Attendance System.

4. Result and discussion

In this chapter, the results of the analysis are presented. Discussion on each segment of results is made especially upon the errors. The sets of data that are going to be presented in this chapter are the results for free-text category of recitation of own name. By verifying PSD method, the best result of minimum Euclidean distance is for PSD2 and PSD3 and the t_{11}, t_{13}, t_{31} and t_{33} for minimum Euclidean distance of 10 subjects verifying their voices by reciting their own names. Figure 4.1 shows the PSD result and Figure 4.2 shows the transition parameter result.

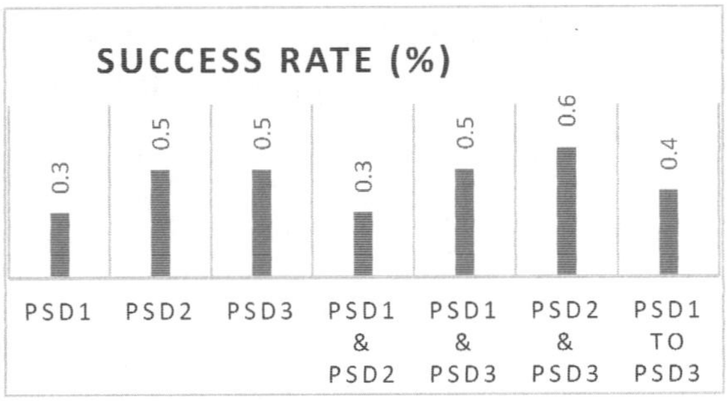

Figure 4.1. PSD result.

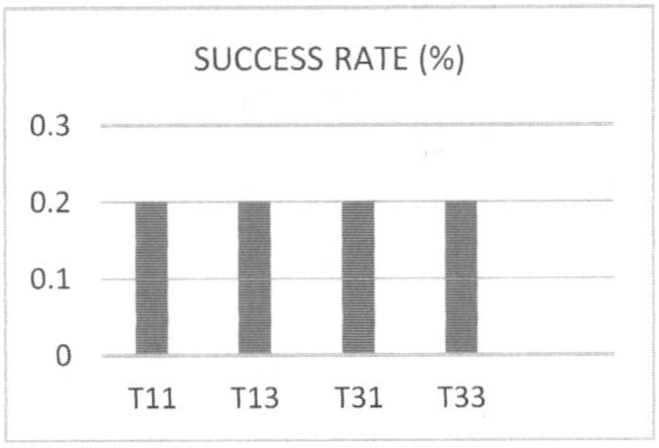

Figure 4.2. Transition parameter result.

The result demonstrated 60 percent or less accurate identification. The low percentage of accuracy is mainly due to the component of speaker variability and speech features as mentioned by [7]. It is to be recommended that the first step of speaker recognition i.e. collection of data is being held in a more reliable and sensitive acts. For example, the session is held in a very closed room that free from acoustical noise. Also, the microphone used must be more quality compare to this research did. Not only that, the prime factor i.e. the speaker or subject itself must be remind to speak in a natural way. The condition of subject's health also must be considered to get a better performance. With this, in shaa Allah the results will be much better. Not only by that, the subjects also must be increase quantitatively. The more subjects, the more data, the better the training.

5. Conclusion

In conclusion, the project went well although the results are not satisfying. This is the part and parcel of research. In addition, this topic is a long and continuous research which means, a set of time is really

IOP Conf. Series: Materials Science and Engineering **260** (2017) 012008 doi:10.1088/1757-899X/260/1/012008

needed to make this project a success. The PSD features and transition parameters features were not effective in identifying the speaker. The effect of different speech in identifying the speaker affected the performance of speaker recognition. Compared to others' related work as mentioned in chapter 2, the result can be improved by using combinations of PSD and transition parameters and other features such as MFCC with different technique of identification.

6. References

[1] Keskinpala HK, Yingthawornsuk T, Wilkes DM, Shiavi RG, Salomon RM. Screening for high risk suicidal states using mel-cepstral coefficients and energy in frequency bands. Eur Signal Process Conf. 2007;(Eusipco):2229–33.

[2] Kekre HB, Athawale A, Desai M. Speaker Identification using Row Mean Vector of Spectrogram. 2011;(Icwet):171–4.

[3] Wu D, Feng Z, Chen L, Xiao Z. Perception Auditory Factor For Speaker Recognition In Noisy Environment. 2016;1916–20.

[4] Agrawal P, Patil HA. Fusion of TEO Phase with MFCC Features for Speaker Verification. 2015;161–6.

[5] Desai KS, Pujara H. Speaker Recognition from the Mimicked Speech : A Review. 2016;2254–8.

[6] Nik Hashim NNW. Analysis of Speech Features as Potential Indicators for Depression and High Risk Suicide and Possible Predictors for The Hamilton Depression Rating (HAMD) And Beck Depression Inventory Scale (BDI-II). Vol. 53. Vanderbilt University; 2014.

[7] Singh N, Khan RA. Underlying Text Independent Speaker Recognition. 2016;6–10.

[8] Theodoridis S and Koutroumbas K 2009 *Pattern Recognition* (London: Academic Press).

Acknowledgments

This work is part of the ongoing project on the development of speech-based multi-level person verification system funded by the Department of Mechatronics Engineering, International Islamic University Malaysia.

IOP Conf. Series: Materials Science and Engineering **260** (2017) 012009 doi:10.1088/1757-899X/260/1/012009

Traffic intensity monitoring using multiple object detection with traffic surveillance cameras

H G Muhammad Hamdan [1] and O O Khalifah [1]

[1] Department of Electrical and Computer Engineering, International Islamic University Malaysia, Jalan Gombak, 53100 Kuala Lumpur, Selangor, Malaysia

Abstract. Object detection and tracking is a field of research that has many applications in the current generation with increasing number of cameras on the streets and lower cost for Internet of Things(IoT). In this paper, a traffic intensity monitoring system is implemented based on the Macroscopic Urban Traffic model is proposed using computer vision as its source. The input of this program is extracted from a traffic surveillance camera which has another program running a neural network classification which can identify and differentiate the vehicle type is implanted. The neural network toolbox is trained with positive and negative input to increase accuracy. The accuracy of the program is compared to other related works done and the trends of the traffic intensity from a road is also calculated. relevant articles in literature searches, great care should be taken in constructing both. Lastly the limitation and the future work is concluded.

1. Introduction

Traffic congestion has become a common plague that is causing trouble for all road users due to uncontrolled urban development, decrease in cost of transportations and growth in population[1]. Nowadays, is almost common for anyone to spend almost on average about 32 hours per year on traffic jams in which he not only spent his time but also caused increased pollution and money for fuel. With the current advancement in computer vision and Internet of Things, it is important to address this issue and try to mitigate it as much as possible. Although there are multiple alternatives such as social messaging services that provide real time congestion information such as Google Maps and Waze, it will be a better option if the systems do not depend on the user data to interpret the congestion as this will require user with internet access.

Figure 1 Road Traffic conditions in Malaysia

In this paper, a method used to identify the traffic intensity to identify the traffic congestion is discussed. The next section will discuss on the related works done by other researchers on this issue and the summary of their works done. In Section 3, the methodology used to develop the solution is elaborated and section 4 will discuss on the result obtained from the implementation. Finally, the future works and conclusion is presented in the last section.

2. Related works

Based on the literature review done, the works that shows similarities with current research cope is discussed in detail in this section. [2] uses blob analysis to track and detect moving cars on the road. Blob analysis uses background subtraction in each frame to generate a blob of moving pixels across the frame. Blob analysis is the simplest way to find moving object in subsequent frames. The author used traffic cameras places in 2 different areas which are cameras facing 3 lanes dubbed as NIPA and TOLL PLAZA with 2 lanes of road. For training the images, Haar Cascade was used to classify the images and XML file was generated using 580 positive images and 1500 negative samples to generate the dataset. Finally, the performance of the system is evaluated using positive detection compared to actual number of vehicles. This paper provided a good platform for this research topic but the cascade used for this research was based on algorithms used for face tracking and result will be more accurate if the features are more distinct in dataset images.

The next paper focused on using foreground-based method to detect moving object across frames. [3]. Image was acquired using traffic cameras and resized into smaller frames. Then the positive images of cars flowing through the traffic is captured into positive and negative frames. The positive images are then ran through the Haar Cascade similar to the work of [2] but the key difference is the sample was not done using the first frames but the first number of frames was used to identify the foreground of the images to increase the detection accuracy and removing unwanted noise in each frames being processed. The image with foreground detection becomes more dilated and easier to detect images that are moving. The advantage of this is the amount of processing required to detect the cars are reduced due to smaller ROI but author does not explain on the key performance details such as frame rates or evaluation on the changes in frames sizes on the accuracy.

Another paper discussing on the same issue is written by [4] where the image capturing method was used to identify the number of cars on the road. The methodology used by the author was background subtraction to identify the moving object on the road and measuring the number of vehicle on the road. Instead of measuring the number of blobs, the author counts the number of pixels in which the image is different form the first image with no congestion. By estimation the range of values in these different congestion rates mapped to range of pixel in each state, the conclusion is reached. Key idea of this paper is the simplicity of the algorithm where the images are processed as is and no post processing is required to obtain data however, the data will be not meaningful if the pattern of cars or vehicle types are something important that need to be identified. The author also does not provide any insight on the measurement for the congestion detection systems.

Neural network are the new research area that is gaining a lot of attention especially in computer vision. [5] developed a system which can theoretically predict the traffic conditions of the road and use congestion control to relief road congestion. The work focused on using Artificial Neural Network to map out number of vehicle on the road and use Jordan's Sequential Network to predict the new condition of the road in real time as well as prediction the future trends. The prediction algorithm could detect traffic congestion with accuracy reaching 98% in some test however, the paper was not able to give any specification on the tools used for the data acquisition or practical implementation of the algorithm in real time.

Edge detection is also a key method in image classification tool as researched by [6] which uses IP cameras places in strategic locations on the road to predict traffic congestion. The object detection process is done using edge detection and the parameters are passed onto a fuzzy logic system for traffic estimation. Edge detection in this paper uses a Sobel filter mixed with Kalman filter to track the moving object. The number of moving object is then passed to a fuzzy logic analyser that considers three parameters such as vehicle density, distance between vehicle and lastly vehicle size. The logic analyser then predicts the traffic into different categories based on the logic table of the fuzzer. This method can accurately predict the traffic intensity of the moving traffic however the region of tracking in these images is not set therefore the ROI tracking of moving vehicles must be done on the whole frames which will increase the processing time.

Table 1 Summary of the related work

Author	Approach	Advantage	Disadvantage
[2]	Haar Cascade combined with blob analysis	The algorithm is easy to implement on the research area	Accuracy measurement of sample was not diverse and no measurement of traffic intensity
[3]	Foreground detection on road	ROI set for vehicle detection is limited to smaller area thus decreasing processing time	Using smaller image will increase noise if there is small change in the camera orientation
[4]	Background subtraction with pixel counting	Simple algorithm to identify the traffic congestion using pixel quantity	Algorithm is unable to tell the difference between the vehicle types and not complete information for traffic intensity
[5]	Artificial neural network to predict the road condition	Very high accuracy to predict the traffic data	High complexity of algorithms requires a large computation engine to run the required processes
[6]	Traffic estimation using fuzzy logic	Clear requirements set for monitoring the traffic flow.	Uses basic blob detection tool to detect the cars therefore the accuracy can be increase is the detection algorithms is improved

To detect the traffic intensity, there are 3 critical approach in the modelling of the traffic flows in roads namely the spatial modelling[7], [8] [9] [10], time prediction modelling [11] [12][13] and non-parametric modelling. Spatial modelling refers to using the physical space on real time to estimate the motion of the vehicles based on the physical parameters such as road distance, car length, road curvature and other key elements. Time prediction models use statistical data collected over time to predict the current trend on the road and update the data on real time if necessary. Non-parametric modelling refers to using computational intensive algorithms such as artificial neural network and work independent from any real-time input but relies on collection of data to synthesize an output.

However, in this implementation the space based modelling technique is used to simplify the result classification

3. Proposed Implementation

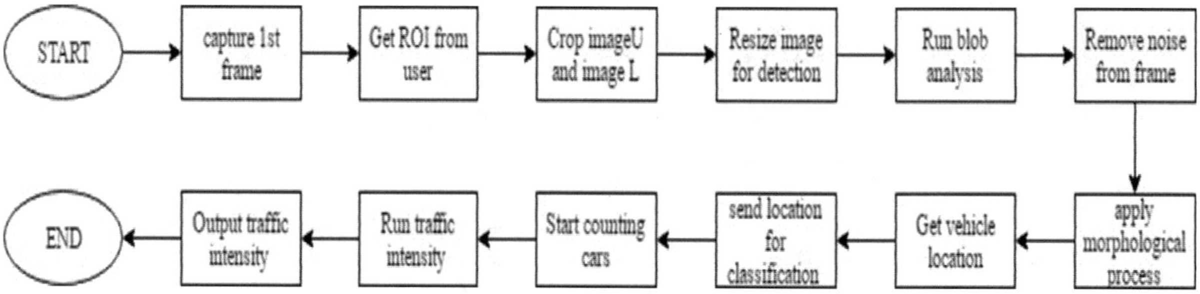

Figure 2 The overall process to identify traffic intensity

The figure shows the overall implementation of the proposed solution. The process starts when the traffic camera is trained on the foreground and background images on an area. Then the program displays the images to the user to identify the Region of Interest(ROI) for the desired location to start counting the cars. The user will choose the lower and upper region of the ROI to determine traffic flow into and out of the link as shown in following figure 3.

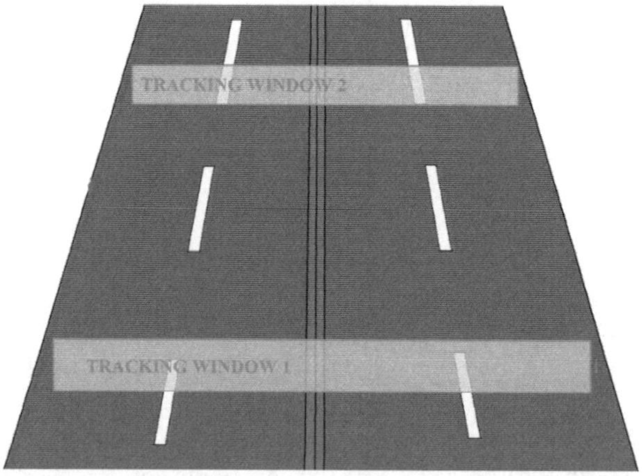

Figure 3 Tracking location of vehicles for traffic intensity

Next the frame captured from camera subtracted from the current frames and the differences are marked using blob analysis and its output is filtered for regions that are larger than 5000 pixels which is dependent on the original video size. Next the centroid of the moving objects is calculated from the blob and it is passed to the classification tool.

The classification tool contains the pretrained object output of images from vehicles given before the program is initialized. The training process involves using positive images of cars and other vehicles and negative image containing background and unrelated random images to allow the computer to differentiate the type of object given to it. The object file is generated using neural network toolbox in Matlab which gives a high accuracy of data for object classification. When the image is passed to this program, it will determine if the centroid given to it contains which vehicle from the trained object.

If the classification program determines if the area from previous program is a car, then it will add the number of cars and other corresponding vehicle types. Once the program runs for a minute, it will

IOP Conf. Series: Materials Science and Engineering **260** (2017) 012009 doi:10.1088/1757-899X/260/1/012009

generate the traffic intensity based on the number of cars entering the link and the cars exiting the link. The equation used to calculate the traffic intensity is based on equation 1 taken from [9] derived from Macroscopic Modelling of Urban transportation model.

$$p_{w,r}(k) = \frac{n_{w,r}(k) - q_{w,r}(k)}{C_{w,r} \cdot l_{veh} - \frac{q_{w,r}(k)}{\rho^{jam}}} \quad (1)$$

This equation translated to the traffic density of oncoming traffic, $p_{w,r}(k)$ and $n_{w,r}(k)$ is the number of vehicle inside the region of interest. The symbol $q_{w,r}(k)$ is the length of the road in field of view set to 30m and $C_{w,r} \cdot l_{veh}$ is the maximum number of cars multiplied with its average length set at 28 and lastly, ρ^{jam} is the jam density measured on average roads which is set to 4.5veh/m/hour . This calculation is done on periods of time for example k will be measured in minutes for a period of 10 minutes and the traffic intensity will be the total intensity within that period.

Finally, the traffic intensity will be displayed using a Graphical User Interface(GUI) that shows the trend of the vehicles moving during different times of the day.

4. Results and discussion

The program is run on a road and the data collected is discussed in this section. The sample location selected was in Penang, Malaysia where a camera is placed on the centre of the road facing the link. The accuracy of the classification program is as shown in table 2.

Table 2. Accuracy of the program

Total number of vehicle	Type of vehicle detected	Number of vehicle detected	Number of vehicle not detected	Accuracy (%)	Average Detection time(ms)
100	Car	85	15	85	150
37	Motorcycle	30	7	80	150

The program can tell with almost 85 percent probability if the detected object is a car and 80 % accuracy of the detected object is a motorcycle. Based on these data, it provides a good reference to optimize the system in the future though is slightly lower than [3] [5] . Next the program is run during the night time to get the accuracy when the condition is dark and table 3 shows the accuracy of the data collected on the same location.

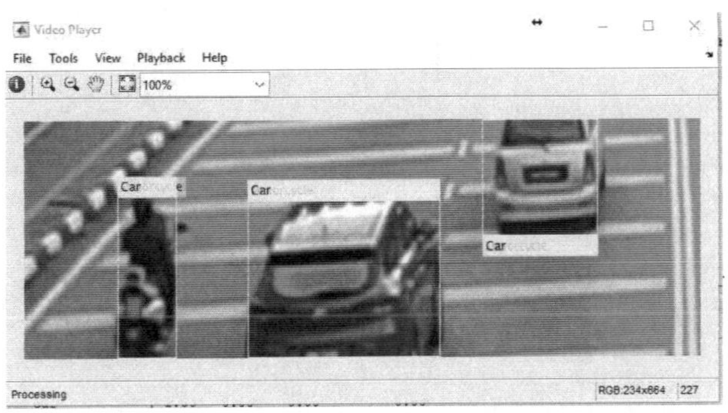

Figure 4 False positive where a motorcycle is detected as a car and a motorcycle

Figure 5 True Positive detection of a car(right)

Table 3 Accuracy of the program during night

Total number of vehicle	Type of vehicle detected	Number of vehicle detected	Number of vehicle not detected	Accuracy (%)	Average Detection time(ms)
32	Car	8	24	25	220
17	Motorcycle	2	15	11	230

The data in table 3 shows that the accuracy has dropped significantly compared to the video captured during the day. These results are not compared to other researches as this test was not done by them. There are a few factors that contributed to this drop in accuracy such as different lighting on the road causing the changes in the shade of car be lower than it should be when the blob analysis code is running. Next, the car at night uses light to illuminate the road thus changing a key feature extracted during training process. The light projected by the car also increased brightness that makes the camera unable to see the actual car shape but only see circle during blob analysis. Lastly, the colours of the light color also change when the car is moving forward and backward as the headlight is white and the break light is red. Next, the traffic intensity during low traffic and high traffic movement is calculated and the result is as shown in table 4 and 5.

Table 4 Number of cars and traffic intensity in low congestion

Time Start	Time end	Number of cars into link(per minute)	Number of car out of link(per minute)	Traffic intensity at time k (no Congestion)
0	1	26	25	0.018367347
1	2	9	10	-0.017307692
2	3	7	6	0.017045455
3	4	3	4	-0.016917293
4	5	22	21	0.018072289
5	6	12	12	0
6	7	4	5	-0.016981132
7	8	11	10	0.017307692
8	9	13	13	0
	Total	107	106	0.019586665

Table 5 Number of cars and traffic intensity during congestion

Time Start	Time end	Number of cars into link(per minute)	Number of car out of link(per minute)	Traffic intensity at time k (Congestion present)
0	1	20	17	0.053359684
1	2	17	14	0.052734375
2	3	22	20	0.036
3	4	19	17	0.035573123
4	5	20	21	-0.018072289
5	6	22	20	0.036
6	7	13	13	0
7	8	23	21	0.036144578
8	9	20	23	-0.05465587
	total	176	166	0.1770836

When there is no congestion, the total traffic intensity in each time is low and the traffic intensity during peak hours are high therefore the program will be able to tell the difference in the traffic intensity clearly. The user can set the threshold where the traffic intensity is high, medium or low depending on the road condition. A typical range of values that has low congestion are below 0.05 and medium congestion at 0.1 to 0.3 and highly congested set above 0.3 but the values maxes out at 0.5 for the initial values set in previous section.

5. Conclusion

In this paper, a method of identifying the traffic intensity of the road is implemented using computer vision. Based on the data acquired, the program can tell accurately if the road is congested based on the equation of the Macroscopic Urban Traffic models. However, using computer vision still has limitations as discussed in the previous section. There are some limitation as the program still uses a straight road to determine moving cars and curved road are still a challenge as of now.

In the future, a better program which collect multiple road statistics and generate a prediction algorithm based on the road trends is needed to be able to tell road users information about their journey on the road.

6. References

[1] A. S. Abdelfatah, M. Z. Shah, and O. C. Puan, "Evaluating the Sustainability of Traffic Growth in Malaysia," *J. Traffic Logist. Eng.*, vol. 3, no. 1, pp. 6–11, 2015.

[2] Y. Hasan, M. U. Arif, A. Asif, and R. H. Raza, "Comparative analysis of vehicle detection in urban traffic environment using Haar cascaded classifiers and blob statistics," *FTC 2016 - Proc. Futur. Technol. Conf.*, no. December, pp. 547–552, 2017.

[3] Q. Wu, W. Kang, and X. Zhuang, "Real-time vehicle detection with foreground-based cascade classifier," *IET Image Process.*, vol. 10, no. 4, pp. 289–296, 2016.

[4] B. Eamthanakul, M. Ketcham, and N. Chumuang, "The Traffic Congestion Investigating System by Image Processing from CCTV Camera," in *2017 International Conference on Digital Arts, Media and Technology (ICDAMT)*, 2017, pp. 240–245.

[5] R. More, A. Mugal, S. Rajgure, R. B. Adhao, and V. K. Pachghare, "Road traffic prediction and congestion control using Artificial Neural Networks," in *2016 International Conference on*

Computing, Analytics and Security Trends (CAST), 2016, pp. 52–57.

[6] A. R. F. Quiros, R. A. Bedruz, A. C. Uy, A. Abad, A. Bandala, and E. P. Dadios, "Machine vision of traffic state estimation using fuzzy logic," in *2016 IEEE Region 10 Conference (TENCON)*, 2016, pp. 2104–2109.

[7] F. Sanchez-Fernandez, P. Brunet, and S. M. Senouci, "Background subtraction for aerial surveillance conditions," *14th Int. Conf. Innov. Community Serv. "Technologies Everyone", I4CS 2014 - Conf. Proc.*, pp. 28–33, 2014.

[8] K. Phalke and R. Hegadi, "Object tracking using Mutiple Hypothesis," in *2016 3rd International Conference on Signal Processing and Integrated Networks (SPIN)*, 2016, pp. 707–710.

[9] F. Yan, F. Tian, and Z. Shi, "An improved traffic network model for model-based predictive control," in *Proceedings of the 33rd Chinese Control Conference*, 2014, pp. 3405–3410.

[10] A. Boulmakoul, L. Karim, M. Mandar, A. Idri, and A. Daissaoui, "Towards Scalable Distributed Framework for Urban Congestion Traffic Patterns Warehousing," *Appl. Comput. Intell. Soft Comput.*, vol. 2015, 2015.

[11] L. Li, "A continuum model of traffic flow on road networks," in *2015 International Conference on Transportation Information and Safety (ICTIS)*, 2015, no. 2, pp. 595–598.

[12] L. Sindhia, K. B. Hariharan, and D. Kumar, "Efficient detection tracking of multiple moving objects in temporal domain," in *2016 International Conference on Recent Trends in Information Technology (ICRTIT)*, 2016, pp. 1–6.

[13] A. Ouessai and M. Keche, "Development of road traffic flow model for nighttime conditions," *3rd Int. Conf. Control. Eng. Inf. Technol. CEIT 2015*, 2015.

Control of cooperative manipulators in holding deformable objects

A A Alkathiri[1] **and N Z Azlan**[2]

[1,2] Department of Mechatronics Engineering, International Islamic University Malaysia, Kuala Lumpur, Malaysia

E-mail: sinnira@iium.edu.my

Abstract. This paper presents the implementation of a control system to control cooperative manipulators to hold deformable objects. The aim is to hold the deformable object without having information on the shape and stiffness of the deformable object beforehand. The prototype of a pair of manipulators has been designed and built to test the controller. A force sensor and a rotary encoder are used to give feedback to the controller, which controls the DC motor actuators accordingly. A position proportional-integral-derivative (PID) controller technique has been applied for one of the manipulators and a PID force control technique is applied to the other. Simulations and experimental tests have been conducted on models and the controller has been implemented on the real plant. Both simulation and test results prove that the implemented control technique has successfully provided the desired position and force to hold the deformable object with maximum experimental errors of 0.34mm and 50mN respectively.

1. Introduction

This paper focuses on the implementation of a control algorithm to handle deformable objects using cooperative manipulators. Deformable objects are objects whose shape change when force is applied [1]. The difficulty of manipulating deformable object is mainly due to the its nonlinear elasticity, friction and parameter variations [2]. Consequently, the manipulation of deformable objects is less researched because of mechanical reasons. On top of that, the modeling cost of deformable objects is high [3].

Despite all the challenges, the research of manipulating deformable object is still of interest due to their significant applications. Deformable objects can be found in petty everyday stuff, such as when folding a bed sheet [4], tying knots, flattening towels and erasing a whiteboard [5]. Handling deformable objects is also important on an industrial level, such as in food processing and recycling [6] [7]. Other significant applications include medical applications [8].

Due to the increase in demand in robots that imitate human arms and hands, the study of cooperative manipulators has been popular [9]. They are also a preferred choice for handling deformable objects [10]. For large objects and objects of heavy payload, such as vehicle chases, cooperative manipulators can be used to share the load and gain stiffness [11].

For the handling strategies of deformable objects, the study in [3] utilizes FEM, which is time-consuming. Thus, a powerful processor is required. The method in [4] is not equipped to handle 3-D deformable objects. In [12], the manipulators have to move slowly due to the online visual estimation method. The general demonstration-based method in [5] faces the challenge of the difficulty of getting the robot to perform the demonstrations. Therefore, the goal of this study to design and implement a controller that is not computationally heavy that at the same time, is capable of holding deformable objects despite unknown stiffness and position. Also of concern are the force and position reaching their desired values without exceeding a time threshold. The difficulty in handling deformable objects is due to the fact that they are challenging to model and simulate. Handling deformable objects requires the robot to exert a particular amount of force to hold the object at the desired position. In [7], dual-motor drive gripper with parallel fingers is designed and parallel position-force algorithm is used.

In this paper, stiffness and the position of the deformable object to be held are unknown. In this sense, a deformable object is one whose shape changes due to an applied force such as paper cup, food and bag filled with sand. At this stage of study, this work focuses on the PID control strategy for cooperative manipulators in holding the deformable objects. The prototype is made of two 1-DOF cooperative manipulators, each powered by a DC motor. This paper focuses on deformable objects such as rings, sponges, paper cups and rubber balls instead of deformable objects of infinite degrees of freedom such as fabric and knots [2].

The rest of the paper is organized as follows: section 2 shows the mechanical design and electrical diagram. Section 3 details the derivation of mathematical models. Section 4 is dedicated to the simulation studies, while section 5 focuses on experimental studies. The conclusion is presented in section 6.

2. Mechanical Design

The scope of this paper is to design a controller that is capable of holding a deformable object in place by exerting desirable amounts of force at two 1-DOF cooperative manipulators. The force is translated from a DC motor to the manipulator through a rack-and-pinion configuration.

To achieve that goal, the system will comprise of an Arduino Mega 2560 microcontroller to interface the control algorithm running on MATLAB'S Simulink with the hardware, DC motors as actuators, a force sensor and an encoder to provide feedback to the controller and rack-and-pinion sets to translate torque generated by the motors into force, in addition to the structure holding these components together. One of the manipulators is assigned as the 'master' and the other is set as the 'slave'. Figure 1 shows the prototype of the master and slave manipulators.

An electrical diagram of the connections of a dual-channel motor driver shield stacked on top of the microcontroller is presented in figure 2. The motors used are brushed DC motors with magnetic quadrature encoder for linear position feedback. For force feedback, the Honeywell FSS1500NST is used due to its low repeatability error of about $1.5mV$ at $300g$ compared to other types of force sensors, such as FSR.

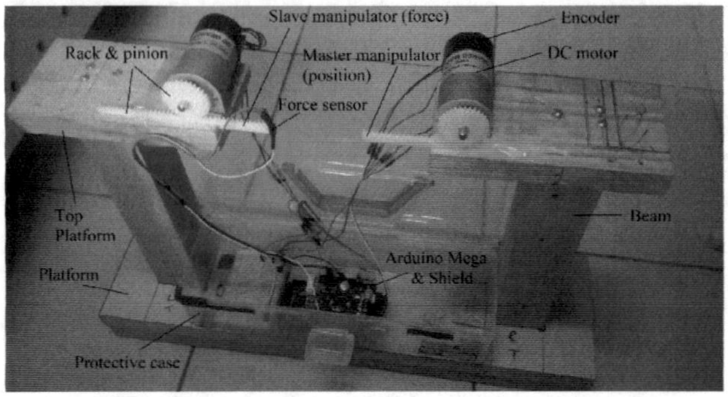

Figure 1. Prototype of the manipulators.

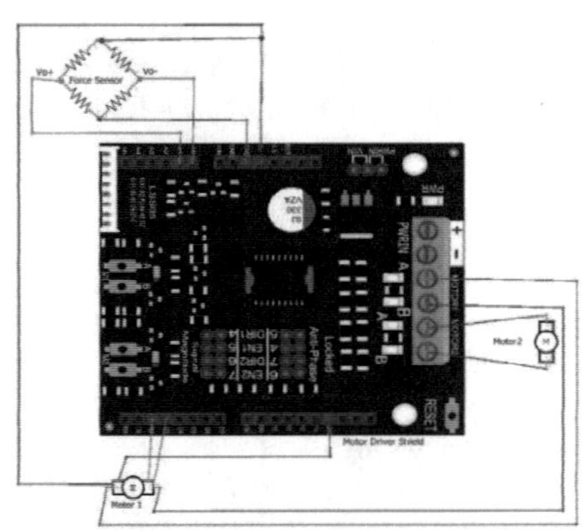

Figure 2. Electrical diagram.

3. Mathematical Modelling

In this section, the relationship between the output voltage of the force sensor and the corresponding force as well as model mathematical models of the plant and controller are presented.

3.1. Force Sensor Calibration

The force sensor is calibrated by taking readings incremental masses up to $500g$. No higher masses were sampled because no higher equivalent normal force is to be exerted on the manipulator. Figure 3 plots the relationship between the mass and voltage. When the relationship is linearized using line of best fit and converted, the force sensitivity amounts to $0.7007N/V$.

3.2. Manipulators

The plant is a 1-DOF manipulator and its free-body diagram is shown in figure 4, where f_{fr} is the friction force and f is the force exerted on the slave manipulator by the motor torque. Therefore, its model can be derived using a force equation:

$$\sum F = ma;\; f - f_e - f_{fr} - b\dot{x}_e = m\ddot{x}_e \tag{1}$$

where m is the mass of the manipulator in kg, b is the damping friction coefficient, f is the force from the actuator in N, f_e is the force exerted on the environment in N, f_{fr} is the friction force in N, and x_e is the position in m, $\dot{x}_e$ and $\ddot{x}_e$ are its first and second derivatives, respectively. For the master manipulator, the term f_e is taken out from (1) since the position control strategy will be applied to this manipulator. If x_e is constant, $\dot{x}_e = \ddot{x}_e = 0$.

The pressure angle of the rack and pinion is 20° and the module is 1. The pitch diameter of the pinion is given by

$$P_d = zm \tag{2}$$

where z is the number of teeth and m is the module. The pinion is of module 1 and has 32 teeth.

One rotation of the pinion will displace the rack by I, which is given by

$$I = \pi P_d \tag{3}$$

The relationship between the exerted torque by the motor and the force is

$$T = fr \sin\theta \tag{4}$$

where r is the distance from the axis of rotation to the point of application of the force in m, and θ is the angle between the force applied and the axis of rotation in degrees, which is 0°, as shown in figure 4. Thus, (4) becomes

$$T = fr \tag{5}$$

3.3. PID controller

A proportional-integral-derivative controller is used for both position control on the master manipulator and force control on the slave manipulator. The PID equation is

$$u(t) = K_p e(t) + K_i \int_0^t e(t)\,dt + K_d \frac{de(t)}{dt} \tag{6}$$

IOP Conf. Series: Materials Science and Engineering **260** (2017) 012010 doi:10.1088/1757-899X/260/1/012010

where K_p, K_i and K_d are the proportional, integral and derivative gains, respectively, and $u(t)$ is the control input to the manipulator, which in this work is the force, f.

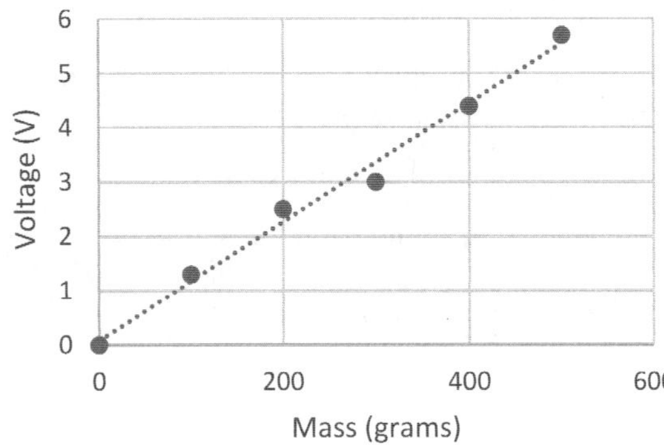

Figure 3. Mass-voltage relationship for the force sensor.

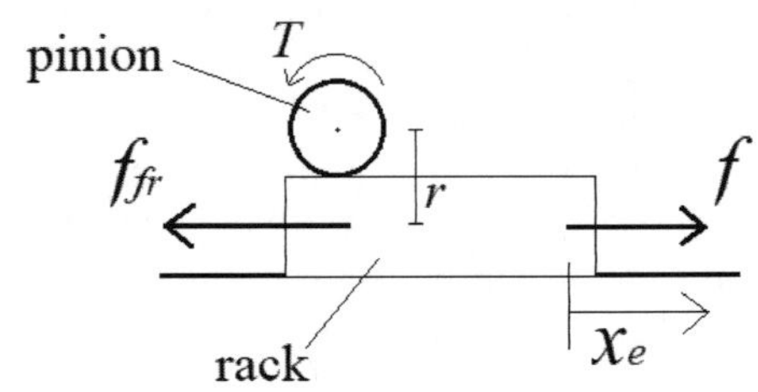

Figure 4. Free-body diagram of the manipulator.

Performing Laplace transform on (6) yields

$$U(s) = K_p + \frac{K_i}{s} + K_d s \quad (7)$$

The error $e(t)$ is the error for the master manipulator is

$$e(t) = e_x = x_d - x_e \quad (8)$$

where x_d is the desired position and x_e is the actual position. On the other hand, the error $e(t)$ is the error for the slave manipulator is

$$e(t) = e_f = f_d - f_e \quad (9)$$

in which f_d is the desired force and f_e is the actual force.

4. Simulation Studies

In this section, a few simulations are run to test the proposed technique. Figure 5 shows the simulation block diagram. The controller used in both cases is PID controller, as in section 3.3. The desired specifications for the controller in both cases are that the percentage overshoot (PO) should be less than 15%, the settling time (T_s) be less than 4 seconds and the steady state error (e_{ss}) be less than 5%. The simulation is run with the following values: $b = 10$, x_e varies between 0 and $2cm$, $F_d = 3N$, $k_e = 2N/cm$, $m = 0.6kg$ and $f_{fr} = 0$ (due to smooth contact surface) into the model in (1). Figures 6 and 7 show the result for the master and slave manipulators, respectively. Figure 6 shows that the manipulator has successfully tracked the desired position, while figure 7 proves that the controller follows the desired force.

To meet the performance specifications above, for the master manipulator controller, the gains values have been manually set to $K_p = 93.275$, $K_i = 139.139$ and $K_d = 7.212$, and for the slave manipulator controller, $K_p = 102.940$, $K_i = 162.104$ and $K_d = 5.077$. Table 1 tabulates the performance of the force and position control simulation.

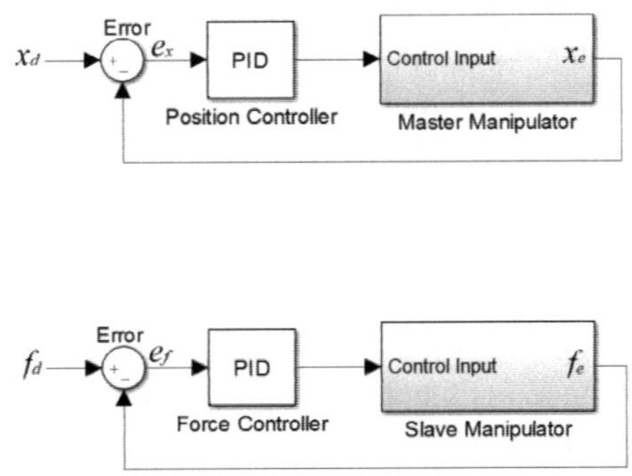

Figure 5. Simulation block diagram.

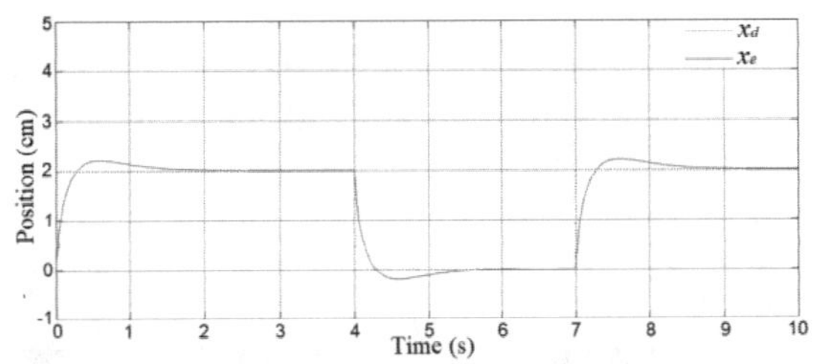

Figure 6. Position control simulation for master manipulator.

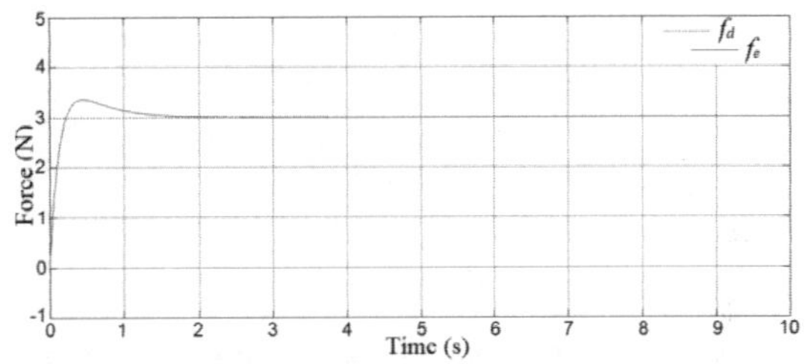

Figure 7. Force control simulation for slave manipulator.

5. Experimental Studies

The results of running the controller on the real plant are shown in figures 8 and 9; figure 8 shows the position tracking on the master manipulator while figure 9 shows the force tracking on the slave manipulator when grasping a tomato, as shown in figure 10. Due to the limited space between the two manipulators, the desired position tracked ranges between 0 and $2cm$. To avoid damaging the tomato, the desired force is set to $3N$.

The values of the gains of the controller have to be tuned manually. Ultimately, to achieve the result shown in figure 8, the values of the gains of the position controller are set to $K_p = 3100$, $K_i = 20$ and $K_d = 150$, and for the force controller in figure 9 the values are set to $K_p = 800$, $K_i = 200$ and $K_d = 500$. Table 1 tabulates the performance of the force and position control simulation.

The master manipulator deals with position, which is independent of the object being held. Figure 8 shows that the actual position of the master manipulator converges toward the desired values. Likewise, for force tracking on the slave manipulator is shown in figure 9. These results validate that the proposed technique is effective in controlling the cooperative manipulators in holding a deformable object.

The K_p, K_i, and K_d values in the implementation are fine-tuned manually since the parameters of the plant are quite different from the values assumed in the simulation. In tuning, having a relatively slower steady-state response is preferred to an overshoot to avoid the object from being squished or damaged.

From Table 1, it can be seen that are some discrepancies between the simulation and experimental results although both results meet the desired specifications. This is due to the unmolded dynamics and noise present in the experiment. It is suggested that a more accurate mathematical modelling to be developed so that the simulation resembles the actual experiment closely.

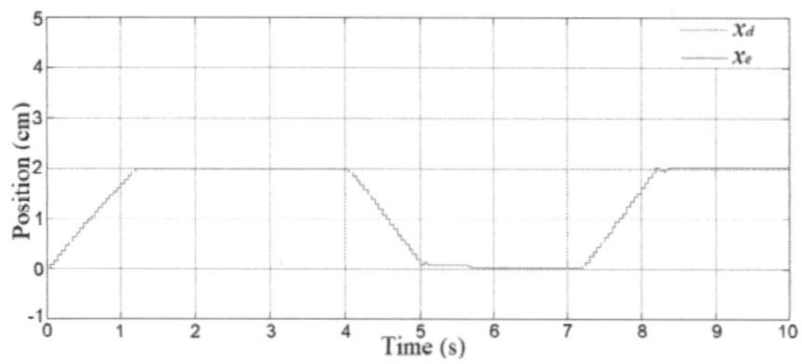

Figure 8. Position tracking of the master manipulation.

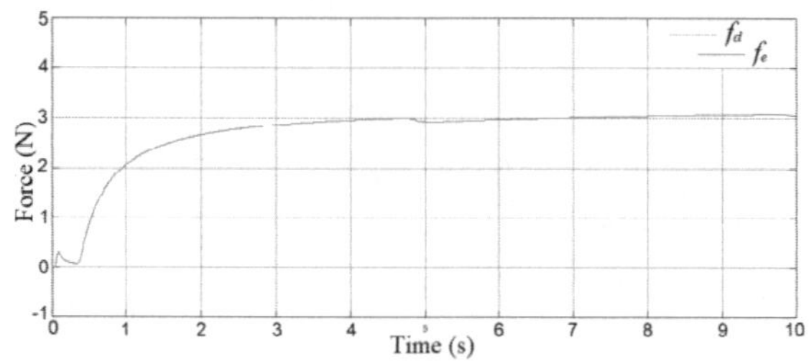

Figure 9. Force tracking of the slave manipulator, holding a tomato.

Table 1. Performance of force and position control comparison.

		e_{ss}	T_s (s)	PO (%)
Simulation	Position	0	0.261	10.484
	Force	0	0.207	12.366
Experiment	Position	$0.323mm$ (3.266%)	3.5	3.266
	Force	$48.387mN$ (4.839%)	1.707	0

Figure 10. Manipulators holding a tomato.

6. Conclusion

The prototype of the cooperative manipulators has been developed and built, and the control system to control the manipulators has been evaluated and implemented successfully. Through simulation and experimental tests, the implemented control technique has successfully held the deformable object with maximum errors of 0.34mm for position, and 50mN for force. The significant difference between the simulation and experimental result indicates the inaccuracy of the initial modelling of the system.

The scope of this paper is limited to the action of holding a deformable object. Due to that focus, 1-DOF manipulators have been utilized in the study. To expand the tasks that the robot can do, the number of DOF can be increased; for example, to allow picking and placing the object. The deformable object to be held in this work is also limited. Currently, this paper regards only objects such as a plastic cup, cardboard boxes, fruits and vegetables. In the future, this study could be extended to handle more complex deformable objects such as fabrics. Moreover, more complex controllers, such as adaptive control could be implemented instead of the PID.

Acknowledgments
The authors would like to thank The World Academic Sciences (TWAS-COMSTECH) for supporting this research under the grant number RG/ENG/AS_C-FR3240288945.

References
[1] Landau D, Pitaevskii L P, Kosevich A M and Lifshitz E M 2012 *Theory of elasticity* (London: Pergamon Press)
[2] Wada T, Hirai, S, Kawamura S and Kamiji N 2001 Robust manipulation of deformable objects by a simple PID feedback *IEEE International Conference on Robotics and Automation* **1** 85-90
[3] Guo F, Lin H and Jia Y 2013 Squeeze grasping of deformable planar objects with segment contacts and stick/slip transitions *IEEE International Conference on Robotics and Automation* 3736-41
[4] Berenson D 2013 Manipulation of deformable objects without modeling and simulating deformation *International Conference on Intelligent Robots and Systems* 4525-32
[5] Lee A X, Lu H, Gupta A, Levine S and Abbeel P 2015 Learning force-based manipulation of deformable objects from multiple demonstrations *IEEE International Conference on Robotics and Automation* 177-184
[6] Hirai S, Tsuboi T and Wada T 2001 Robust grasping manipulation of deformable objects *IEEE International Symposium on Assembly and Task Planning* 411-416
[7] Wang X, Xiao Y, Fan X and Zhao Y 2016 Design and grip force control of dual-motor drive electric gripper with parallel fingers *IEEE Information Technology, Networking, Electronic and Automation Control Conference* 696-700
[8] Cleary K and Nguyen C 2001 State of the art in surgical robotics: clinical applications and technology challenges *Computer Aided Surgery* **6** 312-328
[9] Omar A, Ri P and Zhang Y 2016 Control algorithm trajectory planning for dual cooperative manipulators with experimental verification *MATEC Web of Conferences* **75**
[10] Caccavale F and Uchiyama M 2008 Cooperative manipulators *Springer Handbook of Robotics* ed Siciliano B and Khatib O (Berlin: Springer) 701-718
[11] Aghili F 2013 Adaptive control of manipulators forming closed kinematic chain with inaccurate kinematic model *IEEE/ASME Transactions on Mechatronics* **18** 1544-54
[12] Navarro-Alarcon D, Liu Y, Romero J G and Li P 2013 Model-free visually servoed deformation control of elastic objects by robot manipulators *IEEE Transactions on Robotics* **29** 1457-68

IOP Conf. Series: Materials Science and Engineering **260** (2017) 012011 doi:10.1088/1757-899X/260/1/012011

Dragon Stream Cipher for Secure Blackbox Cockpit Voice Recorder

Fadira Akmal[1], Surya Michrandi Nasution[2], Fairuz Azmi[3]

[123]Security Laboratory Electrical Engineering Faculty
Telkom University Bandung, Indonesia
[1]fadira.akmal@gmail.com [2]michrandi@telkomuniversity.ac.id
[3]worldliner@telkomuniversity.ac.id

Abstract. Aircraft blackbox is a device used to record all aircraft information, which consists of Flight Data Recorder (FDR) and Cockpit Voice Recorder (CVR). Cockpit Voice Recorder contains conversations in the aircraft during the flight.Investigations on aircraft crashes usually take a long time, because it is difficult to find the aircraft blackbox. Then blackbox should have the ability to send information to other places. Aircraft blackbox must have a data security system, data security is a very important part at the time of information exchange process. The system in this research is to perform the encryption and decryption process on Cockpit Voice Recorder by people who are entitled by using Dragon Stream Cipher algorithm. The tests performed are time of data encryption and decryption, and avalanche effect. Result in this paper show us time encryption and decryption are 0,85 seconds and 1,84 second for 30 seconds Cockpit Voice Recorder data witn an avalanche effect 48,67 %.

1. Introduction

The limited ability of aircraft blackbox in the process of sending information is to be a hassle if a aircraft crash occurs and blackbox is difficult to find, because all the facts cause the occurrence of an accident is in the blackbox. Incident Malaysia airlines flight 370(MH70), Phoenix International Holdings, Inc spents 70 days with dives over 5000 meter into the sea to investigate the crash[5]. From the one example of aircraft accident, aircraft blackbox needs the ability to send information but the information submitted must be maintained its authenticity and secrecy. One technique for securing data is cryptography. According to Rinaldi Munir, Cryptography is a field of science that deals with mathematical techniques related to confidentiality information security, data integrity, and authentication[2].
This Dragon algorithm will be implemented on the recording of Cockpit Voice Recorder by calculating the processing time and avalanche effect. Although the system does not discuss the process of sending data, but the ability of process for data security in a short time is important.

2. Related Works

2.1. Cockpit Voice Recorder

Cockpit Voice Recorder (CVR) is part of the blackbox aircraft that works to record all conversations and sounds during the flight. Cockpit voice recorder is active if the plane engine is switched on, and start working from the plane engine is turned on until the engine is turned off. This CVR has four channels consisting of three channels used to record conversations crew of the aircraft,between pilots and co pilots, pilots with air traffic control, flight attendants with passengers, while one more channel is used for environmental sound inside and outside the cabin such as noise, rain or lightning. The CVR can record the entire conversation with a duration of 120 minutes from the four channels[1].

2.2. Dragon Stream Cipher

The dragon algorithm includes the type of stream cipher. In 2005 was submitted to the eSTREAM project in the software profile category and has advanced to phase 3. Incorporation of feedback shift register and non-linear function with memory is the basic design of cipher. This state is initialized with 128- or 256-bit key-IV pairs. Each clock of the stream cipher produces 64 bits of keystream, using simple operations on 32-bit words[3].

a. F Function

F function is reversible processing of 192 bits (six 32-bit words) to 192 bits. It takes six 32 bit words as input and produces six 32-bit words as output. These six 32-bit input words are normally defined as a, b, c, d, e, f. And the output words are a', b', c', d', e', f'. Also the function has other components called G1, G2, G3, H1, H2, H3. The main operations used in the F function are the xor(+) and the addition modulo 2^32[+] divide into 3 layers ; pre mixing, s-box and post mixing[3].

Input $= \{\ a, b, c, d, e, f\ \}$
Pre-mixing Layer:

1. $b = b \oplus a;$	$d = d \oplus c;$	$f = f \oplus e;$
2. $c = c \boxplus b;$	$e = e \boxplus d;$	$a = a \boxplus f;$

S-box Layer:

3. $d = d \oplus G_1(a);$	$f = f \oplus G_2(c);$	$b = b \oplus G_3(e);$
4. $a = a \oplus H_1(b);$	$c = c \oplus H_2(d);$	$e = e \oplus H_3(f);$

Post-mixing Layer:

5. $d' = d \boxplus a;$	$f' = f \boxplus c;$	$b' = b \boxplus e;$
6. $c' = c \oplus b;$	$e' = e \oplus d;$	$a' = a \oplus f;$

Output $= \{\ a', b', c', d', e', f'\ \}$ ... (1)

The G and H functions are non-linear process of 32-bit inputs to 32-bit outputs. Storing a 32×32 s-box requires sixteen gigabytes. Virtual 32-bit s-boxes use small of memory,in the process from 8×32 32 s-boxes. Therefore the G and H functions are processed from two 8×32 s-boxes, S1 and S2[3].

b. Initialization

Key initialization is partition into eight 128-bit words, W0 to W7. Dragon-128 and Dragon-256 have simple two-phase initialization system that different only in the first phase. The 64-bit memory M is loaded with the constant value 0x0000447261676F6E , an ASCII representation of 'Dragon'[3].

IOP Conf. Series: Materials Science and Engineering **260** (2017) 012011 doi:10.1088/1757-899X/260/1/012011

Input $= \{ K, IV \}$
1. $M = 0x0000447261676F6E$
2. $W_0 \parallel ... \parallel W_7 =$

128-bit key and IV

$K \parallel K' \oplus IV' \parallel IV \parallel K \oplus IV' \parallel K' \parallel K \oplus IV \parallel IV' \parallel K' \oplus IV$

256-bit key and IV

$K \parallel K \oplus IV \parallel \overline{K \oplus IV} \parallel IV$

Perform steps 3-8 16 times

3. $a \parallel b \parallel c \parallel d = (W_0 \oplus W_6 \oplus W_7)$
4. $e \parallel f = M$
5. $\{a', b', c', d', e', f'\} = F(a, b, c, d, e, f)$
6. $t = (a' \parallel b' \parallel c' \parallel d') \oplus W_4$
7. $W_i = W_{i-1}, 1 \leq i \leq 7$
8. $W_0 = t$
9. $M = e' \parallel f'$

Output $= \{ W_0 \parallel ... \parallel W_7 \}$ ... (2)

c. Keystreams Generation[3]

The keystream generation generating a keystream using the previously initialized internal state, and the last value of M as input. The internal state is now divided into 32-bit words. During each round of the keystream generation, B0, B9, B16, B19, B30, B31 are used as the F function inputs a, b, c, d, e, f. Output in each round of the keystream generation is a 64-bit length keystream k, and an updated state B and M, which can be used again to generate more keystreams.

Input $= \{ B_0 \parallel ... \parallel B_{31}, M \}$
1. $(M_L \parallel M_R) = M$
2. $a = B_0, b = B_9, c = B_{16}, d = B_{19}, e = B_{30} \oplus M_L, f = B_{31} \oplus M_R$
3. $(a', b', c', d', e', f') = F(a, b, c, d, e, f)$
4. $B_i = B_{i-2}, 2 \leq i \leq 31$
5. $B_0 = b', B_1 = c'$
6. $M = M + 1$
7. $z = a' \parallel e'$

Output $= \{ z, B_0 \parallel ... \parallel B_{31}, M \}$ ... (3)

3. System Design

This system will implement the Dragon stream cipher algorithm, this algorithm will encrypt and decrypt frames per frame from recording of cockpit voice recorder files. Dragon algorithm for this paper using 256 bit key and IV, and use symmetric key for encryption and decryption process.

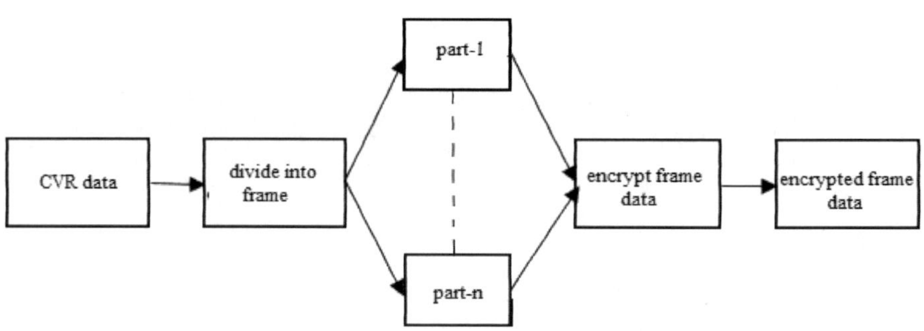

Figure 1. Dragon encryption process

Encryption process start from divide Cockpit Voice Recorder data .wav into frame and also convert into binary file .txt, frame data will be encrypt one by one with keystream from keystream generation processed, then the frame data can send to ground or another place. This paper only discuss about system how to secure data , not discuss about sending data. System for secure cockpit voice recorder on encryption processing time should not take a long time.

IOP Conf. Series: Materials Science and Engineering **260** (2017) 012011 doi:10.1088/1757-899X/260/1/012011

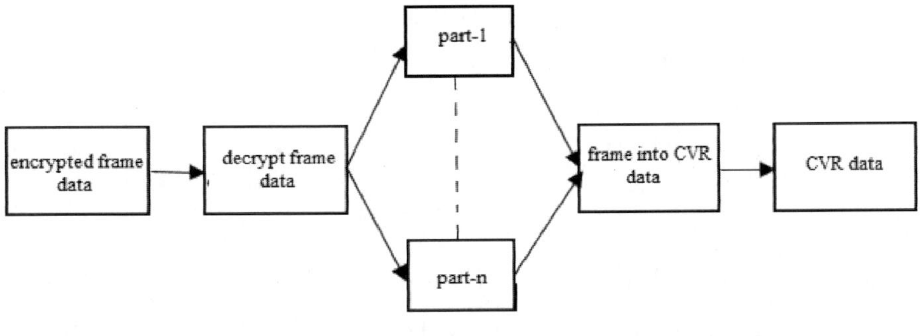

Figure 2. Dragon decryption process

Figure above show the process when frame data on the ground , decryption process will decrypt frame by frame with same keystream on encryption process. After the decryption process is complete, system will concat file frame.txt into CVR data.wav like before. Decryption process may take longer than encryption process, cause there is no process that must be done in the system.

4. Experimental Testing

Experimental testing is performed how long time from encrypt and decrypt cockpit voice recorder files and calculates the avalanche effect of the Dragon algorithm that implemented on the data. This paper testing with 30 seconds cockpit voice recorder duration, with data capacity 840KB.

4.1. Encrypytion and Decryption Time

Table 1. CVR encryption data and time

Testing	200 frame(s)	300 frame(s)
1	0,596096	0,736012
2	0,894002	1,117502
3	0,601053	0,751312
4	0,872117	1,090142
5	1,406586	1,687886
6	0,842201	1,01064
7	0,609878	0,731853
8	1,11603	1,302035
9	0,883208	1,057408
10	0,718711	0.928767

Table 2. CVR decryption data and time

Testing	200 frame(s)	300 frame(s)
1	11,889367	0,845670
2	0,790354	0,790045
3	0,5325493	0,673410

4	0,672110	0,723541
5	0,802844	0,819987
6	0,9124202	1,020
7	0,600498	0,885200
8	1,00603	0,904258
9	0,643214	0,83049
10	0,629417	0,917732

Tables above showed how long waktu encryption and decryption for cockpit voice recorder in 10 times testing with different frame, 200 frames and 300 frames. We can know that the larger number of frames the longer time we need to process CVR. The longest encryption process exists on the 5th test at 300 frames, while the longest decryption process exists on the 1st test at 200 frames.

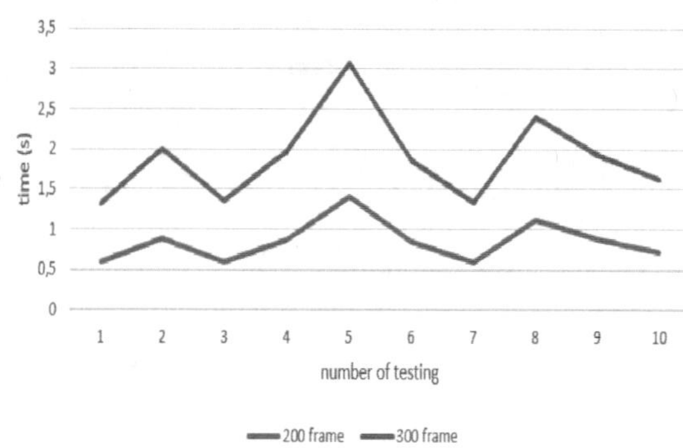

Figure 3a. Graph encryption time

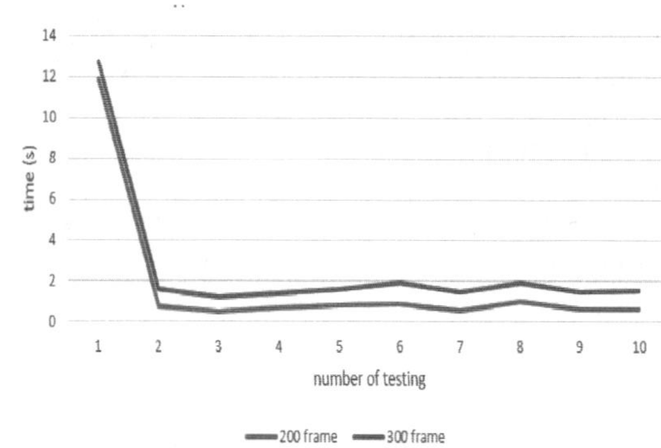

Figure 3b. Graph decryption time

Figure 3a showed encryption process between 200 frames and 300 frames, the graph showed time for encryption process realtively stable, although on 5th test time is uphill. Figure 3b describe decryption process 200 frames and 300 frames, on both cases the 1st test always take longer time than other. So we can calculate the average time for encryption and decryption while divide into 200 frames are 0,85 second and 1,84 seconds. We need longer time for 300 frames CVR in encryption and decryption are 1,05 seconds and 0,84seconds.

4.2. Avalanche effect

Avalanche effect is a way to determine whether or not a cryptographic algorithm. By counting avalanche effect, we will know how much the changes that occur in the ciphertext bits due to the encryption process. The higher value of the avalanche effect will make better the level of security algorithms. An algorithm is said to have a good AE value if only one bit changes in the input results in a change of about half the number of bits in its output[4].

$$Avalanche\ Effect = \frac{\Sigma \text{bit changed}}{\Sigma\, bit\ total} * 100\% \qquad ...(4)$$

Here is an avalanche effect table testing of Dragon Algorithm. The first test is on Table 3 with the different audio frame that showed in table and the key and IV are same for each audio frame.
Key : 00001111 22223333 DDDDDDDD 66667777 88889999 AAAABBBB CCCCDDDD EEEEFFFF
IV : 000011F1 22223333 44445555 68887777 88989999 AAAABBBB CCCCDDDD EEEEFFFF

Table 3. Avalanche effect testing

Audio Frame	Plaintext	Ciphertext	AE(%)
Frame1.txt	010100100100100101000110010001100100011010000000000.....	101001001001001010001100100011001000110100000000000.....	49.0
Frame2.txt	100000001000000010000000100000001000000010000000100.....	100000001000000010000000100000001000000010000000100.....	48.5
Frame3.txt	100000001000000010000000100000001000000010000000100.....	100000001000000010000000100000001000000010000000100.....	48.5

Table 3 is avalanche effect testing for CVR data divide into 200 frame, the average of avalanche effect for 200 frame is 48.67 %. This value including good avalanche effect for security level.
Here the second test is on Table 4 with the same audio frame1.txt that showed in table and different key but IV are same for each audio frame.

Table 4. Avalanche effect testing

Key	Plaintext	Ciphertext	AE(%)
00001111 22223333 DDDDDDDD 66667777 88889999 AAAABBBB CCCCDDDD EEEEFFFF	01010010010010010100011001000110010001101000 00000000.....	10100100100100101000110010001100100011010000 00000000.....	49.0
00001111 22223333 DDDDDDDD 66667777 88889999 AAAABBBB CCCC**5**DDD EEEEFFFF	01010010010010010100011001000110010001101000 00000000.....	10100100100100101000110010001100100011010000 00000000.....	49.8
00001111 22223333 DDDDDDDD 66667777 88889999 AAAABBBB **3**CCCDDDD EEEEFFFF	01010010010010010100011001000110010001101000 00000000.....	10100100100100101000110010001100100011010000 00000000.....	49.5

Table 4 is avalanche effect testing from 1 frame in CVR, the table showed that one change bit in key can change ciphertext(different avalanche effect from three testing). Value of avalanche effect from different key still produce good performance for security level.

5. Analysis

From the Table 1 above, we can see the comparison of the time required by audio with different frame. The 200 frames has duration of time from encryption and decryption is more efficient than the second with 300 frames. The 200 frames needs 0,85 second for encryption and 1,84 seconds for decryption while the 300 frames is needed 1,05 seconds for encryption and 0,84 seconds for decryption.From that comparison, it can be concluded that the number of frames affects the time of encryption and decryption. The more number of frames so the longer the time of the encryption and decryption process
From test results Table 3 above, we can see the average avalanche effect value from the different frame, the same key and IV are 48.67% and Table 4 we can see the changed value of avalanche effect in every different key 49 % - 50% . All result shows that the avalanche effect test produce changes bits is about half the number of bits in its output. Cryptographic algorithm will be difficult to resolve when the key used is not known and the results of the output will be very unique. It means Dragon algorithm has a good performance on security level.

IOP Conf. Series: Materials Science and Engineering **260** (2017) 012011 doi:10.1088/1757-899X/260/1/012011

6. Conclusion

From the implementation and test results, we draw some concluding remarks, as follows :

Result in this paper show us time encryption and decryption are 0,85 seconds and 1,84 second for 30 seconds Cockpit Voice Recorder data with an avalanche effect 48,67 %. The difference of audio frames can affect to time of encryption and decryption. The larger the number of frames, the longer they need time to encrypt and decrypt the audio. The changes bits after testing avalanche effect value is about half of number of bits in its output, it's means algorithm has good security system. Dragon algorithm has a good performance avalance effect range value, 48 % - 50 % on this system.

7. REFERENCES

[1] *Air Transport Series. Product Description. Solid-State Voice/Data&Digital Communications Combined Recording Systems for Air Transport Application.*

[2] Rinaldi Munir. 2006. Kriptografi. Bandung : Penerbit Informatika Bandung

[3] Ed Dawson, Matt Henricken, and Leonic Simpson. 2005. *The Dragon Stream Cipher: Design, Analysis, and Implementation Issues*

[4] Siska Selany. 2015. IEEE *Sosemanuk Algorithm forEncryption and Decryption Video on Demand (VoD)*

[5] Peter K LeHardy and Christopher Moore. 2014. IEEE *Deep Ocean Search for Malaysia Airlines Flight 370*

[6] Paul. R. Schoberg. 2003. *Secure Ground – Based Remote Recording and Archiving Of Aircraft Black Box Data.* Monterey California: AMM: Naval Post Graduate School.

[7] Pardeep Kumar and Hoon Jae Lee. 2009. IEEE *A Secure Data Mechanism for Ubiquitous Sensor Network with Dragon Cipher.* Dongseo University, Busan, South Korea.

Enhancement web proxy cache performance using Wrapper Feature Selection methods with NB and J48

Dua'a Mahmoud Al-Qudah, Rashidah Funke Olanrewaju, Amelia Wong Azman

Faculty of Engineering, International Islamic University Malaysia, Jalan Gombak 53100, Kuala Lumpur, Malaysia;

dm_q66@yahoo.com, frashidah@iium.edu.my, amy@iium.edu.my.

Abstract. Web proxy cache technique reduces response time by storing a copy of pages between client and server sides. If requested pages are cached in the proxy, there is no need to access the server. Due to the limited size and excessive cost of cache compared to the other storages, cache replacement algorithm is used to determine evict page when the cache is full. On the other hand, the conventional algorithms for replacement such as Least Recently Use (LRU), First in First Out (FIFO), Least Frequently Use (LFU), Randomized Policy etc. may discard important pages just before use. Furthermore, using conventional algorithm cannot be well optimized since it requires some decision to intelligently evict a page before replacement. Hence, most researchers propose an integration among intelligent classifiers and replacement algorithm to improves replacement algorithms performance. This research proposes using automated wrapper feature selection methods to choose the best subset of features that are relevant and influence classifiers prediction accuracy. The result present that using wrapper feature selection methods namely: Best First (BFS), Incremental Wrapper subset selection(IWSS)embedded NB and particle swarm optimization(PSO)reduce number of features and have a good impact on reducing computation time. Using PSO enhance NB classifier accuracy by 1.1%, 0.43% and 0.22% over using NB with all features, using BFS and using IWSS embedded NB respectively. PSO rises J48 accuracy by 0.03%, 1.91 and 0.04% over using J48 classifier with all features, using IWSS-embedded NB and using BFS respectively. While using IWSS embedded NB fastest NB and J48 classifiers much more than BFS and PSO. However, it reduces computation time of NB by 0.1383 and reduce computation time of J48 by 2.998.

Keywords: web cache, replacement algorithms, wrapper feature selection, intelligent classifiers.

1. Introduction

As the rapid growth in network usage and significant role for network in our life, there is a need for good service, enough bandwidth and fast response. Web caching mainly depends on storing copies of web objects closer to user. In web caching, if the end user requests a page that founds in the cache, it is sent directly. Meanwhile, if it is not in the cache, fetching page from origin server is required. However, transferring of web object over the networks tends to increase network traffic. Thus, web caching helps in decreasing network traffic and reduce loads on the origin server [1, 2].

Meanwhile, due to limited cache size, appropriate cache replacement algorithms are required to manage web caching. Cache replacement algorithms are a process of evicting pages from cache to make room for new requested page and there is no enough space for it. They have the significant effect in the web caching [3] Although the conventional replacement algorithms such as LRU, LFU, and SIZE are not effective enough in web cache because they consider some features and ignore others which may have a good impact on web cache performance. Furthermore, those algorithms suffer from cache pollution problem where objects will be stored in the cache which may not revisited in future hence, waste of limited resources. Intelligent replacement algorithms are required to effectively utilize the limited resources (web cache) as well enhancing its performance by integrating more than one feature together [4].

extracting subset of features from space of features is called feature selection method. It has a significant impact on improving machine learning techniques performance. Since it is helped in removing noisy, irrelevant and unwanted features. In addition to fastest intelligent systems by reducing computation time [1, 2]. However, features selection method divided into filter methods, wrapper methods and embedded methods.

Chosen features for training intelligent system will strongly influence prediction process. Most strategies that used intelligent systems chose features manually while automated feature selection enhances prediction process by choosing the optimal and best subset of the features that have a significant impact on intelligent systems[1, 2]. Hence, this research study impact of using wrapper feature selection method to choose the best subset of features for training C4.5, NB intelligent classifiers, as well as prevent pollution problem. The paper is organized as follow section two discuss related works, section three provide a theoretical background overview, section four descried and analysis the proposed methods and finally, the conclusion found in section five.

2. Related Works

Integrating of NB classifier with GDS, LRU and DA algorithms were proposed by [4]. The new proposed algorithms divided into two stages that are offline stage with reasonability to train NB classifiers and predict class of object. While online stage integrates NB with replacement algorithms and evaluate the proposed algorithms over conventional algorithms.[5] proposed NB-SIZE and C4.5-Hybrid algorithms which relay on integrating NB classifier and C4.5 decision tree with SIZE and Hybrid conventional algorithms, the proposed approaches enhance web proxy cache performance in terms of HR, BHR and latency. To enhance web cache performance in terms of HR and BHR [6] proposed integration between machine learning algorithms like SVM, NB and C4.5 decision tree with LFU_DA replacement algorithm. However, the feature that selected for training are chosen manually.

Co-operation between J48 classifier and LRU replacement algorithm was proposed by [7] for improving LRU performance and prevent cache pollution problem. The result illustrates that using J48 classifier boost LRU performance. While [8] integrate decision tree classifier with SIZ and Hybrid algorithms to form DT-SIZE and DT-Hybrid new intelligent replacement algorithms. Moreover, Random Forest Tree classifier (RFT) is used to design a novel RFT-LRU and RFT-GDSF intelligent replacement algorithms. RFT classifier used for class prediction purpose if will be revisited again or not. After that, the classifier integrates with LRU, GDSF to improve web cache performance in terms of HR and BHR[9]. A novel web cache replacement method is proposed by [10]. it depends on using SVM for classification and divided the caching probability it will revisited again or not. After that use LFU algorithms when the cache is full. The proposed algorithm is called Hybrid LFU-SVM. The evaluation proves that Hybrid LFU-SVM enhance web cache performance in terms of HR and BHR over SVMLRU, LFU and LRU algorithms However, features that used in training intelligent classifiers that are mentioned before were chosen manually.

Abdalla, Sulaiman and Ali [1], suppose using automated method to extract best subset of features to improve the prediction process in intelligent systems which used in training phase by J48 classifier and NB classifier. Automated prediction scheme caused reducing in the number of features that selected and reducing dimensionality of dataset. Moreover, reduction in computation time is achieved, and improving accuracy of NB, C4.5 using Best First wrapper method compared to manual selection method. However, [2] propose using IWSS embedded NB feature selection method to extract the best subset feature for Naïve Bayes classifier and later integrate Naïve Bayes classifier with Adaptive Weight Ranking Policy via Dynamic Aging (AWRP-DA) to improve the performance of web systems concerning Hit Ratio (HR) and Byte Hit Ratio (BHR).

3. Theoretical Background

3.1. Feature selection

Feature selection is a process of extracting subset of features from features space. It plays a key role in improving and enhancing the performance of machine learning techniques by reducing the effect of noise and irrelevant features. Moreover, it reduces the computational time and storage required to store features. Feature selection methods are classified in three groups, namely: Filter, Wrapper and embedded methods [11]. Hence filter and embedded methods are not considered in this research.

Wrapper methods are general-purpose algorithms they depend on learning algorithms to find best subset of features. Wrapper algorithms adding (forward selection) or deleting features (backward selection) one at a time from feature space until extract the best and optimal feature subset. However, wrapper's disadvantage is the computational cost involved [12]. The main idea of wrapper processes that using introduction algorithm as black box, and then run it over dataset. Partitioning data set into internal training and holdout set are done. However, this research uses best- first method, Incremental Wrapper Subset Selection with naïve bays method and Particle swarm optimization.

3.1.1. The Best First Method (BFS): BFS Searches the space of attribute subsets by greedy hill climbing augmented with a backtracking facility. Setting the number of consecutive non-improving nodes allowed controls the level of backtracking done. Best first can work either by FSS or BSS or Bi-directional. The main goal is to choose the most hopeful node that had been generated so far and that has not already been expanded. Best-first search stop when the goal is reached [13].

3.1.2. Incremental Wrapper Feature Subset Selection with Naïve Bayes (IWSS Embedded NB): It is a wrapper algorithm. But, Naïve Bays classifier embedded with it. Thus, it combines between linear time complexity of filter algorithms and advantages of wrapper methods. IWSSS always attempt to get best ranking features and once a feature is included it maintains them until end of the search. To solve this disadvantage of IWSS method adding NB classifier to it. NB allow algorithm to add new features or interchange new feature with feature that is previously added[14].

3.1.3. Particle Swarm Optimization (PSO): The idea of PSO is derived from social behaviors. Such that a group of birds flocking at promising position for attaining an accurate goal in a multidimensional space. In PSO the problem has been optimized by repeatedly trying to improve a candidate solution. It is like other evolutionary algorithms it has a population of solutions (swarm) of individuals (particles) that characterized depending to their position and velocity and moving them in search space depending on simple mathematical formula. To reach the optimal solution, there are two factors that particles depends on it to change searching direction. That are its own best previous experience (p-best) and the best experience of all other members (g-best) social part. Update that calculate velocity for each particles, Construction which determine particle moving and finally termination that terminated iterations if the iteration reach maximum number of iteration [15, 16].

3.2. Decision Tree

C4.5 is the most know algorithm to build decision tree. And using it is very common in different scopes like marketing, finance, computer science etc. The idea of C4.5 algorithm depends on building a tree start with root node which is the entire given dataset, then it works recursively to divide the data into smaller subsets by testing a given attribute at each node. The process terminated when the subsets are pure [17].

IOP Conf. Series: Materials Science and Engineering **260** (2017) 012012 doi:10.1088/1757-899X/260/1/012012

C4.5 Java code implementation in WEKA data mining tool is J48 classifier. WEKA detailed available at http://www.cs.waikato.ac.nz/ml/weka/. J48 classifier is used to predict output value depending on various features values of training data. Internal node in J48 represent various features. However, branches among nodes represent possible value of features. While output value represents by leaf node. Attribute that need to predicted is called dependent variable since it depends on all other variable values which are called (independent variable) [7].

3.3. Naïve Bayes

Naïve Bayes classifier is a simple supervised machine learning algorithm which depends on the Bayes' probability theory. It can deal with any number of attributes and classes. Moreover, Naïve Bayes technique is considered the fastest machine learning techniques and it is very easy to construct. In addition, its performance is very good in a lot of applications and applied well in many domains [4, 12].

In NB, all features under class label are considered conditionally independent. That leads to ignore any correlation among features. [17]. Discretizing process is required to convert continuous values into discrete values for better performance. Minimum Description Length (MDL) method will be used that described by [18] with default setup in WEKA software to convert continuous values into discrete values. After that, the dataset is ready to train NB. NB classifier trained to classify objects to revisited objects or not within forwarding sliding window. Prior probabilities Pr= (C =cj) and conditional priorities Pr (Ai= ai | C=cj) estimation required by NB to predict object's class. Probabilities calculation is shown in equation 1 and 2. Where Pr refers to probability value, C refers to class and A refers to attribute.

$$\text{Pr}= (C = cj) = \frac{number\ of\ examples\ of\ class\ c}{number\ of\ total\ examples} \ \ldots\ldots\ (1)$$

$$\text{Pr}\ (Ai= ai \mid C=cj) = \frac{number\ of\ examples\ with\ \ Ai = ai\ and\ class\ cj}{number\ of\ examples\ of\ class\ cj} \ \ldots\ldots\ (2)$$

4. Proposed Methods

4.1. Raw Data Collection

proxy log files data had been collected from six proxy servers of IRCache. That collected from BO2, PA, SJ, SV, UC and SD servers that located at Boulder Colorado, Palo Alto California, MAE-West San Jose California, Silicon Valley California (FIX-West), Urbana-Champaign, Illinois San Diego California, between 9 and 10 of January 2007. Data collected in 9-1-2007 is used in training phase. Features that exist in the log trace files are time elapsed, timestamp, object size, client address, log tag with HTTP script, URL, user identification, request method, content type and hierarchy of data and hostname.

4.2. Pre-processing

Data pre-processing is a process of removing noisy, incomplete, outlier and inconsistent data which presented in the dataset. Pre-processing done following three steps that are Parsing, filtering and finalizing.

In parsing boundaries between successive records in log files are determined. In addition to, determining each unique field inside each record. While, Filtering process done by Removing requests that generated by automated programs like web robots, spiders and crawlers. Since they cause traffic to web sites, removing requests for image files associated with requests for particular pages, removing irrelevant records, records that have unsuccessful status. We consider the success status which has 200 codes., removing any request has methods except GET and POST. And removing uncatchable requests that have "?" inside URL and cgi-bin requests. And Finalizing (for simplify) done by replacing each URL with unique random integer number. In addition to, using specified integer numbers to define object type like HTML = 1, Image = 2, Audio = 3, Video = 4, Application =

5 and Other types = zero. Finally, log files will contain the main fields that have impact from which the useful information can be extracted: URL Id, time stamp, elapsed time, object size, type of Web object.

4.3. Feature Extraction

In this state, the features are obtained from log files that play a significant role in web cache performance. The most common features that are extracted and used by researchers are recency, frequency, cost and size since they have an obvious influence in web cache [4, 19-23]. In table 1 features description and extracting procedure are described [1].

Table 1. Features description and extracting procedure[1].

No.	Feature Name	Description	Calculation
1	URL_ID	Identification number for each URL.	Randomly generate unique integer number for each URL.
2	Timestamp	Time when the client socket is closed in millisecond.	Given
3	Elapsed time	Time between accept () and close () of the client socket in millisecond.	Given
4	Size	Number of bytes written to the client in bytes.	Given
5	Object type	Web type object.	Given
6	Frequency	Web object's frequency	Number of frequent occurs of web object.
7	SWL-frequency	Web object's frequency based on sliding window.	Swl-freq = $\begin{cases} swl - freqi - 1 + 1, if\ \Delta t \leq swl \\ 1, otherwise \end{cases}$
8	IR-Frequency	Web objects frequency plus internal request frequency.	IR_Freq = freq + IR
9	Recency	Web object's recency based on sliding window.	Recency= $\begin{cases} max(swl, \Delta t)\ , if\ object\ requested\ before \\ swl\ , otherwise \end{cases}$
10	Time Spend	The time that spent in each page in millisecond.	The different timestamp in millisecond between current page and next one which immediately followed.
11	Mean	The average of time spent in each Page in millisecond.	The sum of time spent in each page for all its frequents divide by its number of frequents.

Sliding window (SWL) was defined as a time before and after the request was made in millisecond. SWL should be around the mean time of object. They use it in calculate object's frequency and recency as shown in table 2. They trained BPNNs with different SWL values between 30 minute and two hours. The results show that when small SWL value uses the training performance is decreased. But, using large training dataset with small SWL value can increase training performance. They conclude that 30 minutes are good enough time periods for large dataset. Sarhan, Elmogy and Ali [21] define internal request as request of another web page originating from root web page. Counting object's frequency with internal request was done by increment frequency of main page with each subpage request. Thus, frequency of root page will be equal to its frequency plus subpages frequency which is called Internal Request Frequency (IR_Frequency).

4.4. Feature Selection

After finalizing data pre-processing and feature extraction. Weka 3.8 is used to select the best subset of features. WrapperSubsetEval is used as attribute evaluator in all data and three different methods used in search method namely: BSF, IWSS-embedded-NB and PSO for both classifier NB and J48 with default Weka attributes for rest parameters. The following tables 3 and 4 present the best subset of features that are chosen by each method. In general, IWSS- embedded NB choose the smallest subset of features while PSO choose larger number of features.

Table 2. The selected best subset for NB classifier in different datasets by different feature selection methods.

NB	BO2	PA	SJ	SV	UC	SD
Best First	Id, Elapsed Time, Type, Freq, FIR, Swl_Freq, Recency, Time Spend, Mean	Id, Type, Freq, FIR, Swl_Freq, Recency	Id,Timestamp, Freq, FIR, Recency, Timespend ,Mean	Id, Timestamp, Type, Freq, FIR, Recency, Mean	Freq, FIR	Id, Type, Freq, Recency, Mean
Number of features	9	6	7	7	2	5
IWSS Embedded NB	Freq, FIR	Freq, FIR	ID, Timestamp, Type, Freq, FIR, Mean, Recency	FIR, ID, Type, Recency	ID, Elapsedtime, Type, Freq, FIR, Mean , Recency, Timespend, Size	Freq, FIR, ID, Type, Mean, Timespend, Elapsedtime
Number of features	2	2	7	4	9	7
PSO	Freq, FIR	Id, Timestamp, Type, Freq , FIR, Recency, Mean	Id, Timestamp, FIR, Swl_Freq, Timespend , Recency, Mean	Id, Type, Freq, FIR, Swl_Freq, Recency	Id, Elapsedtime, Size,Type, Freq, FIR, Swl_Freq, Recency, Mean	Id ,Elapsedtime, Size, Type, Freq, Recency, Mean
Number of features	2	7	7	6	9	7

Table 3. The selected best subset for J48 classifier in different datasets by different feature selection methods.

J48	Bo2	Pa	Sj	Sv	Uc	Sd
Best First	Id, Timestamp, Elapsedtime, Type, Freq, Fir, Recency, Timespend,Mean	Id, Timestamp, Freq, Fir, Type , Recency	Id, Timestamp, Type, Freq , FIR, Swl_Freq, Recency, Timespend, Mean	Id, Timestamp, Type, Freq , FIR, Recency,	Id, Timestamp, Type, FIR, Recency, Mean	Id, Timestamp, Elapsedtime, Type, Freq, FIR, Recency, Timespend, Mean
Number of Features	9	6	9	6	6	9
IWSS embedded NB	Freq, FIR	Freq, FIR	Id, Timestamp, Type, Freq, FIR, Mean , Recency	FIR, Id, Type , Recency	Id, Elapsedtime, Type, Freq, FIR, Mean , Recency, Timespend,	Freq, FIR , Id, Type, Mean, Timespend, Elapsedtime

					Size	
Number Of Features	2	2	7	4	9	7
PSO	Id, Timestamp, Type, Freq , FIR, Recency, Mean	Id, Timestamp, Type, Freq , FIR, Recency, Timespend	Id, Timestamp, Type, Freq , FIR, Recency, Mean, Timespend	Id, Timestamp, Type, Freq, FIR, , Recency	Id, Timestamp, Eapsedtime, Type, Freq, FIR, Timespend , Recency, Mean	Id, Timestamp, Elapsedtime, Type, Freq, FIR, Swl_Freq, Recency
Number of Features	7	7	8	6	9	8

4.5. Training

Training classifiers plays a significant role in replacement algorithm performance. Thus, each proxy datasets are divided into 70% for training and 30% for testing. After that, the data is prepared for training by discretizing it by using Minimum Description Length (MDL) method that suggested by [18] with default setup in WEKA 3.8.

The best subset of features which is chosen by BFS, IWSS embedded NB and PSO are used with the dataset to train and evaluate learning algorithms. Training data is prepared as follow:

1. Defining input features as <X1.....Xn> features set.
2. Setting target output value Y to 1 if object is requested again within forward sliding window. Otherwise, it is setting to zero.
3. Setting input and output in pattern format as <X1, X2... Xn , Y>.

When training pattern prepared in this form any classifier can be trained using this training pattern to predict if object revisited in future or not.Naïve Bayes and J48 classifiers are training and testing four times. Firstly, with all feature extracted which are URL_ID, timestamp, elapsed time, size, type, frequency, IR_frequency, SWL_frequency, recency, time spend and mean. Then using features extracted by Best First method, IWSS embedded NB method and Particle Swarm Optimization method. which are mentioned in detailed in table 3 for NB classifier and in table 4 for J48 classifier. The trained classifiers can be saved for later use.

5. Classifiers Evaluation and Discussion

NB and J48 classifiers are evaluated using Correct Classification Rate (CCR) or accuracy defined in (3), which is the most common performance measures are used. But, the datasets used in this research are imbalance. Thus, CCR or accuracy measures is not enough to assess classifier performance. In imbalance data one class has the majority and another class has minority instances in binary classification .So that sensitivity or true positive rate (TPR) defined in (4), specificity or true negative rate(TNR) defined in (5) and geometric mean (Gmean) defined in 6) are used to assessment classifier performance in addition to CCR [4]. Confusion matrix is presented in table 4.

$$CCR = \frac{(TP+TN)}{(TP+FP+FN+TN)}\% \ldots\ldots (3)$$

$$TPR = \frac{TP}{(TP+FN)}\% \ldots\ldots (4)$$

$$TNR = \frac{TP}{(TN+FP)}\% \ldots\ldots (5)$$

$$Gmean = \sqrt{TPR * TNR}\ \% \ldots\ldots (6)$$

Table 4. Confusion Matrix

	Predicted positive	Predicted negative
Actual positive	True Positive (TP)	False negative (FN)
Actual negative	False Positive (FP)	True negative (TN)

In this study, web object that requested again within forward -looking window SWL which is 30 minutes belongs to the positive class otherwise it belongs to negative class. In this study, minority class was the positive class which is the most important one and the majority represent negative class.

Comparison among six different proxy datasets that are BO2, PA, SJ, SV, UC and SD are presented in evaluation tables for training. Each table presents comparison between NB and J48 classifiers result. In addition, it presents compare between using NB and J48 classifiers with all features and with using features that had been extracted using automated feature selection methods.

Table 5: CCR of different proxy datasets Training phase

Datasets	NB Training				J48 Training			
	WFS	Best First	IWSS embedded NB	PSO	WFS	Best First	IWSS embedded NB	PSO
BO2	91.57%	91.93%	91.93%	91.93%	91.93%	91.93%	91.93%	91.93%
PA	85.53%	85.44%	85.79%	85.44%	87.74%	87.67%	85.80%	87.69%
SJ	77.50%	78.67%	79.38%	81.22%	86.49%	86.49%	86.46%	86.59%
SV	82.38%	82.47%	82.70%	82.47%	88.67%	88.67%	83.19%	88.67%
UC	89.00%	90.20%	90.19%	90.20%	91.91%	91.91%	90.35%	91.91%
SD	92.88%	94.22%	94.16%	94.22%	96.51%	96.51%	94.21%	96.61%
AVG	86.48%	87.15%	87.36%	87.58%	90.54%	90.53%	88.66%	90.57%

Table 5 CCR training table presents that both of NB and J48 give a satisfactory performance. But in the most cases J48 gives a higher accuracy than NB. In average, both classifiers give higher accuracy value using features selected by feature selection methods than without using them. NB training average CCR without using feature selection methods, using Best First, using IWSS embedded NB and PSO are 86.48%, 87.15%,87.36% and 87.58% respectively. whereas, in training J48 the result was as follow 87.58%,90.54%, 90.53%,88.66% and 90.57%. Accuracy results for NB and J48 classifiers indicate that PSO feature selection method boost classifier performance in average more than other feature selection methods.

Table 6 illustrate True Positive Rate (TPR) or sensitivity result for NB and J48 classifiers in training and testing. The result show that TPR for J48 classifier is better than NB classifier. However, both classifier do not have a very high TPR. This is referring to data imbalance since positive class represent the minority. Moreover, SD and UC datasets give the higher sensitivity than other proxy datasets since they have larger number of instances than others. That indicate when the data is large then it will give better indication and better performance for classifiers.

Table 6. TPR of different proxy datasets Training phase

Datasets	NB Training				J48 Training			
	WFS	Best First	IWSS embedded NB	PSO	WFS	Best First	IWSS embedded NB	PSO
BO2	53.2	51.1	51.1	51.1	51	51.1	51	51.6
PA	25.7	27.3	23.2	27.3	39.2	38.6	23	38.6
SJ	24.3	25	33.5	41.1	78.9	78.9	78.8	79
SV	30.5	29.8	29.6	29.8	67.7	67.7	35.8	67.7
UC	49.4	60.6	60.6	60.6	70.3	70.3	61.1	70.3
SD	58.1	79.3	70.4	78.3	82.1	82.1	78.3	82.2
AVG	40.2	45.5166	44.733333	48.0333	64.866	64.783	54.666666	64.9

In contrast to TPR result the NB classifier give higher TNR than J48. And the TNR result is very high compared to TPR as presented in table 7. This is because the negative class represent by majority of instances. Moreover, using PSO feature selection method promote the classifier performance in general.

Table 7. TNR of different proxy datasets Training phase

Datasets	NB Training				J48 Training			
	WFS	Best First	IWSS embedded NB	PSO	WFS	Best First	IWSS embedded NB	PSO
BO2	99	99.9	99.9	99.9	99.9	99.9	99.9	99.9
PA	98.9	98.5	99.8	98.5	98.6	98.7	99.9	98.7
SJ	97.7	99	96.8	96.5	89.4	89.4	89.4	89.4
SV	97.7	98.1	98.2	98.1	94.9	94.9	97	94.9
UC	96.9	96.1	96.1	96.1	96.2	96.2	96.2	96.2
SD	96.1	96	96.9	96	98.1	98.1	96	98.1
AVG	97.716	97.933	97.95	97.5166	96.18	96.2	96.4	96.2

Good classifier the one which achieve high TPR and high TNR, but due to data imbalance Gmean measure is needed. Since it gives mean value between TPR and TNR. Thus, it gives better indication for both together. Gmean measure provides the balanced performance of learning algorithm between two classes positive and negative. Thus, Gmean is used when both classes positive and negative are considered. Table 8 present Gmean result for training NB and J48 classifiers. It can be seen obviously that J48 has higher Gmean than NB. In addition, PSO is the best selection method used to improve classifiers performance. In general, using automated feature selection methods boost classifiers performance.

Table 8. Gmean of different proxy datasets Training phase

Datasets	NB Training				J48 Training			
	WFS	Best First	IWSS embedded NB	PSO	WFS	Best First	IWSS embedded NB	PSO
BO2	72.5727	71.4485	71.448512	71.4485	71.378	71.448	71.378568	71.797
PA	50.4155	51.8560	48.118187	51.8560	62.170	61.723	47.934330	61.723
SJ	48.7248	49.7493	56.945588	62.9773	83.986	83.986	83.932830	84.039
SV	54.5880	54.0682	53.914005	54.0682	80.154	80.154	58.928770	80.154
UC	69.1871	76.3129	76.312908	76.3129	82.236	82.236	76.666942	82.236
SD	74.7222	87.2513	82.593946	86.6994	89.744	89.744	86.699480	89.798
AVG	62.6754	66.7652	66.193881	68.4401	78.987	78.944	72.593847	79.015

A comparison among computation training time for NB and J48 classifiers without using feature selection methods and with using them are illustrated in table 9. It is obvious that NB classifier is faster than. In the average using IWSS embedded NB feature selection method with both classifier fastest classifier training process than using other feature selection methods. Using automated selecting methods decrease the number of features that used by classifier in addition to exclude irrelevant features. Thus, the classifiers run faster. In average, computation time for NB classifier are 0.216666667, 0.086666667, 0.086666667 and 0.096666667 sec for WFS, using Best First, IWSS embedded NB and PSO respectively. For J48 computation time finished with 4.268333333, 2.201666667, 1.27and 2.545 sec using WFS, using Best First and PSO respectively.

At all, the results for each classifier using different feature selection methods was little bit different because the features chosen by them were almost similar differed in one or two features or sometimes they were same. Moreover, some web proxy files provide better result and good indication than other because they have larger number of instances inside them than other. This obvious in SD proxy which gives in general the best result at all since it is the largest log file used. Imbalanced data was the reason behind large diversity between TPR and TNR for each classifier. Clearly, that both classifier give satisfactory performance and using feature selecting methods boost the performance of NB classifier more than j48. In addition, they minimize the number of feature used and fastest the training classifiers operations.

Table 9. Computation Time (sec) of different proxy datasets Training phase

Datasets	NB Training				J48 Training			
	WFS	Best First	IWSS embedded NB	PSO	WFS	Best First	IWSS embedded NB	PSO
BO2	0.18	0.04	0.04	0.04	0.94	0.89	0.31	0.62
PA	0.06	0.03	0.02	0.03	1.29	0.63	0.12	1.19
SJ	0.09	0.07	0.07	0.07	2.66	1.86	1.06	2.14
SV	0.12	0.11	0.03	0.09	2.44	2.85	0.34	2.77
UC	0.34	0.07	0.09	0.07	2.37	2.31	1.75	3.2
SD	0.51	0.2	0.22	0.28	15.91	4.67	4.04	5.35
AVG	0.21666	0.08666	0.0783333	0.09666	4.2683	2.2016	1.27	2.545

6. Conclusion

The training result of NB and J48 classifiers show that using feature selection methods to choose best subset of features enhance classifier performance in terms of accuracy (CCR), sensitivity (TPR) and specificity (TNR). In addition, PSO result show that it is the best selection method used to enhance performance of NB and J48. Meanwhile, computation time for both classifier is decreased using IWSS embedded NB more than BFS and PSO.

The experimental result shows that using feature selection methods enhance classifiers performance to predict class of object accurately. However, this is because it is reduced the number of feature that the classifiers deal with and remove irrelevant feature. In addition, to reduce computation learning time for the classifiers. Moreover, PSO selection method enhance classifiers performance more than BFS and IWSS embedded NB selection methods.

The experimental results show that J48 classifier with PSO achieve higher accuracy (CCR), TPR and Gmean than NB classifier by 2.99%, 16.866 and 10.5749% respectively. However, in term of computation time NB classifier with IWSS embedded NB reduces computation time of NB classifier training by 0.1383 sec. However, IWSS embedded NB speed up J48 classifier by reducing computation time with 2.998 sec. Moreover, NB classifier is faster than J48 classifier. In the future, more feature selection methods would be used with more numbers of classifiers. and integration those classifiers with web cache replacement algorithms should be done.

7. References

[1] Abdalla A, Sulaiman S and Ali W 2015 Intelligent Web Objects Prediction Approach in Web Proxy Cache Using Supervised Machine Learning and Feature Selection *International Journal of Advances in Soft Computing & Its Applications* **7** 146-64

[2] Olanrewaju R F, Alqudah D, Azman A W and Yaacob M 2016 Intelligent web proxy cache replacement algorithm based on adaptive weight ranking policy via dynamic aging *Indian Journal of Science and Technology* **9**

[3] Muralidhar K and Geethanjali D N 2012 Improving the performance of the browsers using fuzzy logic *International Journal of Engineering Research and Technology* **3**

[4] Ali W, Shamsuddin S M and Ismail A S 2012 Intelligent Naïve Bayes-based approaches for Web proxy caching *Knowledge-Based Systems* **31** 162-75

[5] Kumar P V and Reddy V R 2014 Novel Web Proxy Cache Replacement Algorithms using Machine Learning Techniques for Performance Enhancement *INTERNATIONAL JOURNAL OF ENGINEERING SCIENCES & RESEARCH TECHNOLOGY* **3** 339-46
[6] Ali W and Shamsuddin S M 2015 Intelligent Dynamic Aging Approaches in Web Proxy Cache Replacement *Journal of Intelligent Learning Systems and Applications* **7** 117
[7] Ibrahim H, Yasin W, Hamid N A W A and Udzir N I 2014 Intelligent Cooperative Web Caching Policies for Media Objects based on Decision Tree Supervised Machine Learning Algorithm
[8] Kumar P V and Reddy V R 2014 Web proxy cache replacement policies using decision tree (DT) machine learning technique for enhanced performance of web proxy *International Journal on Recent and Innovation Trends in Computing and Communication* **2** 302-9
[9] Benadit J P, Francis S F and Nadhiya M 2014 Enhancement of Web Proxy Caching Using Random Forest Machine Learning Technique *International Journal of Computer Science Issues (IJCSI)* **11** 83
[10] Aimtongkham P, So-In C and Sanguanpong S 2016 A novel web caching scheme using hybrid least frequently used and support vector machine. In: *Computer Science and Software Engineering (JCSSE), 2016 13th International Joint Conference on*: IEEE) pp 1-6
[11] Chandrashekar G and Sahin F 2014 A survey on feature selection methods *Computers & Electrical Engineering* **40** 16-28
[12] Ratanamahatana C A and Gunopulos D 2002 Scaling up the naive Bayesian classifier: Using decision trees for feature selection
[13] Kohavi R and John G H 1997 Wrappers for feature subset selection *Artificial intelligence* **97** 273-324
[14] Bermejo P, Gámez J A and Puerta J M 2014 Speeding up incremental wrapper feature subset selection with Naive Bayes classifier *Knowledge-Based Systems* **55** 140-7
[15] Lin S-W, Ying K-C, Chen S-C and Lee Z-J 2008 Particle swarm optimization for parameter determination and feature selection of support vector machines *Expert systems with applications* **35** 1817-24
[16] Moraglio A, Di Chio C and Poli R 2007 *Genetic Programming*: Springer) pp 125-36
[17] Ali W, Sulaiman S and Ahmad N 2014 Performance Improvement of Least-Recently-Used Policy in Web Proxy Cache Replacement Using Supervised Machine Learning *International Journal of Advances in Soft Computing & Its Applications* **6**
[18] Irani K B 1993 Multi-interval discretization of continuous-valued attributes for classification learning
[19] Ali W, Shamsuddin S M and Ismail A S 2012 Intelligent Web proxy caching approaches based on machine learning techniques *Decision Support Systems* **53** 565-79
[20] Romano S and ElAarag H 2011 A neural network proxy cache replacement strategy and its implementation in the Squid proxy server *Neural computing and Applications* **20** 59-78
[21] Sarhan A, Elmogy A M and Ali S M 2014 New Web cache replacement approaches based on internal requests factor. In: *9th International Conference on Computer Engineering & Systems (ICCES)*: IEEE) pp 383-9
[22] Sathiyamoorthi V and Bhaskaran V M 2012 Web caching through modified cache replacement algorithm. In: *International Conference on Recent Trends In Information Technology (ICRTIT).* (Chennai, Tamil Nadu: IEEE) pp 483-7
[23] Yasin W, Ibrahim H, Udzir N I and Hamid N A W A 2014 Intelligent Cooperative Least Recently Used Web Caching Policy based on J48 Classifier. In: *Proceedings of the 16th International Conference on Information Integration and Web-based Applications & Services*: ACM) pp 262-9

IOP Conf. Series: Materials Science and Engineering **260** (2017) 012013 doi:10.1088/1757-899X/260/1/012013

UAV State Estimation Modeling Techniques in AHRS

Shikin Razali [1, 2, a)] and Amzari Zhahir [1, b)]

[1]Department of Aerospace Engineering, Engineering Faculty, Universiti Putra Malaysia, 43400 UPM Serdang, Selangor, Malaysia.
[2]School of Electrical System Engineering, Universiti Malaysia Perlis 02600 Arau, Perlis, Malaysia

[a)]Corresponding author: shikinrazali01@gmail.com
[b)]amzari@upm.edu.my

Abstract. Autonomous unmanned aerial vehicle (UAV) system is depending on state estimation feedback to control flight operation. Estimation on the correct state improves navigation accuracy and achieves flight mission safely. One of the sensors configuration used in UAV state is Attitude Heading and Reference System (AHRS) with application of Extended Kalman Filter (EKF) or feedback controller. The results of these two different techniques in estimating UAV states in AHRS configuration are displayed through position and attitude graphs.

1. Introduction

Flight translational and rotational states estimation is one of the important parts in autonomous UAV. Estimation of the states depends on the measurement taken by several sensors on board and techniques in processing the sensor signals. Combining multi state sensors compensates the lacking information in individual sensor and reduces the uncertainty in detecting [1-3]. Sensor fusion algorithm or feedback controller is required to perform the combination of these sensor signals as the sensors measurement contains errors and biases.

AHRS configuration includes gyroscope, accelerometer, magnetometer, GPS and pressure sensors to measure UAV states. In AHRS, estimation on UAV position and velocity relies on GPS and pressure sensors. However in estimating attitude, the bias in gyro is first estimated before it is integrated with accelerometer and magnetometer. EKF which is based on linearization of system dynamic is widely used for sensor fusion process in AHRS whereas feedback controller, a compensator based on proportional integral control is a simpler technique. Both techniques differ in gyro biases estimation which result the UAV attitude roll, pitch and yaw states estimation [4].

The objective of this paper is to analyse the result of UAV states estimation using EKF and feedback controller in AHRS when accelerometer and unavailability of GPS signal rate are increased. The state modelling in this paper is based on fixed wing UAV algorithm generated by [4]. As the difference of these two techniques is expected to have more effect in attitude states rather than position and velocity, therefore the discussion below will be focused more on UAV attitude.

IOP Conf. Series: Materials Science and Engineering **260** (2017) 012013 doi:10.1088/1757-899X/260/1/012013

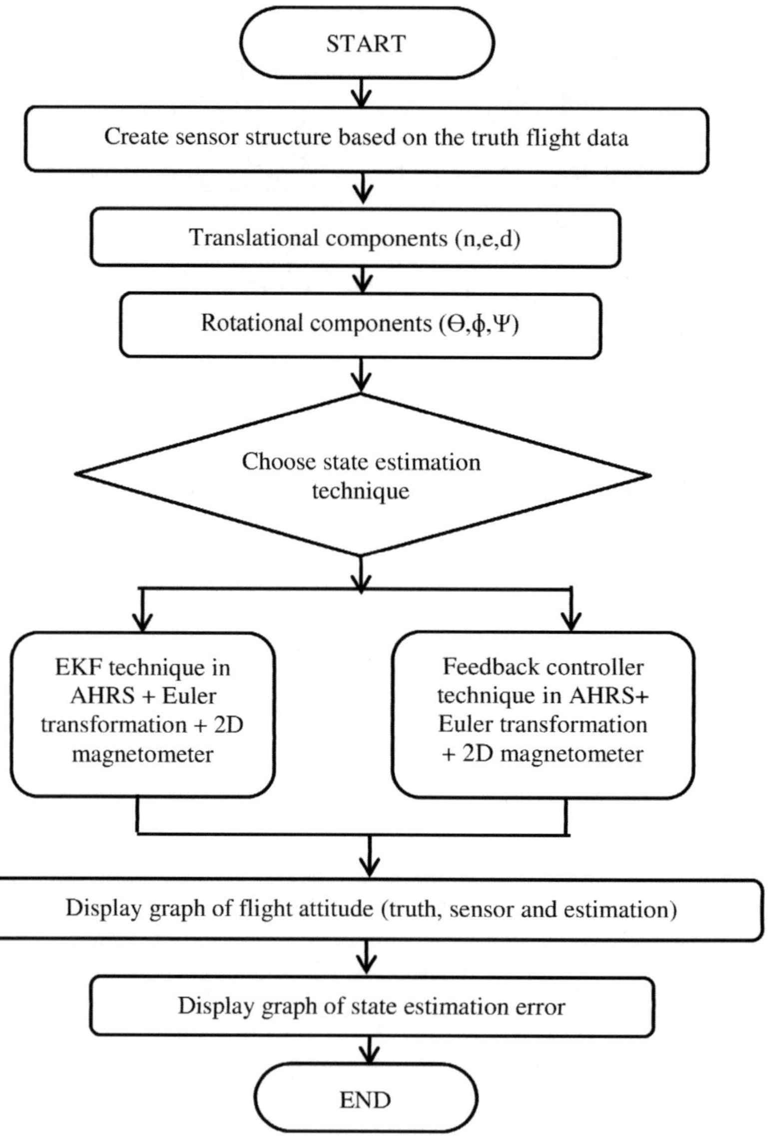

Figure 1: Flow chart of UAV state estimation modeling

2. UAV State Estimation Modeling

Figure 1 shows the process flow of UAV state modelling in MATLAB environment. Sensor structure including errors and biases are generated based on the truth flight data. For this paper, an E route is chosen for the flight map. Table 1 shows the sensors state and related parameters measured in AHRS configuration. Translational states of UAV position and velocity are determined by GPS and pressure sensor. Whereas UAV rotational of yaw (Ψ), pitch (Θ) and roll (ϕ) states are measured by combination of magnetometer, accelerometer and gyroscope. The transformation from translational coordinate of north-east-down to rotational body x-y-z is conducted using Euler angle [5].

IOP Conf. Series: Materials Science and Engineering **260** (2017) 012013 doi:10.1088/1757-899X/260/1/012013

Estimation on UAV states is conducted using EKF or feedback controller. The algorithm development for both techniques in AHRS configuration can be referred in [4]. In EKF technique, observation on gyro biases is tracked over time using the measurement reference from magnetometer, GPS and accelerometer. The de-biased gyro measurements are then integrated to calculate the attitude estimation of pitch, roll and yaw. The measurement and signal processing noises in the model are linearized using Gaussian process. On the other hand, the proportional integral feedback controller technique estimates gyro biases using body frame attitude error vector. Error vector is the expression of angular error and rotation axis between the measurement reference of *North-East-Down* coordinate frame and AHRS-estimated *North-East-Down* coordinate. The north coordinate reference is obtained from magnetometer while down coordinate is derived from accelerometer measurements.

Table 1: Sensor and parameters

State component	Sensor	Paramater	
Translational	GPS	Position	P_n, North position P_e, East position, h_{MSL} height above mean-sea-level
		Velocity	V_n, North velocity V_e East velocity V_d, Downward inertial velocity
Rotational (bank, elevation and heading)	Accelerometer	Specific force vector in body coordinate (f_x, f_y, f_z)	
	Gyroscope	Angular rate in body coordinate (ω_x, ω_y ,ω_z)	
	Magnetometer	Magnetic north in body coordinate (M_x, M_y, M_z)	

3. Result and Discussion

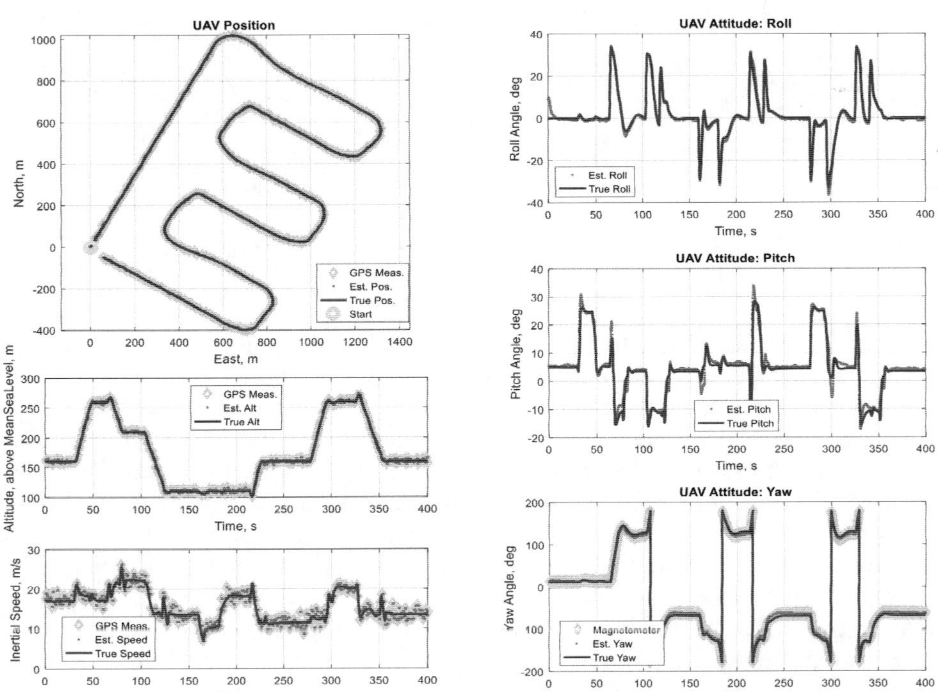

Figure 2: UAV states using EKF in AHRS

IOP Conf. Series: Materials Science and Engineering **260** (2017) 012013 doi:10.1088/1757-899X/260/1/012013

There are 6 graphs in all figures which are parted into left and right column. The left side represents the UAV position, altitude and inertial speed whereas the right column represents the roll, pitch and yaw graphs. Each graph compares truth data, sensor measurement and state estimation. As the GPS signal has minimum errors, both techniques have no problem in estimating UAV position and velocity close to the truth states as in Figure 2 and 3. As well as the result for attitude estimation of roll and yaw. In pitch estimation, however several points are slightly overshot of +/-5^0 from the truth but overall measurements is still accurate. The correct angle estimations of roll, pitch and yaw show that the gyro biases estimation is captured well throughout the time for both techniques.

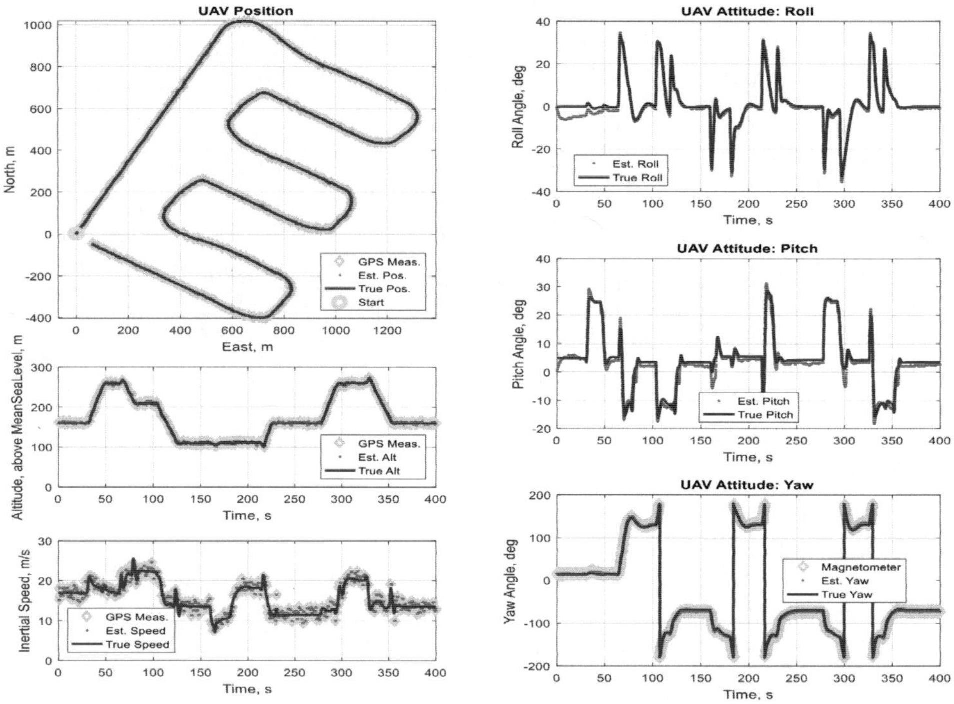

Figure 3: UAV states using feedback controller in AHRS

Figure 4 and 5 show the UAV state estimation by EKF and feedback controller technique when accelerometer and magnetometer noise and GPS dropout signal rate are increased. Unavailability of GPS signal dropout is shown in the graph of UAV position by the empty space between the green dots. The position and velocity graphs for both techniques are quite similar with several estimated points are projected a bit far from the truth states. This is expected as the availability of GPS signal rate is decreased. In Figure 4 of EKF technique, while the estimation of pitch and roll angles are still accurate, the result of attitude estimation of feedback controller in Figure 5 is however different. The roll angle waveform is deteriorated -10^0 and many points in pitch graph are estimated far approximately ±7^0 from the truth states. The deterioration of pitch and roll graphs in feedback controller technique is due to the inaccuracy of attitude estimation as the accelerometer and GPS noises are increased. The existing noises in measurement references caused the inaccuracy of error vector calculation. Thus, the gyro biases are not estimated correctly and further results to deteriorate pitch and roll estimation. The yaw estimation however is not affected as the measurement can be taken directly from the magnetometer reference.

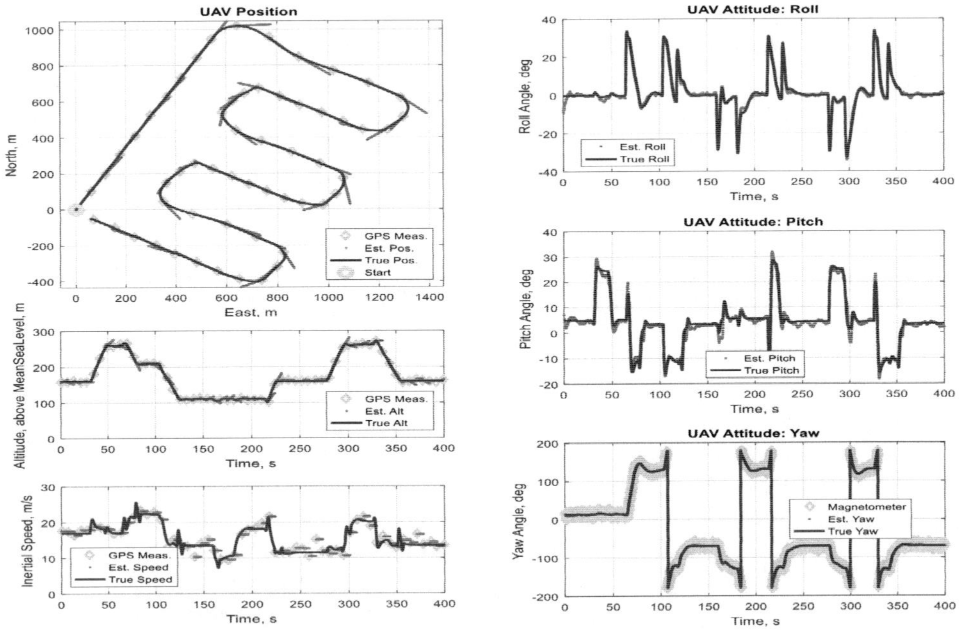

Figure 4: UAV states using EKF when errors are increased

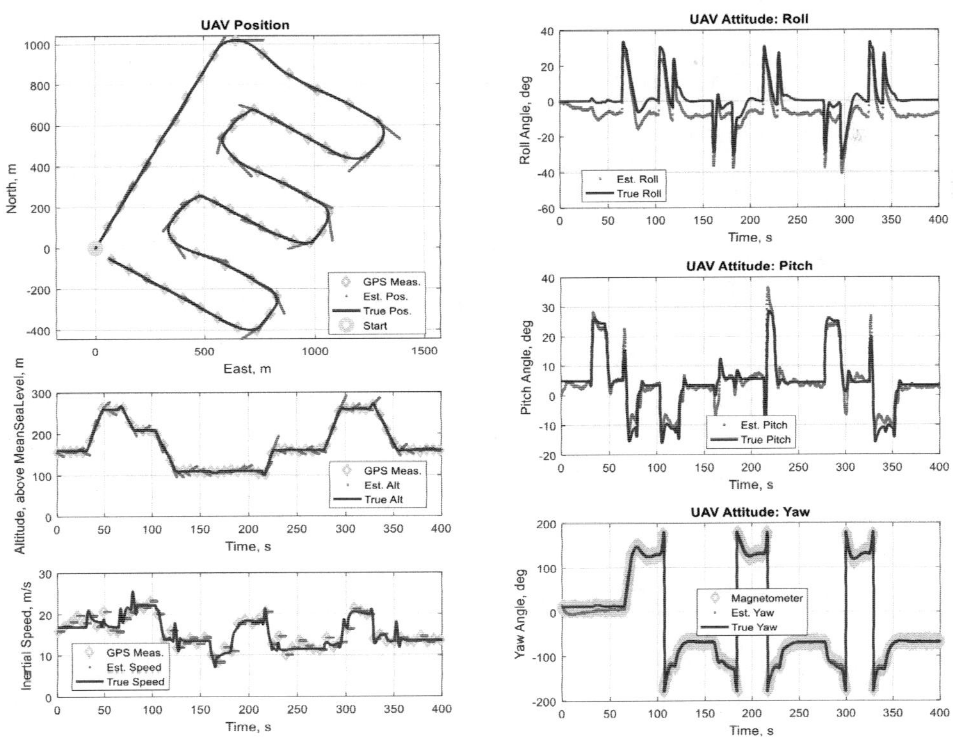

Figure 5: UAV states using feedback controller when errors are increased

4. Conclusion

Based on this modeling, EKF and feedback controller techniques had performed UAV attitude estimation accurately in AHRS configuration. When noises in accelerometer and GPS were increased, attitude estimation in both techniques were affected. However, the effect was more observant in feedback controller rather than EKF with -10^0 and $\pm 7^0$ deterioration of roll and pitch angle from the truth states. In conclusion, although feedback controller is simpler to model in estimating UAV attitudes but it is more sensitive to existing noises in accelerometer and GPS.

Acknowledgement

We would like to thank Universiti Putra Malaysia for supporting this research manuscript under Geran Universiti Putra Malaysia Tahun 2016 with Project Code. GP-IPS/2016/9498300.

References

[1] Rigatos, G. G. (2012). Nonlinear Kalman Filters and Particle Filters for integrated navigation of unmanned aerial vehicles. Robotics and Autonomous Systems, 60(7), 978–995.

[2] Saha, R. K., & Chang, K. C. (1998). An efficient algorithm for multisensor track fusion. IEEE Transactions on Aerospace and Electronic Systems, 34(1), 200–210.

[3] Angelino, C. (2012). UAV position and attitude estimation using IMU, GNSS and camera. 15th International Conference on Information Fusion (FUSION). 735–742.

[4] Barton, J. D. (2012). Fundamentals of Small Unmanned Aircraft Flight, JHU/APL Technical Digest Article 31(2), 132–149.

IOP Conf. Series: Materials Science and Engineering **260** (2017) 012014 doi:10.1088/1757-899X/260/1/012014

Modelling and Simulation Analysis of Rolling Motion of Spherical Robot

N N Kamis[1], A H Embong[2], and S Ahmad[3]

Department of Mechatronics Engineering, Kulliyyah of Engineering, International Islamic University Malaysia, 53100 Gombak, Selangor

[1]nafisahkamis@gmail.com, [2]ehalim@iium.edu.my, [3]salmiah@iium.edu.my

Abstract. This paper presents the findings of modelling, control and analysis of the spherical rolling robot based on pendulum driven within the simulation environment. The spherical robot is modelled using Lagrange function based on the equation of rolling motion. PD-type Fuzzy logic controller (FLC) was designed to control the position of the spherical robot where 25 rules were constructed to control the rolling motion of spherical robot. It was then integrated with the model developed in Simulink-Matlab environment. The output scaling factor (output gain) of the FLC was heuristically tuned to improve the system performance. The simulation results show that the FLC managed to eliminate the overshoot response and demonstrated better performance with 29.67% increasing in settling time to reach 0.01% of steady state error.

1. Introduction

Spherical robot is one type of mobile robot that shows significant advantages over others due to its geometrical shape. It is spherical in shape with driving mechanism located inside the spherical robot. The robot is naturally roll to locate itself and capable to recover from collisions with obstacle. So, the robot can be thrown and dropped without risk to be overturned like other types of mobile robot[1]. This allowed the spherical robot to manoeuvre comfortably in any environments to perform any desired tasks.

This unique features in mobility and safety enable it to be used in many applications such as education, entertainment, security, navigation, monitoring and exploration [2]–[7]. This increases the interest of researchers to explore and develop this area. The number of different mechanical structures has been developing in recent years. Several basic principles have been used to generate the motion of spherical robot. The early version of spherical robot actuated using wheel inside the sphere. The wheel is driven by a motor and can be turned to cause change in direction [8]. Almost similar concept is also been used in [9] where the small remote car is used to provide force to propel the robot. These kinds of mechanisms are easier to be modelled and controlled as long as no accurate tracking control is required. A simple equation of motion has been proposed using linear dynamic model of the system to model the system. Using simple model make the control design become easier but the simple model is unable to capture the nonlinearity and complexity of the system. Therefore, it makes the methods to be unsuitable to be used to model a spherical robot.

Spherical robot also can be actuated by changing the center of the mass of the system. The concept of imbalance mass is used by moving the mechanism like pendulum[10]–[17]. This moving mass will cause unbalance to the system and make the spherical robot to roll. The dynamic model of the pendulum and kinematic model of rolling disk is used to model the system. The complex modelling process can also be simplified with decoupled approach when dealing with more than one degree of freedom [13].

Besides, rolling motion of spherical robot can also be achieved using three independent inertia disc [18]. For this type of design, the system is model based on its geometrical framework and adopting attitude dynamic governed by Poincare equation of motion. Based on this model, a nonlinear state feedback control law is used to achieve asymptotic stabilization of position and reduced attitude. This nonlinear feedback controller id defined from linear state feedback control law that has been transformed using body coordinate velocities.

The kinematic and dynamic model of the spherical robot can also be model using Euler angle and Lagrange equation. Most of the modelling methods discussed before linearized the model to make the controller design easier. LQR with full state feedback controller [15], simple feedback controller [19], gaits control based on motion equation [17] is designed to tackle the stability or tracking the trajectory of spherical robot. But, the linearization cause the elimination of nonlinear part of the system and may cause some behaviour of the system that are related to the unmodelled dynamics, parameter variations, disturbance and uncertainty being neglected. This will affect the design controller.

To overcome this problem, the intelligent controller is introduced and seem to be effective in other mobile robot [20], [21]. In [22], adaptive neuron fuzzy controller where PID controller was combined with fuzzy controller and sliding mode control theory based learning algorithm is introduced to improve the robustness of the system and ensure faster convergence than traditional learning technique.

In this paper, a fuzzy logic controller is designed with 25 fuzzy rules to control the position of rolling spherical robot to follow the reference point. Fuzzy logic controller is design based on the human's heuristic knowledge on the system. This control method is significant in reducing error result from the complex mathematical modelling as the control is expressed in a natural language. With fuzzy logic controller system manage to improve the transient response of the system. The paper is organized as mathematical model based on pendulum driven in next section, and design of fuzzy logic controller in section 3. The simulation result and discussion will be presented in section 4 and finally, the conclusion in section 5.

2. Model of spherical robot

2.1 Kinematic modelling

Kinematic model of the system is derived based on equation of motion of the rolling wheel or disk. In the model ling process, the spherical robot is assumed to roll without slipping, no friction force acting between the spherical robot and horizontal plane. Besides that, the geometric center of the shell and mass of the robot are assumed to be eccentric. This assumptions help to simplify the modelling process. Similar modelling based on pendulum driven had been discussed briefly in [13], [22].

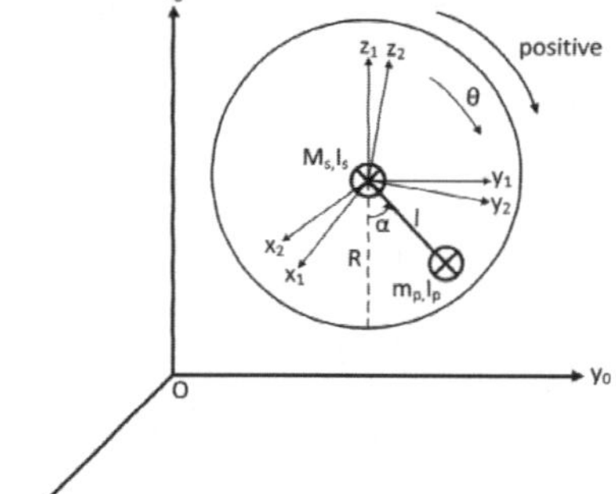

Figure 1. Modelling of the Spherical robot with pendulum driven.[22]

Figure 1 represent the motion of spherical robot in 2D-plane with θ and α as angle of rotation of spherical robot and pendulum around the x-axis respectively. X_O, Y_O, Z_O is the reference frame fixed to the motion ground while X_1, Y_1, Z_1 and X_2, Y_2, Z_2 are the moving frames attached to the center of the spherical robot with radius of R. The kinematic system of the model can be represent with it angular velocity, ω_s and linear velocity, v_s of the spherical robot, the angular velocity, ω_p, and linear velocity, v_p of the mass of the pendulum. Kinematics equations can be writen in vector form as below:

$$\omega_S = -\dot{\theta} i \tag{1}$$

$$V_S = R\dot{\theta} j \tag{2}$$

$$\omega_p = (\dot{\alpha} - \dot{\theta}) i \tag{3}$$

$$V_p = (-R\dot{\theta} + (\dot{\alpha} - \dot{\theta}) l \cos(\alpha - \theta)) j + ((\dot{\alpha} - \dot{\theta}) l \sin(\alpha - \theta)) k \tag{4}$$

2.2 Dynamic model

The dynamic model of the rolling spherical robot can be derived using Lagrangian function of L. L is the subtraction of kinetic energy of linear and rotation system with potential energy of the system. Equation of L is described as below:

$$\begin{aligned} L &= K - P \\ &= \frac{1}{2} M_s \| V_s \|^2 + \frac{1}{2} I_S \| \omega_s \|^2 + \frac{1}{2} m_p \| V_p \|^2 + \frac{1}{2} I_p \| \omega_p \|^2 - m_p g r_p = z \\ &= \frac{1}{2} M_s (R\dot{\theta})^2 + \frac{1}{2} I_s (-\dot{\theta})^2 + \frac{1}{2} I_p (\dot{\alpha} - \dot{\theta})^2 + \frac{1}{2} m_p ((-R\dot{\theta} + (\dot{\alpha} - \dot{\theta}) l \cos(\alpha - \theta))^2 \\ &+ ((\dot{\alpha} - \dot{\theta}) l \sin(\alpha - \theta))^2) - m_p g l \cos(\alpha - \theta) \end{aligned} \tag{5}$$

M_s and m_p representing the masses of the sphere and pendulum. While, mass of moment of inertia, I_s and I_p of sphere and pendulum is derived as equations 6 and 7.

$$I_s = \frac{2}{3} M_s R^2 \tag{6}$$

$$I_p = \frac{1}{12} m_p l^2 + m_p \left(\frac{l}{2}\right)^2 \tag{7}$$

The system also is assumed to have viscous friction acting between the sphere and the surface where the equation can be written in an energy dissipation function of velocities of the system, $\dot{q}_i$ and the damping constant, ξ.

$$S = \frac{1}{2} \xi \dot{q}_i^2 = \frac{1}{2} \xi \left(\dot{\theta}^2 + \dot{\alpha}^2\right) \tag{8}$$

Then, the Euler Lagrange system is written as follow:

$$\frac{d}{dt} \left(\frac{\partial L}{\partial \dot{q}_i}\right) - \frac{\partial L}{\partial q_i} + \frac{\partial S}{\partial \dot{q}_i} = Q_i \tag{9}$$

Where q_1 and q_2 are θ and α respectively are the generalized coordinates. $Q_1 = Q_2 = \tau$ represent the input torque to rotate the pendulum that is rotated through an input torque and a reaction torque about the shaft occurs in the opposite direction. Then, the final rolling motion of spherical robot is written as below:

$$M_{11} \ddot{\theta} + M_{12} \ddot{\alpha} + C_{11} + G_{11} = \tau \tag{10}$$

$$M_{22} \ddot{\alpha} + M_{21} \ddot{\theta} + C_{21} + G_{21} = \tau \tag{11}$$

Where,

$$M_{11} = M_S R^2 + m_p R^2 + m_p l^2 + I_s + I_p + 2 m_p R l \cos(\alpha - \theta)$$

$$M_{12} = M_{21} = -m_p l^2 - I_p - m_p R l \cos(\alpha - \theta)$$

IOP Conf. Series: Materials Science and Engineering **260** (2017) 012014 doi:10.1088/1757-899X/260/1/012014

$$M_{22}= m_p l^2+I_p$$
$$C_{11}=m_p Rl\sin(\alpha-\theta)\left(\dot{\alpha}-\dot{\theta}\right)^2+\xi\dot{\theta}$$
$$C_{21}=\xi\dot{\alpha}$$
$$G_{11}= -G_{21}=-m_p gl\sin(\alpha-\theta)$$

The equations 10 and 11 are used to model the simulink block diagram of spherical robot for rolling motion as in figure 3. The pendulum is actuated using dc motor. The simple model of DC motor is modelled as in equations 12 and 13 where $\theta_m, k_e, k_t, b, R_{m,} V_{m,}, L$ and J are the rotation of motor , moment of inertia of the motor, electromotive force constant, motor torque constant, motor's viscous friction constant, electrical resistance, voltage armature,electrical inductance and moment of inertia of the motor respsctively. The system then was represented in Simulink block diagram for simulation.

$$\ddot{\theta}m = \frac{1}{J}(k_t - b\dot{\theta}m) \quad (12)$$

$$\frac{di}{dt} = \frac{1}{L}(-R_m i + V_m - k_e\dot{\theta}_m) \quad (13)$$

3. Fuzzy logic control (FLC)

Fuzzy logic control is a controller that used the human language and knowledge as an element to control the system. This controller do not used an exact value to determine the input to the plant. Fuzzy logic controller consists of 4 main components; fuzzification, rule based, inference mechanism and defuzzification of the output FLC.

The first step in designing the FLC is to select the input-output parameters of the fuzzy controller. In order to control the position of the system PD type of fuzzy logic is designed. This type of FLC requires error (*e*) and rate of change of error ($\dot{e}$) as the input to the FLC, while producing the voltage of the DC motor as the output to the actuator. The voltage of DC motor will vary the angle of rotation of the pendulum that causes the rotation of the spherical robot.

The inputs and output of the FLC are represented as membership functions as Big Negative (BN), Small Negative (SN), Zero (Z), Small Positive (SP), and Big Positive (BP) using triangular membership function with range [-1 1]. The membership function of inputs and output of the FLC are plotted in figures 2,3 and 4.

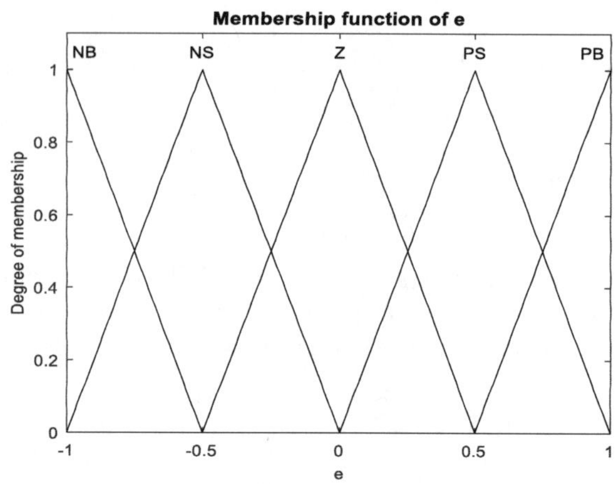

Figure 2. Membership function of error, *e*.

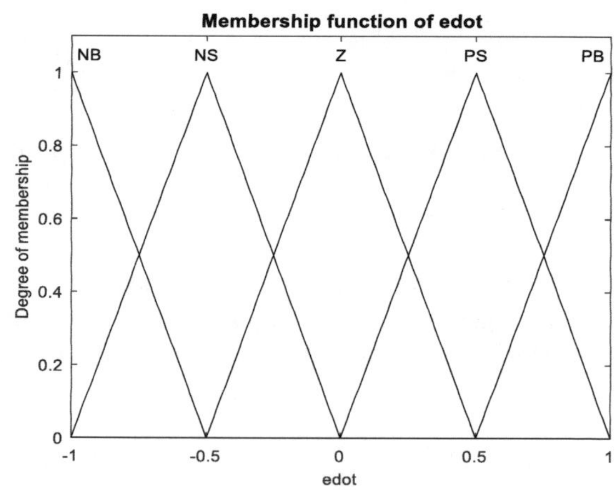

Figure 3. Membership function of rate change of error, $\dot{e}$

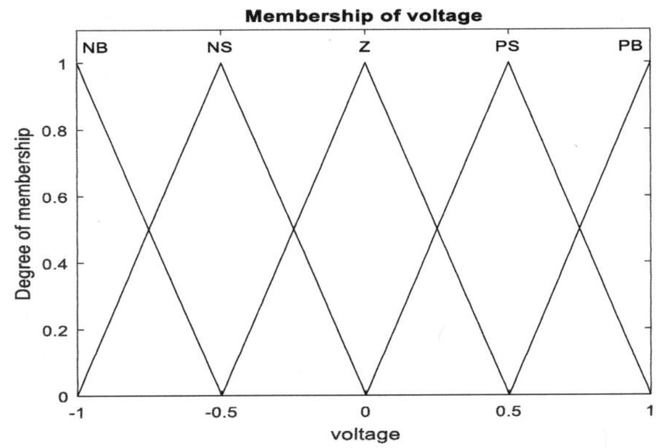

Figure 4. Input membership function of input e, $\dot{e}$, and output voltage.

The 25 rules of membership function are constructing based on the experiences and the understandings of the system. Here, the understanding is gained through its mathematical model that describes its behaviour and experiences of the person. table 1 show the fuzzy rules used in this controller.

Table 1. Rule base of PD type Fuzzy logic controller for position control

Voltage		Change of error($\dot{e}$)				
		BN	SN	Z	SP	BP
Error (e)	BN	BN	BN	SN	SN	Z
	SN	SN	SN	SN	Z	SP
	Z	SN	SN	Z	SP	SP
	SP	SN	Z	SP	SP	BP
	BP	Z	SP	SP	SP	BP

The general form of linguistic rule listed as below:

"**If** premise, **Then** consequent."

As this fuzzy controller takes two inputs; error and rate of change of error, the two inputs are combined using AND. The rule can be understood as "**If** error is big negative (BN) **and** change of error is big negative (BN), **Then**, the voltage supply to DC motor is also big negative (BN)". This rule quantifies the situation where the simulated position of spherical robot is bigger than the desired output and it is moving fast forward (fast changing in error). So, it means the robot move further away from the desired position. Then, the negative voltage is supplied to the DC motor so that it moves on the opposite direction to the desired position.

4. Simulation result and discussion

For the simulation, parameters of the spherical robot have been assumed as I_s = 0.6m, I_p = 0.2m, M_s= 1m, m_p= 0.06kg, R=0.15m, l=0.145m and ξ=1. While the parameters of dc motor are assumed as $k_e = 1, k_t = 1, b = 1, R_m =$ 1ohm, $L =$ 1H and J=1kg.m^2.

The model was first simulated without controller to evaluate the output response of the spherical robot to the input signal. The result to this step input test is shown in figure 5. The transient response of the spherical robot without the controller shows that 6.6% overshoot, rise time, t_s at 4.54s, peak time, t_p at 5.51s and settling time, t_s at 11.29s.

IOP Conf. Series: Materials Science and Engineering **260** (2017) 012014 doi:10.1088/1757-899X/260/1/012014

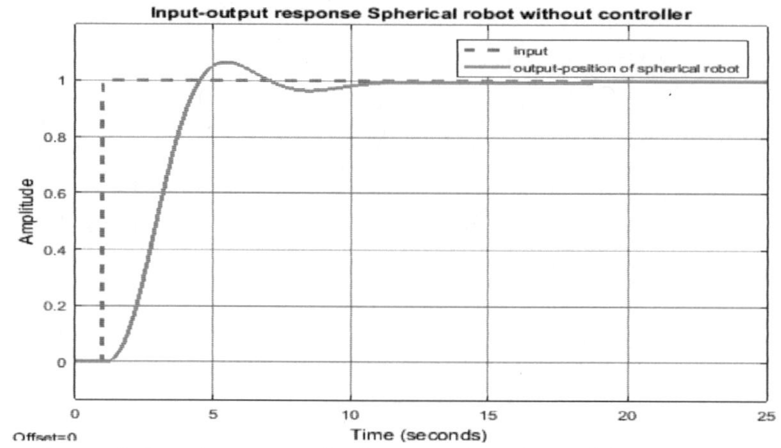

Figure 5. Step input test of spherical robot plant without controller.

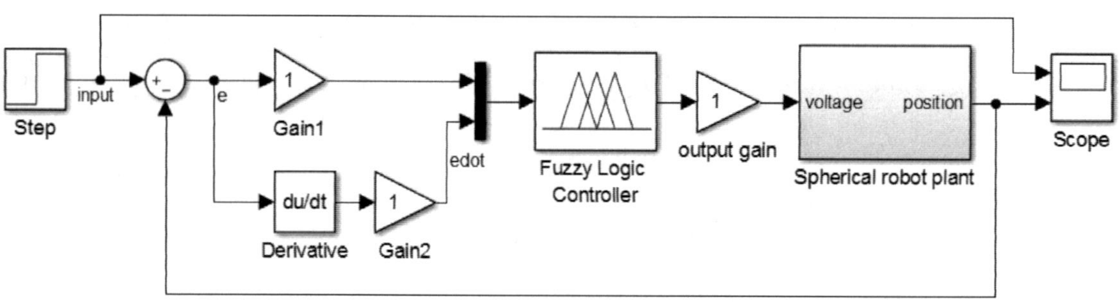

Figure 6. Simulink diagram of fuzzy logic controller of spherical robot.

Once the PD-type fuzzy logic controller is implemented in the simulation as in figure 6, the result shows an improvement in the transient response where the controller manages to eliminate the overshoot problem. To improve the output response, it is necessary to tune the output gain of the FLC, *g*. When the value of output gain FLC, *g* is increased gradually to 6, the settling time is reduced from 11.29s to 7.94s as shown in figure 7. Yet, increasing the value of *g* managed to accelerate the output response but the distortion build in the process can cause the problem to the system. So, the best output gain used to get faster response and least distortion before reach the reference point is 4.

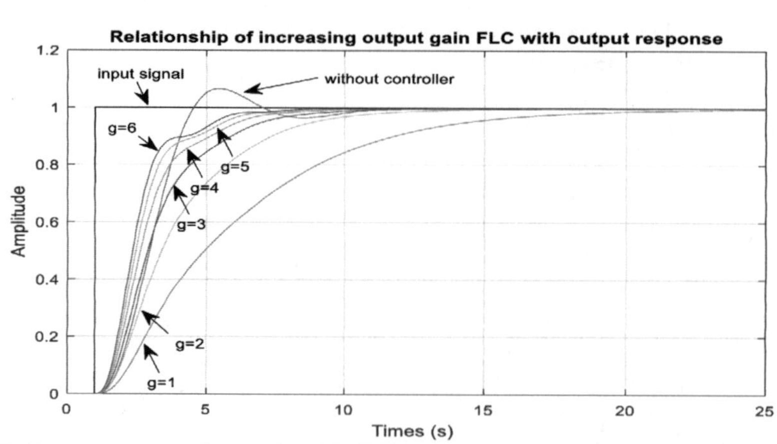

Figure 7. Output response's graph with different values of FLC's output gain, *g*.

Besides, several fuzzy rules are also being designed, tuned and simulated but, the result shows that changing the several rules like BN to SN or vice versa does not give significant impact to the performance of the spherical robot. While, amplifying the input gain of e and $\dot{e}$ also does not improved the performance of the rolling motion.

5. Conclusion

In this study, modelling and Fuzzy Logic controller has been designed and investigated. The controller was focused to control the response of the system to follow the reference input. The modelling of the spherical rolling robot was based on single DOF pendulum driven. The response of the system without controller was compared with the PD-type controller to analyse the effect of fuzzy logic controller. The effect of increasing output gain of controller was analysed. The response manages to reach the reference point faster as the gain increasing, but, the system will suffer to the oscillation before reaching the reference point if the output gain is too high.

References

[1] V. Crossley, "A literature review on the design of spherical rolling robots," *Pittsburgh, PA*, pp. 1–6, 2006.

[2] F. Michaud *et al.*, "Autonomous Sperical Mobile Robot for Child Development Studies," *IEEE Trans. Syst. Man Cybern.*, vol. 35(4), no. 4, pp. 1–10, 2005.

[3] F. Michaud and S. Caron, "Roball-an autonomous toy-rolling robot," *Proc. Work. Interact. Robot. Entertain.*, 2000.

[4] M. Seeman, M. Broxvall, A. Saffiotti, and P. Wide, "An autonomous spherical robot for security tasks," *Proc. 2006 IEEE Int. Conf. Comput. Intell. Homel. Secur. Pers. Safety, CIHSPS 2006*, vol. 2006, no. October, pp. 51–55, 2006.

[5] A. I. C. and P. R. Maxime Meilland, "A spherical Robot Centered Representation for urban navigation," *Intell. Robot. Syst. (IROS), 2010 IEEE/RSJ Int. Conf.*, vol. 1, pp. 5196–5201, 2010.

[6] J. D. Hernández, J. Barrientos, J. del Cerro, A. Barrientos, and D. Sanz, "Moisture measurement in crops using spherical robots," *Ind. Robot An Int. J.*, vol. 40, no. 1, pp. 59–66, Jan. 2013.

[7] B. Li, Q. Deng, and Z. Liu, "A spherical hopping robot for exploration in complex environments," *2009 IEEE Int. Conf. Robot. Biomimetics, ROBIO 2009*, pp. 402–407, 2009.

[8] A. Halme, T. Schonberg, and Yan Wang, "Motion control of a spherical mobile robot," in *Proceedings of 4th IEEE International Workshop on Advanced Motion Control - AMC '96 - MIE*, 1996, vol. 1, pp. 259–264.

[9] J. Alves and J. Dias, "Design and control of a spherical mobile robot," *Proc. I MECH E Part I J. Syst. Control Eng.*, vol. 217, no. 6, pp. 457–467, 2003.

[10] Bo Zhao *et al.*, "Dynamics and motion control of a two pendulums driven spherical robot," *IEEE/RSJ 2010 Int. Conf. Intell. Robot. Syst. IROS 2010 - Conf. Proc.*, pp. 147–153, Oct. 2010.

[11] L. Daliang and S. Hanxu, "Nonlinear sliding-mode control for motion of a spherical robot," *Control Conf. (CCC), 2010 29th Chinese*, pp. 3244–3249, 2010.

[12] M. Zheng, Q. Zhan, J. Liu, and Y. Cai, "Control of a spherical robot: Path following based on nonholonomic kinematics and dynamics," *Chinese J. Aeronaut.*, vol. 24, no. 3, pp. 337–345, 2011.

[13] E. Kayacan, Z. Y. Bayraktaroglu, and W. Saeys, "Modeling and control of a spherical rolling robot: a decoupled dynamics approach," *Robotica*, vol. 30, no. 4, pp. 1–10, 2011.

[14] Y. Cai, Q. Zhan, and X. Xi, "Path tracking control of a spherical mobile robot," *Mech. Mach. Theory*, vol. 51, pp. 58–73, 2012.

[15] F. K. Zadeh, P. Moallem, S. Asiri, and M. M. Zadeh, "LQR motion control and analysis of a

prototype spherical robot," *2014 2nd RSI/ISM Int. Conf. Robot. Mechatronics, ICRoM 2014*, pp. 890–895, 2014.

[16] T. Yu, H. Sun, Q. Jia, Y. Zhang, and W. Zhao, "Stabilization and Control of a Spherical Robot on an Inclined Plane," *Res. J. Appl. Sci. Eng. Technol.*, vol. 5, no. 6, pp. 2289–2296, 2013.

[17] T. B. . Ivanova and E. N. Pivovarova, "Dynamics and Control of a Spherical Robot with an Axisymmetric Pendulum Actuator," pp. 1–14, 2015.

[18] V. Muralidharan and A. D. Mahindrakar, "Geometric controllability and stabilization of spherical robot dynamics," *IEEE Trans. Automat. Contr.*, vol. 60, no. 10, pp. 2762–2767, 2015.

[19] T. Urakubo, M. Monno, S. Maekawa, and H. Tamaki, "Dynamic Modeling and Controller Design for a Spherical Rolling Robot Equipped With a Gyro," *IEEE Trans. Control Syst. Technol.*, vol. 24, no. 5, pp. 1669–1679, 2015.

[20] T. T. Mac, C. Copot, R. D. E. Keyser, T. D. Tran, and T. Vu, "MIMO Fuzzy Control for Autonomous Mobile Robot," pp. 277–282.

[21] H. Omrane, M. S. Masmoudi, and M. Masmoudi, "Fuzzy Logic Based Control for Autonomous Mobile," *Comput. Intell. Neurosci.*, vol. 2016, pp. 1–10, 2016.

[22] E. Kayacan, E. Kayacan, H. Ramon, and W. Saeys, "Adaptive neuro-fuzzy control of a spherical rolling robot using sliding-mode-control-theory-based online learning algorithm," *IEEE Trans. Cybern.*, vol. 43, no. 1, pp. 170–179, Feb. 2013.

DEVELOPMENT OF FUZZY LOGIC CONTROLLER FOR QUANSER BENCH-TOP HELICOPTER

M. H. JAFRI[1], H. MANSOR[2] & T. S. GUNAWAN[3]

Dept of Electrical and Computer Engineering, Kulliyyah of Engineering, International Islamic University Malaysia (IIUM), Jalan Gombak, 53100 Kuala Lumpur, Malaysia.

[1]hazwanjafri@gmail.com, [2]hasmahm@iium.edu.my, [3]tsgunawan@iium.edu.my

ABSTRACT: Bench–top helicopter is a laboratory scale helicopter that usually used as a testing bench of the real helicopter behavior. This helicopter is a 3 Degree of Freedom (DOF) helicopter which works by three different axes wshich are elevation, pitch and travel. Thus, fuzzy logic controller has been proposed to be implemented into Quanser bench-top helicopter because of its ability to work with non-linear system. The objective for this project is to design and apply fuzzy logic controller for Quanser bench-top helicopter. Other than that, fuzzy logic controller performance system has been simulated to analyze and verify its behavior over existing PID controller by using Matlab & Simulink software. In this research, fuzzy logic controller has been designed to control the elevation angle. After simulation has been performed, it can be seen that simulation result shows that fuzzy logic elevation control is working for 4°, 5° and 6°. These three angles produce zero steady state error and has a fast response. Other than that, performance comparisons have been performed between fuzzy logic controller and PID controller. Fuzzy logic elevation control has a better performance compared to PID controller where lower percentage overshoot and faster settling time have been achieved in 4°, 5° and 6° step response test. Both controller are have zero steady state error but fuzzy logic controller is managed to produce a better performance in term of settling time and percentage overshoot which make the proposed controller is reliable compared to the existing PID controller.

KEY WORDS: *fuzzy logic; controller; bench-top helicopter; Quanser*

1. INTRODUCTION

Fuzzy logic theory is a mathematical reasoning that compute "degrees of truth" that works between value 0 and 1 [1]. Fuzzy logic controller usually determined by some set of rules which later these set of rules will be implemented in the system. Since the numerical variables have been converted to the linguistic variables, mathematical modeling is not required in this fuzzy logic. Fuzzy logic consists of three parts which are fuzzification, interference engine and defuzzification. Fuzzy logic consists of input and output which are represented in form of membership function. This membership function can be represented in various shapes such as triangular, bell, trapezoidal and other shape. The membership function affect the output result of the system. To design fuzzy logic controller, there are several design stages need to be performed such as input and output membership function and fuzzy rules.

Fuzzy logic controller has been used widely in control system. Some of the researchers have introduced the fuzzy logic controller because of its ability in state of stability and better responses compared to other controller [2,3,4]. Other than that, fuzzy logic also has been introduced to be implemented with other controllers such as PID in controlling the aerial vehicles to produce better response of the system [5,6]. Fuzzy logic controller fuzzy logic controller will act as a controller to overcome weaknesses that has been faced by other controllers because of its ability to work with non-linearity system such as a helicopter. Since the designed controller need to be worked with the bench-top helicopter which is a system with high non-linearity and complexity, fuzzy logic has been proposed to overcome weakness by other controller. Therefore, fuzzy logic will improve the performance of the system to produce a better result.

In this research, fuzzy logic has been proposed as a controller to control Quanser bench-top helicopter. There are several researches that have proposed different type of controllers for Quanser bench-top helicopter. One of the techniques that have been developed is quantitative feedback theory which has been proposed by researcher to achieve a desired robust design over a specified region of

IOP Conf. Series: Materials Science and Engineering **260** (2017) 012015 doi:10.1088/1757-899X/260/1/012015

plant uncertainty [7]. Fuzzy logic is a type of controllers that cover a wider range of operating conditions. Other than that, fuzzy logic controller is suitable to deal with nonlinearities and uncertainties. Thus, in this research, fuzzy logic will be introduced as a type of alternative controller to control the bench-top helicopter and its behavior will be identified throughout the research. Other than that, robust adaptive LQR controller also has been developed for Quanser bench-top helicopter [8]. LQR controller is proposed by researcher because it simply creates a stable system without explicitly optimizing anything and it is also straightforward to use for multivariable systems where the design procedure is essentially the same as for single-input-single-output systems.

2. METHODOLOGY

2.1 Bench-Top Helicopter

Quanser bench-top helicopter is a laboratory helicopter model to represent the actual model of helicopter [9,10]. Bench-top helicopter works by moving at 3 Degrees of Freedom (3DOF) which are elevation axis, pitch axis and travel axis. Figure 1 shows the 3-DOF bench-top helicopter while Figure 2 shows the free body diagram for this 3-DOF helicopter system.

Figure 1 : 3-DOF Bench-Top Helicopter [11]

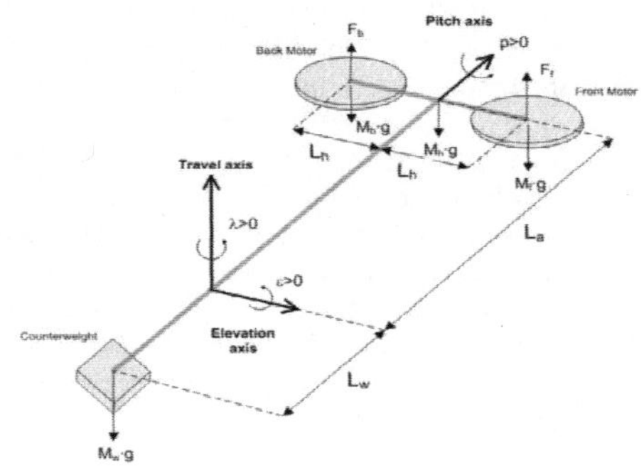

Figure 2: Free body diagram of 3-DOF Helicopter system [11]

Parameters of 3-DOF helicopter are described as follows:

M_h	Mass of the helicopter
M_w	Mass of counter weight
M_f	Mass of front propeller assembly
M_b	Mass of back propeller assembly
L_a	Distance between travel axis to helicopter body
L_h	Distance between pitch axis to helicopter body
L_w	Distance between travel axis to the counter weight
g	Gravitational constant
F_b	Back force
F_f	Front force
p	Pitch

Two electrical DC motor of bench-top helicopter are attached to the body and making two propellers turn. Aerodynamic has caused the total force *F* to make the total system turn around an angle measured by encoders. The arm has counterweight that helps propellers lift the body weight. With 3-DOF, elevation angle, pitch angle and travel angle can be measured by an absolute encoder and controlled by any kind of controllers such as PD controller [11]. This research will focus on developing a fuzzy logic controller to control elevation angle at certain degree.

This research will identify the response of the Quanser bench-top helicopter by using Matlab & Simulink software. Therefore, in this research, simulation results of bench-top helicopter that has been implemented with the fuzzy logic controller will be presented and discussed.

2.2 Fuzzy Logic Controller Design

The early step in designing the fuzzy control system is to set the fuzzy rules. In this part, set of fuzzy rules consists of two inputs which are *Error* and *Change in Error* while the output is *response* of the system. In order to design the fuzzy logic controller, the subset of input and output will be determined. In this case, each input will consist subset {NEG, Z, POS} while subset of output contain {VS, S, M, H, VH}. Based on the subset of each inputs and output that have been determined, the fuzzy rules will be set.

Table 1 shows the representation of fuzzy rules set. Based on Table 1, there are nine fuzzy rules that have been set for the bench-top helicopter system. Later, this rule will be used in designing the fuzzy logic controller.

Table 1: Fuzzy Rules of Bench-Top Helicopter Response

	ERROR		
DEL ERROR	**NEG**	**Z**	**POS**
NEG	*VS*	*S*	*M*
Z	*S*	*M*	*H*
POS	*M*	*H*	*VH*

List of abbreviation:
NEG - Negative
Z - Zero
POS - Positive
VS - Very Slow
S - Slow
M - Medium
H - High
VH - Very High

In order to perform simulation results for the Quanser bench-top helicopter system response, fuzzy rules will be set in FIS editor in Matlab. FIS editor is needed to set the fuzzy rules and determine the membership function for the fuzzy logic controller. 9 fuzzy rules that have been discussed before will be used in designing fuzzy logic controller. The FIS editor will be set by using "and" method which is consider the min value and the defuzzification by using centroid. After inputs and an output have been determined, fuzzy rules that have been discussed earlier will be set in fuzzy logic controller designation by using rule editor in FIS editor.

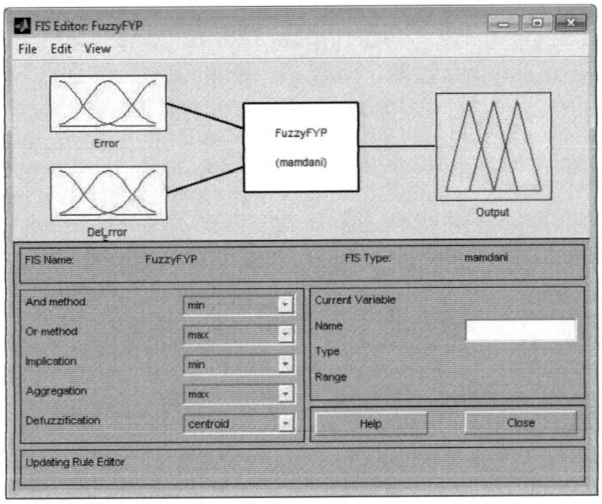

Figure 3: FIS Editor

Figure 3 shows FIS editor in Matlab that has shown two input membership functions which are *Error* and *Change in Error* while an output membership function is the *Response* of the system. In this part, membership functions have been tuned and suitable range of three membership functions has been identified. Other than that, shape of the membership functions also have been determined where are triangular and trapezoidal shape will be used. In order to identify a best response, this method works by changing range of each membership function and observe the output produced. Other than that, position and parameters value of membership functions also will be adjusted. Membership functions will be tuned until the best output response has been identified.

2.3 3-DOF Quanser Bench-Top Helicopter Simulations

In this part, a simulation file named *s_heli3d* that has been supplied by Quanser Inc. will be opened in Simulink in order to determine the behavior of output response regarding with the implementation of controllers. Figure 4 shows Simulink block diagram for Quanser 3-DOF Bench-top Helicopter Simulation.

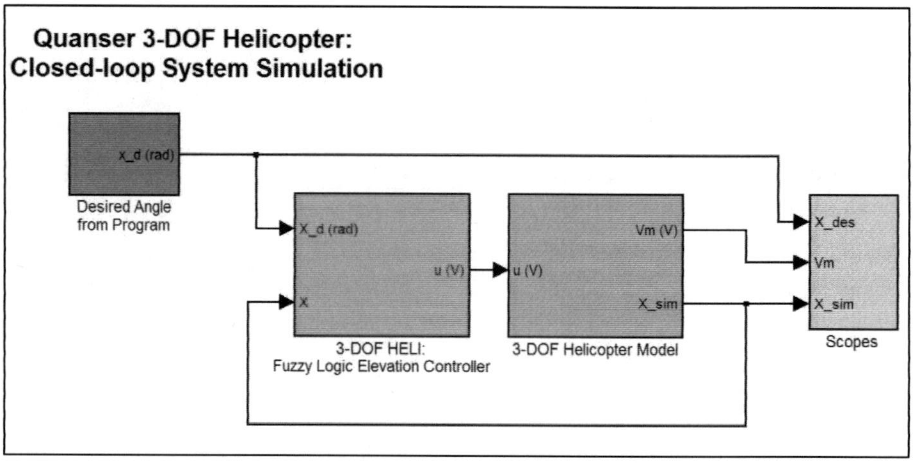

Figure 4: Simulink block diagram for Quanser 3-DOF Bench-top Helicopter Simulation

Based on Figure 4, *s_heli3d* represents the Quanser 3-DOF Helicopter: Closed-loop System Simulation. In this system, there are four blocks that will be used to run the bench–top helicopter system simulation. *Desired Angle from Program* block has been used to set the desired angle that needed to be simulated. In *PID Controller* block, it consist a controller that needed to be used to control the desired angle. In this project, fuzzy logic controller has been developed to replace the PID controller. Then, *3-DOF Helicopter Model* block contains designed parameters for helicopter model to work. This block will receive signals from controller to control certain part of the helicopter. Lastly, in *Scopes* block will display responses of the Quanser 3-DOF Helicopter: Closed-loop system simulation that contains four scopes which are elevation, pitch, travel and motor voltage.

Figure 5 shows the implementation of fuzzy logic elevation controller that has been designed by replacing the existing PID controller. The PID controller has been optimized in controlling the elevation angle by considering the elevation error and travel error. Other than that, the PID controller also will depend on the gain value of the bench-top helicopter. Fuzzy logic controller will consider elevation error and travel error to control the elevation angle where these errors will be added and summation produced will be acted as inputs for fuzzy logic controller in term of error and change in error.

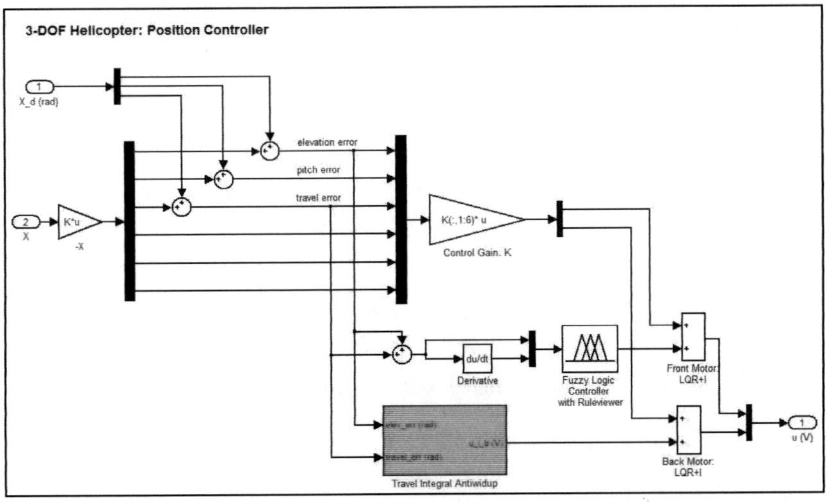

Figure 5: Fuzzy Logic Controller Implementation for Elevation Control

For Quanser bench-top helicopter hardware test, the system that has been used is quite similar to Figure 4 and the controller implementation can be seen in Figure 5. However, the system named Quanser 3-DOF Helicopter: Closed-loop Actual System will be used to connect to the real Quanser bench-top helicopter and it can be run by using Matlab and Simulink by interconnecting the hardware with q4 data acquisition card that attached to the computer. The steps for hardware connection can be referred in Quanser 3-DOF Helicopter Reference Manual [11].

3. RESULT AND ANALYSIS

3.1 Membership Function

In this section, result from the project that has been conducted will be analyzed and discussed. The important part of fuzzy logic controller is the designation and adjustment of membership functions that have been designed in Matlab. In design part, membership functions are designed and the behavior of output response will be observed. In this project, fuzzy logic controller is designed to control the output of bench-top helicopter. Thus, fuzzy logic controller is designed to control the desired elevation angle that will be set at 5°. The best membership functions that have been designed can be seen in Figure 6, 7 and 8 which represent error membership function, change in error membership function and output membership function respectively.

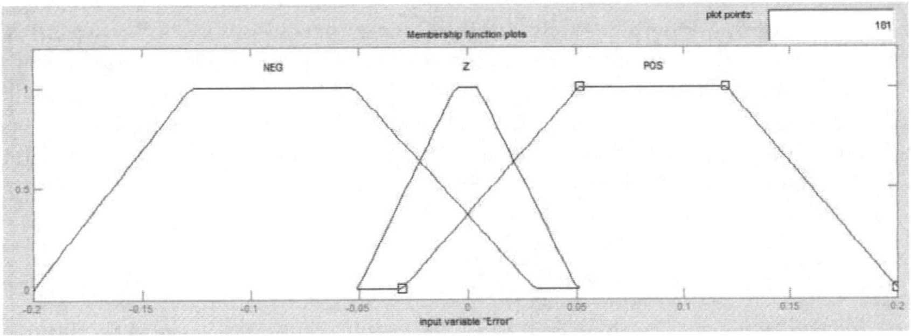

Figure 6: Error Membership Function

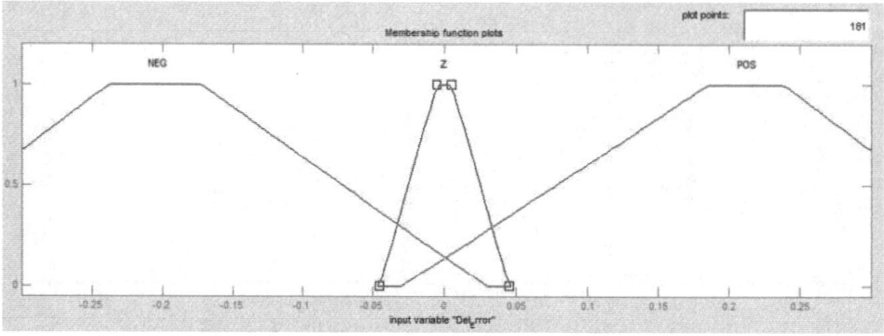

Figure 7: Change in Error Membership Function

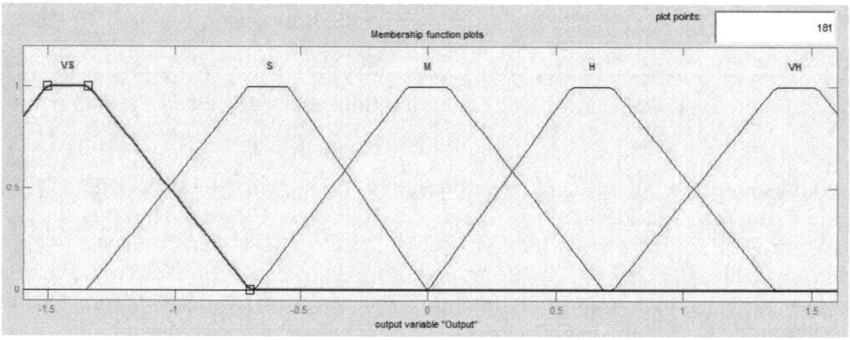

Figure 8: Output Membership Function

Based on the membership functions that have been designed, it can be seen that every membership functions have different range which are [-0.2,0.2], [-0.3,0.3] and [-1.6,1.6] for *Error*, *Change in Error* and *output* membership function respectively. It also can be observed that each of the membership function has symmetrical position. It can be said that the position of membership function does affect the output response of the system.

3.2 Quanser Bench-Top Helicopter Simulation

In this part, 3-DOF Quanser bench-top helicopter simulation has been conducted. Simulation is performed to determine the response of bench-top helicopter before it can be tested on real hardware. Therefore, fuzzy logic controller that has been designed is implemented in Quanser bench-top helicopter system to observe its ability in the Quanser bench-top helicopter system controlling. To determine the performance of fuzzy logic controller test on Quanser bench-top helicopter simulation test, three different angles have been chosen which are 4°, 5° and 6°. These angles is chosen because the fuzzy logic controller that has been designed can be optimized by tuning the membership functions to determine which membership functions is working well with these angles. In order to make sure that the fuzzy logic controller works with different angle, the membership functions need to be tuned.

3.2.1 Fuzzy Logic Elevation Control Test

The designed fuzzy logic controller has been tested with three different step input tests to determine its behavior. The first test is step response test at 4° elevation angle. The output response is obtained from the scope block shown in Figure 4. Based on Figure 9, it can be seen that the fuzzy logic elevation control response has faster settling time which is 3.4s and its percentage overshoot is recorded at 6.25%.

Test 1: Step Response (4°)

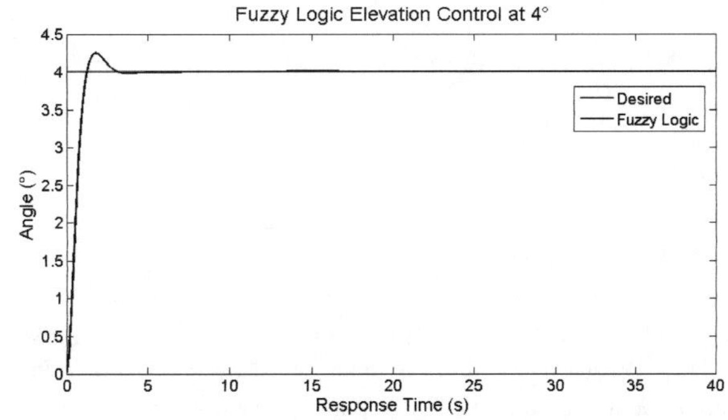

Figure 9: Fuzzy Logic Elevation Control at 4°

The second test is step response test at 5° elevation angle. Based on Figure 10, it can be observed that settling time about 3.3s with percentage overshoot of 8%.

Test 2: Step Response (5°)

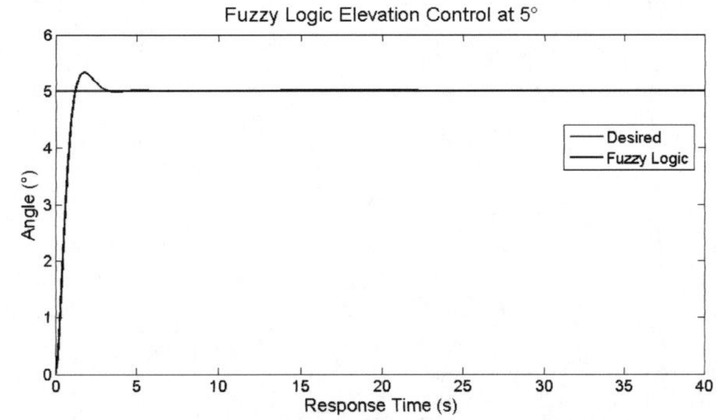

Figure 10: Fuzzy Logic Elevation Control at 5°

Then, the last test of step response is set at 6° elevation angle. It can be observed from Figure 11 that it settling time is recorded at 4.2s with percentage overshoot of 7%.

Test 3: Step Response (6°)

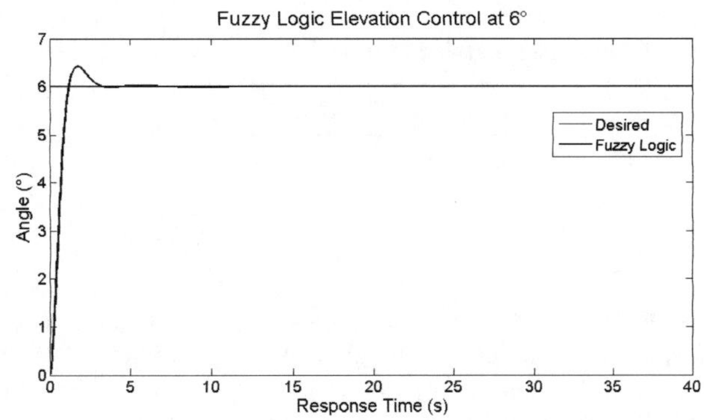

Figure 11: Fuzzy Logic Elevation Control at 6°

Based on the simulation results in Figure 9, 10 and 11, its show that fuzzy logic controller that has been designed and being implemented in the system is succeed in controlling elevation angle at 4°, 5° and 6°. It can be seen that the steady state error for these three angles are 0°. Table 2 shows

IOP Conf. Series: Materials Science and Engineering **260** (2017) 012015 doi:10.1088/1757-899X/260/1/012015

the summary of three test results that has been conducted on the Quanser bench-top helicopter simulation system.

Table 2: Fuzzy logic controller elevation response for 4°, 5° and 6°

Test	P.O	Ts	s.s.e
1	6.25%	6.2s	0
2	8%	3.3s	0
3	7%	4.2s	0

Based on Table 2, it can be seen that the steady state error for these three angles are 0° it is a smaller value in term of percentage overshoot and settling time. Therefore, fuzzy logic controller that has been designed has gave a good response for bench-top helicopter system in controlling elevation angle based on simulation that has been conducted.

3.2.2 Comparison between Fuzzy Logic Controller and PID Controller

After simulation has been conducted, the output behavior of fuzzy logic elevation control will be compared with output of PID elevation control. Figure 12 shows comparison between PID controller and fuzzy logic controller that has been designed for elevation control at 5°.

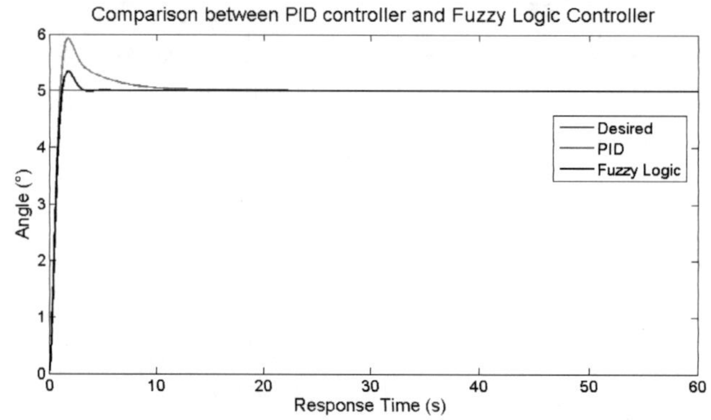

Figure 12: Comparison between PID Elevation Controller and Fuzzy Logic Elevation Controller

Table 3: Comparison between Fuzzy Logic Controller and PID Controller

Controller	P.O	Ts	s.e.e
PID	18%	14s	0
Fuzzy Logic	8%	3.3s	0

Table 3 shows the transient response characteristics of both PID and fuzzy logic controller. Based on Table 3, it can be seen that fuzzy logic controller provides a better response compared to PID controller in state of percentage overshoot and settling time. Meanwhile, steady state error shows that elevation control of both controllers is completed at 5°. Figure 13 shows the simulation result of the system when function generator of 0.04Hz has been implemented in the system. It can be seen that fuzzy logic controller is provides a better output response in term of settling time when the output time is settled at 2.9s compared to PID which is about 11.04s. Other than that, when the output response is observed, it can be seen that fuzzy logic elevation control is managed to reduce percentage overshoot which is 15.2% compared to the PID elevation control which is 36.4%.

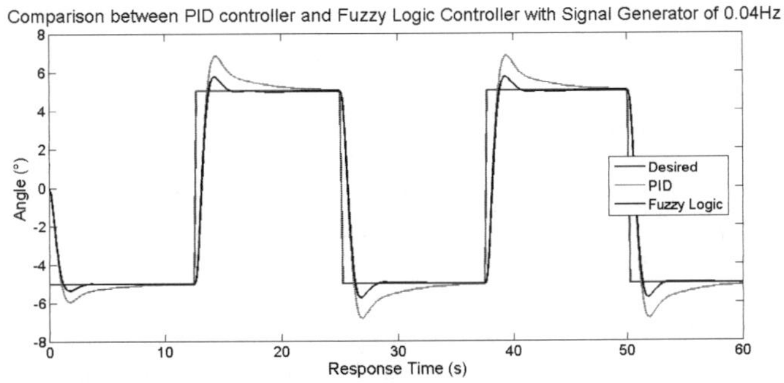

Figure 13: Comparison between PID Elevation Controller and Fuzzy Logic Elevation Controller with Signal Generator

3.3 Quanser Bench-Top Helicopter Hardware Test

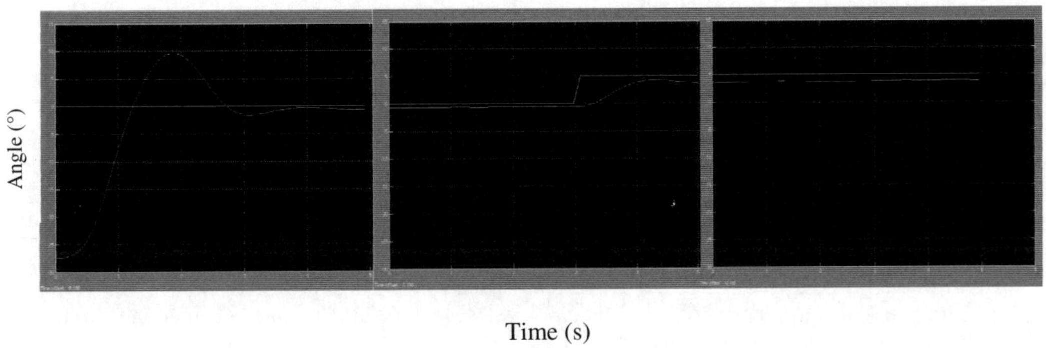

Figure 14: Quanser Bench-top Helicopter Hardware Test (a) 0 – 5 seconds (b) 5 – 10 seconds (c) 10 – 15 seconds

Based on Figure 14, it can be seen that fuzzy logic elevation control has been tested on hardware which is the Quanser bench-top helicopter located at Spacecraft, Guidance, Navigation and Control Lab at E5, Kulliyyah of Engineering, IIUM. In the Figure 14, yellow line indicates the desired response and purple line indicates actual response. Figure 14 (a) shows the response of bench-top helicopter from 0 to 5 seconds. Initially, the bench-top helicopter is located at -27.5°. When the system is started and elevation angle is set at 0°, it can be seen that the system response is trying to follow the 0° desired angle until 5° step response is applied at 8 seconds in Figure 14(b). The remaining time shows the response is trying to follow the 5° desired angle. This test shows that the system is managed to elevate at 3.9° angle with 22% of steady state error and takes about 1.8 seconds to achieve the settling time.

Fuzzy logic is expected to give a better performance compared to PID controller not only for simulation test but hardware test as well. In this case, membership function can be tuned in future work in order to make sure that fuzzy logic controller will give a better response for hardware test compared to the existing PID controller.

4. CONCLUSION

Fuzzy logic controller has been successfully designed and implemented in the system of Quanser bench-top helicopter. The objectives of this project are to design the fuzzy logic controller and to verify its performance over the existing PID controller. In this research, it can be seen that fuzzy

IOP Conf. Series: Materials Science and Engineering **260** (2017) 012015 doi:10.1088/1757-899X/260/1/012015

logic controller that has been designed for elevation control is working well for 4°, 5° and 6° for simulation system. Other than that, the controller that has been designed are managed to elevate the Quanser bench-top helicopter nearly to 5° for hardware test. Then, the behavior of simulation results show that elevation control of fuzzy logic controller has smaller percentage overshoot compared to PID controller. In addition, the simulation also shows that the settling time for fuzzy logic controller is also faster than the PID controller. It can be concluded that fuzzy logic controller give a better performance when the simulation results have been analyzed.

REFERENCES

[1] A. P. Sharma, J. B. Thakur, S. V. Surve, D. R. Singh and S. P. Sawant, "Comparison of PI and Fuzzy logic controller implemented in an APF for renewable Power generation," *2016 International Conference on Energy Efficient Technologies for Sustainability (ICEETS)*, Nagercoil, 2016, pp. 521-526.

[2] H. Chang, "Design of Fuzzy Controller for Temperature of Bag Filter," *2015 International Conference on Computational Intelligence and Communication Networks (CICN)*, Jabalpur, 2015, pp. 1258-1260.

[3] M. Negnevitsky, "Design of a fuzzy controller: some experience," *Fuzzy Systems, 2001. The 10th IEEE International Conference on*, 2001, pp. 912-915 vol.3.

[4] M. Chowdhury, JunbinGao and R. Islam, "Fuzzy logic based filtering for image de-noising," *2016 IEEE International Conference on Fuzzy Systems (FUZZ-IEEE)*, Vancouver, BC, Canada, 2016, pp. 2372-2376.

[5] M. Mehndiratta, E. Kayacan and T. Kumbasar, "Design and experimental validation of single input type-2 fuzzy PID controllers as applied to 3 DOF helicopter testbed," *2016 IEEE International Conference on Fuzzy Systems (FUZZ-IEEE)*, Vancouver, BC, 2016, pp. 1584-1591.

[6] J. Ma and R. Ji, "Fuzzy PID for Quadrotor Space Fixed-Point Position Control," 2016 Sixth International Conference on Instrumentation & Measurement, Computer, Communication and Control (IMCCC), Harbin, 2016, pp. 721-726.

[7] A. H. Mohd Hairon, H. Mansor, T. S. Gunawan, and S. Khan, "Travel angle control of Quanser bench-top helicopter based on quantitative feedback theory technique," Indonesian Journal of Electrical Engineering and Computer Science, 2016, vol. 1, no. 2, pp. 310-318.

[8] R. I. Boby, H. Mansor, T. S. Gunawan and S. Khan, "Robust adaptive LQR control of nonlinear system application to 3-Dof flight control system," *Smart Instrumentation, Measurement and Applications (ICSIMA), 2014 IEEE International Conference on*, Kuala Lumpur, 2014, pp. 1-4.

[9] S. A. S. Yazan, H. Mansor, T. S. Gunawan and S. Khan, "Development of robust Quantitative Feedback Theory controller for Quanser bench-top helicopter," *Smart Instrumentation, Measurement and Applications (ICSIMA), 2014 IEEE International Conference on*, Kuala Lumpur, 2014, pp. 1-4.

[10] Hasmah Mansor, Teddy S. Gunawan, Sheroz Khan, N. I. Othman, N. Tazali, R. I. Boby, and S. B. Mohd-Noor. 2014. Self-Tuning Dead Beat PD Controller for Pitch Angle Control of a Bench-Top Helicopter. In *Proceedings of the 2014 International Conference on Computer and Communication Engineering* (ICCCE '14). IEEE Computer Society, Washington, DC, USA, 119-122.

[11] Quanser. Quanser 3-DOF Helicopter Reference Manual London: Quanser Incorporasted. 2010

IOP Conf. Series: Materials Science and Engineering **260** (2017) 012016 doi:10.1088/1757-899X/260/1/012016

Design and analysis of planar spiral resonator bandstop filter for microwave frequency

S. M. A. Motakabber and Muhammad Shaifudin Suharsono
Department of Electrical and Computer Engineering
Faculty of Engineering, International Islamic University Malaysia
53100 Kuala Lumpur, Malaysia

Email: amotakabber@iium.edu.my, shaifudin.suharsono@gmail.com

Abstract. In microwave frequency, a spiral resonator can act as either frequency reject or acceptor circuits. A planar logarithmic spiral resonator bandstop filter has been developed based on this property. This project focuses on the rejection property of the spiral resonator. The performance analysis of the exhibited filter circuit has been performed by using scattering parameters (S-parameters) technique in the ultra-wideband microwave frequency. The proposed filter is built, simulated and S-parameters analysis have been accomplished by using electromagnetic simulation software CST microwave studio. The commercial microwave substrate Taconic TLX-8 has been used to build this filter. Experimental results showed that the -10 dB rejection bandwidth of the filter is 2.32 GHz and central frequency is 5.72 GHz which is suitable for ultra-wideband applications. The proposed design has been full of good compliance with the simulated and experimental results here.

1. Introduction

An electrical or mechanical resonator is a physical structure that displays resonance behavior. Usually, it has a natural oscillation frequency, which can either be an electromagnetic or mechanical oscillation in nature. At this frequency greater amplitude of oscillation is provided than the other frequencies. The resonators are insensitively used to make waves at a specific frequency or to select a band of a specific frequency from a signal. This characteristic enables electronic devices to produce oscillations of the appropriate frequency. Dielectric resonators are effectively used at microwave frequencies. The electrical behavior of one or more multi ports, the electrical network can be performed over a narrow range of frequencies by using S-parameters analysis. A planar transmission line is commonly used as a resonator to make a microwave filter [1][2]. It is possible to construct the filter precisely by using microwave printed board materials or substrate and photolithography and etching processes.

The microstrip planar components are widely used in the microwave frequency band [3][4]. A cross sectional view of a microstrip line is shown in Figure 1. Generally, these comprise of a low loss dielectric substrate having one or two layers of the metal sheet. The signal and ground currents are flowing on the separate conducting metal surface. The microstrip planar resonator can any structure which is able to resonate at least one oscillating electromagnetic field. Basically, a microstrip is a printed geometric structure of metal strip over a ground plane or conducting metal sheet. This structure has a tendency to radiate electromagnetic waves and it increases with increasing the space between the ground plane and the strip. However, it is possible to diminish the radiation without making the strip width too narrow by using a substrate of very low thickness compared to the wavelength of the operating frequency. The microstrip transmission lines are extensively used in the RF and microwave circuits, due to its planar nature, easy integration of active and passive components, good mechanical support, excellent heat sinking and easy to construct using photolithography and etching processes.

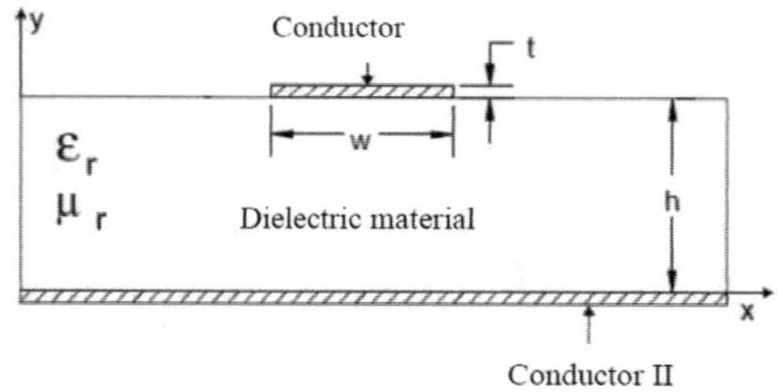

Figure 1. Cross sectional view of basic structure of a microstrip line

A large signal S-parameter analysis has been used for analyzing the proposed filter performance study. When an n-port network is interconnected into a microwave transmission line the propagating waves that are scattered or reflected can be analyzed by S-parameter. From the S-parameter analysis, it is possible to predict the circuit behavior to the designed level of precision. In addition, it can also provide a means of illustrating the quantitative and qualitative measures and behavior of the circuit elements as well. To obtain an accurate design and calculation, only computer based test system could reasonably handle the type of measurement and mathematics involved. The system consists of N5230A PNA-L Vector Network Analyzer with S-Parameter test set that is able to measure the frequency range of interest. Besides, a software program that is capable of performing vector error correction, computes an equivalent circuit parameter, setting up measurements and calibrating the measurement system is also implemented.

2. Methodology

The design was modeled according to a specific dimension by calculating the required parameters using the formula. The simulation and design process was conducted using the CST Microwaves Software. Then, the design was fabricated and the values were analyzed using N5230A PNA-L Network Analyzer. This validity test was conducted to confirm that the simulation through software would give approximately the same result as the fabrication built. The design parameters were, gap between turns, $s = 0.3$ mm, width of turns, $w = 0.80$ mm, distance between line and resonator, $d = 0.20$ mm, width of transmission line, $W_L = 2.26$ mm, width of the resonator, $W_T = 5.20$ mm and length of the resonator, $L_T = 8.00$ mm. The PCB was fabricated using Taconic TLX-8 as the substrate with its thickness of 0.50 mm. The dielectric constant was 2.55 and the material used was copper with a thickness of 0.035 mm. Table 1 shows the summary of the dimensions of the filter elements.

Table 1: Dimensions of the Filter Elements

Filter parameter	Value(mm)	Filter parameter	Value(mm)
S	0.3	L_R	8.0
W	0.8	W_T	2.26
d	0.2	L_T	17.75
W_R	5.2	- -	- -

Design with less than 0.2 mm width might cause a problem during the fabrication process. Therefore, some of these parameters were chosen for the practical purpose during fabrication while other

IOP Conf. Series: Materials Science and Engineering **260** (2017) 012016 doi:10.1088/1757-899X/260/1/012016

parameters were obtained using formula Equation 1, Equation 3 and Equation 3. The design, layout and the photograph of the prototype filter are shown in Figure 2.

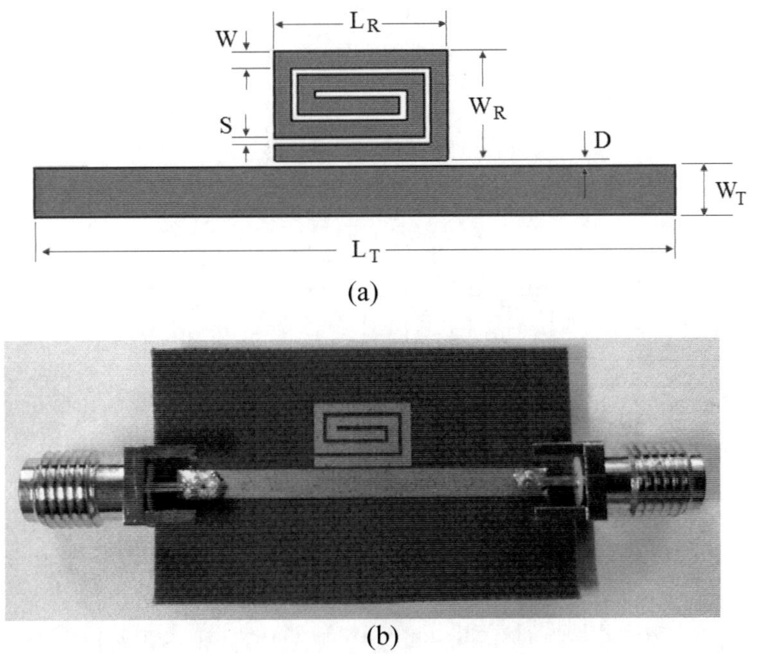

Figure 2. Proposed filter, (a) layout and (b) developed filter with SMA connectors

Using the same design parameters, a real spiral resonator bandstop filter prototype was fabricated on the Taconic TLX-8 substrate. For microwave circuit, PCB enables the transmission line to be laid out in a planar form. Briefly, the design was printed out onto a glossy paper and was transferred onto the PCB via heat compressor. After that, the unwanted materials were removed using a specific combination of solutions. The finished product was then soldered with Sub Miniature version A (SMA) connector on both sides of the ports. Specifically, there were using two female PCB type for the PCB. The finished prototype was analyzed by Vector Network Analyzer model N5230A PNA-L as shown in Figure 3.

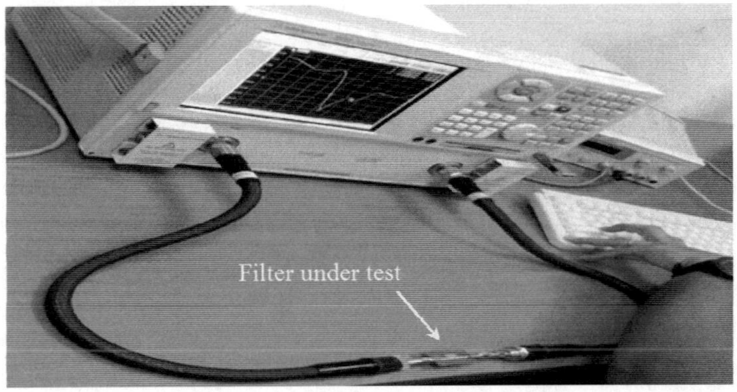

Figure 3. Photograph to measure the data of S11 and S21 by using Vector Network Analyzer

Dimensions of the filter are calculated by using the proposed formulas [5][6]. The approximate of the effective dielectric constant of a microstrip line can be calculated by using Equation 1.

$$\varepsilon_e = \frac{\varepsilon r + 1}{2} + \frac{\varepsilon r - 1}{2} \frac{1}{\sqrt{1 + 12d/W}} \tag{1}$$

The characteristic impedance can be calculated by Equation 2.

$$Z_0 = \begin{cases} \frac{60}{\sqrt{\varepsilon_e}} \ln\left(\frac{8d}{W} + \frac{W}{4d}\right) & \text{for } \frac{W}{d} \le 1 \\ \frac{120\,\pi}{\sqrt{\varepsilon_e\left[\frac{W}{d} + 1.393 + 0.667 \ln\left(\frac{W}{d} + 1.444\right)\right]}} & \text{for } \frac{W}{d} \ge 1 \end{cases} \tag{2}$$

The W/d ratio can be calculated by characteristic impedance Z_0 and dielectric constant ε_r.

$$\frac{W}{d} = \begin{cases} \frac{8\,e^A}{e^{2A} - 2} & \text{for } \frac{W}{d} \le 2 \\ \frac{2}{\pi}\left[B - 1 - \ln(2B - 1) + \frac{\varepsilon_r - 1}{2\,\varepsilon_r}\left\{\ln(B - 1) + 0.39 - \frac{0.61}{\varepsilon_r}\right\}\right] & \text{for } \frac{W}{d} \ge 2 \end{cases} \tag{3}$$

The constant parameters *A* and *B* can be calculated by Equation 4(a) and Equation 4(b) respectively.

$$A = \frac{Z_0}{60}\sqrt{\frac{\varepsilon_r + 1}{2} + \frac{\varepsilon_r - 1}{\varepsilon_r + 1}\left(0.23 + \frac{11}{\varepsilon_r}\right)} \tag{4.a}$$

$$B = \frac{377\,\pi}{2\,Z_0\sqrt{\varepsilon_r}} \tag{4.b}$$

3. Result and Discussion

The simulation and experimental graphical results are described from Figures 4 to 8.

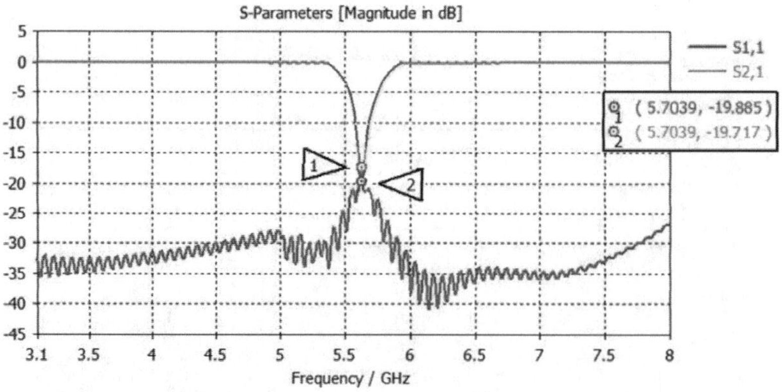

Figure 4. Simulated S-parameters of S11 and S21

To study the S-parameter characteristic of the circuit can help to find out the resonant frequency of the spiral resonator. Based on the result, it is observed that the resonant frequency is 5.7039 GHz.

Figure 5 shows both the S11 and S21 properties of the filter circuit. The graphical results show that it is a bandstop filter characteristic and there is a high attenuation of the signal at the resonant frequency. At resonant, S11 exhibit its reflective signal characteristics where the signal passes as high as possible while S21 exhibit its transmitted signal characteristics, the signal goes as low as possible.

The CST software simulation of surface current is shown in Figure 5. From the figure, it shows that there are almost no current flows through the transmission line at frequency 5.7039 GHz. Since at that frequency, all the current absorbed by the spiral resonator.

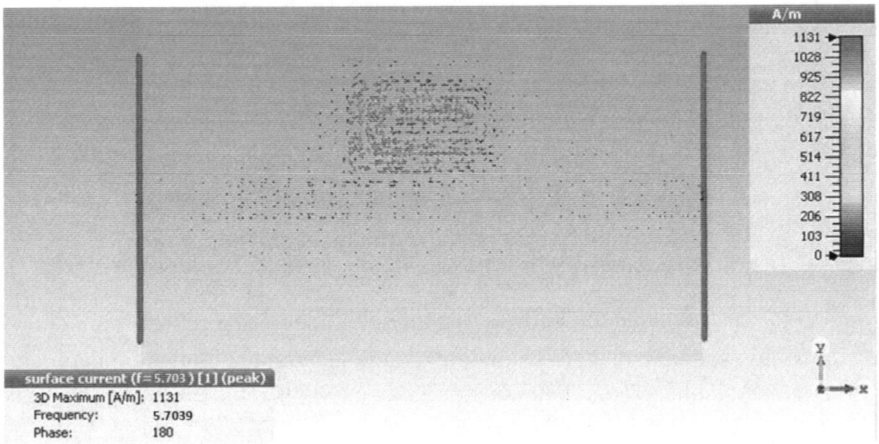

Figure 5. The surface current of the spiral resonator at frequency 5.7039 GHz.

The result from VNA (Vector Network Analyzer) is shown in Figure 6. The measured value shows that the resonant frequency is at 5.726 GHz. Nevertheless, the experimental result shown is close to the simulation result with a difference of 0.023 GHz (5.726 GHz – 5.7039 GHz = 0.023GHz). It is believed that this disagreement exists due to conductor loss which was not accounted in the simulation. Moreover, there might be an error in the fabrication process and some losses occur in the connection during measurement. From the article reviewed, MMIC technology can overcome these types of problems, if the resonator was designed and fabricated using this technology.

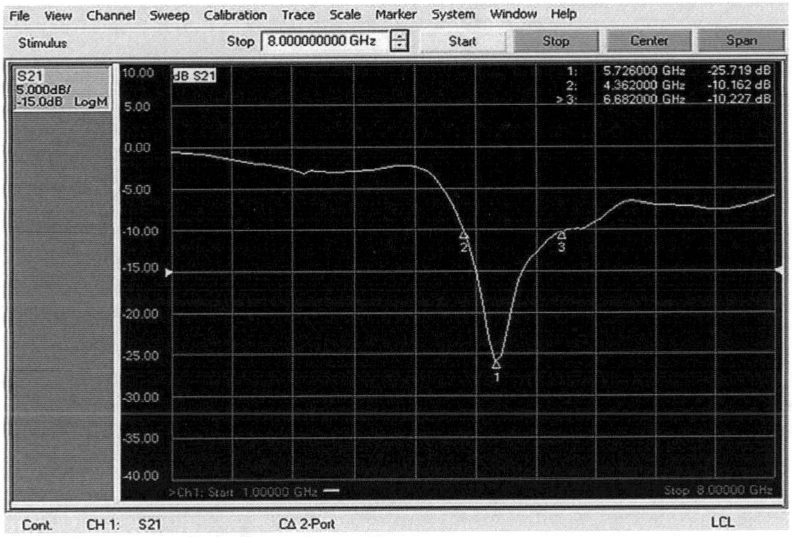

Figure 6. S-Parameter of S21 port from VNA.

To compare the simulation with the experimental results, graphs are combined in Figure 7. From Figure 6 it is found that the -10 dB bandwidth of the filter is 2.32 GHz. The simulation and experimental results of S21 are well agreed as shown in Figure 7.

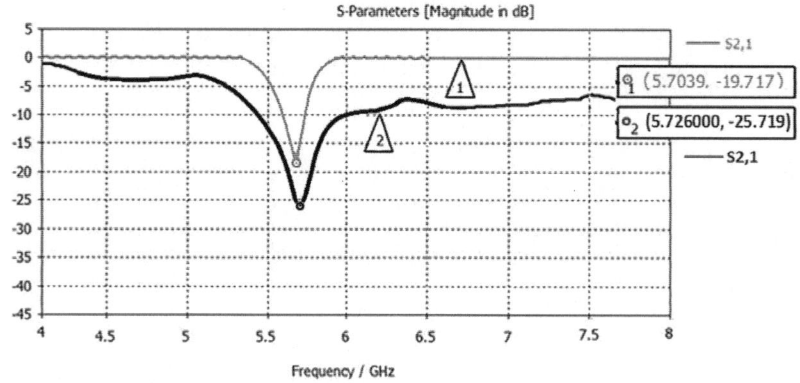

Figure 7. S-parameters of S21 port from VNA is the black line and CST simulation is green line

4. Conclusion

Although the size of this spiral resonator is small, the experimental result proves that this resonator has a -10 dB bandwidth of 2.32 GHz. It also shows spiral resonator can be used to construct a small bandstop filter. The spiral resonator also corresponds closely to the simulated results thus validating the theory from simulated data. It is a simple design procedure can also reduce the cost of fabrication. This project shows that each parameter is interconnected when designing a spiral resonator. Thus, understanding the theories and formula related to it is essential for filter designer. The S-parameter provides great insight into the design problem and it can work at high frequencies. This shows that S-parameter is important knowledge in microwave design. Moreover, when deciding on the dimension, it is important to consider practicality when fabricating as an error in fabrication can affect the experimental result. The simulation and experimental result should not disagree with each other as disagreement shows the problem in the design. However, a small disparity can be tolerated as there is no perfect PCB that can be fabricated.

Acknowledgement

This research has been supported by the Malaysian Ministry of Science and Technology through the eScienceFund under the project ID: SF14-010-0060.

References

[1] A. K. M. Z. Hossain, S. M. A. Motakabber and M. I. Ibrahimy, "Microstrip Spiral Resonator for the UWB Chipless RFID Tag," *23rd International Conference on Systems Engineering*, 19-21 August 2014, USA.

[2] S. Lavanya, M. U. Maheshwari et al, "Planar Ultra-Wideband Bandpass Filter Using Defected Ground Structure," *IEEE Sponsored 2nd International Conference on Innovations in Information Embedded and Communication Systems*, 2015.

[3] S. Kumari, M. Kumar, "Ultra-Wideband Planar Bandpass Filter with Dual Notch Behavior," *IEEE International Conference on Advances in Engineering & Technology Research*, August. 2014, Unnao, India.

[4] A. K. M. Z. Hossain, M. I. Ibrahimy and S. M. A. Motakabber, "Spiral Resonator for Ultra-Wide Band Chipless RFID Tag," *5th International Conference on Computer and Communication Engineering*, 23-25 September 2014, Malaysia.

[5] L. Sandrolini, U. Reggiani, and G. Puccetti, "Analytical Calculation of the Inductance of Planar Zig-Zag Spiral Inductors," *Progress in Electromagnetics Research*, vol. 142, pp. 207–220, 2013.

[6] S. Kumari and M. Kumar, "Design of Planar Bandpass Filter for Ultra-Wideband Applications," *Fourth International Conference on Communications Systems and Network Technologies*, pp. 38-40, 2014.

IOP Conf. Series: Materials Science and Engineering **260** (2017) 012017 doi:10.1088/1757-899X/260/1/012017

Real-time Human Activity Recognition

N.Albukhary and Y.M. Mustafah

Department of Mechatronics Engineering, International Islamic University Malaysia (IIUM), Jalan Gombak, 53100 Kuala Lumpur, Malaysia

Email: mira.bukhary@gmail.com, yasir@iium.edu.my

Abstract. The traditional Closed-circuit Television (CCTV) system requires human to monitor the CCTV for 24/7 which is inefficient and costly. Therefore, there's a need for a system which can recognize human activity effectively in real-time. This paper concentrates on recognizing simple activity such as walking, running, sitting, standing and landing by using image processing techniques. Firstly, object detection is done by using background subtraction to detect moving object. Then, object tracking and object classification are constructed so that different person can be differentiated by using feature detection. Geometrical attributes of tracked object, which are centroid and aspect ratio of identified tracked are manipulated so that simple activity can be detected.

1. Introduction

Security camera systems, such as Closed-circuit Television (CCTV) are widely used and can be found in places where monitoring is needed. [1] The cheaper price and the increasing use of CCTVs in many areas are making it to be easily deployable. However, this traditional CCTV requires human to monitor the scene continuously [2]. It is very costly and inefficient to have human monitoring the scene. Monitoring activity is dull in nature and the process of analyzing series of videos afterward can also be time-consuming. Therefore, a system with human activity recognition is needed for this purpose.

This paper presents a visual human activity recognition, by retrieving information from one single camera. The scenario of human activity at an open space, considering the space is uncrowded. Firstly, object detection from video collected is done before proceeding to object classification and activity analysis.

There are growing interests on recognizing simple activity generated from human such as recognizing if someone is walking or running which can be seen from [3], or a combination of few activities [4], [5], [6]. The higher level includes recognizing the activity of several interacting people. [7], [8].

According to [9], object detection, object tracking and object classification need to be done before proceeding to activity analysis. Object detection is for detecting moving object [10], object tracking is the capability to track moving object as they move around the scene and object classification assign label to tracked object [9].

Activity analysis can be done by recognizing the posture of tracked object. The object can be characterized by points inside region of interest. In [4], centroid position is used as one of the features to classify activity of tracked object. Next, object can be represented by their geometric shapes. Bounding box is used as object tracking in [4], [5], [8], and [11]. Object can also be tracked using silhouette which is represented by the area inside the contour. [12] and [13] use contour to get human

IOP Conf. Series: Materials Science and Engineering **260** (2017) 012017 doi:10.1088/1757-899X/260/1/012017

appearance representations. Another representation for tracked object is by using skeletal model object which is used to model articulated and rigid object. Skeletal model can be found in [14] and [15].

2. Proposed Method

This research was developed in three parts which includes moving object detection, moving object classification and activity classification. Figure 1 below shows process overview of the research.

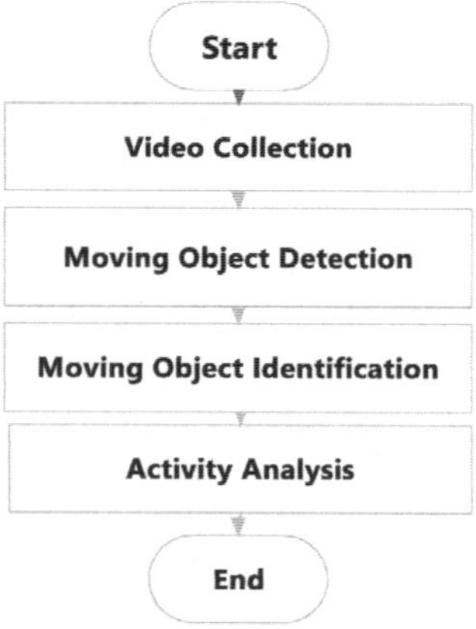

Figure 1. Process overview

2.1. Moving object detection

Background subtraction is done to get moving object in the scene. Foreground object is obtained by differencing pixel by pixel in the current frame to the reference background image.

Due to environmental factors that might affect the foreground image detection, noise from the image acquired should be expected. Image portion might become distorted due to noise. To suppress noise, morphological operation of opening and closing is used which includes a combination of dilation and erosion. The Figure 2 below shows the blob of tracked object that is successfully detected.

Figure 2. Moving object detection

2.2. Moving object identification and tracking

To successfully track and identify a moving object, correct identity should correctly be assigned to that certain track throughout the scene. For this purpose, feature-based tracking can be used for object classification since the same object shared common features. The feature used in this paper includes geometrical attributes which is the distance between tracked object across frame, colour information as well as the texture of tracked object. The centroid of tracks, $p(x, y)$ is located at the center of the bounding box as shown in Figure 3 below. Given the track in the previous frame has centroid position $p_i(x_i, y_i)$ and the track in the current frame with centroid $p_n(x_n, y_n)$, The centroid and distance between the current and previous frame is computed by using equation (1) and (2) below.

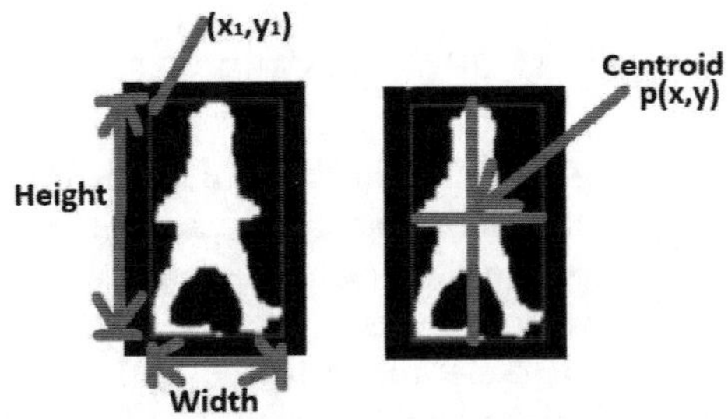

Figure 3. Geometrical attributes using bounding box.

$$x = x_l + \frac{width}{2} \text{ and } y = y_l + \frac{height}{2} \tag{1}$$

$$d_p(\mathrm{p_n}, \mathrm{p_i}) = \sqrt{|x_n - x_i|^2 + |y_n - y_i|^2} \tag{2}$$

Next, feature is described based on colour. The colour property used in hue and RGB. Hue is taken from HSV colour based. The similarities of hue between the current and previous frame is computed. For RGB, the colour feature was extracted by comparing the intensity values of Red, Green and Blue. The largest intensity values were assigned to the pixels. After that, colour histogram was computed and probabilities of each colour is made. similarity of RGB can be get by getting the difference between probabilities in the current and previous frame.

Feature detection for colours need to be used whenever new track entered the scene, merging between the tracks occur or when tracks leave the scene. Else, if none of this happens and the object seems to be close with each other, same label will be given to tracks. This is decided after comparing number of bounding box for previous with number bounding box in current frame.

Three condition of comparison between number of bounding box number between current frame and previous frame is analyzed to know is there is any track leaving or entering the scene, or if any tracks are merging or splitting. This can be done by firstly comparing if number of bounding box is the same for both current and previous frame, or if the current frame has greater or less tracks compared to previous. If needed, for example splitting condition, feature detection will be done to ensure correct labelling is given after splitting. Figure 4 below shows how object identification and tracking is done.

Figure 4. In (a) there are two tracks approaching each other, (b) merging between two tracks, (c) splitting of tracks.

2.3. Activity Analysis

This paper concentrates in differentiating five simple activity which includes walking, running, sitting, standing and landing.

For activity analysis, the position and aspect ratio of tracked object is calculated before analysing the activity with basic assumptions that the activities that need to analysed can be ascertained by measuring these aspects. The centroid and distance can be computed by using formula (1) and formula (2). Equation to calculate aspect ratio is in (3) below.

$$Aspect\ ratio = \frac{Height\ (pixels)}{Width\ (pixels)} \quad (3)$$

While human is walking, or running, the motions between both arms and both legs change the shape region dynamically [4]. However, the height of object is usually quite constant. Therefore, human shape is used as one of the characteristics to determine activities such as walking and running. For walking and running, direction is classified by positive and negative. If tracks move towards the right side, positive symbol will be given. And if tracks move to the left, negative symbol will be given. It is important to get to know the direction since any changes towards direction can be notified. Using ratio proposal itself is not enough to clarify human activity. Therefore, centroid position is also used to differentiate human activity.

3. Results

For moving activity like walking and running, the centroid will change dynamically per frame. Difference in centroid is much larger in running since the speed of tracks is higher when running. This is different with stationary or almost stationary. For the other three actions, which is standing, sitting and landing, there won't be really any different between centroid between frames. The activity can be considered by using aspect ratio. If the aspect ratio is high, the track can be considered as standing. If aspect ratio is almost one, the track can be considered as sitting and if aspect ratio is low, tracks is assumed to be landing. The result for activity analysis can be seen in Figure 5.

Table 1 shows the average calculation of distance centroid per frame based on pixel difference, bounding box ratio and direction label.

Figure 5. Position walking (+w, -w) , running (+r, -r) , sit (s) , stand (st) and landing (l)

Table 1. Average calculation of distance centroid per frame, bounding box ratio and direction label

	Walk	**Run**	**Standing**	**Landing**	**Sit**
Distance centroid per frame	1 to 4	> 4	< 1	< 1	< 1
Bounding box ratio (height/width)	> 1.3	> 1.3	> 1.3	< 1	0.7 to 1.3
Direction label (-ve,+ve)	/	/	NA	NA	NA

For the system to work in real-time, image spatial resolution was limited to 384x586. For a resolution of 384x586, with up to 3 objects in the frame, the processing speed can go up to 61ms per frame which is 16 frames per second, acceptable for the system to work in real time [16]. However, as more objects appears in the frame, the processing time will increase further. Our future improvement would involve studying the trade-off between the number of object, spatial resolution and speed of the system.

4. Conclusion

A method to recognize human action using centroid and aspect ratio by comparing frame is presented. Position information of human in the scene is estimated by introducing the distance calculation between frames. The walking and running direction is also presented. The advantage of the system is built from a single camera using geometrical properties of image and the average time taken per cycle is 61ms.

References

[1] R. Arroyo, J. J. Yebes, L. M. B. Bergasa, I. G. . Daza, and J. Almazán, "Expert video-surveillance system for real-time detection of suspicious behaviors in shopping malls," *Expert Syst. Appl.*, vol. 42, no. 21, pp. 7991–8005, 2015.

[2] S. Rho, G. Min, and W. Chen, "Engineering Applications of Artificial Intelligence Advanced issues in artificial intelligence and pattern recognition for intelligent surveillance system in smart home environment," *Eng. Appl. Artif. Intell.*, vol. 25, no. 7, pp. 1299–1300, 2012.

[3] C. Y. C. Yam, M. S. Nixon, and J. N. Carter, "Automated person recognition by walking and running via model-based approaches," *Pattern Recognit.*, vol. 37, no. 5, pp. 1057–1072, 2004.

[4] J. Hariyono and K.-H. Jo, "Centroid based Pose Ratio for Pedestrian Action Recognition," in *Industrial Electronics (ISIE), 2016 IEEE 25th International Symposium*, 2016, pp. 895–900.

[5] D. Gerónimo and H. Kjellström, "Unsupervised Surveillance Video Retrieval based on Human Action and Appearance," in *International Conference on Pattern Recognition*, 2014, pp. 4630–4635.

[6] H. Zhao and Z. Liu, "Shape-Based Human Activity Recognition Using Edit Distance," in *2009. 2nd International Congress on Image and Signal Processing, CISP*, 2009.

[7] M. S. Ryoo, "Human activity prediction: Early recognition of ongoing activities from streaming video," in *2011 IEEE International Conference on Computer Vision (ICCV),* 2011, pp. 1036–1043.

[8] M. Raptis and L. Sigal, "Poselet Key-framing : A Model for Human Activity Recognition," in *IEEE Conference on Computer Vision and Pattern Recognition (CVPR)*, 2013, pp. 2650–2657.

[9] K. H. S, G. S. R, K. S. B, and P. Sangmesh, "Smart Video Surveillance," *Int. J. Emerg. Eng. Res. Technol.*, vol. 3, no. 1, pp. 109–112, 2015.

[10] C. L. Devasena, R. Revathí, and M. Hemalatha, "Video Surveillance Systems – A Survey," *IJCSI Int. J. Comput. Sci. Issues*, vol. 8, no. 4, pp. 635–642, 2011.

[11] J. M. Chaquet, E. J. Carmona, and A. Fernández-caballero, "A survey of video datasets for human action and activity recognition," *Comput. Vis. Image Underst.*, vol. 117, no. 6, pp. 633–659, 2013.

[12] M. Dragan and I. Mocanu, "Human Activity Recognition in Smart Environments," in *19th International Conference on Control Systems and Computer Science (CSCS)*, 2013.

[13] J. Cai, X. Tang, and G. Feng, "Learning pose dictionary for human action recognition," in *2014 22nd International Conference on Pattern Recognition*, 2014.

[14] M. Magno, F. Tombari, D. Brunelli, L. Di Stefano, and L. Benini, "Multimodal Video Analysis on Self-Powered Resource-Limited Wireless Smart Camera," vol. 3, no. 2, pp. 223–235, 2013.

[15] E. S. L. Ho, J. C. P. Chan, D. C. K. Chan, H. P. H. Shum, Y. Cheung, and P. C. Yuen, "Improving posture classification accuracy for depth sensor-based human activity monitoring in smart environments," *Comput. Vis. Image Underst.*, vol. 0, pp. 1–14, 2016.

[16] Y. M. Mustafah, R. Noor, H. Hasbi, and A. W. Azma, "Stereo vision images processing for real-time object distance and size measurements," in *2012 International Conference on Computer and Communication Engineering, ICCCE 2012*, 2012, pp. 659–663.

IOP Conf. Series: Materials Science and Engineering **260** (2017) 012018 doi:10.1088/1757-899X/260/1/012018

Optimized Vibration Chamber for Landslide Sensory and Alarm System

[1]Eliza Sabira Binti Ismail, [1]Mohamed Hadi Habaebi, [2]Jamal I. Daoud and [1]Md Rafiqul Islam

[1]Electrical and Computer Engineering Department, [2]Science in Engineering Department, International Islamic University Malaysia

elizasismail@gmail.com; habaebi@iium.edu.my; jamal58@iium.edu.my; rafiq@iium.edu.my

Abstract. Landslide is one of natural hazard that is not unfamiliar disaster in Malaysia. Malaysia has experienced this disaster many times since 1969. This natural hazard has become a major research concern for Malaysian government when many people were injured badly and even had been killed. Many previous research works published in the open literature aimed at designing a system that could detect landslide in early stage before the landslide becomes catastrophic. This paper presents the early works on a major work-in-progress landslide early warning system for Malaysian environment. The aim of this system is to develop the most efficiently reliable cost-effective system in which slight earth movements are monitored continuously. The challenge this work aims at is to work with a low budget system that produces efficient performance. Hence, the material used is off-the-shelf. Early design optimization results of the vibration sensor used is quite promising detecting the slightest faint tremors, which are amplified using the best vibration chamber available. It is shown that the choice of proper pipe length and diameter dimensions in combination to a gravel to exaggerate the produced higher sensitivity level noise of 5 dB.

1. Introduction

Landslide also known as slope failure, landslips or slums can be defined as movement of rock or debris down a slope. In other words, landslide is the changes of the condition of the slope from stable to unstable. Landslide can happen due to natural phenomenon and human activities. The factors that causes landslides naturally are earthquakes, groundwater, erosion and volcanic eruptions. Whereas, acts by human that result in landslides are deforestation, construction, blasting, earthwork and vibrations from machinery. In Malaysia, 1969 was the earliest landslide occurred and it was in Pahang. There were seventeen landslides were recorded in total now and the latest was the last year in Selangor. People were injured and killed in this natural hazard and had caused tremendous damages to the environment infrastructure. These would have cause a lot of money for the recovery. The purpose of invention of landslide early warning system is to reduce all the risks mentioned above. In fact, many lives could be saved from becoming the victim of this natural hazard. Many research papers were published regarding this landslide early warning system from time to time to develop the most reliable and efficient system in order to alert people in the prone area about this hazard while it is still early. This paper aims to design a vibration sensor and to investigate its different parameters to improve its capability in detecting landslides using commercial off-the-shelf materials. The optimum sensing structure dimensions were investigated for sensor casing for maximum acoustic density. The type of sensor that can detect ground movement is identified and it was decided to use vibration sensor SW-420. Next, as an effort to increase the accuracy and intensity of the vibration, gravels were placed inside a pipe before buried in the ground. In fact, several experiments were carried out to investigate this matter and the results obtained is positive. Lastly, in finding the optimum dimension of pipe, numerous experiments were conducted for three different lengths of pipes with four different diameters for each length. The paper is organized as follows. Section 2 briefly surveys the different projects used recently for landslide monitoring. Section 3 introduces the methodology. Section 4 presents the results and discussion. Section 6 concludes the paper.

2. Literature review

The development of landslide sensor has been evolving every day in order to produce the most reliable and effective system in detecting landslide hence alert the people in the prone area at the same time. A lot of studies regarding landslide sensor can be found today and generally, they are all different in two important aspects which are first, the type of sensor or sensors utilized in their projects and second is the data transmission from sensor node to base unit. Given these points, in this chapter, those two important aspects will be reviewed thoroughly coupled with the outcomes from their studies. Landslide can be described as a movement of soil, debris or rock down a slope caused by earthquake, rainfall or rapid snow melt [1,2]. Since 50 years ago, the speed of the movement in the ground have been changing from slowest to very slow [3]. To put it another way, the speed of the landslide is getting higher from time to time. Every year, landslides has affected many countries over the world in fact has endanger even killed many human life and destroyed the environment. Therefore, data transmission in real-time regarding the status of stability of slope in the prone areas has become a major concerned in research area in the hope that an efficient landslide early warning system could be developed. Landslide monitoring system is basically requiring experimental studies to test the sensors together with the data transmission from the sensor node to the base unit before deploy those sensors in the prone area. However, for more precise work, landslide mapping can be done before conducting any experiments. Landslide mapping helps to identify the exact area of susceptible landslide hence determine the right place to deploy sensor. A thorough investigation regarding landslide and its complex interaction on control factors such as lithology, elevation and slope are very beneficial in predict landslide both in space and in time [4]. Data transmission is actually a communication from sensor node to base unit. It lets the base unit to know when there is changes in the ground. Nowadays, wireless sensor node (WSN) has been widely used because of its capability to transmit data in real-time. Once a data is received at the base unit that indicates there is potential landslide, early warning system (EWS) plays a major role to alert the people to save their lives. EWS can be defined as to instantly spread meaningful warning information so that people nearby that are exposed to a hazard can prepare and act accordingly to reduce the possibility of danger or loss [5]. EWS is not just merely flashing red light on computer whenever a threshold value is exceeded. The people could be warned by SMS or a siren and when that happens, these people should know and understand what kind of hazard they are threaten to and how to act appropriately [5]. This alarm system is very important because this is the only way to alert the people in the prone area to that a natural hazard is about to happen and their lives are in danger. By the time the alarm is generated, the hazard is still in early phase and before it become catastrophic, the people could have enough time to save their lives and valuable belongings such as identification card, bank card, mobile phone also important documents. Since the focus of this paper is on the sensor itself, we will limit our discussion to it.

Type of sensors: Different types of sensors have been introduced for landslide detection in every research paper. The first type of sensor is called vibration sensor and it is used to sense any movement in the ground. Biansoongnern, Plungkang, & Susuk, (2016) propose to place SHOCK-801S which is a

IOP Conf. Series: Materials Science and Engineering **260** (2017) 012018 doi:10.1088/1757-899X/260/1/012018

vibration sensor in a stainless steel pipe [6] . The pipe then is buried in the ground together with the sensor. The utilization of this pipe can be the solution to prevent the sensor from broken during the landslide phenomenon. On the other hand, Woo [2] claims that landslide can be detected by ADXL 202 biaxial accelerometer which measures both inclination angle and acceleration of ground movement. Besides that, Woo (2016) has conducted several experiments in order to confirm the ability of this sensor to detect any motion in the ground [2]. Figure 1 shows different types of sensors.

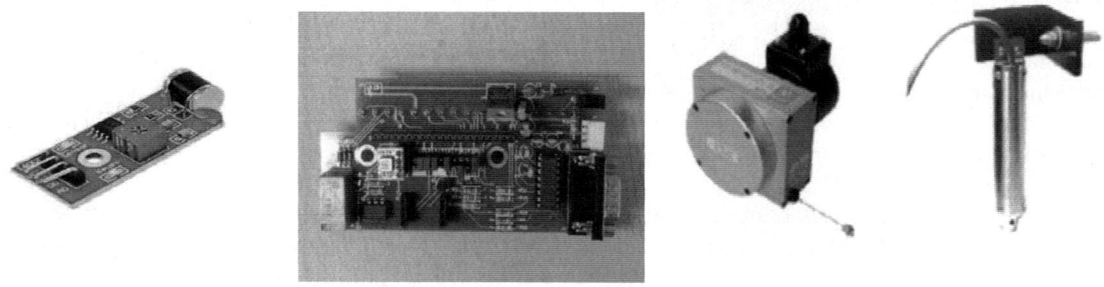

Figure 1. From left to right: Vibration Sensor (SHOCK-801S) [6], Accelerometer sensor (ADXL 202 biaxial accelerometer) [2], Extensometer and Tilt Sensor.

However, extensometer can be used to measure the slope surface displacement while inclinometer can be used to measure the movement of the slope. Above all, for slow and sudden ground movement, tiltmeter is the best choice [7,1]. Apart from gravity, rainfall also could trigger any ground movement or landslide [6]. Landslide rainfall-induced is caused by either high intensity rainfall or prolong rainfall with medium intensity [1] . For this reason, other parameters that are influenced by ground movement such as pore water pressure, rainfall intensity, soil pressure and surface displacement are considered in landslide monitoring system [7]. As suggested by [1,7], sensors like osmometer, dielectric moisture sensor, pore pressure piezometer, strain gauge and rain gauge are very useful to measure the parameters that is to say pore pressure and moisture content of the soil.

Table 1. Sensors and their functions.

Sensor	Function
Vibration	To detect debris vibration in the ground.
Accelerometer	To obtain the acceleration of the ground movement.
Extensometer	To measure the displacement of the slope.
Inclinometer	To measure vertically and horizontally the movement of the slope.
Tiltmeter	To measure slow and sudden movement of soil layer
Osmometer	To measure pore water pressure.
Dielectric moisture sensor	To measure the permitivity or dielectric constant of the soil.
Pore pressure piezometer,	To measure pore water pressure.
Strain gauge	To measure the deflection of soil layer when the ground has a movement.
Rain gauge	To measure the accumulative of rainfall.
Geophone	To measure vibration.

3. Experiments and Methodology

IOP Conf. Series: Materials Science and Engineering **260** (2017) 012018 doi:10.1088/1757-899X/260/1/012018

The type of sensor used to detect ground movement is vibration sensor, SW-420 known as highly sensitive sensor. When there is no vibration, the output signal is low and the LED will not turn on. Whereas, when there is vibration the output signal is high and the LED will turn on. As shown in figure 2, the vibration module consists of the sensor itself, an LM 393 chip and a potentiometer. LM 393 chip is a comparator and it is used to detect vibration if it is exceeded the threshold. This threshold can be adjusted by the potentiometer. Turning the potentiometer in clockwise direction will decrease the sensitivity while anticlockwise direction will increase the sensitivity. A PVC pipe is used to place the sensor and several experiments were conducted to determine the optimum dimensions in terms of length and diameter in producing the highest vibration intensity. As for the real-time data transmission, it is very important to transmit data from the sensor to the base unit before this natural hazard become catastrophic. Therefore, wireless sensor network (WSN) is the best choice to transmit data from the sensor to the base unit instantly. Wi-Fi module, ESP 8266 is used to read the value from the sensor and to transmit the data to the Base Unit. An efficient alarm system is needed in order to alert the people nearby when there is potential landslide. Many lives and valuable properties could be saved if these people receive this alarm.

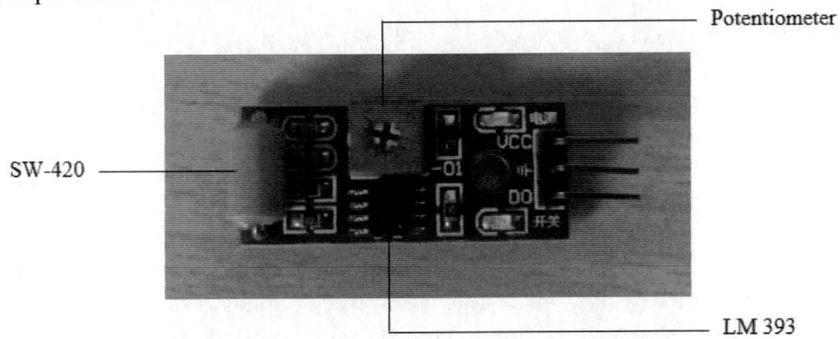

Figure 2. Vibration sensor module SW-420.

Off-the-shelf PVC pipes of 1 meter, 2 meters and 3 meters length and diameter (mm) of 15, 32, 50 and 80 for each length are used and tested which one is the optimum dimension for placing the sensor with some gravels before buried in the ground. When there is a movement in the ground, the gravels inside the pipe is moving as well and hit the inner wall of the pipe hence produce vibration thus sensed by the vibration sensor. Other than that, other advantage of using this pipe is, it can protect the sensor during the landslide so that it can continuously function during the landslide phenomenon. There are three different experiments were conducted. The first experiment was conducted to test the functionality of the sensor. The objective of this experiment is to test the functionality of the sensor. Suppose the sensor obtain a measurement when there is motion applied to the pipe and not record any measurement when there is no motion applied to the pipe. The second experiment was conducted to identify which position of pipe produce higher vibration measurement also to test if the gravel inside the pipe can increase the vibration measurement. The third experiment was conducted to find the optimum dimension of the pipe in terms of length and diameter. Components used in the experiments are: Vibration module sensor (SW-420), LED, Arduino UNO board, Jumper wire, PVC pipes, gravel and laptop.

Table 2. Dimensions of pipe.

Length (m)	Diameter (mm)			
1	15	32	50	80
2	15	32	50	80
3	15	32	50	80

IOP Conf. Series: Materials Science and Engineering **260** (2017) 012018 doi:10.1088/1757-899X/260/1/012018

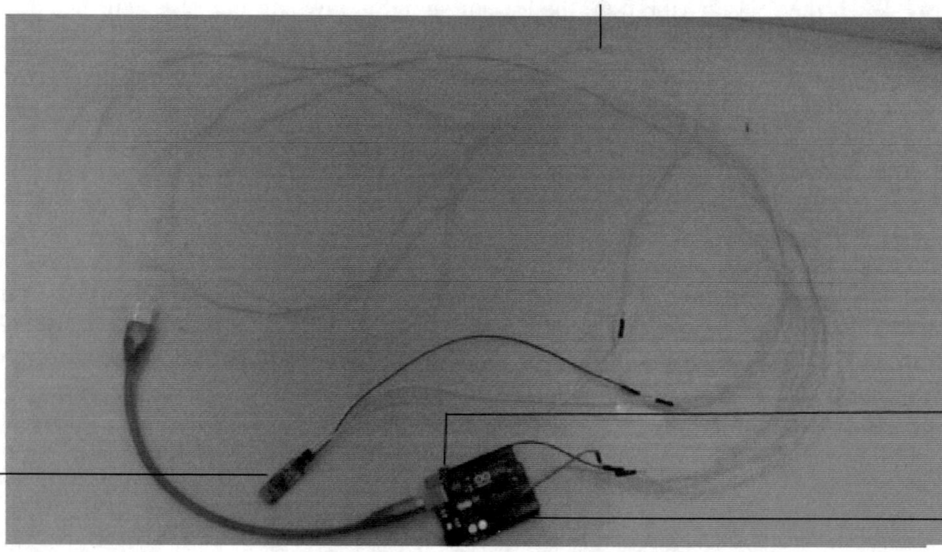

Figure 3. Components used in the experiment (SW-420, jumper, LED, Arduino Uno board.

In order to test the functionality of the sensor, the assembly is connected to the laptop and the sensor is placed inside the 2m pipe with diameter 15mm. The code is compiled and uploaded by using Arduino software. A serial monitor on Arduino software is clicked to obtained the measurement of the sensor and the measurement is recorded for ten seconds. Secondly, the vibration intensity was tested from the pipe with gravel and without gravel inside the pipe when the pipe is laid horizontally and vertically and for different pipe lengths and diameters.

Figure 4. The experiment set-up for Experiment 1 and Experiment 2 for pipe laid horizontally.

The vibration was induced vertically and horizontally by shaking the pipe in back/forth and up/down motions for ten seconds. To find the optimum dimension of pipe in terms of length and diameter, 10 pieces of gravels are put inside the different pipe to be tested to induce stronger acoustic vibration.

Figure 5. The experiment set-up for Experiment 2 for pipe held vertically.

4. Results and Discussion

All the measurement recorded were presented in graph form for better visualization and comparison.

Experiment 1: To test the functionality of the sensor.

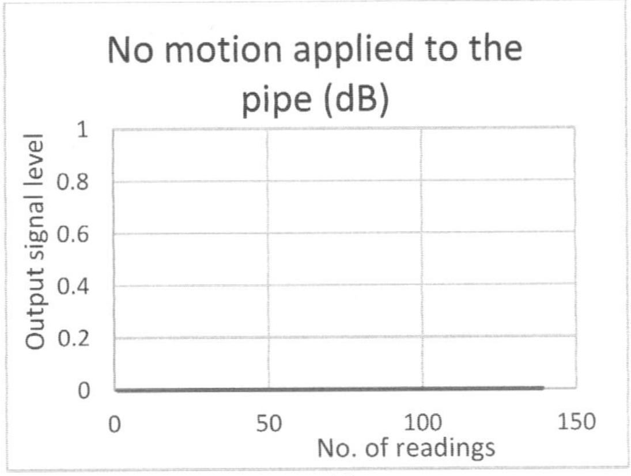

(a)

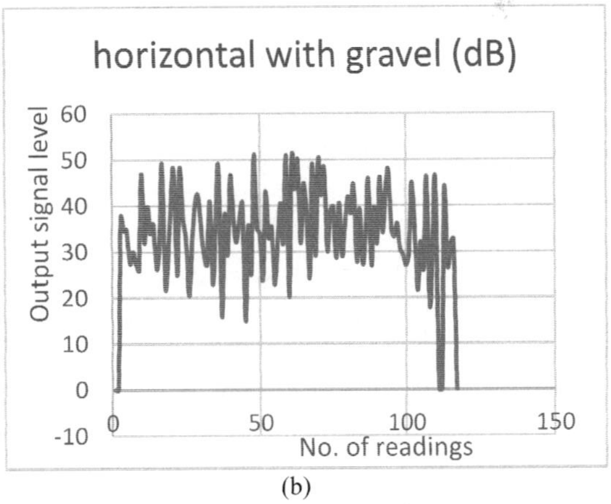

(b)

Graph 1. (a)Vibration measurement when no motion applied to the pipe and (b) vibration measurement when motion applied to the pipe.

Based on graph 1(a) there was no measurement recorded in Experiment 1 since there was no motion applied to the pipe. When there is motion applied to the pipe, there was measurements recorded as shown in graph 1(b). Therefore, it can be confirmed that this sensor is working perfectly hence it can be used for the next experiments.

Experiment 2: To test the vibration intensity with gravel and without gravel inside the pipe when the pipe is in horizontal and vertical position.

Table 3. Comparison of data for horizontal and vertical position of pipe.

HORIZONTAL		VERTICAL	
With gravel		With gravel	
Average	15096.709	Average	4208.340741
Variance	759911221	Variance	135802130.4
Standard deviation	27566.487	Standard deviation	11653.41711
Without gravel		Without gravel	
Average	8039.777	Average	1748.291667
Variance	305353399	Variance	20824466.71
Standard deviation	17474.364	Standard deviation	4563.383252
With gravel (dB)		With gravel (dB)	
Average	35.197631	Average	28.94987849
Variance	71.56194	Variance	67.12980137
Standard deviation	8.4594291	Standard deviation	8.193277816
Without gravel (dB)		Without gravel (dB)	
Average	31.997333	Average	24.80375801
Variance	89.095055	Variance	63.220448
Standard deviation	9.4390177	Standard deviation	7.951128725

While for Experiment 2, the value of mean, variance is obtained from those measurements as shown in table 3. From these values, it can be seen that the pipe with the gravels inside has higher value compared to pipe without gravels inside for both horizontal and vertical position of the pipe. This means, the gravels produced vibration as well when the pipe was moved hence increased the vibration intensity. For this reason, it can be confirmed that the gravels could increase the accuracy in detecting landslide movement. These measurements are important in setting the threshold value in which will determine the occurrence of ground motion.

The comparison between the horizontal vibration measurement and the vertical vibration measurement can be seen in table 3. It clearly shows that when the pipe is laid horizontally with the gravels inside the pipe produce the most vibration intensity. This is because, when the pipe is moved back and forth horizontally, the were reflection occurred between the gravels and the inner wall of the pipe. Less reflection occurred when the pipe is held vertically and is moved up and down as the gravels are moved up and down as well and not hit the wall much. Therefore, it can be concluded that this sensor can functioned perfectly when the pipe is laid horizontally and can be utilized in this project together with the gravels.

Experiment 3: To find the optimum dimension of pipe in terms of length and diameter.

Comparison of pipes in terms of length (m) and diameter (mm) based on average value from the output signal level.

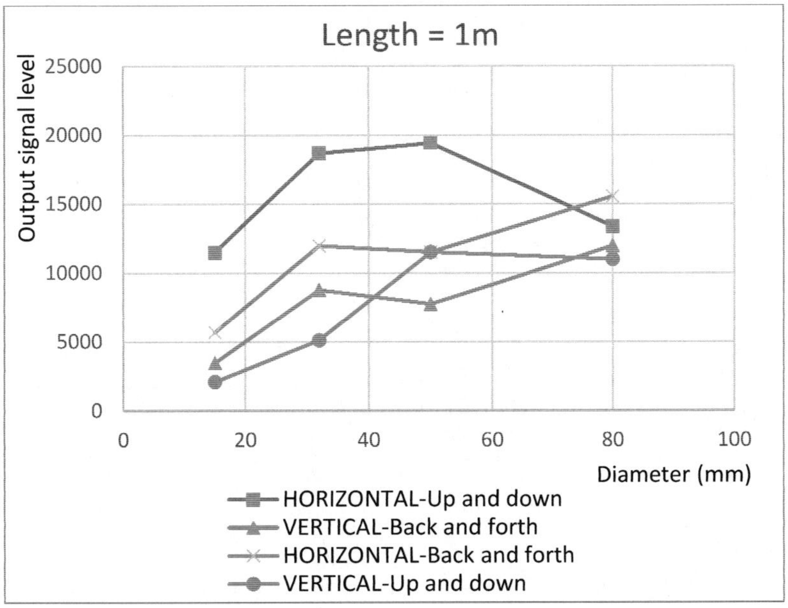
Length = 1m
Output signal level
25000
20000
15000
10000
5000
0
0
20
40
60
80
100
Diameter (mm)
HORIZONTAL-Up and down
VERTICAL-Back and forth
HORIZONTAL-Back and forth
VERTICAL-Up and down

(a)

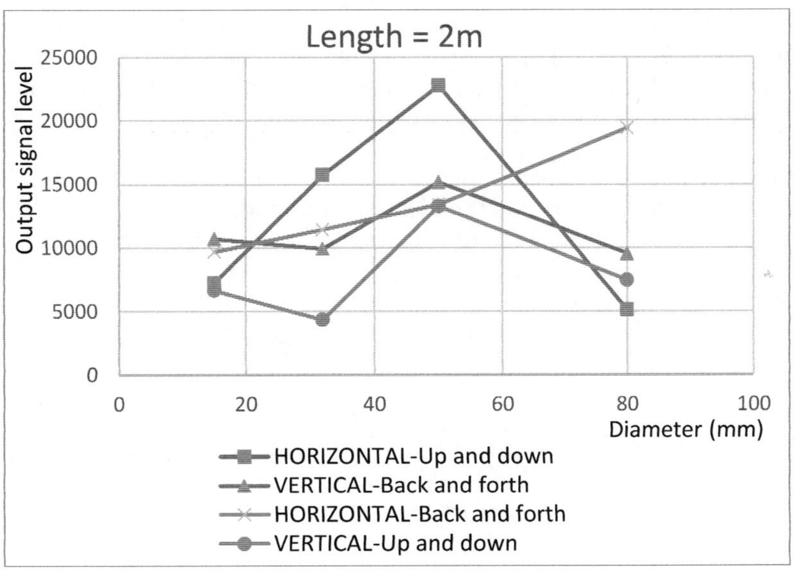
Length = 2m
Output signal level
25000
20000
15000
10000
5000
0
0
20
40
60
80
100
Diameter (mm)
HORIZONTAL-Up and down
VERTICAL-Back and forth
HORIZONTAL-Back and forth
VERTICAL-Up and down

(b)

IOP Conf. Series: Materials Science and Engineering **260** (2017) 012018 doi:10.1088/1757-899X/260/1/012018

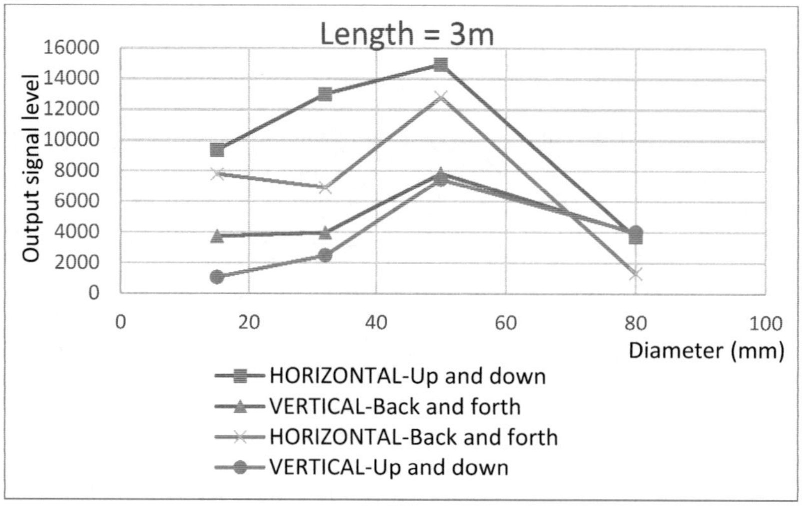

(c)

Graph 2. Vibration intensity for diameter 15mm, 32mm, 50mm and 80mm when pipe is (a) 1m length, (b) 2m length and (c) 3m length. Square indicates the pipe is laid horizontally and is moved up and down continuously. Triangle indicates the pipe is held vertically and is moved back and forth. Cross indicates the pipe is laid horizontally and is moved back and forth. Circle indicates the pipe is held vertically and is moved up and down.

Moreover, from Experiment 3, we can see in those graphs clearly that, when the pipe is laid horizontally, the vibration intensity is much higher compared to when the pipe is held vertically. In graph 2, it can be confirmed that the optimum dimension of the pipe is 2m length and 50mm diameter.

Table 4. Average values from dominant output signal level for (a) pipe length is 1m, (b) pipe length is 2m and (c) pipe length is 3m.

(a)

Diameter (mm)	HORIZONTAL-Up and down	VERTICAL-Back and forth
15	11477.15929	3447.957746
32	18674.85106	8761.093023
50	19415.27586	7725.948529
80	13349.20952	11968.59091

(b)

Diameter (mm)	HORIZONTAL-Up and down	VERTICAL-Back and forth
15	7213.244275	10701.80992
32	15707.84259	9907.015625
50	22741.28889	15162.71304
80	5122.768595	9528.75

(c)

Diameter (mm)	HORIZONTAL-Up and down	VERTICAL-Back and forth

15	9366.97479	3732.014184
32	13015.28462	3964.978102
50	14940	7846.96063
80	3697.046296	3956.823129

5. Conclusion

As can be seen, the first objective for this project has achieved successfully that is to investigate different types of sensors in detecting landslide. Through the reading of research papers had help a lot in understanding the concept of this project. Every research papers have introduced different approaches with different types of sensor. However, their goals are the same that is to detect any ground movement at early phase before the natural hazard become catastrophic and of course this is very important so that people can save their lives and valuable belongings. After conducted several experiments, the vibration measurements produced from vibration sensor SW-420 has clarify that, PVC pipe of length 2m and diameter 50mm with gravel inside the pipe (laid horizontally) is the most favourable. Hence, our future test-bed will use this information into the design consideration.

Acknowledgement

This work is partially sponsored by International Islamic University Malaysia Grant RIGS16-362-0526.

References

[1] Ramesh M 2014 Design, development, and deployment of a wireless sensor network for detection of landslides *Ad Hoc Networks* **13** 2-18.

[2] Kim H W 2016 Development of wireless sensor node for landslide detection *Proceedings of the Asia-Pacific Advanced Network* **42** 56-60.

[3] Mihali S, Arbanas Z and Krka M 2013 Comprehensive landslide monitoring system: The Konstanjek lansdslide, Croatia.

[4] Bayrak T, Eyo E, Basoglu N, Musa T and Akpee D 2015 Development of an alternative low-cost landslide monitoring method using data from tusaga-aktif gnss network *Boletim de Ciencias Geodesicas* **21(3)** 610-23.

[5] Thiebes B and Glade T 2016 Landslide early warning systems-fundamental concepts and innovative application *Landslides and Engineered Slopes: Experience, Theory and Practice* 12-19.

[6] Biansoongnern S, Plungkang B and Susuk S 2016 Development of low cost vibration sensor network for early warning system of landslides *Energy Procedia* **89** 417-20.

[7] Qiao G, Lu P, Scaioni M, Xu S, Tong X, Feng T, Wu H, Chen W, Tian Y, Wang W and Li R 2013 Landslide investigation with remote sensing and sensor network: from susceptibility mapping and scaled-down simulation towards in-situ sensor network design *Remote Sensing* **5(9)** 4319-46.

IOP Conf. Series: Materials Science and Engineering **260** (2017) 012019 doi:10.1088/1757-899X/260/1/012019

Miniaturized Water Flow and Level Monitoring System for Flood Disaster Early Warning

SALAMI IFEDAPO ABDULLAHI , MOHAMED HADI HABAEBI, TEDDY SURYA GUNAWAN AND MD RAFIQUL ISLAM

Dept of ECE, Fac. of Eng., International Islamic Univ. Malaysia (IIUM), Jalan Gombak, 53100 Kuala Lumpur, Malaysia.

abdusalife@gmail.com; habaebi@iium.edu.my; rafiq@iium.edu.my

Abstract: This study presents the performance of a prototype miniaturised water flow and water level monitoring sensor designed towards supporting flood disaster early warning systems. The design involved selection of sensors, coding to control the system mechanism, and automatic data logging and storage. During the design phase, the apparatus was constructed where all the components were assembled using locally sourced items. Subsequently, under controlled laboratory environment, the system was tested by running water through the inlet during which the flow rate and rising water levels are automatically recorded and stored in a database via Microsoft Excel using Coolterm software. The system is simulated such that the water level readings measured in centimeters is output in meters using a multiplicative of 10. A total number of 80 readings were analyzed to evaluate the performance of the system. The result shows that the system is sensitive to water level rise and yielded accurate measurement of water level. But, the flow rate fluctuates due to the manual water supply that produced inconsistent flow. It was also observed that the flow sensor has a duty cycle of 50% of operating time under normal condition which implies that the performance of the flow sensor is optimal.

KEY WORDS: *Flood Monitoring, Early Warning System.*

1. Introduction

Flooding is usually a result of overflow of river over the river banks due to high water rise contributed by the intense rainfall. Apart from rainfall, a number of factors are responsible for this phenomenon such as melting of ice at the North and South pole, tidal waves and seasonal wind and direction of flow [2]. According to [1,8], flood accounts for 40-50 percentage of global disaster with huge consequences on lives, economy and infrastructure. In the year 2014, Malaysia lost 21 lives to flooding and assets worth of 12.685 million US dollar [4,5]. Flood have resulted to a number of other geohazard disasters such as landslide and mudflow in Cameroon Highland and other locations in Malaysia leading to many lives been lost and properties destroyed. For instance in 2010, Kedah(Malaysia) an estimated 5615 residents were affected by flood and lost vast area of agricultural land[5].

Flood can hardly be controlled, but understanding flood prone areas and availability of timely information about water discharge especially during heavy rainfall can help mitigate the impact of this event[15]. A number of studies have been done by the geospatial engineering experts to map areas that are prone to flooding in different parts of the country [6,7,8,9]. Likewise, a number of systems have been put in place to monitor river water flow and level to provide early warning to adjoining communities on the imminence of the event [10,11,12,13]. Despite these efforts, the situation persist yearly with enormous social and economic distruption. Can we say the systems are not working? Far from it. The challenges with the existing systems is the fewer number of monitoring stations and the cost of installation. Available systems are heavy, costly and difficult to be deployed at any particular location of choice. Therefore, it is essential to have a miniaturised system that is less expensive, portable and capable of been deployed to any location along the river course.

The aim of this present study is to design a prototype water flow and level sensor and to test its performance under controlled laboratory environment. Experimental apparatus is designed with the water inlet positioned vertically which allows natural gravity pull water down the pipe. Since the gravity has a constant acceleration, the velocity of the water moving downward is also assumed to be constant. However, at the elbow joint the pressure is expected to drop because, naturally, elbow joints cause friction loss resulting to drop in pressure [18]. This is expected to increase the rate of water flow into the reservoir in reference to Bernoulli's principle which states that an increase in fluid flow only occurs when there is a decrease in pressure or potential energy [19]. The effect of frictional loss to elbow piping is not significant at this stage to affect the accuracy of the result. The success of this study will allow densification of existing systems and make data more available for precise flood prediction, adequate for early warning and disaster preparedness. The paper is organization starts with the materials and method section followed by the result and discussion section and finally the concluding section.

2. Materials and Method

2.1 Selection of microcontroller and sensor

The microncotroller used in this project is an Iteaduino, a predecessor of Arduino uno, which makes serial communication with the USB port of a computer. Arduino gets its power via the USB port or altrnatively it can be powered with Litium ion battery. Iteaduino has both digital and analog pins that enables it communicate with the sensors. The microcontroller powers the sensor with between 3.3 and 5 volts. Furthermore, it comes with its own integrated design environment (IDE) called Arduino IDE that allows programming codes and graphical sketches and their output to be read on serial monitor [17]. The microcontroller works with contact type water level sensor and Saier SEN-HZ21WA flow sensor. The sensor employs a float switch mechanism with a magnet inside the floating blob that detects the opening and closing circuit. For the flow sensor, three wires red, black and yellow, for deignated functions. The redwire connects the 5 volts power of the Arduino pin, the black to the ground, and the yellow connects the digital pin. This connection powers the sensor up and gets it ready for sketching and water flow rate measurement. The flow sensor utilizes a hall effect sensor that detects magnetic pulses installed on a fan blade inside the casings[17]. The flow sensor gets the reading of the flow rate passing through the inlet of the flow sensor using the expression in equation (1) [16].

$$Q = V \times A \tag{1}$$

where V is the volume flow, V is the velocity and A is the cross sectional area.

IOP Conf. Series: Materials Science and Engineering **260** (2017) 012019 doi:10.1088/1757-899X/260/1/012019

2.2 System design and laboratory setup

Functional code is the heart of the system; programming was done in Arduino IDE to record the flow and water level readings. The Digitalwrite Mode code reads the water height by observing whether the pinMode of the siensor it is high or low. An interrupt and digitalread code enables the flow sensor to sense the pinMode of the magnetic pulse designed to count in every 4.5 seconds that read and calulates the flow rate of the water. The code function in such a way that each float switch reads water level at the instant when the pin mode is high. Severity of rising level is indicated using different thresholds for water level and flow rate. The degree of severity is categorised as safe at the lowest flow and water level, caution level when it is averagely filled and dangerous when high water level and flow rate reach maximum.

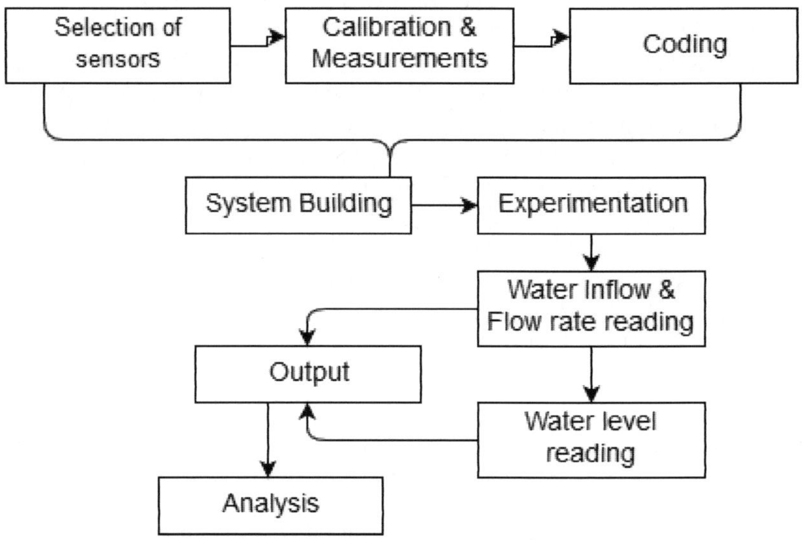

Figure 1:Workflow diagram

For the experimental setup, five float switches were installed on a PVC pipe at approximately 34 cm in length and 2.8 cm diameter (Figure 2). The laboratory experiment was conducted using a plastic container, hereafter called the reservoir. The reservoir is about 18 cm high with average diameter of 25 cm and has a water retaining capacity of 5.8 litres. Built in the reservoir is PVC pipe of about 2.8 cm. On the pipe is attached five float switch sensor arranged at 5 cm interval along the vertical direction to record the water level. Another 20 cm PVC pipe of the same dimeter was fixed some to the reservoir as water inlet. The inlet supplies water into the reservior at approximately 2 cm from its base. The flow sensor is placed right on-top of the water inlet beneath a handmade funnel to read the flow rate of the water coming into the reservoir.

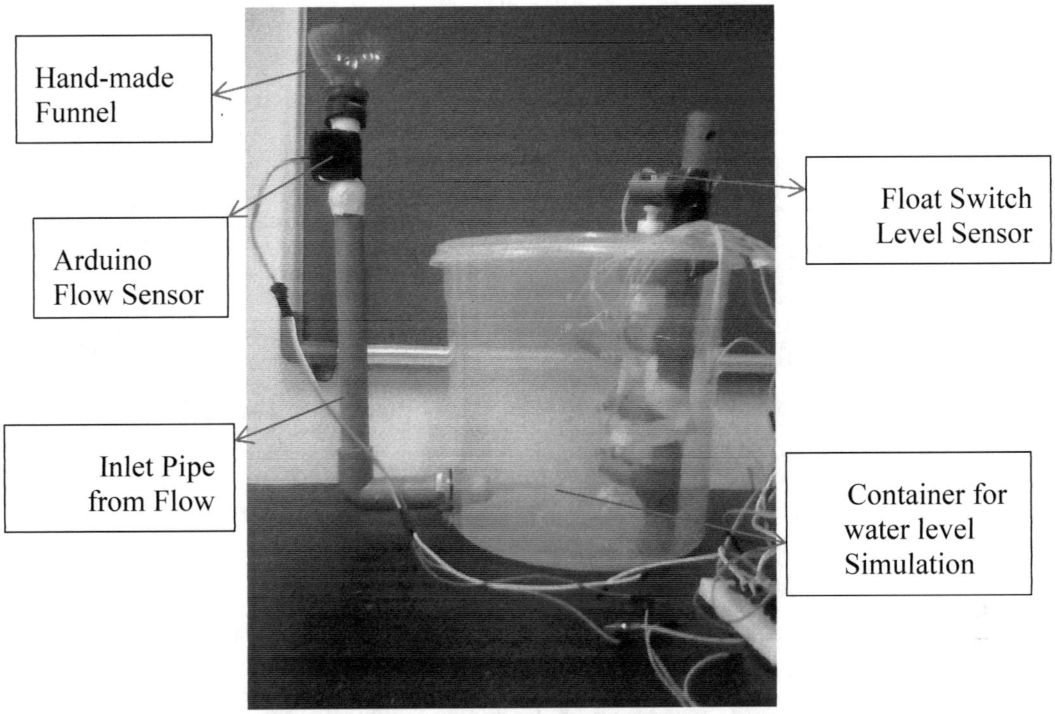

Figure 2: Display of Integrated System for Simulation

To compare the efficiency of the design above, a printed circuit board (PCB) (Figure 3) level sensor was also utilized. PCB is a simple process because the design involves making a drawing of the lines that represent the circuit. But rather than a loop circuit, linear and parallel circuit was designed in this project with adjacent lines decreasing by few centimeters. The steady decrease of the tracing lines is to reduce resistivity between each line and to allow reasonable amount of current to activate the PCB level sensor. The design was followed by printing of the sketch on a copper plate and subsequently a photo-platter was used to print the circuit onto the copper plate. After that, the circuit was flashed under flourescence for about 180 seconds to imprint it onto the PCB. The imprint was later etched to remove the unused copper plate. Finally, the ink used to cover the copper tracing was removed using acetone solution and all necessary wire connections soldered to their respective components for water level measurement.

Figure 3: Fabricated PCB Water Level Sensor

2.3 Flow and water level measurement

The sensor has been pre-calibrated to measure between 1litre to 30 litres per minutes; however, for accuracy the PCB water level was calibrated using the microcontroller and a DC multimeter. The DC multimetre measured the voltage difference between the reference PCB and the copper tracing to obtain current for each copper tracing. The amount of input voltage needed to be decided so that sufficient current can be generated to activate the transistors on the signal conditioning circuit (Figure 4). The wire from the tracing copper was connected to the analog pins of the Arduino to get the

IOP Conf. Series: Materials Science and Engineering **260** (2017) 012019 doi:10.1088/1757-899X/260/1/012019

calibration of the PCB water sensor ready.The analogRead code reads the voltage value that is detected for each copper tracing. This voltage readings are recorded and used to set boundaries to print out designated water level at the instant the rising water touches the tracing copper.

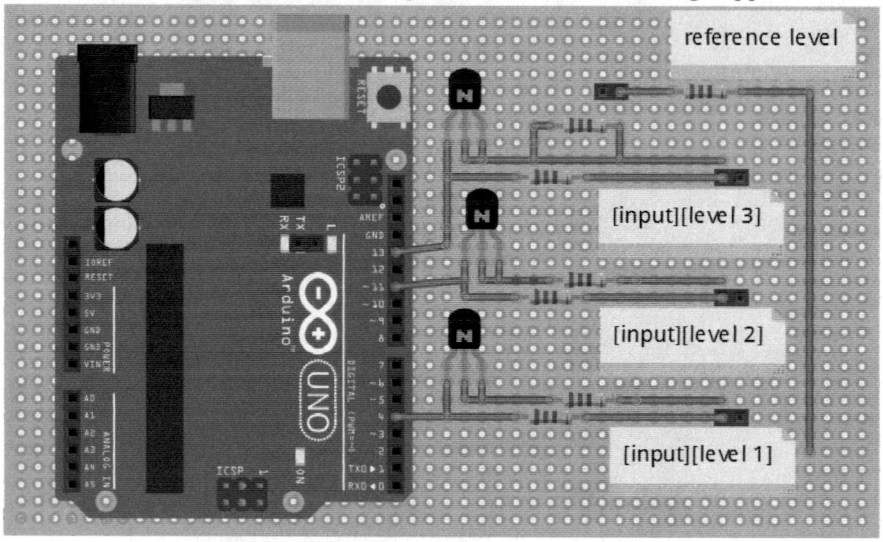

Figure 4: Signal conditioning circuit diagram for PCB level sensor

The system was tested in the laboratory by taking experimental readings. This was done by manually supplying water into the reservoir through the funnel. As the water passes throught the flow sensor, the instrument record the flow rate. Similarly, the water level sensors record the rising water level at each sensor position. The result to the output water level was measured was multiply by 10 to assume our readings as metre. Both the flow rate and water level readings are automatically logged into Microsoft Excel using Coolterm software as numerical values. A total number of 80 flow rate reading were taken and their corresponding water levels observed continuously untill the reservior was filled up. Meanwhile, the water was passed through the inlet pipe using a glass baker. Two experiments were conducted with each taking approximately 10 minutes. The flow sensor shows a quick response to the rate of flow of water. The total time delay between consequetive flow rate reading is approximately 1ms. The flow rate increases as water rushes through the water inlet and enters the opening of the flow sensor. The same concept exists from the simulation that was done in the Matlab Simulink which shows motor response when there is a voltage applied across it. The mechanism works with the movement of the fan blade which creates a voltage surge and caused transient reponse when the fan blade rotates. Some basic statistical analyses were done to evaluate the result obtained.

3. Results and Discussion

Sensitivity of each line of the PCB copper tracing representing water levels was evaluated by comparing the voltage output of each line (Figure 5a). The result shows unstable pattern with average voltage of 3.6V. Between water level of 0 m and 1.5 m, voltage reading displays an undulating wave-like graph that rises steadily between each level around 3.59 – 3.65V. The voltage pattern changed between water level 1.5 m and 2 m dropping to approximately 3.15V. This happens because of insufficient voltage received by the signal conditioning circuit at that level and did not pass through a filter that can regulate the voltage readings and assign a specific voltage to each copper tracing [15]. This indicates that float switch is much better for water level sensor than PCB level sensor because it produces consistent output, usually converting the voltage to 0s and 1s (Figure 5b). In addition, it has a simpler circuit construction.

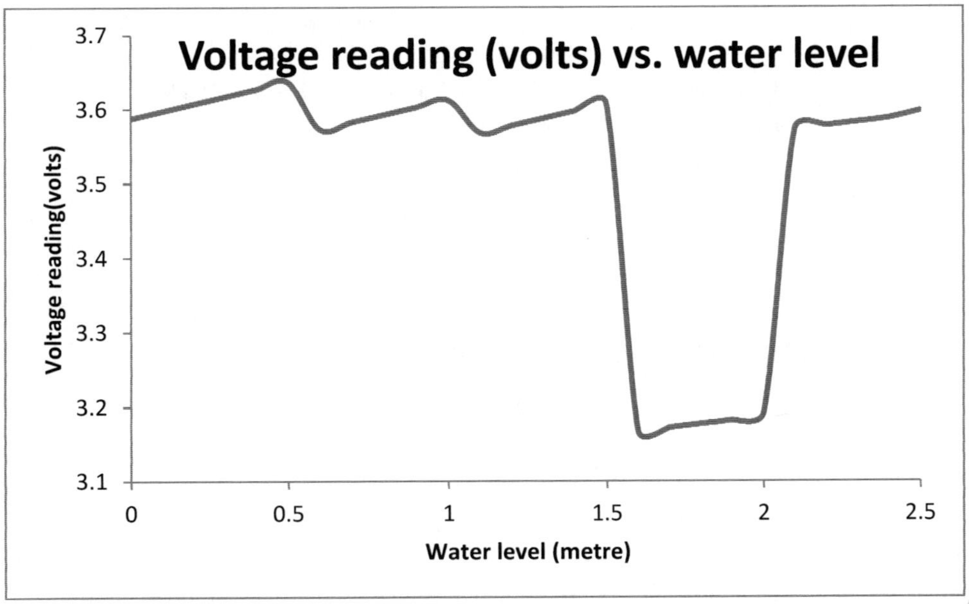

Figure 5a: Graph of Sensitivity of the PCB Water Level Sensor

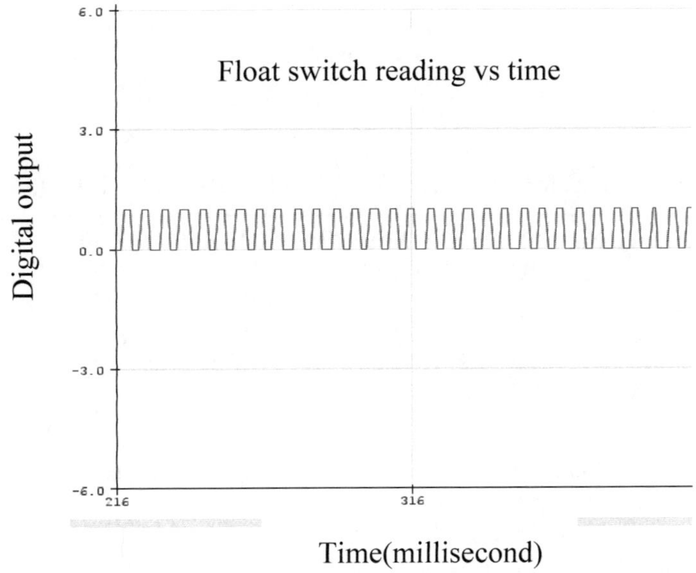

Figure 5b: Float Switch Reading from Serial Plotter

The flow rate and the corresponsing water level readings were automatically recorded and stored as a database in excel (Table 1). It can be observed that the system record water level reading in a decreasing order of frequency as the water level rises. The descriptive statistics of the readings is presented in Table 2.

IOP Conf. Series: Materials Science and Engineering **260** (2017) 012019 doi:10.1088/1757-899X/260/1/012019

Table 1: experimental recorded data

Flow rate	**Water level**	**Flow rate**	**Water level**	**Flow rate**	**Water level**	**Flow rate**	**Water level**	**Flow rate**	**Water level**
1	0.5	5	0.5	5	1	4	1.5	6	2
6	0.5	5	0.5	4	1	6	1.5	4	2
6	0.5	6	0.5	6	1	4	1.5	7	2
6	0.5	6	0.5	6	1	5	1.5	6	2
6	0.5	6	0.5	7	1	7	1.5	6	2
6	0.5	5	0.5	7	1	7	1.5	7	2
6	0.5	6	0.5	3	1	7	1.5	3	2
7	0.5	7	0.5	4	1	7	1.5	5	2
7	0.5	7	0.5	5	1	7	1.5	6	2
7	0.5	3	0.5	4	1	7	1.5	3	2
7	0.5	4	0.5	5	1	7	1.5	5	2
6	0.5	4	0.5	7	1	3	1.5	4	2
6	0.5	3	0.5	6	1	6	1.5		
1	0.5	4	0.5	7	1	6	1.5		
1	0.5	5	0.5	7	1	6	1.5		
6	0.5	6	1	6	1.5	6	2		
5	0.5	4	1	5	1.5	5	2		

Table 2: Quantitative Analysis of the Flow Rate and Water Level

Variable	**Obsn**	**Min**	**Max**	**Mean**	**Std. dev**
Flow Rate	80	1	7	5.400	1.498
Water level	80	0.5	2	1.081	0.571

XLSTAT, an add on to Microsoft Excel, was used to to produce the graph of flow rate against water level and flow rate and water level against time was plotted from the readings (Figure 6). The water level steadilly increases with time while the flow rate fluctuates similar to the study of [10].The flow rate produce unstable wave because the water was manually discharged through the inlet. Although this does not affect the level reading, but it certainly makes the consistency difficult to valuate since water discharge, volume and rate of discharge account for different stages of flooding [10].

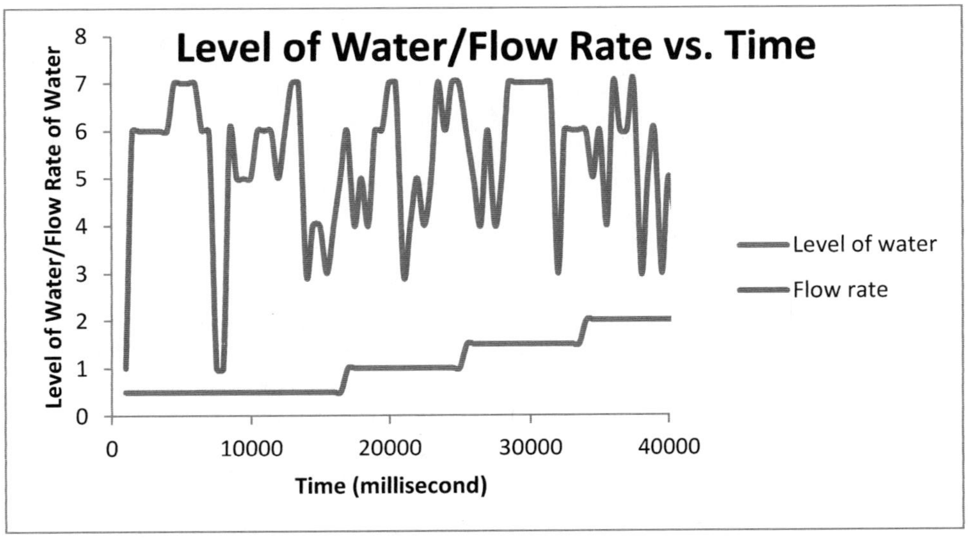

Figure 6: Flow Rate and Water Level vs. Time

Linear regression and correlation of the flow rate and water level were examined (Table 3). The Pearson correlation matrix produced a value of 0.073 which shows relativity of independent to each other; however, the closer the value to 1 the stronger relationship. Similarly, the coefficient of determination of this test yields R^2 of 0.005 indicative of how closely the points are fitted. The inconsistency arises from the manual water supply during the experiment and insufficient variablility of the experimentation process.

Table 3: Determination of statistical relationship

Correlation Matrix			Co-efficiency of Determination(R^2)	
Variables	Water level	Flow Rate	Water level	Flow Rate
Water level	**1**	0.073	**1**	0.005
Flow Rate	0.073	**1**	0.005	**1**

4. Conclusion

This study presents the initial stage of the design of a miniaturized prototype of a water level sensor aimed at providing an integrative early warning system for flood emergency response. The system, under laboratory experiment, produces a promising system that can efficiently measure water flow rate and water level rise. However, several other factors are still required to ascertain its capability in real-life situation. For instance, the size of the experimental setup and manual supply of water affects the accuracy and consistency of the result obtained. Upon completion of this study, it is expected that causality arising from flood disaster will be reduced to minimum in Malaysia and elsewhere because of the density of installation and the volume of information arising therefrom for precise evaluation and analysis. Future work will use controlled water input mechanism by varying the pressure of the inlet to assess the consistency. Besides, the system will be tested under different conditions, including real-life deployment in an experimental natural flowing river. For outdoor useage, the system will be mounted on fixed poles and the sensors firmly strapped to them to avoid being washed away by water current.

ACKNOWLEDGEMENT

The authors thank Mr Mohammed Oludare Idrees of the Geospatial Information Research Centre, Universiti Putra Malaysia for his contributions and guidance in providing the idea of this project, and helping with the manuscript editing and proof reading. This work is partially sponsored by International Islamic University Malaysia Grant RIGS16-362-0526.

References

[1] Shaari, M. S. M., Karim, M. Z. A., & Hasan-Basri, B. (2016). Flood disaster and GDP growth in Malaysia. *European Journal of Business and Social Sciences*, *4*(10), 27-40.

[2] Subramaniam, S. K., VIGNESWARA, R. G., Subramonian, S., & Hamidon, A. H. (2010). Flood level indicator and risk warning system for remote location monitoring using Flood Observatory System. *Universiti Teknikal Malaysia Melaka*, 160.

[3] Shafiai, S., & Khalid, M. S. (2016). Flood Disaster Management in Malaysia: A Review of Issues of Flood Disaster Relief During and Post-Disaster. *The European Proceedings of Social & Behavioural Sciences EpSBS.*

[4] Mokhtar, E. S., Pradhan, B., Ghazali, A. H., & Shafri, H. Z. M. (2017). Comparative assessment of water surface level using different discharge prediction models. *Natural Hazards*, *87*(2), 1125-1146.

[5] *"Floods kill 21 in Malaysia, waters recede". Agence France-Presse. AsiaOne. 31 December 2014. Retrieved 31 December 2014.*

[6] Pradhan, B., Abokharima, M. H., Jebur, M. N., & Tehrany, M. S. (2014). Land subsidence susceptibility mapping at Kinta Valley (Malaysia) using the evidential belief function model in GIS. *Natural hazards*, *73*(2), 1019-1042.

[7] Pradhan, B. (2010). Flood susceptible mapping and risk area delineation using logistic regression, GIS and remote sensing. *Journal of Spatial Hydrology*, *9*(2).

[8] Tehrany, M. S., Pradhan, B., & Jebur, M. N. (2014). Flood susceptibility mapping using a novel ensemble weights-of-evidence and support vector machine models in GIS. *Journal of Hydrology*, *512*, 332-343.

[9] Kia, M. B., Pirasteh, S., Pradhan, B., Mahmud, A. R., Sulaiman, W. N. A., & Moradi, A. (2012). An artificial neural network model for flood simulation using GIS: Johor River Basin, Malaysia. *Environmental Earth Sciences*, *67*(1), 251-264.

[10] Bolshakov, V. (2013). Regression-based Daugava River Flood Forecasting and Monitoring. *Information Technology and Management Science*, *16*(1), 137-142.

[11] Hashim, N. M. Z., Hamdan, N. B., Zakaria, Z., Hamzah, R. A., & Salleh, A. (2013). Flood Detector Emergency Warning System. *International Journal Of Engineering And Computer Science (IJECS)*, *2*(8), 2332-2336.

[12] Sunkpho, J., & Ootamakorn, C. (2011). Real-time flood monitoring and warning system. *Sonklanakarin Journal of Science and Technology*, *33*(2), 227.

[13] Basha, E. A., Ravela, S., & Rus, D. (2008, November). Model-based monitoring for early warning flood detection. In *Proceedings of the 6th ACM conference on Embedded network sensor systems* (pp. 295-308). ACM.

[14] Roy, J. K., Gupta, D., & Goswami, S. (2012, December). An improved flood warning system using WSN and Artificial Neural Network. In *India Conference (INDICON), 2012 Annual IEEE* (pp. 770-774). IEEE.

[15] Reverter, F., Li, X., & Meijer, G. C. (2007). Liquid-level measurement system based on a remote grounded capacitive sensor. *Sensors and Actuators A: Physical*, *138*(1), 1-8.

[16] Suresh, N., Balaji, E., Anto, K. J., & Jenith, J. (2014). Raspberry Pi Based Liquid Flow Monitoring And Control. *International Journal of Research in Engineering and Technology*, *3*(07).

[17] Waghmare, A., & Naik, A. A. (2015, November). Water velocity measurement using contact and Non-contact type sensor. In *Communication, Control and Intelligent Systems (CCIS), 2015* (pp. 334-338). IEEE.

[18] Ma, Z. W., & Zhang, P. (2012). Pressure drops and loss coefficients of a phase change material slurry in pipe fittings. *international journal of refrigeration*, *35*(4), 992-1002.

[19] Bashiri-Atrabi, H., Hosoda, T., & Shirai, H. (2016). Propagation of an air-water interface from pressurized to free-surface flow in a circular pipe. *Journal of Hydraulic Engineering*, *142*(12), 04016055.

The Effect of Electrical Stimulation in Improving Muscle Tone (Clinical)

M F Azman* and A W Azman

Department of Electrical and Computer Engineering,
Kulliyyah of Engineering, International Islamic University Malaysia, Kuala Lumpur, Malaysia
*paan4256@gmail.com, amy@iium.edu.my

Abstract. Electrical stimulation (ES) and also known as neuromuscular electrical stimulation (NMES) and transcutaneous electrical stimulation (TES) involves the use of electrical current to stimulate the nerves or nerve endings that innervate muscle beneath the skin. Electrical stimulation may be applied superficially on the skin (transcutaneously) or directly into a muscle or muscles (intramuscularly) for the primary purpose of enhancing muscle function. The basic theoretical premise is that if the peripheral nerve can be stimulated, the resulting excitation impulse will be transmitted along the nerve to the motor endplates in the muscle, producing a muscle contraction. In this work, the effect of mere electrical stimulation to the muscle bulk and strength are tested. This paper explains how electrical stimulation can affect the muscle bulk, muscle size, muscle tone, muscle atrophy and muscle strength. The experiment and data collection are performed on 5 subjects and the results obtained are analyzed. This research aims to understand the full potential of electrical stimulation and identifying its possible benefits or disadvantages to the muscle properties. The results indicated that electrical stimulation alone able to improve muscle properties but with certain limits and precautions which might be useful in rehabilitation programme.

Keywords: ES, Muscle Atrophy, Muscle Contraction

1. Introduction

Electrical muscle stimulation (EMS) or otherwise named as neuromuscular electrical stimulation (NMES) is the elicitation of muscle compression using electrical driving forces or impulses [1]. Transcutaneous electrical nerve stimulation (TENS) focus the electrical stimulation on nerve rather than muscle. While TENS is focus on relieving chronic and acute pain [2], EMS has a different role which will be discussed further later in this paper.

People who involved in accidents or serious illness sometimes led to broken limbs or bedridden due to comma. These types of injuries usually associated with immobility can last for months. Prolong immobility will lead to muscle atrophy, which is the state where the muscle lost its mass, size and its functions degrades, which can be seen in Fig. 1. Thus, rehabilitation process is needed where a specialized healthcare is perform to improve or restore mobility, physical strength and cognition

[3]. From the literature, there are many published work that indicate electrical stimulation to play a major role as well as becoming an essential rehabilitation tool to give back patients the control of their muscle function through Functional Electrical Stimulation (FES) and improve their mobility [4]. A normal healing process without electrical stimulation intervention which normally takes almost a year can now be done in average of three month with the aid of electrical stimulation.

However a certain degree of effectiveness of electrical stimulation in rehabilitation or healing process directly related to muscle still remains unclear. The most significant limitation is that a rehabilitation process with electrical muscle stimulation (EMS) must come with functional task to ensure the effectiveness of the therapy. In this paper, the effect of EMS on muscle tone, size and strength was examined to find out the limits and extend of EMS in improving all of these parameter.

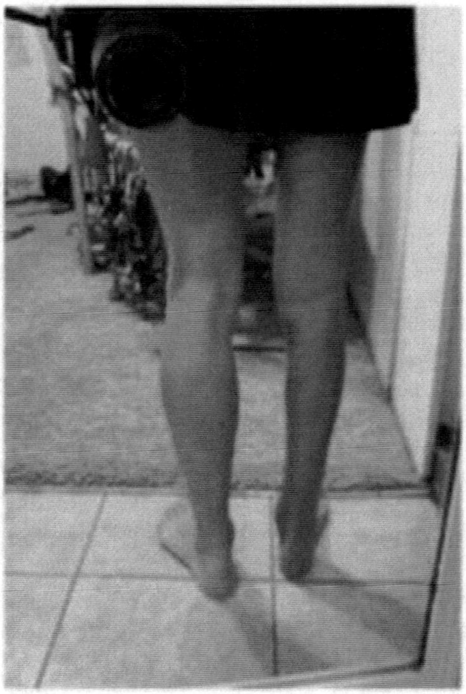

Fig. 1: Example of muscle atrophy patient. The right leg seems smaller than the left one. Retrieved from https://www.slideshare.net/hallerbakk/the-muscular-system-38480382

IOP Conf. Series: Materials Science and Engineering **260** (2017) 012020 doi:10.1088/1757-899X/260/1/012020

2. Related Work

In utilizing the EMS, the frequency, pulse width, ramp time, duty cylcle and amplitude are among the vital parameters of EMS that will be reviewed in this research work. Frequency refers as the pulses produced per second during stimulation and in the units of Hertz. Most clinical regiment's uses frequency between 20 to 50Hz to obtain optimum results, although, it also differs accordingly to the intervention, intention or objectives at that time [5]. Higher frequencies have proven to be more comfortable because the force response is smoothed and only given out tingling effect, whereas, lower frequencies give out a tapping effect which distinguished the individual pulses [6]. The pulses produce by the electrical stimulator device usually are represented by a geometrical shape either square, sine wave or peaked wave. The time taken for each single pulse is known as pulse duration or pulse width, the example is shown in Fig. 2. Besides pulse width, another time variable parameter is also consider vital that is the ramp time. It is refers as the time taken for the stimulation to turned on and reach the desirable frequency depending to the actual onset [7]. This 1 to 3 seconds time period is essential when a subjects muscle tone increase to create resistance when stimulation occur.

This time variable is define as the ratio of time taken for each stimulation against the rest time. A ratio of 1:2 means that it on for 10 seconds and off for another 20 seconds. The purpose of this duty cycle is demonstrated where the muscle tissue was able to recover more faster and produce greater torque when the periods of force development were interrupted with silent periods [8]. Usually a duty cycle of 1:3 is considered standard for clinical purpose. This however, can still be change to accommodate specific patient's conditions. Measured in ampere, this is the value of the maximum current being administrated during the stimulation. A higher amplitude of EMS will activates a larger number of muscle fibers for contraction force to occur, however overdosing it will caused antidromic transmission which is when the both motor and sensory impulses are block to travel to the spinal motor pool causing less activation of the central nervous system [9].

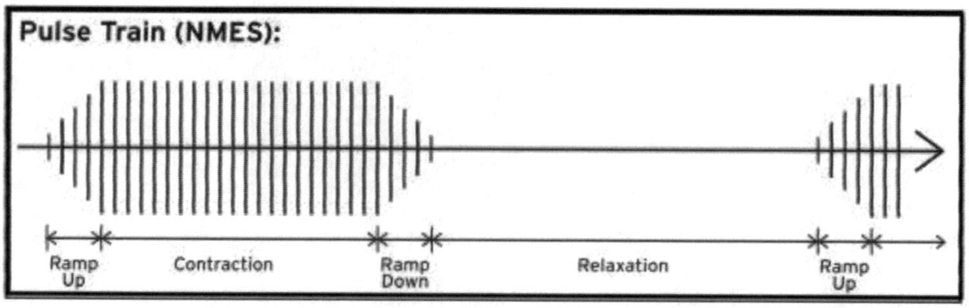

Fig. 2: Pulse Train of EMS/NMES. Retrieved from https://lh3.googleusercontent.com/0X44ujeA2pcv906ljv8lQYf2QHv7lmhL5kw3OF6gbOyCG7abWa-a18tdiIBOw-t4q7E9=s170

For a home based therapy purposes, either for rehabilitation, chiropractic or even physiotherapy, a simple and small portable stimulator device which is normally battery operated is used. A transcutaneous surface electrodes that adhere to the skin is used to deliver the current from the device to the muscle. The standard pre-gelled electrode's has adhesive to ease the removal and application of the electrode to the skin. It is important to highlight it here that the outcomes can be dependable to the quality of the skin-electrode interface and the consistency of the electrode placement in the tested area [1]. Moreover, the repeatability of the placement spot of the electrode can affect the outcomes of the stimulation.

In NMES therapy, a suggested electrode placement guides or charts are usually provided by the manufacturer, included with the purchase of the device. This is for getting the optimum result and utilizing the positive and negative terminal of the stimulator device. In a research, the larger the surface area of the electrodes, the more of the muscle tissue will be activated, though it will also disperse the current in a wider surface area, making the current density decreasing. Whereas a smaller electrodes will focus more current densities, giving a full focal concentration of current which lessen the chance of the stimulation to migrate to unwanted nearby muscle, however this dense current can also increase feeling of discomfort or even worse, pain [10]. It is also advice to not share the same adhesive electrodes with other test subjects for hygiene purpose and keep in mind the repeatability of using the electrodes which normally last for 5 to 20 times of use, according to the material use for the electrodes.

The use of electrical stimulation (ES) as muscle atrophy rehabilitation intervention had been discuss in various research works. Johnsen et al. [2] proposed integrating surface electrode into a modular polymer cast design to introduce EMS to the forearm muscle groups without removing the cast for patient undergoing cast immobilization due to forearm bone fracture. Their target patient for the therapy regiment were the patients who had been cast for 3-4 weeks, where they introduce their research intervention up till the cast was completely remove.

A research from both by Ali & Qin [11] and Dupont Salter et al. [12] focus on using animal model for their works. Ali & Qin [11] approach the use of different frequencies ranging from 1 to 20 Hz on a rat model. While Dupont Salter [12] works also involved a rat model, they approach by introducing a lower frequency at a range of 2 to 10 pulses per second.

In Lo et al. works [13], a short-term propulsion of ES was introduced accompanied by a functional leg-cycling task to patient with stroke. They hypothesized that this intervention can reduce the spasticity of the affected leg muscle. Another work by Keller et al. [14] use a novel approach of using a multi-channel stimulation which perform a real-time spatial and temporal variations of the electrical current density on the surface and deep tissue layer of the skin. This approach was hypothesized to generate a better muscle selectivity and improved muscle activation patterns.

The result and discussion of every work and research mentioned above show a positive feedback but still bound to several limitations and challenges. In the work by Johnsen et al. [2], a throughly safety measures must be planned for their prototype. The use of a microcontroller, and some hardware and software programming introduce a complex intervention. A simple mistake in the coding part will render the the entire system inoperable. Nevertheless, this intervention is a good progress in rehabilitation engineering as it can reduce the post-recovery time necessary to restore normal strength and motor function of the affected muscle beside preventing muscle

IOP Conf. Series: Materials Science and Engineering **260** (2017) 012020 doi:10.1088/1757-899X/260/1/012020

atrophy. The research by Ali & Qin [11] and Dupont Salter et al. [12] show a contradiction in their result. Ali & Qin [11] stated that the use of 20 Hz give the highest percentage of soleus muscle mass and the trabecular bone volume recovery of the animal model. Thus, suggested that this frequency is the most optimum frequency for the therapeutic treatment for disuse induced muscle atrophy. Dupont Salter [12] works on the other hand stated that a frequency of 2 pulses per second is better than higher-frequency stimulation for the level of atrophic changes the muscle experienced. The 2 pulses per second frequency give the least percentage of mean atrophy in the slow soleus muscle.

Lo et al. [13] works although give a promising results, the research introduced a functional task using their leg-cycle wheelchair prototype. This prototype will be costly and need supervision by the expert and not a do-it-yourself intervention for the patient that undergo home-based rehabilitation. The effect also varied according to the muscle tone of the patient which is the higher the tone of the muscle, the lower the spasticity level of the muscle was reduced. Although, this intervention promise the quickest therapy regiment with high effectivity. The multi-channel stimulation approach by Keller et al. [14] shows a more throughly intervention involving the integration of stimulation hardware into a garment sleeve. The experiment on the patient with stroke shows a positive outcome as they can generate selective finger and wrist extension movement in the affected area. The challenges will be once again the integration of this new intervention to ease the use, stimulation and wear comfort. Nevertheless, it can be a better intervention that the art ES system which operate with predetermined electrode positions.

3. Design and Methodology

A normal human body can only withstand a certain amount of electrical current before the body cells are damaged. At 100mA, a severe contraction of muscle is occurring and between 100 to 300mA of current causes electrocution to the body [15]. Therefore, a typical EMS stimulator produce 9 to 15mA according to the resistor used. The main core of the work here is the electrical stimulation. The area tested is the upper limb of the hand, focusing on the 3 main muscle that is the biceps, triceps and deltoid, as shown in Fig. 3. For this work, the left, undominant hand is undergoing the ES treatment, while the right, dominant hand is kept constant as control variable. Thus, a total of 5 healthy subjects of a college-aged (21-24 years old), weight around 50 to 60kg and did not commit to any heavy exercise routine, especially in the upper body section for the past 3 month before the therapy is done voluntarily participated in this research. Prior to the experiment, all of the test subjects have signed the ethical as well as the consent form. The parameters of the tested muscle that are focus on this work is the muscle tone, muscle size and muscle strength.

IOP Conf. Series: Materials Science and Engineering **260** (2017) 012020 doi:10.1088/1757-899X/260/1/012020

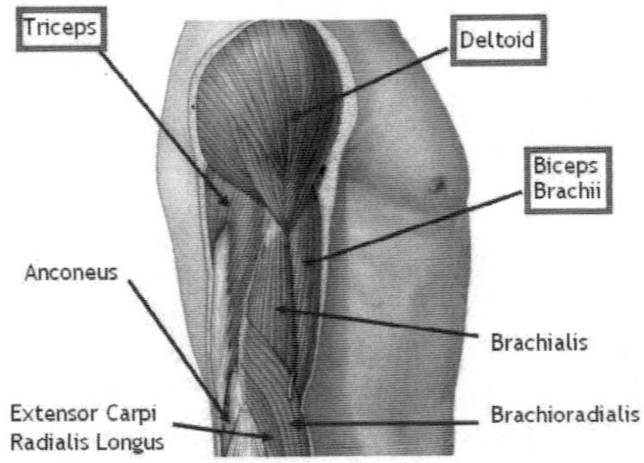

Fig. 3: Muscle anatomy (only biceps, triceps and deltoids were tested). Retrieved from http://thebettyrocker.com/sexy-arms-5-great-moves-for-your-biceps-and-triceps-no-equipment-required/

Fig. 4: The MediStim XP. Retrieved from http://www.neurotechgroup.com/us/products/avivastim-xp

The EMS stimulator used in this work is the MediStim XP from the Neurotech as shown in Fig. 4, commonly used by medical expertise for rehabilitation. It offers up to 9 pre-set programs according to the condition and indication needed, from rehabilitation process to acute pain management. The specification of this stimulator is shown in Table. 1. For the therapy work, the program number 2 was chosen as the parameters listed in Table. 2 is the same as the one needed for this work.

Table 1: The specification of MediStim XP

Nominal output voltage / power			
Parameter	500Ω	1kΩ	1.5kΩ
Output RMS voltage (RMSV)	7.5 V	12.32 V	13.74 V
Output RMS current (RMSA)	15 mA	12.3 mA	9.16 mA
Output frequency	4 – 99 Hz	4 – 99 Hz	4 – 99 Hz
DC Component	0 C	0 C	0 C
Pulse width	80 – 400 μs	80 – 400 μs	80 – 400 μs

Table 2: The parameter of program no. 2 of the MediStim XP

Program	Frequency (Hz)	Pulse Width (μs)	Ramp Up (s)	Ramp Down (s)	On time	Off time	Burst or Trigger
1	50	300	1	1	5	5	Trigger
2	50	300	1	1	5	10	Trigger
3	50	300	1.5	1.5	10	20	Trigger
4	35	400	1	1	5	5	Trigger
5	10	300	1	0.5	5	5	Trigger
6	Ch. 1: 50 Ch. 2: 10	300	1	0.5	5	5	None
7	35	350	1	0.5	5	5	None
8	8	80	1	0.5	5	5	Trigger
9	4 – 99	150	Continuous (TENS)				Trigger

The therapy lasted for the duration of 3 months. Each week, all the 5 test subjects undergo 30 minutes per day of electrical stimulation therapy on the targeted muscles for 4 days on every Monday, Wednesday, Friday and Sunday. Each targeted muscles spend 10 minutes of therapy then add up to 30 minutes per session for each biceps, triceps and deltoid. Test subjects was advised to wear loose shirt only during the therapy. During the therapy, a series of pre-test is conducted to alert the changes in the targeted muscle parameters. The muscle size is noted by measuring the arm circumference using a measuring tape as shown in Fig. 5. For the muscle strength, a digitalized approach was used.

Fig. 5: Measuring the circumference of the left hand using a measuring tape.

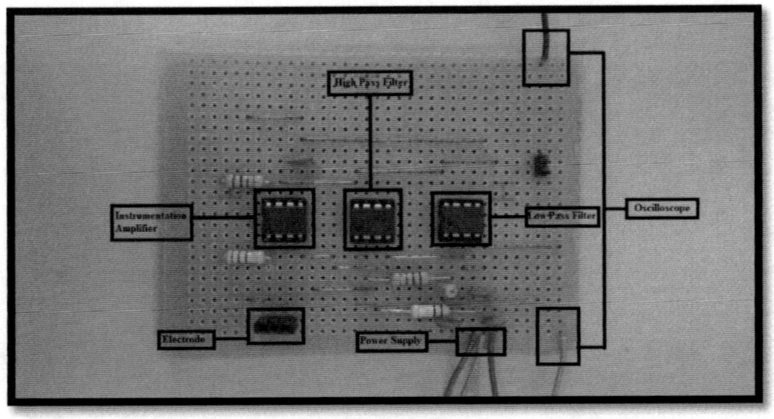

Fig. 6: The EMG conditioning circut. Retrieved from F.A. Kamal, Final Year Project, ECE Department, IIUM.

With the use of National Instrument NI-USB 6216, a data acquisition (DAQ) module, EMG signal conditioning circuit in [16, Fig. 6], and the LabVIEW software, it produce a clean EMG signal and graph that shows the change in muscle strength parameter. Both the right and left hand was tested to compare the strength between a dominant and non-dominant hand side. All the test subject flexed their arm to see the muscle contraction. Thus, a changes in muscle contraction force was depicted through EMG signal and graph. After 3 months of therapy, the final measurement was taken and a series of question was asked to the test subjects to note any respond from them personally after the therapy.

4. Result and Analysis

4.1 Muscle bulk

It is identified that the muscle size of all test subjects increase throughout the 3 months therapy. Test subjects 3 and 5 have a 4.35% increase, while test subject 2 and 4 have 3.88% increment and the last, test subjects 1 has an increase of 4.74% in muscle bulkiness. The factor influencing the rate of increment is the level of thickness of the subcutaneous fat tissue and the muscle condition. The 2 test subjects of 3.88% muscle bulk increment have a thicker subcutaneous fat tissue, therefore decreasing the rate of current density flowing to the muscle [17]. The only test subjects of 4.74% muscle bulk development was influenced by the therapy the test subjects had taken during the part 1 of this work. Thus, given the condition of a trained muscle to the test subjects, resulting in higher rate of muscle bulk expansion. as the subcutaneous fat tissue thickness, the electrode size and the duration of the therapy. Moreover electrical stimulation do not give a permanent effect in the muscle bulkiness as it shown a descending rate of muscle bulk after a long period of break, suggesting that electrical stimulation must be done periodically to ensure the positive impact. The results were summarized in Table 3.

IOP Conf. Series: Materials Science and Engineering **260** (2017) 012020 doi:10.1088/1757-899X/260/1/012020

Table 3: The result of muscle bulk measurement of the left hand side

Date/TS	TS 1	TS 2	TS 3	TS 4	TS 5
25 Sept	25.3cm	25.8cm	24.3cm	26.4cm	25.3cm
2 Oct	-	-	-	-	-
9 Oct	-	-	-	-	-
16 Oct	25.6cm	26.0cm	24.6cm	26.6cm	25.6cm
23 Oct	25.7cm	***break***	***break***	***break***	***break***
30 Oct	25.8cm	26.1cm	24.7cm	26.7cm	25.7cm
6 Nov	25.9cm	26.2cm	24.8cm	26.8cm	25.8cm
13 Nov	26.0cm	26.3cm	24.9cm	26.9cm	25.9cm
20 Nov	26.1cm	26.4cm	25.0cm	27.0cm	26.0cm
27 Nov	26.2cm	26.5cm	25.1cm	27.1cm	26.1cm
4 Dec	26.3cm	26.6cm	25.2cm	27.2cm	26.2cm
11 Dec	26.4cm	26.7cm	25.3cm	27.3cm	26.3cm
17 Dec	26.5cm	26.8cm	25.4cm	27.4cm	26.4cm

TS = test subject

4.2 Muscle strength

The raw EMG signal read from the muscle will undergo filtering and amplification before obtaining a clean EMG signal. The EMG signal is then projected in the LabVIEW software to be analysed. The value to be analysed from the signal obtained were the root mean square (RMS) value and the signal amplitude. From the graph, it can be seen that the signal intensity of right hand side from Fig. 8, which is the dominant hand was still higher than the left hand side, shown in Fig. 7. This indicates that EMS alone cannot afford to develop the muscle strength or isometric muscle contraction. This was due to the incapability of EMS in increasing the fast-twitch myofibrils, which is the muscle fibre responsible in giving the muscle its strength and power characteristic. To efficiently increase the rate of isometric muscle contraction, a series of functional activity or exercise must be done, while EMS aids in giving faster and efficient effect.

However, from the test conducted, the relationship between EMG signal and muscle strength was proven as the linearity of the relationship was recorded in the graph obtain. The RMS value shows a higher reading when the test subject flex their arm, indicating the isometric muscle contraction rate increasing. Thus, mark the increase of the muscle strength when flexing compared to when resting. A research by Heloyse et al. [18] pointed out that the magnitude of the EMG signal is directly proportional to muscle strength but with some limitations to be considered. The factors were the EMG signal cross-talk, non-fixed electrodes recording placement and the involvement of other muscle fibre in force generation.

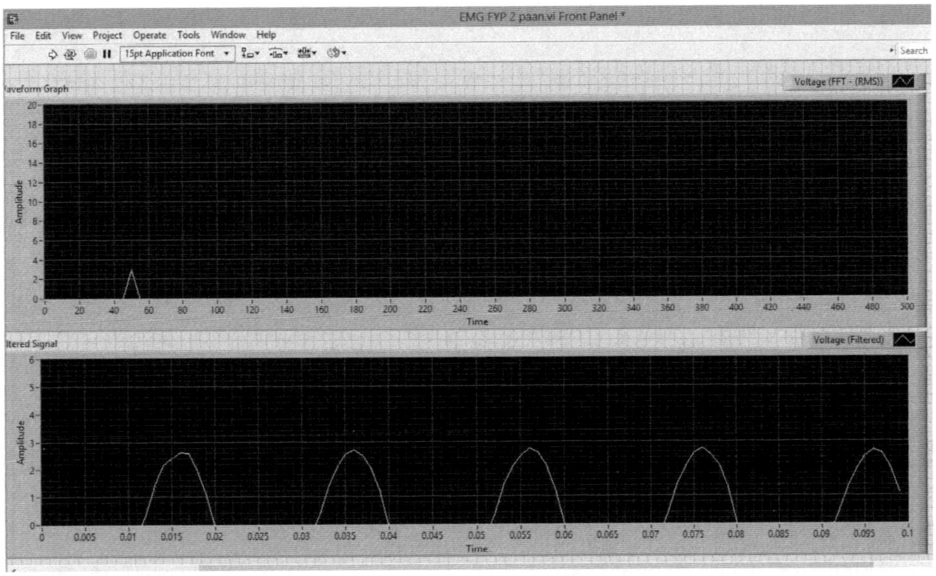

Fig. 7: RMS and amplitude value of the EMG signal for the left handside.

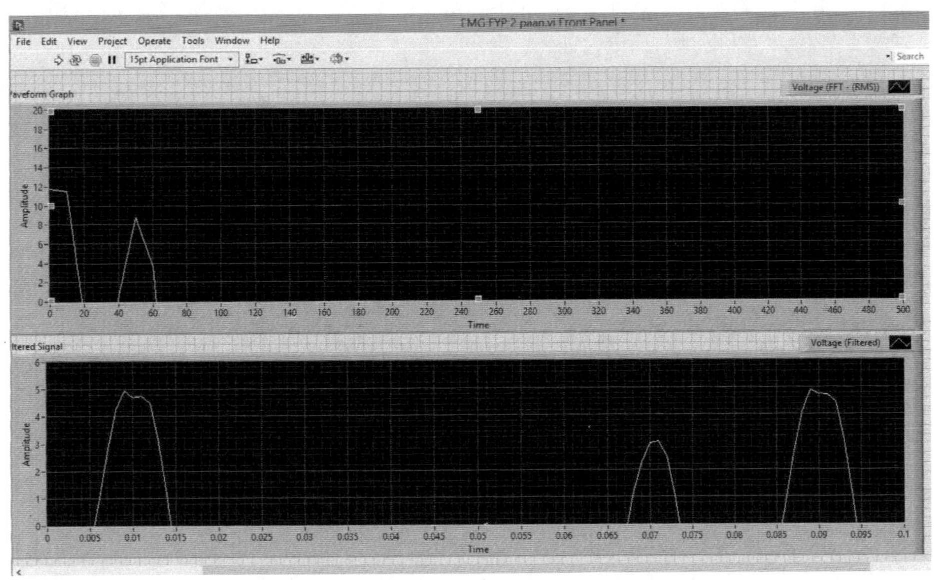

Fig. 8: RMS and amplitude value of the EMG signal for the right handside.

IOP Conf. Series: Materials Science and Engineering **260** (2017) 012020 doi:10.1088/1757-899X/260/1/012020

5. Conclusion

From this research work, it is evident that electrical muscle stimulation can promote and develop muscle bulk along side continuous therapy, although, lacking the capability to increase muscle strength. Factors that can influences the rate of increment of muscle bulk by EMS were the level of subcutaneous fat tissue thickness the electrode size, the duration of the therapy and the condition of muscle whether trained or not. As for the muscle strength, EMS alone cannot increase the fast-twitch myofibrils to give a higher output in muscle strength development. It must come with functional activities or exercise routine to trained the muscle. However, the relationship between EMG signal and isometric muscle contraction force were proven linear. Based on the outcome, EMS prove to be a promising rehabilitation intervention to avoid muscle atropy, especially to the patients experiencing injury due to sports.

Acknowledgement

The authors would like thank all the test subjects who participated and cooperated in the electrical muscle stimulation therapy session. A warm gratitude to Dr. Nadia Mohd Mustafah from the Universiti Teknologi MARA, Sungai Buloh, for sharing valuable input to the design of the methodology of this work. Thank you also to Assoc. Prof. Dr. Ahmad Aidil Arafat Bin Dzulkarnain from the Kulliyyah of Allied Health Sciences for providing us with information on ethical clearence to conduct all of this experiment.

IOP Conf. Series: Materials Science and Engineering **260** (2017) 012020 doi:10.1088/1757-899X/260/1/012020

References

[1] Bergquist A, Clair J, Lagerquist O, Mang C, Okuma Y and Collins D 2011 Neuromuscular electrical stimulation: implications of the electrically evoked sensory volley *Eur J Appl Physiol* pp 2409–2426

[2] Deyo R A, Walsh N E, Martin D C, Schoenfeld L S and Ramamurthy S 1990 A controlled trial of transcutaneous electrical nerve stimulation (TENS) and exercise for chronic low back pain *New Engl J Med.* pp 1627–1634

[3] Johnsen E, DiMeo G, Franco B, Wilson R 2011 Design and construction of an electrical muscle stimulation system to decrease Forearm Muscle Atrophy post-fracture *2011 IEEE 37th Annual Northeast Bioengineering Conf. (NEBEC) Troy, NY* pp 1-2

[4] Willis A 2015 How Does Electrical Muscle Stimulation Work?

[5] De Kroon J R, Ijzerman M J, Chae J, Lankhorst G J, Zilvold G 2005 Relation between stimulation characteristics and clinical outcome in studies using electrical stimulation to improve motor control of the upper extremity in stroke *J Rehabil Med.* pp 65–74

[6] Sluka K A and Walsh D 2003 Transcutaneous electrical nerve stimulation: basic science mechanisms and clinical effectiveness *J Pain.* pp 109–121

[7] Baker C, Wederich D, McNeal C, Newsam R and Waters R 2000 Neuromuscular Electrical Stimulation: A Practical Guide *4th edition. Downey, CA: Los Amigos Res. & Edu. Inst.*

[8] Boom H B K, Mulder A J and Veltink P H 1993 Fatigue during functional neuromuscular stimulation *Prog Brain Res.* pp 409–418

[9] Sheffler L R, Hennessey M T, Naples G G and Chae J 2007 Improvement in functional ambulation as a therapeutic effect of peroneal nerve stimulation in hemiplegia: two case reports *Neurorehabil Neural Repair.* pp 366–369

[10] Sha N, Kenney L P J, Heller B W, Barker A T, Howard D and Moatamedi M 2008 A finite element model to identify electrode influence on current distribution in the skin *Artif Organs.* pp 639–643

[11] Ali A M and Qin Y X 2007 Effects of electrically stimulated muscle contractions on bone loss & muscle atrophy *2007 IEEE 33rd Annual Northeast Bioengineering Conf. , Long Island, NY*, pp 37-38

[12] Dupont Salter A C, Richmond F J R and Loeb G E 2003 Prevention of muscle disuse atrophy by low-frequency electrical stimulation in rats *IEEE Transactions on Neural Syst and Rehabil Eng* vol. **11** no. 3 pp 218-226

IOP Conf. Series: Materials Science and Engineering **260** (2017) 012020 doi:10.1088/1757-899X/260/1/012020

[13] Lo H C, Tsai K H, Tsu F C, Chang G L and Yeh C Y 2009 Effects of a functional electrical stimulation-assisted leg-cycling wheelchair on reducing spasticity of patients after stroke *J Rehabil Med* vol. **41** pp 242-246

[14] Keller T, Lawrence M, Kuhn A and Morari M 2006 New multi-channel transcutaneous electrical stimulation technology for rehabilitation *2006 International Conference of the IEEE Eng in Med and Bio Society, New York, NY* pp 194-197

[15] Carr and Joseph J 1997 Safety for electronic hobbyists *Popular Electronics*

[16] Kamal F A 2014 Analysis of EMG signals as control signal inputs in system design". Final year project, ECE Dept, International Islamic University Malaysia

[17] Doheny E P, Caulfield B M, Minogue C M and Lowery M M 2008 The effect of subcutaneous fat thickness on the efficacy of transcutaneous electrical stimulation *2008 IEEE Eng Med and Bio Society Int. Conf.*

[18] Kuriki H U, Azevedo F M, Takahashi L S O, Mello E M, Filho R F N and Alves N 2012 The relationship between electromyography and muscle force University of Estadual Paulista and University of de São Paulo, Brazil

[19] Bracciano A G 2008 Physical Agent Modalities *Bethesda, MD*

[20] Moe J and Post H 1962 Functional electrical stimulation for ambulation in hemiplegia *J Lancet.* pp 285–288

IOP Conf. Series: Materials Science and Engineering **260** (2017) 012021 doi:10.1088/1757-899X/260/1/012021

An Experimental Design of Bypass Magneto-Rheological (MR) damper

MM Rashid, Mohammad Abdul Aziz*, Md. Raisuddin Khan
International Islamic University Malaysia, Jalan Gombak, 53100 Kuala Lumpur, Selangor, Malaysia

E-mail*: abdul.aziz07me@gmail.com

Abstract. The magnetorheological (MR) fluid bypass damper fluid flow through a bypass by utilizing an external channel which allows the controllability of MR fluid in the channel. The Bypass MR damper (BMRD) contains a rectangular bypass flow channel, current controlled movable piston shaft arrangement and MR fluid. The static piston coil case is winding by a coil which is used inside the piston head arrangement. The current controlled coil case provides a magnetic flux through the BMRD cylinder for controllability. The high strength of alloy steel materials are used for making piston shaft which allows magnetic flux propagation throughout the BMRD cylinder. Using the above design materials, a Bypass MR damper is designed and tested. An excitation of current is applied during the experiment which characterizes the BMRD controllability. It is shown that the BMRD with external flow channel allows a high controllable damping force using an excitation current. The experimental result of damping force-displacement characteristics with current excitation and without current excitation are compared in this research. The BMRD model is validated by the experimental result at various frequencies and applied excitation current.

1. Introduction

Magnetorheological (MR) dampers is a smart device for controlling vibration using electro-magnetic field. MR damper contains smart fluids (MR fluid) that has a controllability for semi active passive system and required less power supply. The properties of MR damper can be changed by changing applied current[1]. MR damper have been chosen over a decade for vehicle suspension system[2-4], helicopters seat[5], civil structure base isolation [6-8], and railway transport[9, 10]. Semi-active MR damper is especially applicable for unpredictable motion conditions and capable large payload condition. In the MR damper, MR fluid has an ability to create controllable damping force and this controllable force can be creating by applying electromagnetic flux distribution through the whole MR damper body[11]. The modern design method of MR damper valve is familiar at all and bypass valve design requires an exact design configuration. During the experimental operation, the bypass valve is configured with accurate flow mode operation. Jin-Hyeong Yoo has proposed meso-scale valve of MR damper which produces higher ground relying yield stress of MR fluid. Whereas, MR valves are controlled by analyzing axial flow rate and pressure drop in the valve[12, 13]. Xian-Xu Bai developed a series of inner, annular, bidirectional and radial bypass system. In the inner dispersion system offers maximum dynamic range to the minimum off stroke damping force range[14]. In the twin tube inner dispersion system added MR damper hydro-mechanical higher damping range and lower stroking load [15].

Other study analyzes an annular inner bypass with higher dampening force and lower field off-state stroking load. Bingham-plastic materials models removing minor losses of MR damper velocity profile[14]. The basic advantages of his works to optimize the MR damper performance and two concentric tubes, bobbin piston arrangement are used for MR fluid flow as well as shock and vibration control[16]. An annular radial duck process for series flow of MR fluid in the gap which increases velocity under controllable condition [17]. However, Mao et al. evaluated single ended MR Bifold valve for higher dynamic capacity which includes larger piston speed for damping operation. His proposed model demonstrates two MR damper valve model and firstly, piston head inside valve and secondly, bifold MR valve at MR damper both end[18]. Ryan m. robinson et al. and eugene cook et al. proposed porous media (variation of porosity and tortuosity) dispersion system of MR damper. In porous media (compacted spheres and cylinder beds) valve tests includes maximum force, damping coefficient, hydraulic piston cylinder cross-sectional area and yield stress production by varying magnetic field efficiency [19, 20].

In this research an experimental evaluation of Bypass MR damper with different stroke modes are analysed and it also illustrates the design technique of BMRD shear mode operation.

2. Experimental setup of Bypass MR damper

Bypass MR damper test components consists of bypass MR damper, universal testing machine (UTM) called Shimadzu. UTM staircase steps excitation analyze the bypass MR damper. However, for different types of test experiments compressor, grippers and upper and lower pulley are attach during the experiments. In the top and bottom of bypass MR damper, two rectangular aluminum plate have attach during the experiments. This is done due to higher pressure and tensile strength. Hydraulic actuator moves due to movable head in which TRAPEZIUMX software is used to generate signal. In the lower head, a load cell is connected for measuring applied force to the bypass damper. A linear variable differential transformer (LVDT) sensor is used to measure the displacement of the BMRD piston. A DC voltage source is used to supply for controlling current in the damper inner coil and piston coils. Moreover, force-displacement, force-time, and stress-strain, time displacement graphs are produced during the experiment. In addition to that Stroke rate, stroke length and input currents is varied due to characterization of MR damper and performance. The total length of stroke from 10mm to 20mm were done during the experiment, which is shown in Figure 1. In figure 1 flow chart presents the experimental setup.

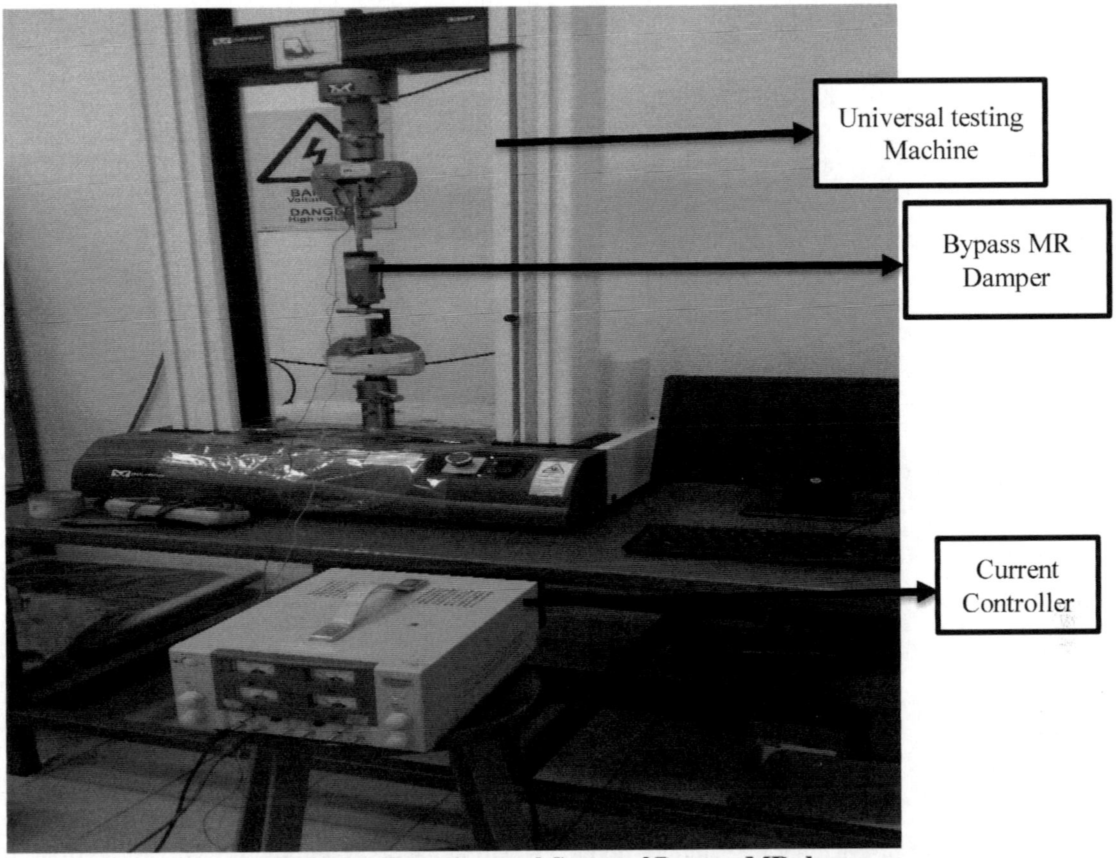

Figure 1. Experimental Setup of Bypass MR damper

3. Bypass MR Damper Damping Constant Calculation for Shear Mode Operation

In figure 2 presents the MR damper piston cylinder arrangement in shear mode operation.

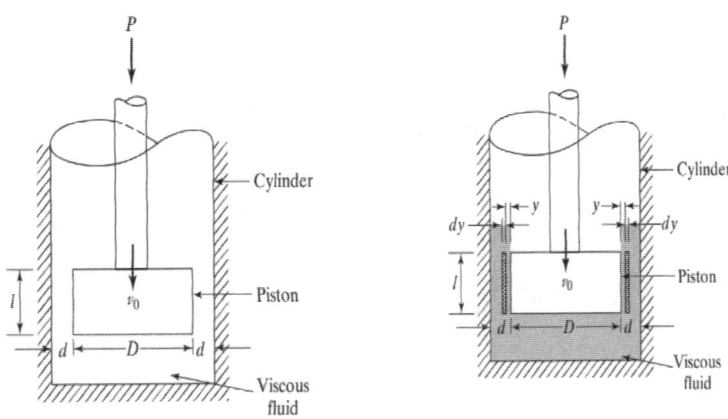

Figure 2. MR damper piston cylinder arrangement in shear mode[29].

The viscous force for this annular damper [29]

IOP Conf. Series: Materials Science and Engineering **260** (2017) 012021 doi:10.1088/1757-899X/260/1/012021

$$F = \pi D l d\tau = \pi D l \frac{d\tau}{dy} \mathrm{dy} \quad (1)$$

The expression of shear stress

$$\tau = -\mu \frac{dv}{dy} \quad (2)$$

The negative sign is expressed for decreasing velocity gradient

$$F = \pi D l dy \mu \frac{d^2 v}{dy^2} \quad (3)$$

The pressure difference causes due to force applied on the piston and calculated by the following equation

$$p = \frac{p}{\frac{\pi D^2}{4}} = \frac{4p}{\pi D^2} \quad (4)$$

Or

$$\frac{d^2 v}{dy^2} = \frac{4p}{\pi D^2} \quad (5)$$

After two times integration using boundary condition $v = -v_0$ at $y = 0$ and $v = o$ at $y = d$, the equation becomes

$$v = \frac{2p}{\pi D^2 l}(yd - y^2) - v_0(1 - \frac{y}{d}) \quad (6)$$

The flow rate through the element after integration the equation becomes and the limits are $y = 0, y = d$

$$Q = \int_0^d v\pi D dy = \pi D \left[\frac{2pd^3}{6\pi D^2 l\mu} - \frac{1}{2} v_0 d\right] \quad (7)$$

The velocity of the piston will be equal to the rate of flow divided by the piston area

$$v_0 = \frac{Q}{\pi/_4 D^2} \quad (8)$$

$$v_0 = \frac{Q\pi D \left[\frac{2pd^3}{6\pi D^2 l\mu} - \frac{1}{2} v_0 d\right]}{\pi/_4 D^2} \quad (9)$$

From equation 1 &2

$$p = \left[\frac{3\pi D^3 l \left(1 + \frac{2d}{D}\right)}{4d^3}\right] \mu v_0 \quad (10)$$

If the $p = cv_0$, then the damping constant

$$c = \mu \left[\frac{3\pi D^3 l \left(1 + \frac{2d}{D}\right)}{4d^3}\right] \quad (11)$$

Finally, the equation of motion can be rewrite as

$$m\ddot{x}(t) + \mu\left[\frac{3\pi D^3 l\left(1+\frac{2d}{D}\right)}{4d^3}\right]\dot{x}(t) + kx(t) = f(t) \tag{12}$$

4. Damper configuration and testing setup

A bypass channel is developed and damping performance i.e. controllability is experimentally assessed using BMRD verification techniques. The bypass channel was constructed using a high strength alloy steel with an inner diameter of 10mm and length of 99 mm. The diameter of this piston rod is 10mm and the piston is 38mm and the applied current for sinusoidal displacement excitations ranging from 0 to 1.5 A. In figure 3 presents the BMRD structure and in table 5, the BMRD components are shown.

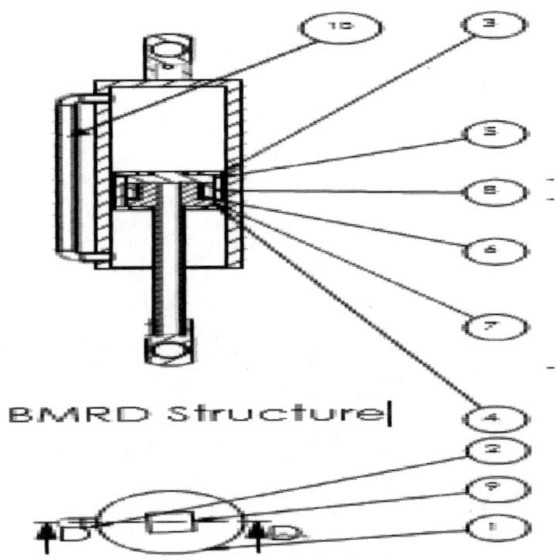

Figure 3. Bypass MR damper structure

Table 1: BMRD parts

Item no	Part Number	Quantity
1	Mono Tube Cylinder	1
2	Pipe	1
3	Piston Cover	1
4	Piston Cover Cap	2
5	Piston Cap	1
6	Piston Coil Cylinder	1
7	Plastic Coil Case	1
8	Copper	1
9	Top Rod	2
10	Bypass pipe	10

5. Bypass MR damper characterization

In figure 4 & 5 it is seen that the damping force rapidly increases during the beginning stroke for all the stroke rates applied during the experiment. After 4mm of stroke length, the damping force show steady behavior. In this curves least trend force is seen at 100mm where highest is seen 200 mm rate. The other rates are as 300,400 & 500 mm shows force values between the 100 & 200 mm rates. Due to the applied load forces increases sharply and reached its peak. However, for the following 1 mm displacement, forces decrease slowly and stable at 10mm stroke. However, in figure 5 after 5mm of stroke length, the damping force show steady behavior. In this curves least trend force is seen at 500 mm where highest is seen 100 mm rate. The other rates are as 200,300, 400 & 500 mm shows force values between the 100 & 500 mm rates. In addition, the steady force under different stroke rate do not maintain any trend for both 10 mm and 20 mm strokes.

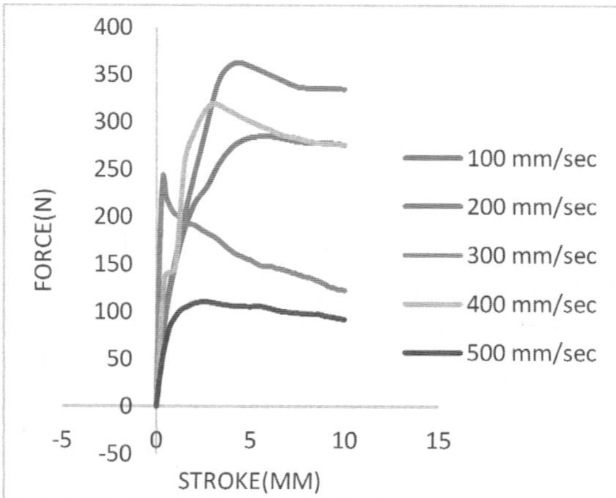

Figure 4. Force vs Stroke curve for 10mm down stroke at zero current for different stroke rate.

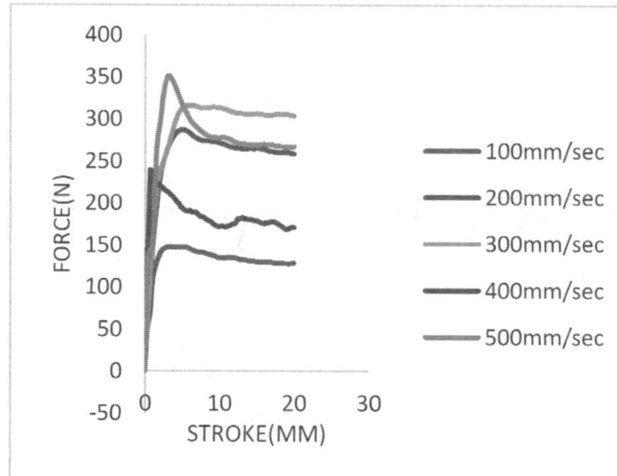

Figure 5. Force vs Stroke curve for 20mm down stroke at zero different stroke rate.

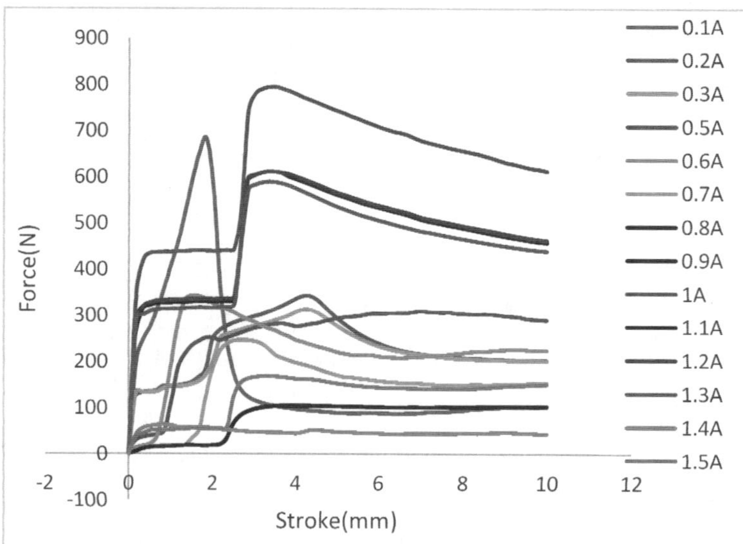

Figure 6. Force vs Stroke curve for 10mm down stroke at different current supply for a stroke rate of 100mm/sec.

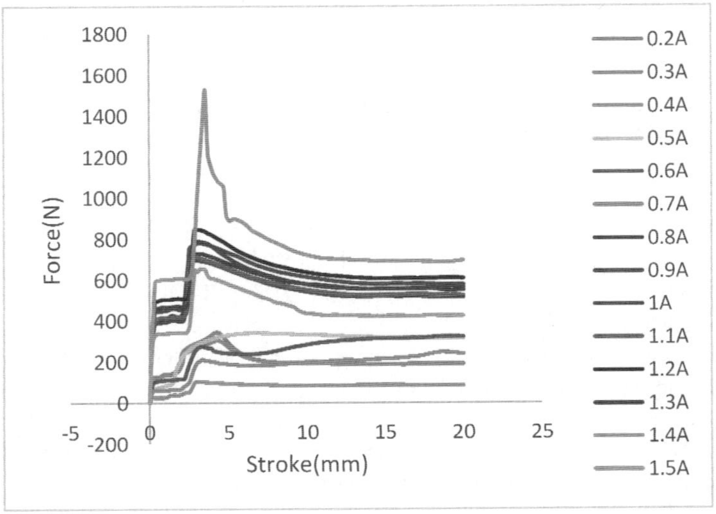

Figure 7. Force vs Stroke curve for 20mm down stroke at different current supply for a stroke rate of 100mm/sec.

Like the down stroke, the upstroke forces have been observed for different stroke length and stroke rate, and current was varied from 0.1A to 1.5 A. The applied force for this experiment from 100mm/sec to 500mm/sec. In figure 6 it is seen that the damping force rapidly increases during the beginning stroke for all the stroke rates applied during the experiment. After 4mm of stroke length, the damping force show steady behavior. However, the steady force under different stroke rate do not maintain any trend. In this curves least trend force is seen at 1.4 A where highest is seen 1A current. The other currents are as 0.1A,0.2A,0.3A,0.4A,0.5A,0.6A,0.7A,0.8A,0.9A,1.1A,1.2A,1.3A,1.5A shows force values between the 1 & 1.4 A currents.

In figure 7 it is seen that the damping force rapidly increases during the beginning stroke for all the stroke rates applied during the experiment. After 4mm of stroke length, the damping force show steady behavior. However, the steady force under different stroke rate do not maintain any trend. In this curves least trend force is seen at 0.4A where highest is seen 0.2A current. The other currents are as 0.1A,0.3A,0.5A,0.6A,0.7A,0.8A,0.9A,1.1A,1.2A,1.3A,1.5A shows force values between the 0.2 & 0.4 A currents.

6. Conclusion

Magnetorheological damper with a bypass channel was developed to produce higher damping force. This external bypass method provides extra passage for flowing MR fluid and the flow of MR fluid controlled by current flow. The BMRD was tested for 10-20mm peak to peak amplitude and currents of 0 to 1.5 amps were applied to the coil. The damper was found to show sharp increase of damping force at the beginning of stroke for different stroke rates and then maintain steady force onwards. The lowest and highest forces found from shear mode experiments are 85.51 N and 1525.67 N respectively. It is also seen that the damping force increases with the increase of excitation current from the range of 0.2 ampere to 1.4 ampere. The design method was validated using experimental results.

7. Reference

[1] H. Fujitani, H. Sodeyama, K. Hata, Y. Komatsu, N. Iwata, K. Sunakoda, et al., "Dynamic Performance Evaluation of 200kN Magneto-rheological Damper," Technical Note of National Institute for Land and Infrastructure Management, vol. 41, pp. 349-356, 2002.

[2] S.-B. Choi, M.-H. Nam, and B.-K. Lee, "Vibration control of a MR seat damper for commercial vehicles," Journal of Intelligent Material Systems and Structures, vol. 11, pp. 936-944, 2000.

[3] D. Simon and M. Ahmadian, "Vehicle evaluation of the performance of magneto rheological dampers for heavy truck suspensions," TRANSACTIONS-AMERICAN SOCIETY OF MECHANICAL ENGINEERS JOURNAL OF VIBRATION AND ACOUSTICS, vol. 123, pp. 365-375, 2001.

[4] A. H.-F. Lam and W.-H. Liao, "Semi-active control of automotive suspension systems with magneto-rheological dampers," International Journal of Vehicle Design, vol. 33, pp. 50-75, 2003.

[5] W. Hu and N. M. Wereley, "Magnetorheological fluid and elastomeric lag damper for helicopter stability augmentation," International Journal of Modern Physics B, vol. 19, pp. 1471-1477, 2005.

[6] H. Jung, B. Spencer Jr, Y. Ni, and I. Lee, "State-of-the-art of semiactive control systems using MR fluid dampers in civil engineering applications," Structural Engineering and Mechanics, vol. 17, pp. 493-526, 2004.

[7] M. Liu, V. Sethi, G. Song, and H. Li, "Investigation of locking force for stay cable vibration control using magnetorheological fluid damper," Journal of Vibration and Acoustics, vol. 130, p. 054504, 2008.

[8] F. Gordaninejad, X. Wang, G. Hitchcock, K. Bangrakulur, S. Ruan, and M. Siino, "Modular high-force seismic magneto-rheological fluid damper," Journal of structural engineering, vol. 136, pp. 135-143, 2010.

[9] C. Guo, X. Gong, L. Zong, C. Peng, and S. Xuan, "Twin-tube-and bypass-containing magneto-rheological damper for use in railway vehicles," Proceedings of the Institution of Mechanical Engineers, Part F: Journal of Rail and Rapid Transit, vol. 229, pp. 48-57, 2015.

[10] Y.-J. Shin, W.-H. You, H.-M. Hur, J.-H. Park, and G.-S. Lee, "Improvement of ride quality of railway vehicle by semiactive secondary suspension system on roller rig using magnetorheological damper," Advances in Mechanical Engineering, vol. 6, p. 298382, 2014.

[11] G. H. Hitchcock, F. Gordaninejad, and X. Wang, "A new by-pass, fail-safe, magneto-rheological fluid damper," in Proceedings of SPIE Conference on Smart Systems for Bridges, Structures, and Highways, 2002, pp. 345-351.

[12] J.-H. Yoo and N. M. Wereley, "Design of a high-efficiency magnetorheological valve," Journal of Intelligent Material Systems and Structures, vol. 13, pp. 679-685, 2002.

[13] N. M. Wereley and L. Pang, "Nondimensional analysis of semi-active electrorheological and magnetorheological dampers using approximate parallel plate models," Smart Materials and Structures, vol. 7, p. 732, 1998.

[14] X.-X. Bai, W. Hu, and N. M. Wereley, "Magnetorheological damper utilizing an inner bypass for ground vehicle suspensions," Magnetics, IEEE Transactions on, vol. 49, pp. 3422-3425, 2013.

[15] X.-X. Bai, N. M. Wereley, W. Hu, and D.-H. Wang, "A Bidirectional-Controllable Magnetorheological Energy Absorber for Shock and Vibration Isolation Systems," in ASME 2012 Conference on Smart Materials, Adaptive Structures and Intelligent Systems, 2012, pp. 485-495.

[16] X.-X. Bai, N. M. Wereley, and W. Hu, "Maximizing semi-active vibration isolation utilizing a magnetorheological damper with an inner bypass configuration," Journal of Applied Physics, vol. 117, p. 17C711, 2015.

[17] X.-X. Bai, D.-H. Wang, and H. Fu, "Principle, modeling, and testing of an annular-radial-duct magnetorheological damper," Sensors and Actuators A: Physical, vol. 201, pp. 302-309, 2013.
[18] M. Mao, W. Hu, Y.-T. Choi, and N. M. Wereley, "A magnetorheological damper with bifold valves for shock and vibration mitigation," Journal of Intelligent Material Systems and Structures, vol. 18, pp. 1227-1232, 2007.
[19] R. M. Robinson, W. Hu, and N. M. Wereley, "Adaptive Dampers Employing Magnetorheological Valves Filled with Porous Media," in 51st AIAA/ASME/ASCE/AHS/ASC Structures, Structural Dynamics, and Materials Conference 18th AIAA/ASME/AHS Adaptive Structures Conference 12th, 2010, p. 3068.
[20] E. Cook, W. Hu, and N. M. Wereley, "Magnetorheological bypass damper exploiting flow through a porous channel," Journal of Intelligent Material Systems and Structures, vol. 18, pp. 1197-1203, 2007.

PREDICTION OF BECK DEPRESSION INVENTORY (BDI-II) SCORE USING ACOUSTIC MEASUREMENTS IN A SAMPLE OF IIUM ENGINEERING STUDENTS

Muhamad Fikri Zanil[1], Nik Nur Wahidah Nik Hashim[2] and Huda Azam[3]

[1-3] Department of Mechatronics Engineering, International Islamic University Malaysia

Abstract. Psychiatrist currently relies on questionnaires and interviews for psychological assessment. These conservative methods often miss true positives and might lead to death, especially in cases where a patient might be experiencing suicidal predisposition but was only diagnosed as major depressive disorder (MDD). With modern technology, an assessment tool might aid psychiatrist with a more accurate diagnosis and thus hope to reduce casualty. This project will explore on the relationship between speech features of spoken audio signal (reading) in Bahasa Malaysia with the Beck Depression Inventory scores. The speech features used in this project were Power Spectral Density (PSD), Mel-frequency Ceptral Coefficients (MFCC), Transition Parameter, formant and pitch. According to analysis, the optimum combination of speech features to predict BDI-II scores include PSD, MFCC and Transition Parameters. The linear regression approach with sequential forward/backward method was used to predict the BDI-II scores using reading speech. The result showed 0.4096 mean absolute error (MAE) for female reading speech. For male, the BDI-II scores successfully predicted 100% less than 1 scores difference with MAE of 0.098437. A prediction system called Depression Severity Evaluator (DSE) was developed. The DSE managed to predict one out of five subjects. Although the prediction rate was low, the system precisely predict the score within the maximum difference of 4.93 for each person. This demonstrates that the scores are not random numbers.

1. Introduction

Based on The National Health and Morbidity Survey on 2015, 29 percent of Malaysians had depression which is 17 percent higher than the statistic in 2011[1]. The percent increment shown by this statistic demonstrated that this issue has become very alarming and shows that this project have very high potential to help this problem. Depression does not only has effect on the individual, it also adds problems to the development of country as each individual has a value to the country.

However, we still having difficulties in solving this problem since it is very challenging for clinician to identify the depression severity of a patient. Researchers have struggled to find a way to identify the severity of depression since long ago, as a result, Aaron T. Beck has published an instrument to measure the severity of depression named Beck Depression Inventory. It is actually a self-report, which contain 21 questionnaires and each one have four options which indicate the score from 0 (not present) to 3 (severe). Each of the item or question have its own category, including sadness, pessimism, past failure, loss of pleasure, suicide thought, crying, agitation, loss of interest, indecisiveness, worthlessness, loss of energy, change in sleep pattern, irritability, change in appetite, concentration

IOP Conf. Series: Materials Science and Engineering **260** (2017) 012022 doi:10.1088/1757-899X/260/1/012022

difficulty, fatigue and loss interest in sex respectively [2]. The severity of depression can be determined by the rating of overall BDI-II scores as in the Table 1.

Table 1. Depression indication according to scores range [2]

BDI-II Scores	Depression Indication
10 and lower	Normal
11 to 16	Sad
17 to 20	Mild
21 to 30	Moderate
31 to 40	Severe
40 and more	Need Serious treatment

Nowadays, the advancement of technologies has changed our daily lives. Many things have been invented and at the same time encourage researchers to look at new approaches in identifying the severity of the depression. Some use imaging device like camera to do image processing and some use audio device to do audio processing and some even do both. This new approach has the goal to accurately measure the severity of depression instantaneously.

As for this project, we focus on audio processing using the reading speech of the subject. It is good to review others works related to this project. Based on other research by others [3] [4] [5], most of them extract Mel-Frequency Ceptral Coefficients (MFCC) speech features, but the approach to predict the scores were different. They try to predict using Gaussian Mixture Model [3], Gaussian Staircase Regression [4] and Multiple Linear Regression Model [5]. The best method to apply is using Linear Regression as it gave the minimum mean absolute error.

2. Data

Data is an important requirement to complete this project as this data are needed to train the prediction system. The subject need to fill the BDI-II Inventory and then the need to be interviewed and read a passage which will be recorded using Tascam DR-05 with sample rate at 44.1 kHz. All the recorded session is done in a closed room. After the whole data collection is done, the subject will receive RM10.00 as honorarium.

The subjects were male and female with the age range 18-29 which are engineering students at International Islamic University Malaysia (IIUM). All of the subjects do not have any respiratory problem and not under the influence of alcohol. The recording session was done in a closed room, but not fully soundproof to make sure the robustness of the system.

During the recording session the subject reads a passage. The passage should be a rainbow passage, but since there is no Bahasa Malaysia passage that is equivalent to the rainbow passage, we decided to go with any controlled passage.

3. Speech Features

Several types of speech were collected to be tested to identify the best combination of speech features to be implemented in DSE system. The speech features include 3 Power Spectral Densities (PSD), 13 MFCC, 6 Transition Matrix (TM), 2 formant, maximum and minimum pitch which in total of 26 features. This system, however only applies PSD, MFCC and TM since the best combination of speech features only contain these 3 types of speech features.

IOP Conf. Series: Materials Science and Engineering **260** (2017) 012022 doi:10.1088/1757-899X/260/1/012022

3.1. Power spectral Density (PSD).

Three features are collect from PSD. PSD is actually a power intensity of a signal in frequency or for a sound waves it is loudness. Most energy in a speech of an adult contain in a rage 0 to 2000Hz. To minimize the number of features, the range of 0Hz to 2000Hz is divided into 4 segments with equal width. Thus, the first band would be 0Hz to 500Hz and the second band from 500Hz to 1000Hz and the third band would be 1000Hz to 1500Hz. Only the first three bands are collected as the fourth band can be predicted by the first three bands.

3.2. *Transition Matrix.*

This feature is a set of probability of state transition as shown in figure 1. Each block represent 0.04 second segment and categorized into number 1, 2 and 3 which represent voiced, unvoiced and silence respectively. The unvoiced speech segments are basically a noise like signal which have high frequency components. Whereas the voiced signal have low frequency components. Butterworth Filter was used to categorize this signal. However, silence speech sound wave also has low frequency components like voice speech sound wave. In order to categorize between them, threshold is needed. [6]

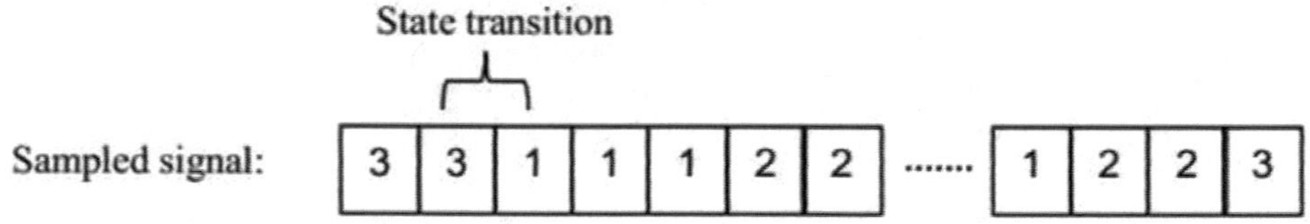

Figure 1. Graphical representation of state transition [5]

3.3. Mel-Frequency Ceptral Coefficients (MFCC)

Mel-Frequency Cepstral Coefficients are quite popular speech features and commonly used for speech analysis. MFCC represents the logarithmic perception of the human auditory system. MFCC also can represent the spectrum of an audio signal in the Mel scale. It acts like human hearing system.

4.Methodologies

The system development was divided into two parts which are identifying the best speech features for part 1 to achieve the first two objectives and develop Depression Severity Evaluator (DSE) to achieve the third objective.

In part 1 the flowchart is shown in Figure 2. This process is very important in order to continue to part 2. The goal of this process is to find the combination of speech features that give the lowest mean absolute error in predicting the BDI-II score using the data that have collected. Too many features may result in high error, so the Sequential Forward Selection (SFS) method was used with linear regression to find the predicted scores. Each feature was selected one-by-one and evaluated using linear regression until all features are evaluated. Feature that gives the least mean sum of absolute error was selected and add it to the empty set one at a time.

Jackknife method also used in this process to train the system. The implementation of the jackknife method in this process is on the basis of leave-one-out. The process involves discarding out one student's score data from the data set and develops a training data set with the remaining N-1 students. The excluded data is then tested. This process is repeated by excluding the next students from the overall set of data until all students have been chosen to be excluded as testing data.

The combination of speech features that give the least mean absolute error were selected as the best and used to find the model coefficient, a using the formula of linear regression $b = D \times a$ where b is the column vector of the actual value of BDI-II scores, D is the input vectors (combination of speech features). Note that the model coefficient, a, is in matrix form.

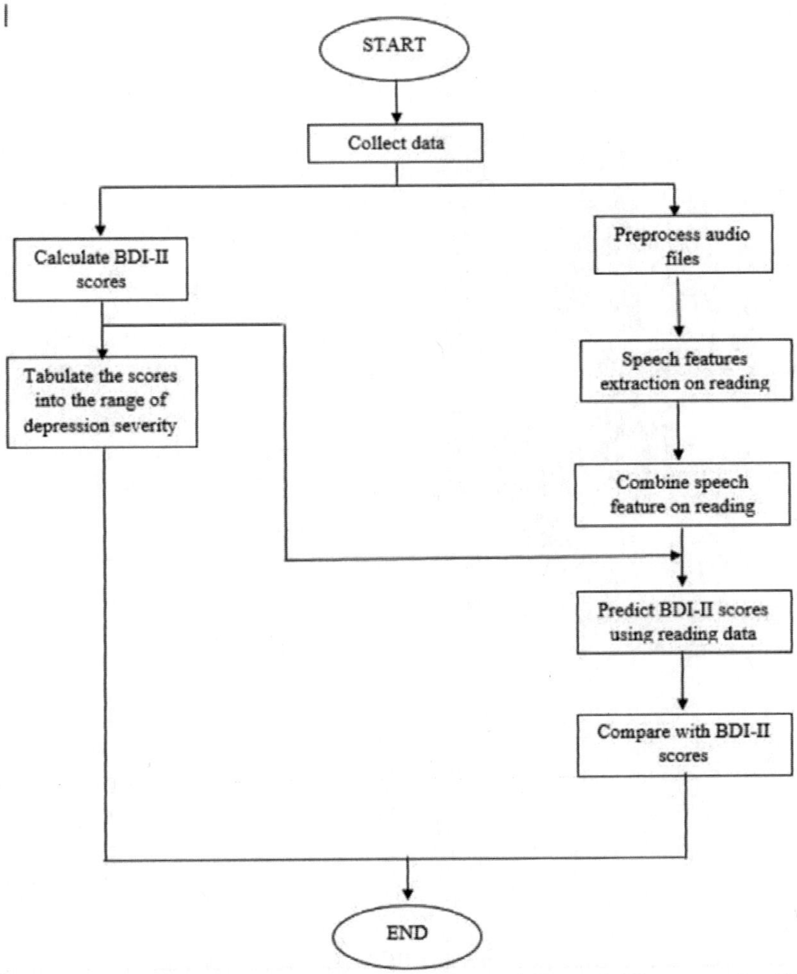

Figure 2. Flowchart of identifying best combination of Speech Features for BDI prediction.

Once we have the model coefficient, **a**, from the best combination of speech features, we can continue to part 2 to build the Graphical User Interface (GUI) of the DSE. The simple block diagram of DSE is shown in figure 3.

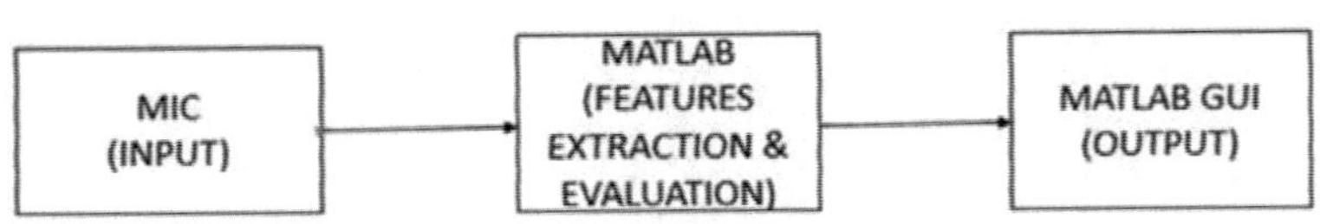

Figure 3. Simple Block Diagram of DSE

IOP Conf. Series: Materials Science and Engineering **260** (2017) 012022 doi:10.1088/1757-899X/260/1/012022

Graphical User Interface(GUI)

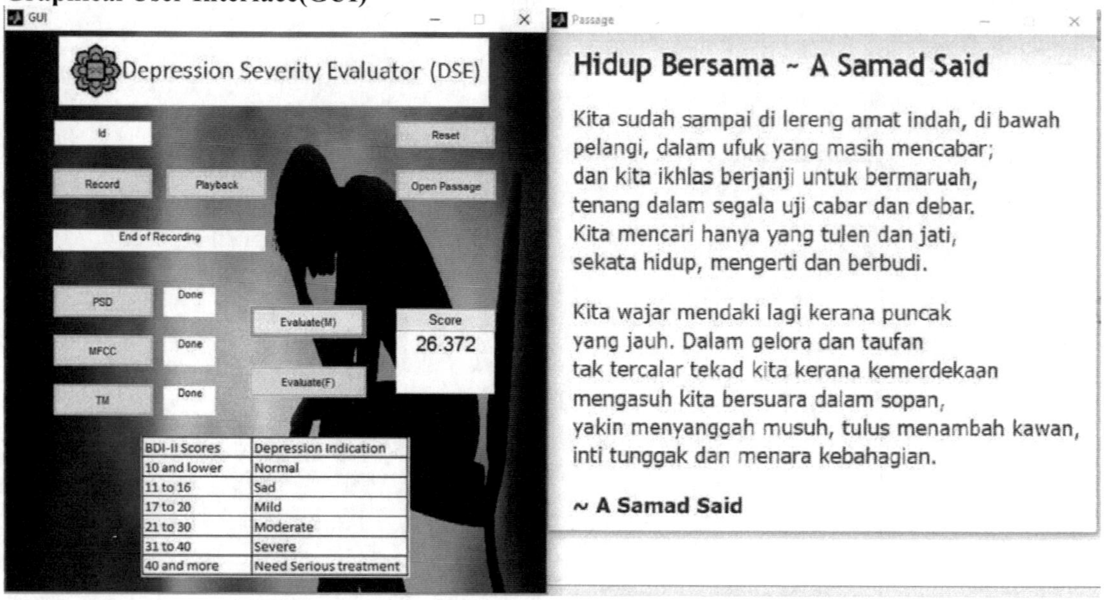

Figure 4. Depression Severity Evaluator (DSE) GUI

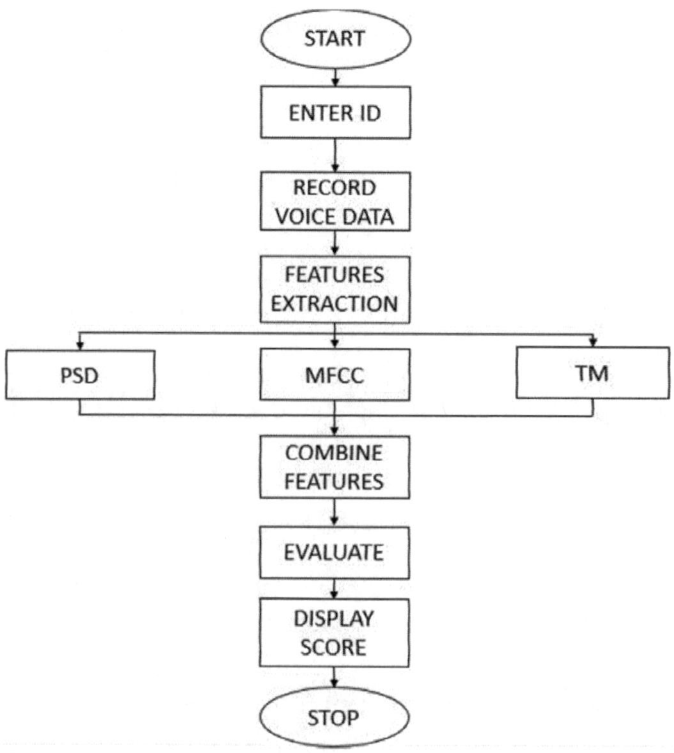

Figure 5. Depression Severity Evaluator (DSE) operational flowchart.

The process of this system almost the same with the process in part 1. Only few changes were done on the feature extraction. The major changes in this system compare to part 1 is on evaluating process. Where the model coefficient, **a**, in part 1 were used to multiply with the best combination of speech features which shown in table 2 for male and female.

Table 2. Best speech features combinations

Male		Female	
Number of features	List of feature	Number of features	List of feature
6	mfcc7, mfcc1, tm9, mfcc9, mfcc11, tm1	6	mfcc2, mfcc11, mfcc8, mfcc7, mfcc10, psd3

5.Result and discussion
For part 1, the result was in term of mean absolute error (MAE). As the process depended on gender separately same goes on the result. For male reading, it can be seen that (SFS) method give the most minimum MAE which is 0.098437 with 6 features selected. The minimum MAE for (SBS) method, however could not outperform the SFS method with MEA of 1.95651 with 8 features left.

For female reading, again, the Sequential Forward Selection method gives the lowest mean absolute error with also 6 speech features. The mean absolute error is quite low which at 0.4096. Whereas, the MEA for SBS method is 4.261898 with 7 features left.

Thus, the best combination of speech features for male reading are mfcc7, mfcc1, tm9, mfcc9, mfcc11, tm1 whereas for female reading mfcc2, mfcc11, mfcc8, mfcc7, mfcc10, psd3. Quite interesting that the remaining best features are MFCC, PSD and TM. MFCC and PSD are a spectrum based measure where spectrum based, in a sense that it involves the energy information, whereas for Transition Matrix, more like speech pattern.

As for part 2, the DSE GUI was tested by using the train data to validate the system as few changes have been made. The result was pretty good, we can say that the system can predict the score accurately within the trained data. Then, the system was tested on a new subject where the subject need to fill BDI-II survey as an actual score and read a passage two times to get scores using DSE as predicted scores. All the result were evaluated and tabulated in table 3. The percentage error were calculated by using actual score and the average of test score 1 and 2. Whereas, the test score difference shows the different between test score 1 and 2. Despite the unsatisfactory on percentage error, the test score difference shows the potential of DSE.

Table 3. DSE analysis result

Subject	Test1	Test2	Test3	Test4	Test5
Actual Score	17	13	4	13	34
Test Score 1	4.29	4.25	25.36	27.63	40
Test Score 2	5.36	2.87	28.03	26.96	35.07
% Error	71.62	72.62	567.38	107.35	37.53
Test Score Difference	1.07	1.38	2.67	1.35	4.93

6.Conclusion
As a conclusion, the relevant speech features from Bahasa Malaysia's speech were obtained. The combination of speech features for BDI-II score prediction also managed to be evaluated and the prediction system Depression Severity Evaluator has developed. Even though the DSE could not accurately measure the severity of depression, the score is not just a random numbers which have

IOP Conf. Series: Materials Science and Engineering **260** (2017) 012022 doi:10.1088/1757-899X/260/1/012022

potential to be improved. Thus, it highlights the potential of this project in identifying the severity of depression.

7.References

[1] Ministry Of Health Malaysia 2015 *National Health and Morbidity Survey of 2015: Non-Communicable Diseases, Risk Factors & Other Health Problem* **2** pp187-184

[2] Simin H and Zahra G 2011 *A survey Beck test in university students & its relationship between some related risk factor.*

[3] Lopez-Otero P, Dacia-Fernandez L and Garcia-Mateo C 2013 *A Study Of Acoustic Features for Depression Detection*

[4] Williamson J R, Quatieri T F, Helfer B S, Ciccarelli G and Mehta D D 2013 *Vocal Biomarkers of Depression Based on Motor Incoordination and Timing*

[5] Nik Hashim NNW 2014 *Analysis Of Speech Features As Potential Indicators For Depression And High Risk Suicide And Possible Predictors For The Hamilton Depression Rating (HAMD) And Beck Depression Inventory Scale (BDI-II)* 113-41

[6] Ozdas A 2001 *Analysis of Paralingguistic properties of speech for near-term suicidal risk assessment* 51-48

Acknowledgments

This work is part of the ongoing project on the development of BDI-II score prediction system funded by the Department of Mechatronics Engineering, International Islamic University Malaysia.

IOP Conf. Series: Materials Science and Engineering **260** (2017) 012023 doi:10.1088/1757-899X/260/1/012023

Prediction of dry ice mass for firefighting robot actuation

M T Ajala[1,2], Md R Khan[1], A A Shafie[1], MJE Salami[1], M I Mohamad Nor[3]

[1]Department of Mechatronics Engineering, International Islamic University, Kuala Lumpur, 53100 Malaysia (e-mail: ajala.mosud@live. iium.edu.my).

[2] Department of Industrial Maintenance Engineering, Yaba College of Technology, PMB 2011, Yaba, Lagos State, Nigeria

[3] Department of Chemical Engineering, University of Malaya, Kuala Lumpur, 53100 Malaysia

Abstract. The limitation in the performance of electric actuated firefighting robots in high-temperature fire environment has led to research on the alternative propulsion system for the mobility of firefighting robots in such environment. Capitalizing on the limitations of these electric actuators we suggested a gas-actuated propulsion system in our earlier study. The propulsion system is made up of a pneumatic motor as the actuator (for the robot) and carbon dioxide gas (self-generated from dry ice) as the power source. To satisfy the consumption requirement (9cfm) of the motor for efficient actuation of the robot in the fire environment, the volume of carbon dioxide gas, as well as the corresponding mass of the dry ice that will produce the required volume for powering and actuation of the robot, must be determined. This article, therefore, presents the computational analysis to predict the volumetric requirement and the dry ice mass sufficient to power a carbon dioxide gas propelled autonomous firefighting robot in a high-temperature environment. The governing equation of the sublimation of dry ice to carbon dioxide is established. An operating time of 2105.53s and operating pressure ranges from 137.9kPa to 482.65kPa were achieved following the consumption rate of the motor. Thus, $8.85m^3$ is computed as the volume requirement of the CAFFR while the corresponding dry ice mass for the CAFFR actuation ranges from 21.67kg to 75.83kg depending on the operating pressure.

1. Introduction

Power systems for firefighting robot (FFR) applications are dominated by engines or electric motors (dc motor). The electric actuators are efficient in performance in ordinary temperature environment but are unreliable in a high-temperature environment [1-2] which characterizes an indoor fire emergency. This is because electric actuators are not designed to function in severe temperature in addition to the loss of their magnetic and piezoelectric properties at such temperatures [2].

Research trends on firefighting robot that can operate in a high-temperature environment (fire emergency) are limited in literature because the majority of the existing literature is inclined towards the preventive task of firefighting. Thus, development of fire detection systems and algorithm take the central stage of studies in firefighting robot applications [3] and little attention is paid to mobility in the high-temperature environment.

Up till now, water-hydraulic-powered and gas-powered actuators have emerged as alternatives to electric-powered actuators in FFR applications. The water-powered actuator is the heart of a water-hydraulic propulsion system (WHPS) proposed in the studies of Zhang et al. [1] and Liljeback, Stavdahl, & Beitnes [4]. The WHPS is composed of a hydraulic motor (as the actuator) and water as the powering fluid instead of oil. As Zhang et al. [1] adopted the WHPS for an indoor firefighting robot which is remotely controlled, Liljeback, Stavdahl, & Beitnes [4] implemented the WHPS on a snake firefighting robot in a tunnel fire (outdoor) application. The FFR developed in both studies lack self-power ability required for efficient operation of an FFR in a fire emergency.

Gas-powered actuators have been applied majorly in the area of material handling of equipment. Examples include linear actuators that are developed to function under high temperatures [2],[5],[6]. The application of gas-powered actuators in firefighting robot led to the concept of the gas-actuated propulsion system (GAPS). Thus, GAPS is composed of a pneumatic actuator (linear or rotary) and a compressed air or gas as its power source. While Ono & Kato [7] introduced a rubber bellow based pneumatic actuator as a GAPS for the mobility of a pipe inspection robot (an outdoor application), Ajala, Khan, Shafie, & Salami [8] proposed a carbon-dioxide gas powered autonomous firefighting robot (CAFFR) for firefighting inside a high temperature fire environment.

Implementing the GAPS for firefighting application as suggested in our previous study [8] requires that the gas is self-generated within the FFR and produced in quantity enough to sustain the operation of FFR inside the fire. Consequently, it becomes critical to determine the volume of carbon dioxide gas as well as the mass of dry ice that will provide sufficient power to drive and actuate the gas driven motor. This is because the CAFFR, adopting GAPS concept, is projected to self-generate the carbon dioxide gas from dry ice within the robot. The schematic of achieving the power for the CAFFR actuation is shown in figure 1.

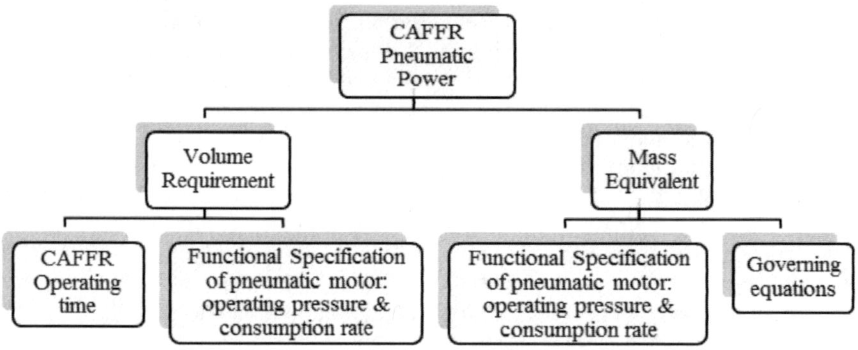

Figure 1. Schematic diagram of power required for CAFFR actuation

Figure 1 shows the schematic of producing the required power for CAFFR actuation. From the figure, the volume of carbon dioxide gas and corresponding mass of dry ice required for the CAFFR performance rest on functional specifications such as the consumption rate of the pneumatic (air) motor and the operating pressure of the motor. Also, the volume requirement can be achieved once the operating period of the CAFFR is known. Additionally, the corresponding mass of dry ice can be determined when the governing equations are established. As a result, this paper focuses on determining analytically the volume of carbon-dioxide gas as well as the mass of dry ice required to sustain the actuation of a CAFFR. The rest of this report is organized as follows: Section 2 specifies the operating time and the volume requirement of the robot. Analysis and modeling equation for predicting the mass of dry ice required to operate the CAFFR are presented in section 3 while the summary and conclusion are given in section 4.

2. Required Volume of CO_2 gas for the CAFFR Operation time

2.1. Specification for the CAFFR Operating time

In this paper, the burning time for the projected fire scenario will be used to describe the operating time of the CAFFR. So, this research adopted the fire scenario model used in the study of Lai, Tsai, & Chen [9]. A modified furnished office room dimension was used. Consequently, the office room dimension was specified as 6m length x 5m wide x 2.4m high, with a 0.6m wide x 1.8m high door. A fire load (E_f) of 11892MJ amounting to an approximate fire load density e_f of 390MJ/m^2 [9] is specified. Equation (1), adopted from Buchanan [10], was used to determine the burning time of the fire scenario. This time was thereafter employed as the robot operating time in the fire. A burning time of 2105.53s (approximately 35mins) resulted from the calculation. Thus, the operating period of the CAFFR shall be for thirty-five minutes (35mins).

$$t_b = \frac{E_f}{\Delta H_c}\left[0.18A_v\left(\frac{H_v W}{D}\right)^{1/2}\left(1 - e^{-0.036\Omega}\right)\right]^{-1} \quad (1)$$

where, t_b = burning time (s), E_f = Fire (fuel) load (MJ), ΔH_c = Heat of combustion, A_v = Area of door opening (m^2), H_v =Door Height (m), W = Room width (m), D = Depth of room (m), and Ω = Opening factor. The details of A_v and Ω are given in the appendix

2.2. *Required Volume to actuate the CAFFR*

We have proposed an FFR that can successfully operate in a high-temperature fire environment which will have the capability to return for fueling/ refueling. The selected air motor (Model No LG30FT) requires a gas supply (consumption rate) of 0.004248m^3/s [11]. Based on this consumption rate, the volume of the carbon dioxide gas that can drive the air motor for selected operating periods is computed and the result is tabulated in Table 1.

Table 1. Computed Volume requirement of $CO_{2\,(g)}$ to power a pneumatic motor.

Operating Time (s)	600	1200	1800	2400	3000	3600
Volume (m^3)	2.55	5.10	7.65	10.20	12.74	15.29

The required volume was plotted as a function of the operating time of the CAFFR and the resulting graph is shown in figure 2. A linear model presented in equation (2) was selected to establish the relationship between the volume of the carbon dioxide gas required to actuate the CAFFR and the operating time of CAFFR in the fire spot. The model has a slope (coefficient of t_b = 0.0042) which represents the consumption rate of the pneumatic motor, a coefficient of determination (R^2) value of 1, and a standard deviation of 0.00324. With a *p-value* of <0.0001, the operating time had a significant linear effect on the volume of carbon dioxide required to power CAFFR. The model was subsequently used to predict the volume of carbon dioxide gas at the calculated operating time of 2105.53s. Thus, a volume of 8.85m^3 of carbon-dioxide gas will be required for the mobility of the CAFFR inside the fire spot.

$$V_{CO_{2(g)}} = 0.0042t_b + 0.0047 \quad (2)$$

where, $V_{CO_{2(g)}}$ is volume of $CO_{2\,(g)}$ (m^3), and t_b is now referred to as the operation time (s).

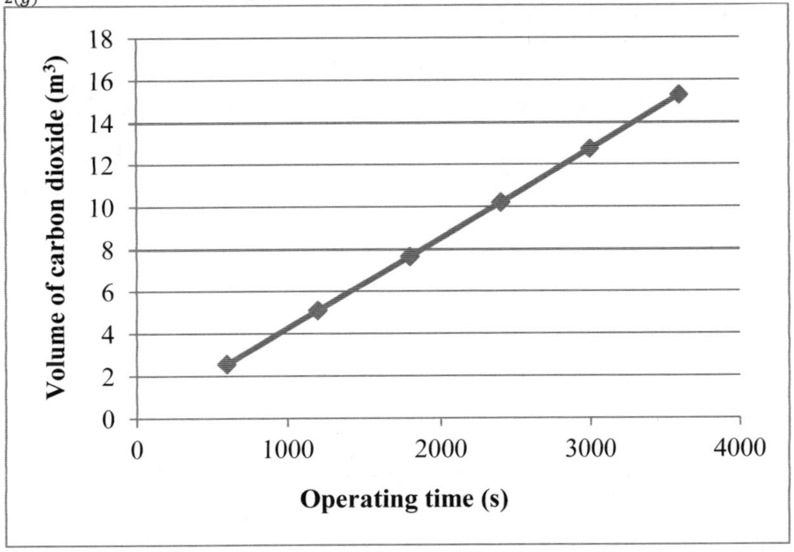

Figure 2. The plot of the volume of carbon dioxide required to power CAFFR

3. Equivalent Mass of Dry Ice

3.1. Governing Equation

A literature search on the solid-gas relationship of dry ice revealed that existing data from literature and dry ice manufacturers lack details as displayed in Table 2. The available data either covers a limited area or parameter (like the temperature of operation) is missing which makes the information inadequate. Thus, the computation of the equivalent mass of dry ice that will produce a given volume of carbon dioxide gas is examined and benchmarked with the existing data.

Table 2. Existing information.

S/N	Mass of Dry Ice (kg)	Volume of $CO_{2(g)}$ (m^3)	Temperature (°C)	Source of Information
1	0.430	0.2180	25	Wu, Kitagawa, & Tsukagoshi [12]
2	0.454	0.2500	Not Specified	DryIceProduction [13]
3	0.454	0.2500	Not Specified	Caldwell, Lewis, Shaffstall, & Johnson [14]; Airgas [15]
4	1	0.5457	21.1	AIRPRODUCT [16];

Computational analysis of different hypothetical dry ice masses at temperatures commonly used in the literatures (15.6°C, 20°C, 21.1°C, 25°C and 27°C) was performed. It was discovered that the ideal gas equation displayed in equation (3) can approximate the relationship between dry ice mass and the volume of carbon dioxide gas produced. The result of the computation was presented in Table 3.

$$PV = nRT \tag{3}$$

where, P = Pressure (kPa), n = No of moles (mol), R = Universal gas constant, T = Temperature (K)

Table 3. The volume of Carbon dioxide gas produced from a given mass of dry ice.

Mass (kg)	Volume[1] at selected temperatures				
	15.6°C	**20°C**	**21.1°C**	**25°C**	**27°C**
0.005	0.0027	0.0027	0.0027	0.0028	0.0028
0.01	0.0054	0.0055	0.0055	0.0056	0.0056
0.015	0.0081	0.0082	0.0082	0.0083	0.0084
0.02	0.0108	0.0109	0.0110	0.0111	0.0112
0.044	0.0237	0.0241	0.0242	0.0245	0.0246
0.1	0.0539	0.0547	0.0549	0.0556	0.0560
0.43	0.2316	0.2352	0.2360	0.2392	0.2408
0.45	0.2424	0.2461	0.2470	0.2503	0.252
0.454	0.2446	0.2483	0.2492	0.2525	0.2542
0.5	0.2693	0.2734	0.2745	0.2781	0.2800
1	0.5387	0.5469	0.5489	0.5562	0.5599

[1]Volumes are computed using equation (4) for P = 101.325 kPa, R = 8.3145 $Pa m^3 K^{-1} mol^{-1}$

Observations of the data in Table 3 show that 0.43kg (430g) mass of dry ice will sublime to give 0.2392 m^3 of CO_2 gas (green color legend). This gives a value of 0.0212 higher than the experimental volume recorded in the study of Wu, Kitagawa, & Tsukagoshi (2005). The sublimation temperature which was not specified in Caldwell, Lewis, Shaffstall, & Johnson, (2006) and Airgas (2002) was discovered to be 20°C (blue color legend). Also, the computed volume of the gas is recorded as 0.2482 m^3 with a 0.0018 difference to the existing value. A 1kg mass of dry ice is computed to produce 0.5466m^3 of carbon dioxide gas at 21.1°C (yellow color legend).

Table 4 shows the comparison between the existing and the predicted volume of carbon dioxide. The predicted values are very close to the existing values and root mean squared error of 0.00123 was recorded.

Table 4. Comparison of existing and computed volume of CO_2 gas from a given mass of dry ice.

S/N	Mass of Dry Ice (kg)	Temperature (°C)		Volume (m^3)		RMSE
		Existing	Predicted	Existing	Predicted	
1	0.430	25	25	0.218	0.2392	
2	0.454	Not Specified	20	0.250	0.2482	0.0123
3	1	21.1	21.1	0.5457	0.5466	

Predicting the actual mass of dry ice that will power the air motor for the operation time of the CAFFR requires further analysis of Table 3 in terms of the varying operating pressure of the pneumatic motor. This is because there are speed and torque variation in the real operation of CAFFR. Thus, the mass of dry ice required to produce the 8.85m^3 volume of carbon dioxide is dependent the operating pressure. Hence, a computation of the mass of dry ice at different operating pressures of the air motor is required.

3.2. Specification for CAFFR Operating Pressures

Pneumatic motors are powered with pressures ranging from 137.9kPa to 1034.25kPa though 620.55kPa remains the optimum operating pressure of a pneumatic motor. Also, carbon dioxide gas phase diagram (figure 3) revealed that the critical pressure of the gas is 517.13kPa (triple point pressure). At this pressure, the gas exists in its three phases. But the liquid is not desirable for this work, so our design pressure should be less than 517.13kPa which will make the interaction remains as solid-gas expansion. As a result, the operating pressures of the CAFFR range from 137.9kPa to 482.65kPa. This pressure range is therefore used to determine the mass equivalent. Hence, the equivalent mass is computed by using equation (4). Equation (4) was realized by expressing equation (3) in terms of the operating pressure and a given mass of dry ice. The result is displayed in Table 5.

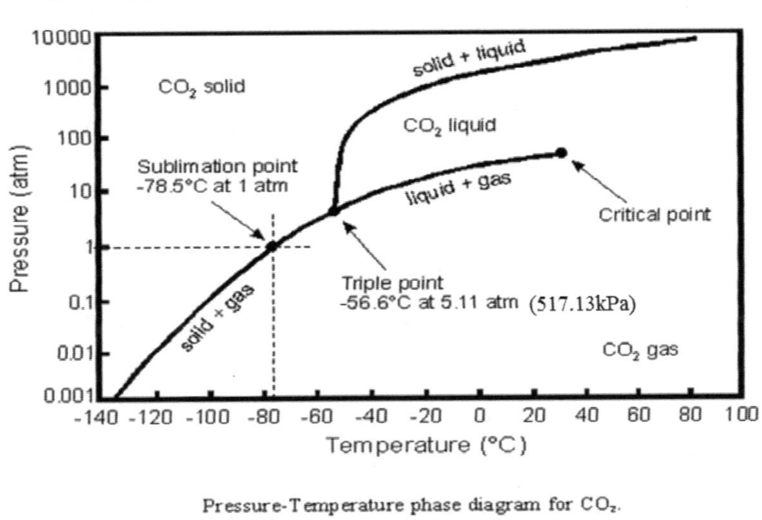

Figure 3. Phase diagram of carbon dioxide (Photo reprinted from Stevengoddard) [17].

IOP Conf. Series: Materials Science and Engineering **260** (2017) 012023 doi:10.1088/1757-899X/260/1/012023

$$\frac{P}{m_d} = k \quad (4)$$

where, m_d (kg) is a given mass of dry ice, and P is operating pressure (kPa)

Table 5. Computation of the mass equivalent of the required volume of dry ice for different operating pressures @ 25°C

Pressure (kPa)	**Mass of dry Ice (kg)**
137.90	21.67
206.85	32.50
275.80	43.33
344.75	54.17
413.70	65.00
482.65	75.83

The result of the plot of dry ice mass as a function of the operating pressure is displayed in figure 4. A linear model represented by equation (5) was selected to establish the relation between the mass of dry ice (m_d) required to actuate the CAFFR and the operating pressure of the pneumatic actuator (P_{motor}). The operating pressure linear effect on the dry ice mass was quite significant with a *p-value* of <0.0001. The linear model had an R^2 value of 1, and a standard deviation of 0.00293. The inverse of the slope of the graph (coefficient of P_{motor} = 0.1571) represents the actual value of *k* in equation (4). It was discovered that for every 68.95kPa (10psi) rise in pressure an approximate 10.83kg (68.95 × 0.1571) mass of dry ice is consumed.

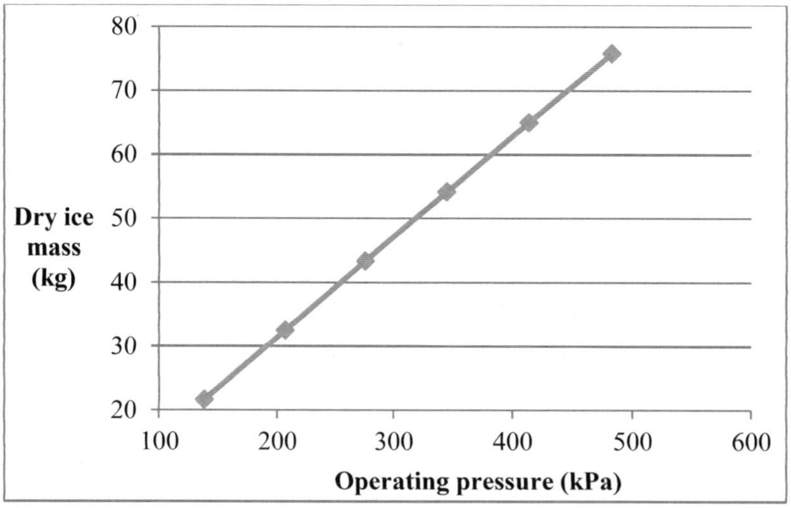

Figure 4. The plot of dry ice mass required to power an FFR for different operating pressures.

$$m_d = 0.1571 P_{motor} - 1 * 10^{-13} \quad (5)$$

Validation of the model equation was achieved by comparing the computation using equation (4) and predictions using the linear model in equation (5). The result is displayed in Table 6. An RSME value of 0.00816 was obtained. Thus, the dry ice mass required to power CAFFR at the selected operating pressures range from 21.67kg to 75.83kg.

Table 6. Comparison of computed and predicted mass of dry ice required to power CAFFR.

Pressure (kPa)	Mass (kg) using equation (4)	Mass Prediction from the model equation	RMSE
137.9	21.67	21.66	
206.85	32.50	32.50	
275.8	43.33	43.33	0.00816
344.75	54.17	54.16	
413.7	65.00	64.99	
482.65	75.83	75.82	

4. Conclusion

The amount of carbon-dioxide gas that would make a firefighting robot, CAFFR, designed to operate successfully in an indoor fire emergency is discussed. It was established that the ideal gas equation can be used to determine the volume of carbon-dioxide gas that can be produced from a given mass of dry at any temperature. Different masses of dry ice were examined using this equation and the corresponding volumes of carbon dioxide gas were calculated. The computed volumes were validated using the existing information on mass to volume sublimation of dry ice. Base on the consumption rate of the motor, operating pressures ranging from 137.9kPa to 482.65kPa was specified for the power of the CAFFR. Furthermore, a mathematical model was developed for the actual mass of dry ice that can conveniently power the CAFFR for the operation time of 2105.53s and the different operating pressures. It was discovered that the mass equivalent depends upon the operating pressure. Thus the actual dry ice mass ranges from 21.67kg to 75.83kg depending on the operating pressure used.

Appendix

$$A_v = H_v \times W_d; \tag{2.1}$$

$$\Omega = \frac{A_t - A_v}{A_v (H_v)^{1/2}} \tag{2.2}$$

$$A_t = 2(l_1 l_2 + l_1 l_3 + l_2 l_3); \tag{2.3}$$

A_t= Total internal surface area of the office room (m^2), l_1 = Length of room (m), l_2 = Width of room (m), and l_3 = Height of room (m), W_d = Width of door (m)

References

[1] Zhang L, Kitagawa A, Tsukagoshi H. A Study on Water Hydraulic Fire-Fighting Robot. Proc JFPS Int Symp Fluid Power. 1999;1999(4):727–32.

[2] Suzumori K, Matsuoka H, Wakimoto S. Novel Actuator Driven with Phase Transition of Working Fluid for Uses in Wide Temperature Range. In: In Intelligent Robots and Systems (IROS), 2012 IEEE/RSJ International Conference on. IEEE; 2012. p. 616–21.

[3] AlHaza T, Alsadoon A, Alhusinan Z, Jarwali M, Alsaif K. New Concept for Indoor Fire Fighting Robot. Procedia - Soc Behav Sci [Internet]. Elsevier B.V.; 2015;195:2343–52. Available from: http://linkinghub.elsevier.com/retrieve/pii/S1877042815036708

[4] Liljebäck P, Stavdahl Ø, Beitnes A. SnakeFighter - Development of a Water Hydraulic Fire Fighting Snake Robot. In: Control, Automation, Robotics and Vision, 2006 ICARCV '06 9th International Conference on. IEEE; 2006. p. 1–6.

[5] Matsuoka H, Suzumori K. Gas / Liquid Phase Change Actuator for Use in Extreme

Temperature Environments. Int J Autom Technol. 2014;8(2):140–6.
[6] Matsuoka H, Suzumori K, Kanda T. Development of a gas/liquid phase change actuator for high temperatures. ROBOMECH J [Internet]. Springer International Publishing; 2016;3(1):1–7. Available from: "http://dx.doi.org/10.1186/s40648-016-0041-7
[7] Ono M, Kato S. A study of an earthworm type inspection robot movable in long pipes. Int J Adv Robot Syst. 2010;7(1):85–90.
[8] Ajala MT, Raisuddin K, Shafie AA, Salami M-JE. Development of a New Concept for Fire Fighting Robot Propulsion System. In: International Conference on Material, Industrial and Mechanical Engineering (ICMIME2016 - ICONTES2016). Kuala Lumpur, Malaysia: ICONTES2016; 2016. p. 90–1.
[9] Lai CM, Tsai MJ, Chen CJ. FULL-SCALE ANALYSIS ON FIRE CHARACTERISTICS OF A FURNISHED OFFICE ROOM.
[10] Buchanan AH. Structural Design for Fire Safety. John Wiley & Sons Ltd; 2001. 421 p.
[11] TONSON. M3-LG30. TAIWAN: TONSON AIR MOTORS MFG CORP; 2015.
[12] Wu H, Kitagawa A, Tsukagoshi H. DEVELOPMENT OF A PORTABLE PNEUMATIC SOURCE USING PHASE TRANSITION AT THE TRIPLE POINT. In 2005. p. 310–5.
[13] DryIceProduction. All About Dry Ice [Internet]. Dry Ice Cold Jet. [cited 2016 Mar 2]. Available from: http://www.coldjet.com/minisites/dryiceproduction/en/dry-ice.php
[14] Caldwell DC, Lewis RJ, Shaffstall RM, Johnson RD. The Sublimation Rate of Dry Ice Packaged in Commonly Used Quantities by the Air Cargo Industry [Internet]. FEDERAL AVIATION ADMINISTRATION. OKLAHOMA CITY OK: CIVIL AEROMEDICAL INST.; 2006. Available from: http://oai.dtic.mil/oai/oai?verb=getRecord&metadataPrefix=html&identifier=ADA461451
[15] Airgas. Carbon Dioxide MSDS. Radnor PA: AIRGAS INC.; 2002. 1-12 p.
[16] AIRPRODUCT. Material safety data sheet Carbon dioxide. Air Products & Chemicals Inc.; 1993 p. 1–8.
[17] CO2 phase diagram [Internet]. [cited 2016 Jan 22]. Available from: https://stevengoddard.wordpress.com/2010/09/05/the-freezing-point-and-the-dew point-part-2/

IOP Conf. Series: Materials Science and Engineering **260** (2017) 012024 doi:10.1088/1757-899X/260/1/012024

Modified Ashworth Scale (MAS) Model based on Clinical Data Measurement towards Quantitative Evaluation of Upper Limb Spasticity

A Ahmad Puzi[1, a], S N Sidek[1, b], H Mat Rosly[1, c], N Daud[2, d], H Md Yusof[1, e]

[1]Department of Mechatronics Engineering, International Islamic University Malaysia, Kuala Lumpur, Malaysia
[2]Faculty of Allied Health Sciences International Islamic University Malaysia, Kuantan, Malaysia

[a]asmarani8487@gmail.com

Abstract. Spasticity is common symptom presented amongst people with sensorimotor disabilities. Imbalanced signals from the central nervous systems (CNS) which are composed of the brain and spinal cord to the muscles ultimately leading to the injury and death of motor neurons. In clinical practice, the therapist assesses muscle spasticity using a standard assessment tool like Modified Ashworth Scale (MAS), Modified Tardiue Scale (MTS) or Fugl-Meyer Assessment (FMA). This is done subjectively based on the experience and perception of the therapist subjected to the patient fatigue level and body posture. However, the inconsistency in the assessment is prevalent and could affect the efficacy of the rehabilitation process. Thus, the aim of this paper is to describe the methodology of data collection and the quantitative model of MAS developed to satisfy its description. Two subjects with MAS of 2 and 3 spasticity levels were involved in the clinical data measurement. Their level of spasticity was verified by expert therapist using current practice. Data collection was established using mechanical system equipped with data acquisition system and LABVIEW software. The procedure engaged repeated series of flexion of the affected arm that was moved against the platform using a lever mechanism and performed by the therapist. The data was then analyzed to investigate the characteristics of spasticity signal in correspondence to the MAS description. Experimental results revealed that the methodology used to quantify spasticity satisfied the MAS tool requirement according to the description. Therefore, the result is crucial and useful towards the development of formal spasticity quantification model.

1. Introduction

Muscle control disability is resulted from imbalanced signals from the central nervous system (CNS) which are composed of brain and spinal cord to the muscles. This factor causes restriction for the disables to perform their Activities of Daily Living (ADL). Upper limbs are part of the human body that is used in many tasks such as eating, writing and other prehensile actions. Thus, the affected subjects have to undergo rehabilitation training program in order to recover from the disabilities.

Stretching the muscle repeatedly may potentially increase the muscle tension and resistance to passive movement [1]. Therefore, subjects need the assistance from expert therapist to perform the training. Three common causes of spasticity are due to stroke, cerebral palsy and traumatic brain injury [2]. According to National Stroke Association Malaysia (NASAM), stroke is the second largest cause of death after heart disease, while cerebral palsy is high amongst the children. It is important for subject to have regular training and therapy in order to reduce the stiffness or tightness of the muscles in order to recover muscle control.

Before undergoing training, subject need to be assessed by therapist to determine his muscle spasticity severity. In the current assessment methodology, this is done purely on qualitative basis and subjected to therapist's jurisdiction and experience [3, 4]. Thus, the assessment is susceptible to variation and this could pose a challenge to monitor the progress of the subject effectively especially if the training sessions are conducted by different therapists. There exist quite a number of assessment tools that were used in clinical practice to measure the level of spasticity. These tools are going to be discussed in the next section.

In this paper, clinical data measurement was elaborated towards the development of upper limb quantitative assessment model. It is composed of four sections where in section 2, spasticity

IOP Conf. Series: Materials Science and Engineering **260** (2017) 012024 doi:10.1088/1757-899X/260/1/012024

assessment tools are briefly described. The subsequent sections explain on the system description, the methodology of clinical data measurement and, the result with discussion.

2. Spasticity Assessment Tools

Management of spasticity is required for clinical evaluation of the subject. Clinical assessment of spasticity should be based on a validated assessment system. Most of conventional clinical therapies refer to certain assessment tools such Fugl-Meyer Assessment (FMA), Modified Tardieu Scales (MTS) and Modified Ashworth Scale (MAS).

Fugl-Meyer Asessment tool is used to assess motor control of the patient. It is created to assess motor function, balance, sensation and joint function in subjects with post-stroke hemiplegia [5, 6]. It is applied clinically and in research to determine the severity level of the disease, describe motor recovery, and to plan and assess treatment [7]. However, this tool is inadequate for spasticity assessment due to its main function more on motor control of particular affected part assessment.

The Tardieu scale and the modified Tardieu scale [8, 10] are the tool that takes into account resistant to the passive movement at both slow and fast speeds. It measures quality of muscle reaction to passive movement and angle of muscle reaction at the point of resistance to the fastest movement speed [10]. Several researches indicated that MTS can be used as reliable measurement for quantifying spasticity of lower limb [11, 12] due to high inter rater validity results.

The more famously used tool found in the practice is the Ashworth scale or the modified Ashworth scale [13, 15].The scale measures the resistance and spasticity (or catch) in the joint during passive movement. The resistance is scored from 0 to 4 (Ashworth scale) and from 0 to 5 (modified Ashworth scale). The score is based on the classification and description as shown in Table 1. According to [14], there is high inter-rater reliability when using the MAS compared to MTS to assess the upper limb spasticity [14, 15].

The shortcomings of all of these tools are that they are subjective and dependent on the therapist impression of spasticity on passive movement. Previous studies have shown that spasticity scales provide insufficient information about muscles involved in spastic movement and the reliability and validity of the scales has been questioned in clinical practice [16, 17]. Thus, introducing quantitative measurement based on the MAS may help to overcome the inadequacy of current method.

Table 1 Modified Ashworth Scale (MAS) Scoring System

Grade	Description
0	No increase in muscle tone
1	Slight increase in muscle tone, manifested by a catch and release or by minimal resistance at the end of the range of motion when the affected part(s) is moved in flexion or extension
1+	Slight increase in muscle tone, manifested by a catch, followed by minimal resistance throughout the remainder (less than half) of the ROM
2	More marked increase in muscle tone through most of the ROM, but affected part(s) easily moved
3	Considerable increase in muscle tone, passive movement difficult
4	Affected part(s) rigid in flexion or extension

3. System Description

The measurement of spasticity level is done with the help of a mechanical system. It is used to collect data of the passive flexion movement of upper extremity. As shown in Figure 1, the system has a kinematic structure of 1-DOF system which consists of a single revolute joint. This preliminary system is composed of four main parts which are the lever, platform, torque sensor and potentiometer. Platform was placed along X to be revolute within fixed axis at angle θ_1.Torque sensor and potentiometer are attached to the pivotal point of the system. The function of each part is described in Table 2.

Table 2 System Desription

Part/Component	Function
Lever (manual handle)	To make passive movement of forearm (flexion) and control by human (therapist).
Platform	To hold subject's forearm during passive movement.
Torque Sensor (custom made)	To measure torque exert by the subject's forearm.
Potentiometer (981 HE)	To measure angle between upper and lower limb and to detect the catch position.

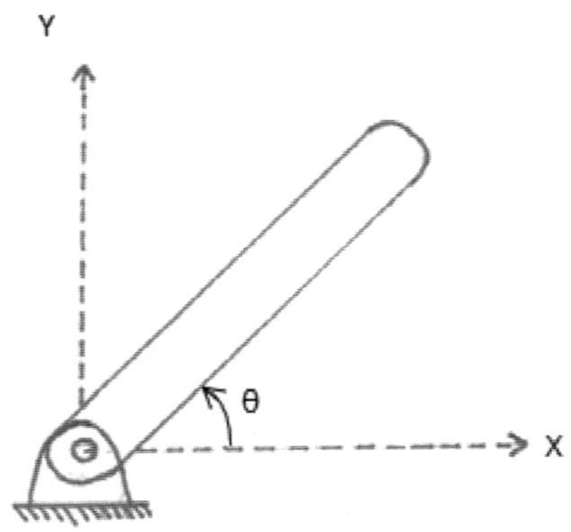

Figure 1 Kinematic Model

4. Methodology

Before the start of data collection process, an ethical clearance has been obtained from the IIUM Research Ethics Committee (IREC) [Ref No: IREC 659]. The selection of subjects is based on certain criteria that are they must have abnormality in motor control with hemiparesis condition and aged over 18 years old. The work is only focused on the upper limb part of the subjects. The subject was brief before the data collection process started and his consent was obtained. The clinical data collection was conducted at the Physical Rehabilitation Centre, IIUM Kuantan and the ambient temperature of the room where the experiment took place was set to 24^0C.

4.1 Data collection set-up

The therapist explained to the subject on the system and the process of data collection together with safety issues. The therapist was assisted by a technical team on the procedure in handling the system during measurement process.

Torque sensor was placed along the pivot point of the elbow joint to measure the spasticity experienced by the subject during passive movement induced by the therapist based on lever mechanism. A single turn potentiometer was used to measure range of motion (ROM) and angle where the catch happened. Catch is defined as a sudden appearance of increased muscle activity in response to a fast passive stretch, which leads to an abrupt stop or sudden increased resistance during the movement, at a certain angle before ROM is reached[18,19]. Both the torque sensor and potentiometer are connected to a DAQ system and the data was recorded using Labview software.

Before the measurement of spasticity by using the mechanical system was conducted the level of muscle spasticity for each subject was accessed by therapist using manual method which is based on the Modified Ashworth Scale tool. Then, the subject was asked to place his arm in the supine position (relax position) and his forearm placed on the platform with the elbow joint parallel to the pivot point of the system. The length of the forearm was measured and recorded. The platform was then moved against the subject's forearm as it flexed at slow and fast speed by the therapist using the lever throughout the full range of motion. The slow movement is to determine the accessibility range of the motion of the affected upper limb. Meanwhile the fast movement is to detect the catch position throughout the full range of motion. Based on the previous study, the average active range of motion (ROM) of elbow flexion was 130 degree [20]. The data measured by the sensors were measured and recorded.

The process is repeated three times for each subject to get the best estimated model of the muscle spasticity. The flow of the data collection process is captured in Figure 2.

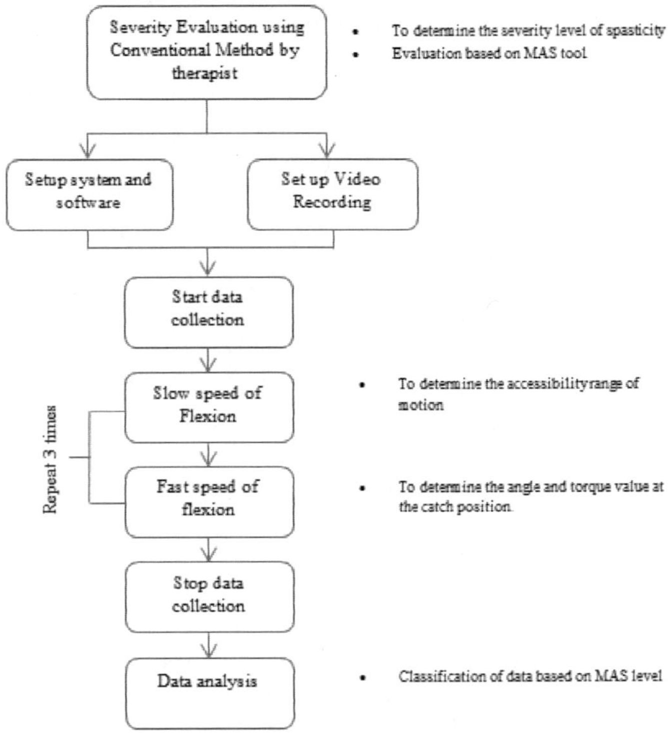

Figure 2 Flowchart of experiment protocol

5. Result and Discussion

The data for torque representing muscle spasticity together with the flexion angle of the forearm were recorded for all the sessions. Figure 3 and 4 depicts the signal measured for subject 1 and 2 respectively. The torque value is plotted against the flexion angle in order especially to identify the catch position.

Based on Figure 3, it shows that the full ROM is 120 ^{0}and catch happen at 59 ^{0} which is less than half of the ROM with a minimal release. From the MAS description, the condition satisfies MAS level 2 as more marked increase in muscle tone is seen through most of the ROM. The catch position is detected at the peak point of the torque which happens at 59 ^{0} flexion angle.

In Figure 4, the full ROM is shown to be 135 ^{0} and the resistance keeps increasing throughout the full ROM. Prior to that, there is no catch and release appears. The graph satisfies the description of MAS level 3 as a considerable increase in muscle tone with passive movement difficulty is captured. Table 3 summarizes the description of the subjects with their respective level of spasticity.

Table 3 Results of clinical assessment based on the preliminary system

Subject Number	1	2
Gender	Male	Male
Spastic Arm	Left	Right
Cause of symptom	Traumatic Brain Injury	Stroke
MAS level	2	3
Catch position (degree)	59	None

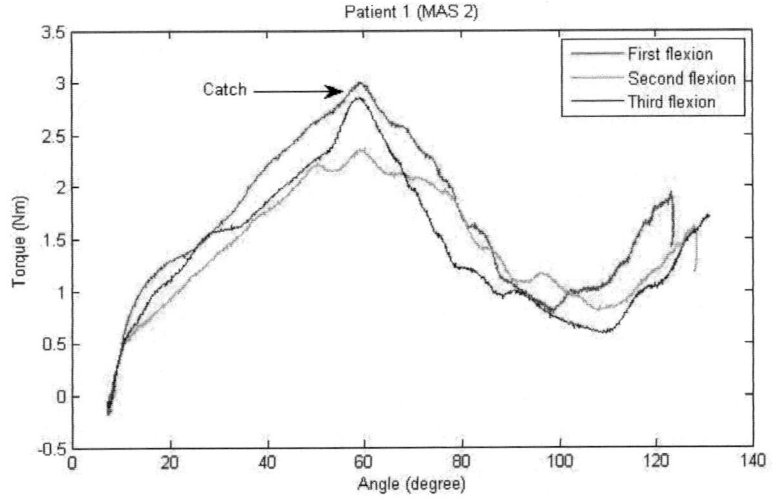

Figure 3 Clinical data for Subject 1 with MAS level 2

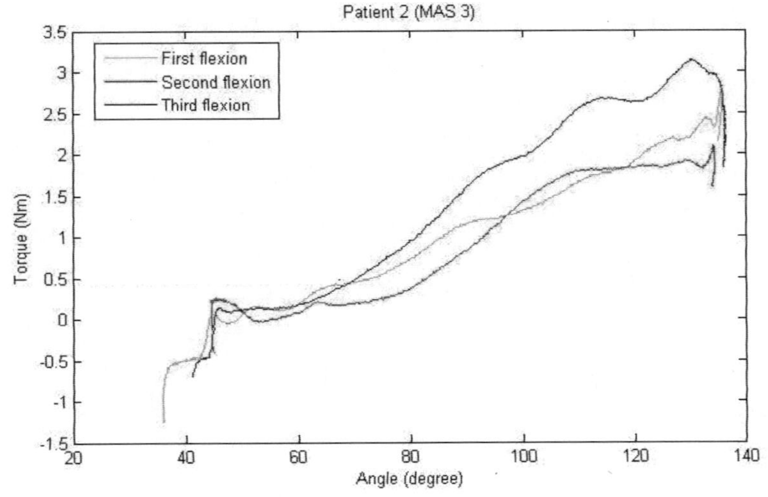

Figure 4 Clinical data for subject 2 with MAS level 3

6. Conclusion

In a nutshell, the paper proposes the use of a mechanical system to measure muscle spasticity of upper limb. The methodology leverages on a standard mean in data collection of muscle tone during fast and slow movement which can be analyzed quantitatively. The preliminary experimental result also suggests that the method for spasticity quantification satisfies the MAS tool requirement based on its description. We have planned in the future to collect more sample data with larger sample size in order to generalize the model.

Acknowledgment

The authors wish to gratefully acknowledge the grant funding from the Ministry of Science, Technology and Innovation (MOSTI) Malaysia (SF15-015-0065).

References

[1] Woods, K., Bishop, P., & Jones, E. (2007). Warm-up and stretching in the prevention of muscular injury. *Sports Medicine*, *37*(12), 1089-1099.

[2] Nair, K. P., & Marsden, J. (2014). The management of spasticity in adults. *bmj*, *349*, g4737.

[3] Priebe, M. M., Sherwood, A. M., Thornby, J. I., Kharas, N. F., & Markowski, J. (1996). Clinical assessment of spasticity in spinal cord injury: a multidimensional problem. *Archives of physical medicine and rehabilitation*, *77*(7), 713-716.

[4] Blackburn, M., van Vliet, P., & Mockett, S. P. (2002). Reliability of measurements obtained with the modified Ashworth scale in the lower extremities of people with stroke. *Physical therapy*, *82*(1), 25-34.

[5] Fugl-Meyer, A. R., & Jääskö, L. (1979). Post-stroke hemiplegia and ADL-performance. *Scandinavian journal of rehabilitation medicine. Supplement*, *7*, 140-152.

[6] Gladstone, D. J., Danells, C. J., & Black, S. E. (2002). The Fugl-Meyer assessment of motor recovery after stroke: a critical review of its measurement properties. *Neurorehabilitation and neural repair*, *16*(3), 232-240.

[7] Opheim, A., Danielsson, A., Murphy, M. A., Persson, H. C., & Sunnerhagen, K. S. (2014). Upper-limb spasticity during the first year after stroke: stroke arm longitudinal study at the University of Gothenburg. *American Journal of Physical Medicine & Rehabilitation*, *93*(10), 884-896.

[8] Tardieu, G. (1954). A la recherche d'une technique de mesure de la spasticite. *Rev Neurol*, *91*, 143-144.

[9] Fosang, A. L., Galea, M. P., McCoy, A. T., Reddihough, D. S., & Story, I. (2003). Measures of muscle and joint performance in the lower limb of children with cerebral palsy. *Developmental medicine and child neurology*, *45*(10), 664-670.

[10] Mehrholz, J., Wagner, K., Meißner, D., Grundmann, K., Zange, C., Koch, R., & Pohl, M. (2005). Reliability of the Modified Tardieu Scale and the Modified Ashworth Scale in adult patients with severe brain injury: a comparison study. *Clinical rehabilitation*, *19*(7), 751-759.

[11] Azarnia, S. (2017). Inter-rater reliability of the modified Tardieu scale for the assessment of knee extensor spasticity in patient with multiple sclerosis. *Koomesh*, *19*(1), 220-226.

[12] Ansari, N. N., Naghdi, S., Hasson, S., Rastgoo, M., Amini, M., & Forogh, B. (2013). Clinical assessment of ankle plantarflexor spasticity in adult patients after stroke: inter-and intra-rater reliability of the Modified Tardieu Scale. *Brain injury*, *27*(5), 605-612.

[13] ASHWORTH, B. (1964). Preliminary trial of carisoprodal in multiple sclerosis. *Practitioner*, *192*, 540-542.

[14] Bohannon, R., Smith, M. (1987). Interrater reliability of a modified Ashworth scale of muscle spasticity. Physical Therapy 67(2): 206.

[15] Li, F., Wu, Y., & Li, X. (2014). Test-retest reliability and inter-rater reliability of the Modified Tardieu Scale and the Modified Ashworth Scale in hemiplegic patients with stroke. *European journal of physical and rehabilitation medicine*, *50*(1), 9-15.

[16] Malhotra, S., Pandyan, A. D., Day, C. R., Jones, P. W., & Hermens, H. (2009). Spasticity, an impairment that is poorly defined and poorly measured. *Clinical rehabilitation*, *23*(7), 651-658.

[17] Fleuren, J. F., Voerman, G. E., Erren-Wolters, C. V., Snoek, G. J., Rietman, J. S., Hermens, H. J., & Nene, A. V. (2010). Stop using the Ashworth Scale for the assessment of spasticity. *Journal of Neurology, Neurosurgery & Psychiatry*, *81*(1), 46-52.

[18] Mayer, N. H. (1997). Clinicophysiologic concepts of spasticity and motor dysfunction in adults with an upper motoneuron lesion. *Muscle & nerve*, *20*(S6), 1-14.

[19] Scholtes, V. A., Becher, J. G., Beelen, A., & Lankhorst, G. J. (2006). Clinical assessment of spasticity in children with cerebral palsy: a critical review of available instruments. *Developmental medicine and child neurology*, *48*(1), 64-73.

[20] Gates, D. H., Walters, L. S., Cowley, J., Wilken, J. M., & Resnik, L. (2016). Range of motion requirements for upper-limb activities of daily living. *American Journal of Occupational Therapy*, *70*(1), 7001350010p1-7001350010p10.

IOP Conf. Series: Materials Science and Engineering **260** (2017) 012025 doi:10.1088/1757-899X/260/1/012025

Network-Based Real-time Integrated Fire Detection and Alarm (FDA) System with Building Automation

F Anwar[1], R I Boby[1], M M Rashid[2], M M Alam[1] and Z Shaikh[2].

[1] Electrical and Computer Engineering, Kulliyyah of Engineering, International Islamic University Malaysia, 53100, Gombak, Kuala Lumpur, Malaysia.
[2] Mechatronics Engineering, Kulliyyah of Engineering, International Islamic University Malaysia, Gombak, Kuala Lumpur, Malaysia.

Abstract. Fire alarm systems have become increasingly an important lifesaving technology in many aspects, such as applications to detect, monitor and control any fire hazard. A large sum of money is being spent annually to install and maintain the fire alarm systems in buildings to protect property and lives from the unexpected spread of fire. Several methods are already developed and it is improving on a daily basis to reduce the cost as well as increase quality. An integrated Fire Detection and Alarm (FDA) systems with building automation was studied, to reduce cost and improve their reliability by preventing false alarm. This work proposes an improved framework for FDA system to ensure a robust intelligent network of FDA control panels in real-time. A shortest path algorithmic was chosen for series of buildings connected by fiber optic network. The framework shares information and communicates with each fire alarm panels connected in peer to peer configuration and declare the network state using network address declaration from any building connected in network. The fiber-optic connection was proposed to reduce signal noises, thus increasing large area coverage, real-time communication and long-term safety. Based on this proposed method an experimental setup was designed and a prototype system was developed to validate the performance in practice. Also, the distributed network system was proposed to connect with an optional remote monitoring terminal panel to validate proposed network performance and ensure fire survivability where the information is sequentially transmitted. The proposed FDA system is different from traditional fire alarm and detection system in terms of topology as it manages group of buildings in an optimal and efficient manner.Introduction

1. Introduction

Fire detection process for trading and housing places is a main requirement to reduce destruction of personal property due to fire incident both natural and induced. Combining the network with controller approach, an innovative improvement has been made in building automation. The real-time surveillance system not only monitors the building automation ministrations but also control the FDMS by gathering, observing and storing related information. Among the basic technologies, networking had been used for improved building automation controller construction. Research notes that the concept of networking based controller used in many building automation process are more effective in fire detection and protection [1-3].

It has been noted that, the integration of networking, detection modules, sensors and a server work as a system to transmit data for fire detection. The communication method of exchanging information takes place via a digital communication medium. The exclusive design of the detectors spot the exact region of the fire effortlessly. Also, the workstations are organized to monitor the fire detectors at regular intervals in order to estimate any repair needed in the system. Most of the network-based fire detection systems are recently manufactured and traded with their proprietary network protocols [4-6].

The design criteria of any fire alarm systems; should differentiate between a false and genuine alarm, allowing quick rescue and implementing accurate evacuation process, thus aiding the fire defenders. Furthermore, the fire alarm system should have the ability to reduce energy intake, in comparison with the existing systems. Hence, an intelligent fire system is required. These innovations will make up a new openings and occasions in marketing fields for fire detection, safety and alarm systems [7-8]. Researcher also designed fire detection system unified with smoke detection sensors, health monitoring devices, display module, wireless communication modules, multi detectors sensors, distant approach through networking module, temperature sensors and liner heat sequences monitoring system. The main ideology to combine these techniques is to develop trustworthiness, reduce the occurrence of fake alarm, discriminate between a natural or induced fire smoke and most importantly increase the life of property by fire protection systems. The conventional fire detection system requires much improvement and plenty of research nowadays to increase detection sensitivity of any sign of smoke at a particular area [9-11]. Similarly, the cost-effective sensor network system have more advantages then conventional to enhance duribility and reliability of an intelligent automation system [13].

A system has also been developed with wireless sensor based neural-network for fire detection. The neural-network was used to increase the life expectancy of the network. The researcher used simulation process to show the performance [14]. They also validate that the simulation process is efficient to reduce communications between sensor nodes. Another research group developed grey-fuzzy controller based neural-network to detect the fire detection in the Tibet region. The design was developed to monitor signal values of fire detection sensors to calculate the value of smoke, temperature and CO_2 density in air during the detection period, where the 30 seconds dealy detection technique was used. Finally, the data was used as prediction result input to ANFIS system to determine the detecting probability [15]. A communication network of a system was proposed to extend by an intermediate-layer or dependability layer. The layer should be transparent to operate on seamless intricated network. Thus, the group addressed reliability in terms of fault tolerance. They basically offered a redundant network in the system and named as heartbeat mechanism. Where the system acknowledgment and specific message format will hold the safety requirements [16]. A Network feedback based Internet of things (IOT) system and method was proposed for home and small business. The main motive was to connect, control and monitor more than one appliances together. The proposed system is used to understand electrical consumption, make decisions on fault detection, prompt to save energy, upgradation of the system, Internet based monitoring and alarm for the system as well as prevention of device over-heated or frozen, damaged device detection, possibility of fire losses and loss of life. The proposed system also includes Internet communication based software and data communication, hardware interfaces on safety issue in communication line and optional wireless outlet [17].

A stand-alone security system was transformed into intelligent network based on computer controlled automatic building automation system. The purpose of the research was to design a pilot models for security using wired network technology. web based virtual instrument was introduced to monitor and control the base from long distance. It was claimed that, they can be re-usable with little customization and increasing some more nodes. Though the system shows more potential but it is not reliable in the long run, because of maintenance and non-automatic plug and play service [18]. Another research group proposed wirelessly node detection using FPGA based management. It was a Zigbee technology using LabVIEW software. LabVIEW developed FPGAs are used for high performance system, where this system will be highly reliable [19]. An intelligent alarm system was designed with photoelectric materials to detect smoke and temperature. The system use AT89C51 unit as hardware control and monitoring. The controller unit converts the analogue signal to digital signals and produce machine output. The system is simple and low cost [20]. Wireless Sensors Network for FDS also became more popular because, detection message will be delivered through wireless sensors. The wireless sensor nodes include catalytic gas sensors, controller and wireless transceiver using BACnet protocol but the weakness is noise and false alarm. To overcome this problem controller area network (CAN) was develop through a radio protocol known as wireless protocol [21-23]. A system was developed to reduce the false alarm in the train. The system was developed by smart sensors and microcontroller. It will automatically detect fire and use mobile network to share information. The nodes use pattern matching system where a predefined set of rules were use as packet of traffic. To reduce false alarm, the signals are re-analysed with the predefine packet of traffic and detect the signal as bad traffic. But, the main drawback of the network is that, the ratio of the false alarm depends on the performance of detection engine. To reduce this problem signature-based traffic methode was developed [24-25].

Researcher also developed UV and infrared stereovision sensors to detect fire in low visibility range. These sensors can be used for long term fire detection system using 3D coordination effect of fire. Visual servo system was used to target the fire location and place the suppuration module towards the fire to remove fire immediately. The system perfectly worked on visibility condition of 1.62 meter. The surrounding environment was developed with simulated wind and smoke effect [26]. A probabilistic classification method was developed using thermal detection method. The algorithm automatically locates fires by thermal imaging technique, where visibility is zero. This algorithm was also stablished through test data set [27]. Again, a signal traffic management system for FDS was proposed for smart cities. Where, a robust fire detection method was proposed using dictionary method. The method computes each patch using linear SVM [28-31].

Optical fiber sensors are proven to be used in industrial applications because of long durability. Common distributed sensors performance is exploited through optical sensors because of fire detection special resolution. The sensors can detect long distance with higher accuracy [32]. A distribution network using optical fiber for close room environment is highly effective for fire detection system. The problem of these sensor was that, it can work only in close room. Though, the method demonstrated that it represents a robust and reliable technique to localize the fire place [33]. The communication system for fire detection system should be more reliable and durable. The system will be a bi-directional highspeed distribution network and the data will be carried out through data bus signal [34-35].

For optical fiber network all possible network combination is suitable because of high data transfer rate. In the next generation all network architecture designing on optical fiber, however P2P network are different from each other. For AON (active optical network) and wavelength division multiplexing passive optical network (WDM PON) fiber is the best choice. Even future 5G network also designing on fiber network due to the data transfer capacity. Also, cloud based network excess relay on optical fiber network [36-38].

The proposed framework of powerful fire alarm detection system depends upon Ring Topology. In this work, the focus is set to develop an efficient fire alarm detection system, applying a topology which depends on algorithmic approach. This framework enlightens currently available process of powerful fire detection system by minimizing the time to activate an alarm system. Moreover, by integrating the FDA system into the buildings, it will allow the system to react to variations in real time, thus able to generate alarm during the state of an emergency.

Table 1. Summary of the important research developments.

Methods	Years	Advantages	Disadvantages
Network feedback based IOT	2014	1. connect, control and monitor more than one appliances together through IOT.	1. Constant monitoring and services is required.
Wired Building	2015	1. pilot models with wired	1. Mainte-nance required.

Automation		network. 2. Web based instrument.	2. Automatic nodal updates not possible.
Network-based FDS via CAN	2004	1. Very short response time. 2. Direct data accusation for accuracy.	1. Data receive directly from the fire detectors any interruption means system loss.
FPGA based smart system.	2016	1. Smart system. 2. Auto update.	1. Use for stand-alone system.
BACnet-based FDS	2014	1. Monitoring system using MS/TP protocol. 2. reference model is operated by wireless-BACnet for flexibility.	1. Signal is directly dependent upon the traffic load. 2. Inter-operability of different vendors' field devices makes problems.
Address-able FACS	2015	1. Broad range of device connecting capabilities.	1. Standard capacity of maximum 127 points.
Dedicated Computer Controlled system	2016	1. Added power, Internet connectivity and remote service capabilities	1. Failure of dedicated computer or connection problem can fail the system. 2. Required maintenance.
Address-able VAD (Visual Alarm Device)	2010	1. Visual Alarm Device.	1. For a small plant. 2. Extension not possible.
Pattern matching process	2015	1. Use predefine pattern matching process. 2. Signature based traffic.	1. The performance depends on detection engine.
Mechatro-nic approach	2016	1. Autonomous fire-suppression module was used. 2. Versatile device addable.	1. The autonomous system was used is expensive.
Optical fiber sensors	2014	1. Long durability. 2. Long distance detection.	1. Problem of hot air.

2. Network-Based Fire Detection System

As it is known that a number of existing technologies are implemented in the past for effective fire detection system in a real-time environment which faces a major drawback in the transmission of data over a large distance. A long-distance data transmission using normal communication cable can lead to a drop-in signal, hence making it inconvenient for long distance data transfer. If all the different buildings are made to connect to one network, and by implementing a suitable network topology, that will be able to take the shortest path to transmit the monitored data, then the stability of the system can be enhanced. The network topology selected in this work is based on the prime interest which is being able to connect different buildings in one network. From many available network topologies like bus topology, star topology, ring topology, mesh topology, tree topology, and hybrid topology, we have chosen the ring topology. In this situation as compared to others, since communication between long distance is the key factor in this work. The inter-building communication is normally done using

IOP Conf. Series: Materials Science and Engineering **260** (2017) 012025 doi:10.1088/1757-899X/260/1/012025

communication cable where signals tend to drop significantly over long distance data transfer. Fig. 1 shows a typical ring network topology setup. Here, a token or a tiny data packet is transmitted around the network and whenever there is a need to transmit a data packet, the device keeps the token for the next trip around and fixes its data packet to it. This set up makes it extremely easy to manage the peer to peer network. While the transmission is intended for a short distance, an effective and high quality communication is ensured. This set up is perfectly suitable to handle traffics of high volume in short distance. Furthermore, the ring topology uses dual rings as a backup in case of one link failure with less wiring.

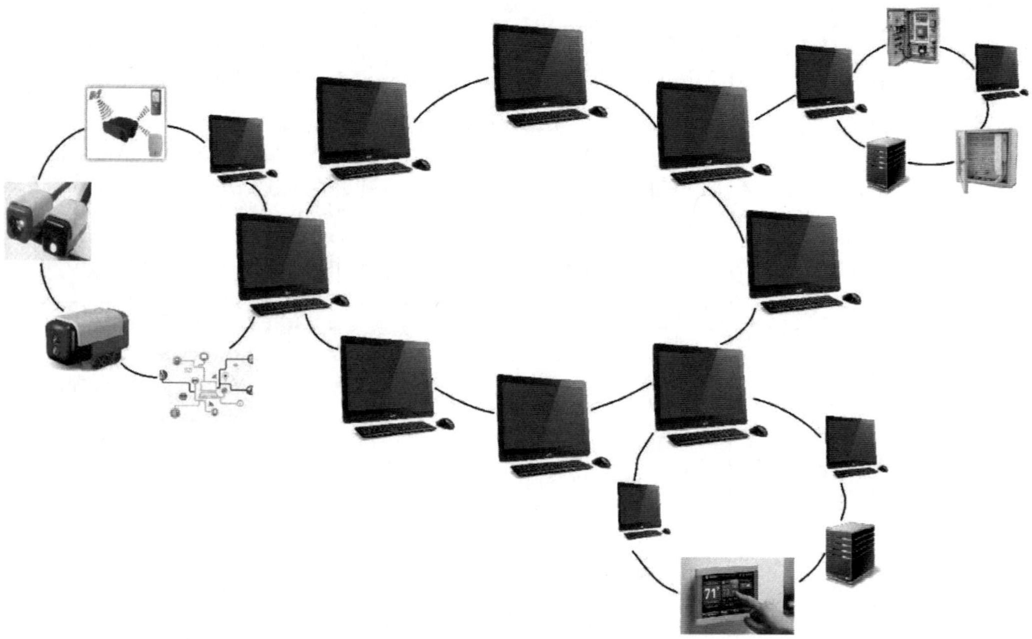

Figure 1. Ring topology setup.

3. Proposed Algorithmic Approach of Fire Alarm Detection System

To upgrade the capability of currently running fire detection and alarm system for long distance, this algorithmic was proposed, to reducing noise and signal drop problem while delivering an authentic alarm in the network. Optical fiber and optical module were proposed in building nodes where internal sensors connection will be the same with normal network cables. By using optical network procedure during a crucial time, the fire affected buildings could be saved with minimum losses. A prototype simulation model system, based on real world situation (having the same parameters) was implemented / designed for fire detection system and fire alarm, as like, the FDA system screens sequence of buildings in a restricted area, such as in industrial environment, some buildings located closely in an area. All the buildings and CCR was connected to each other by means of a ring network. The framework would share the information in peer to peer method, connecting the remotely situated control panels. To supply an alarm signal and give proclamation of fire state, the panel connected in the network would produce alarm signal on network state to send information to the other panels or CCR. Because of signal analysis and signal conversion it may few times, But the real-time communication time depends on the number of connected building nodes. Considering this problem, the framework will work on shortage path calculation and proclaim its fire state to CCR. Again, if the communication is somehow interrupted the longest path will be considered. The shared information

IOP Conf. Series: Materials Science and Engineering **260** (2017) 012025 doi:10.1088/1757-899X/260/1/012025

work in a consecutive manner from one building to another. Fig. 2 illustrates the flow chart of the algorithm used in this work.

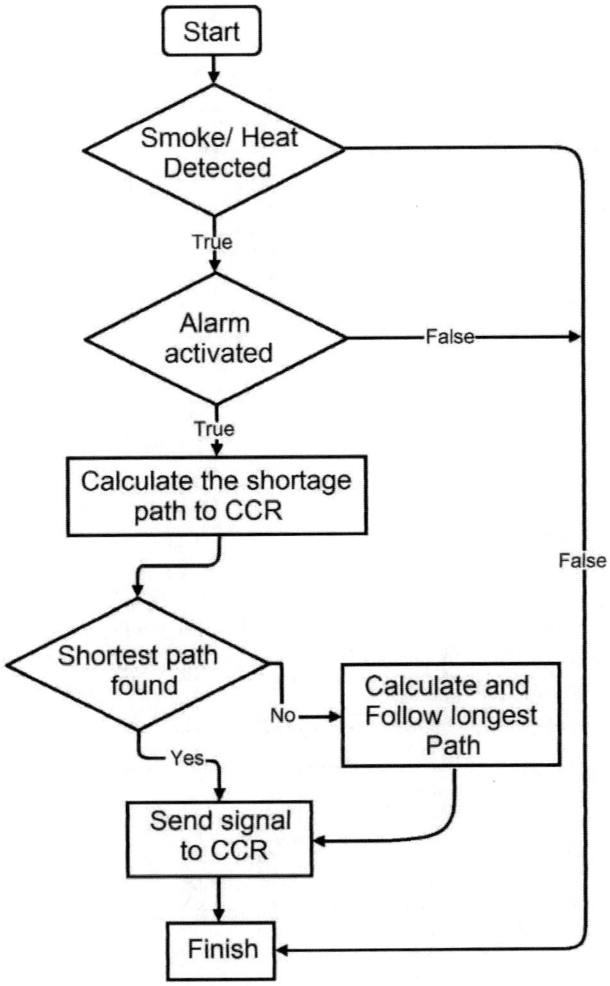

Figure 2. Flow chart of proposed algorithm.

4. Fire Detection System with Ring Topology Building Automation System

In this following section, it is explained how the least and highest time is calculated in a building during a fire incident for different scenarios. The work conducted is intending to use three building blocks. Blocks A, B, and C are assumed to be situated in an industrial area. Each block is surrounded by six other buildings within a distance of 500 meters (as seen in Figure 3), fire alarm panels are affiliated to fiber optic converter to form ring topology when all buildings are joined in a group. The management of transmission between the buildings is made effortless through ring -based approach. The Block A consist of building number 1, 2, 3, 4, 5 and 6. Similarly, Block B and C consist of building numbers 7, 8, 9, 10, 11, 12 and 13, 14, 15, 16, 17, 18 respectively. These blocks are connected with Gytc8s single optical fiber 8-layer outdoor and its respective signal conversion module. The inter block ring network also can work if any inter block connection failure happens and signal will flow through any short or long possible way.

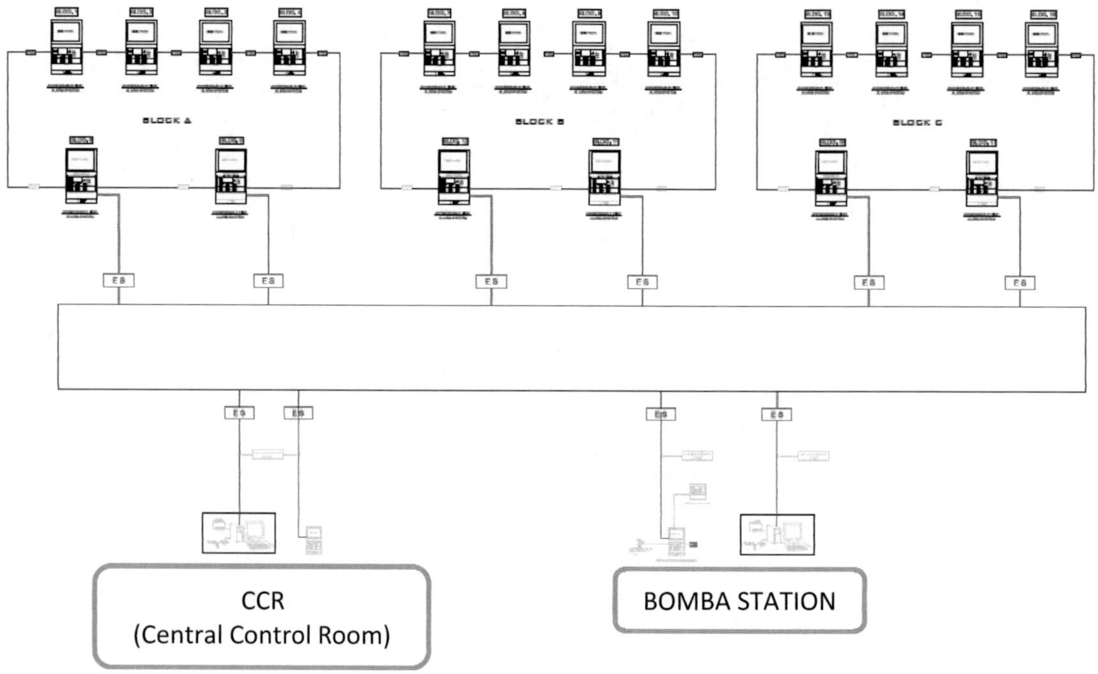
CCR
(Central Control Room)
BOMBA STATION

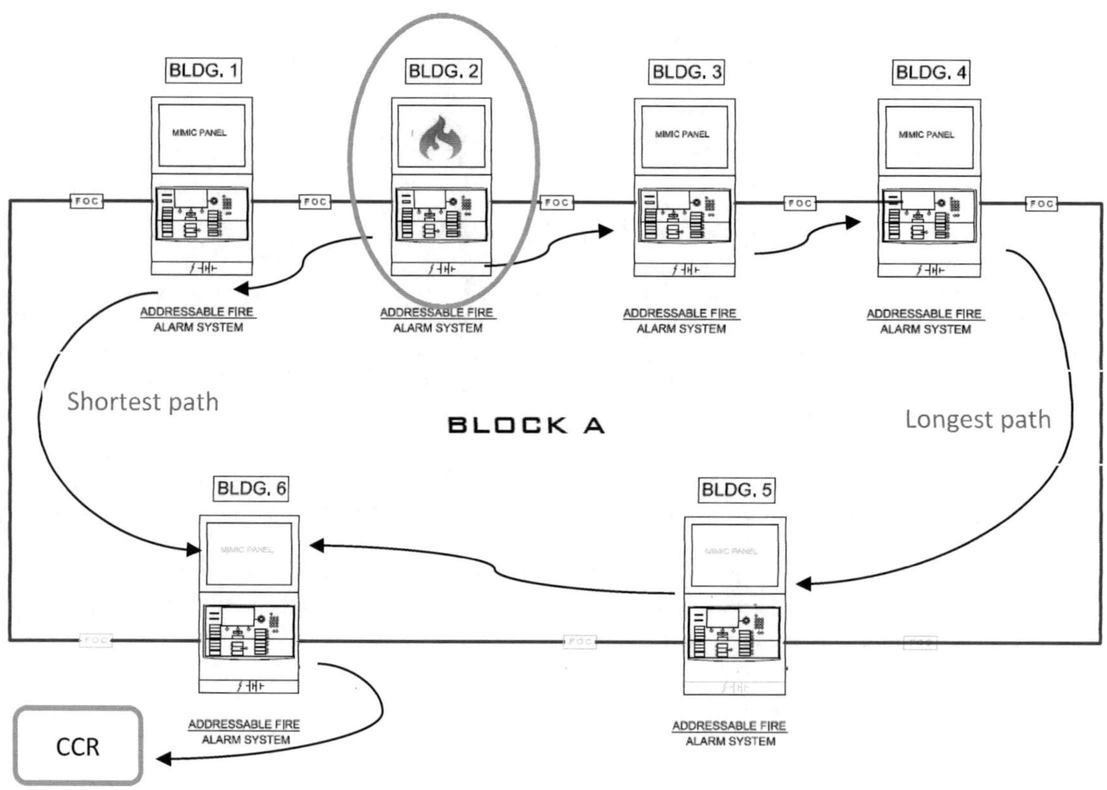
BLDG. 1
BLDG. 2
BLDG. 3
BLDG. 4
MIMIC PANEL
FOC
ADDRESSABLE FIRE ALARM SYSTEM
Shortest path
BLOCK A
Longest path
BLDG. 6
BLDG. 5
CCR

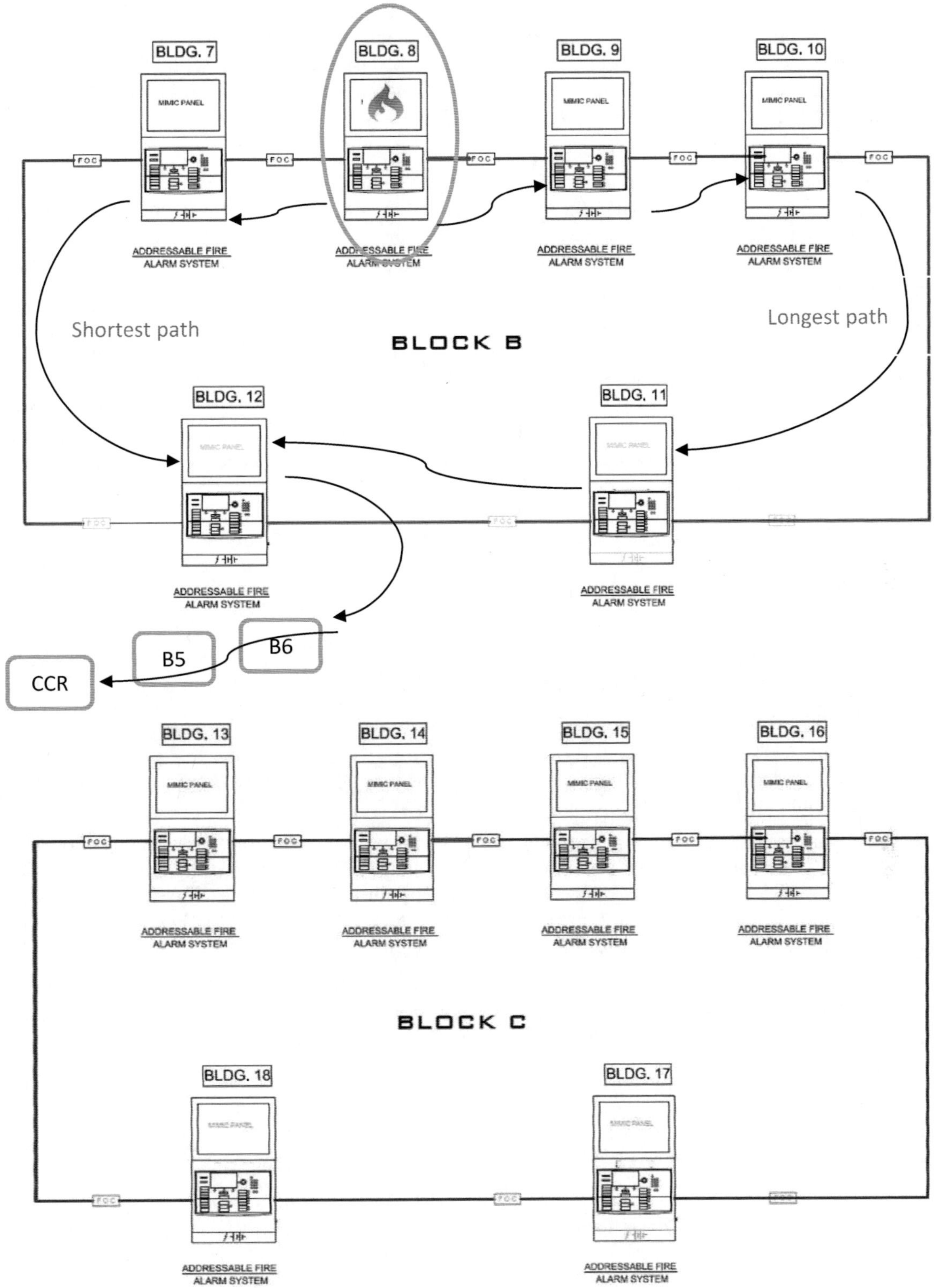

Figure 3. Projected network tropology for the buildings with the control rooms (CCR) and fire rescue department (Bomba Station) includind Block A, Block B and Block C.

The buildings 5 and 6, 11 and 12, 17 and 18 were contacted in same network fashion by Gytc8s outdoor use cable to share information in CCR loop network. The reason for selecting these buildings to be connected with the core is that they would have the shortest distance from the central control building and the fire rescue department in the network.

The core communication line connects control units and fire rescue department with rest of buildings in block. Building 6 in block A is in most immediate reach to the control unit. And for the fire rescue department it is 17. Distance of the control unit from the fire rescue department is about 500 meters as shown in Fig. 3. The fire alarm situation in the network is observed and monitored through communication among each other. They maintain contact with other buildings in blocks through Ethernet switch by using cate STP cable.

The function of the central control room (CCR) and fire rescue department is to supervise and regulate the communication. On the contrary when either CCR or fire rescue department is deviated from its function due to certain circumstances then still one can control it by alternative ways. Further, information can be sent or received among each other. This makes more persistent connection between topology of building and these control units. Under worst events where both control units fail, the buildings connected in a loop network can work as standalone network and during fire detection the loop network can perform their operation individually by switching local alarm or visual devices, sending information through passive proclamation gadgets, releasing the magnetic lock, AHU management, governing of sprinkler and elevator alarm declaration activation, releasing relatively non-secure door links, switching gas control unit, locking the highly secure area those are far away from fire detection place, CO2 gas unit interfacing and switching the main power module to off etc..

5. Discussion and result analysis

In this work JAVA script was used for coding and execution of the algorithm. The results of shortest and longest paths are discussed.

A case scenario is taken to understand how the FDA system works. It was assumed that building 2 caught fire in Block A as shown in Fig. 3, shortest time taken for sending signal to control room would be through building 1 to 6 to CCR.

In the same way, only using building 3 the longest path can send information to CCR. The short and long paths results are shown in table 1 and table 2 respectively. It is obvious that the longest path (B2, B3, B4, B5, B6, CCR) consumes more time than shortest path (B2, B1, B6, CCR). The topology of buildings frames out the reason behind the shortest path. Figure 4 and Figure 5 are showing the result of building 8 block A for the shortest path and the longest path calculation.

For shortest path calculation in block A when building 2 catches fire the shortest path has been calculated as a minimum number of buildings from affected building to control unit through B1 as shown in Table 1 and Figure 4. According to time calculation, it took 3 seconds to reach control unit where only two buildings are crossed from source to destination.

Table 2. Shows the time and other parameters calculation of shortest path results in Block A.

No. of Path	**Building Under Fire**	**Shortest Path To CCR**	**Distance Covered to CCR (Meter)**	**Time to Reach Signal (Second)**	**No. of Buildings Crossed**
1	Building 2	B2, B1, B6 & CCR	1500	3	3

```
Total Number Of Buildings:6
------------------------------
Building1
Building2
Building3
Building4
Building5
Building6
------------------------------
BUILDING 2 GOT FIRE!!!!:
------------------------------
CALCULATING SHORTEST PATH TO SEND INFORMATION TO CONTROL CENTER:
------------------------------
Path: [Building2, Building1, Building6, CentralControlRoom]
------------------------------
Path Calculation Information
------------------------------
Number Of buildings crossed:3
Total Distance Covered to CentralControlRoom: 1500.0 m.
Total Time : 3 sec
```

Figure 4. Receiving signal from building no.8 through shortest path to the Central Control room.

In the case of the longest path, the route from building B2 to CCR is through B3. Here the distance covered is 2500 meters and time to reach the destination is 5 seconds as shown in Table 2 and Fig. 5. The buildings crossed are 4 as compared to 2 in shortest path. Here if shortest path is not feasible due to some reasons then the communication can be completed by longest path.

Table 3. Shows the time and other parameters calculation of longest results in Block A.

No. of Path	Building Under Fire	Longest Path to CCR	Distance Covered to CCR (Meter)	Time to Reach Signal (Second)	No. of Buildings Crossed
2	Building 2	B2, B3, B4, B5, B6 & CCR	2500	5	5

IOP Conf. Series: Materials Science and Engineering **260** (2017) 012025 doi:10.1088/1757-899X/260/1/012025

```
Total Number Of Buildings:6
-------------------------------
Building1
Building2
Building3
Building4
Building5
Building6
-------------------------------
BUILDING 2 GOT FIRE!!!!:
-------------------------------
CALCULATING LONGEST PATH TO SEND INFORMATION TO CONTROL CENTER:
-------------------------------
Path: [Building2, Building3, Building4, Building5, Building6, CentralControlRoom]
-------------------------------
Path Calculation Information
-------------------------------
Number Of buildings crossed:5
Total Distance Covered to CentralControlRoom: 2500.0 m.
Total Time : 5 sec
```

Figure 5. Receiving signal from building no.8 through longest path to the Central Control room.

Let's suppose building 8 caught fire in block B as shown in Fig. 3. The minimum time required for the signal from building 7 to 12 can be calculated. In the same way pathways of 9 to 12 to Control room is estimated. shortest and longest paths results are shown in Table 3 and Table 4 and in Fig. 6 and Fig. 7 respectively.

In the case of block B if building 8 caught fire then the shortest path is from B7 to CCR where it took 5 seconds to send signals to control unit. The total distance covered is 2500 meters and number of building crossed is 5.

Table 4. Shows the time and other parameters calculation of shortest path results in Block B.

No. of Path	Building Under Fire	Shortest Path to CCR	Distance Covered to CCR (Meter)	Time to Reach Signal (Second)	No. of Buildings Crossed
1	Building 8	B8, B7, B12, B5, B6 & CCR	2500	5	5

```
Total Number Of Buildings:8
-------------------------------
Building5
Building6
Building7
Building8
Building9
Building10
Building11
Building12
-------------------------------
BUILDING 8 GOT FIRE!!!!:
-------------------------------
CALCULATING SHORTEST PATH TO SEND INFORMATION TO CONTROL CENTER:
-------------------------------
Path: [Building8, Building7, Building12, Building5, Building6, CentralControlRoom]
-------------------------------
Path Calculation Information
-------------------------------
Number Of buildings crossed:5
Total Distance Covered to CentralControlRoom: 2500.0 m.
Total Time : 5 sec
```

Figure 6. Receiving signal from building no.8 through shortest path to communicate with the Central Control room.

In the case of block B if building 8 got fire then the longest path is from B9 to CCR as shown in Table 4 and Fig. 7 where it took 7 seconds to send signals to control unit. The total distance covered is 3500 meters and number of building crossed is 7.

Table 5. Shows the time and other parameters calculation of Longest results in Block B

No. of Path	Building under Fire	Longest Path To CCR	Distance Covered to CCR (Meter)	Time to Reach Signal (Second)	No. of Buildings Crossed
2	Building 8	B8, B9, B10, B11, B12, B5, B6 & CCR	3500	7	7

IOP Conf. Series: Materials Science and Engineering **260** (2017) 012025 doi:10.1088/1757-899X/260/1/012025

```
Total Number Of Buildings:8
------------------------------
Building5
Building6
Building7
Building8
Building9
Building10
Building11
Building12
------------------------------
BUILDING 8 GOT FIRE!!!!:
------------------------------
CALCULATING LONGEST PATH TO SEND INFORMATION TO CONTROL CENTER:
------------------------------
Path: [Building8, Building9, Building10, Building11, Building12, Building5, Building6, CentralControlRoom]
------------------------------
Path Calculation Information
------------------------------
Number Of buildings crossed:7
Total Distance Covered to CentralControlRoom: 3500.0 m.
Total Time : 7 sec
```

Figure 7. Receiving signal from building no.8 through longest path to the Central Control room.

6. Conclusion and Future Work

To conclude, in this work, an advanced Fire Detection and Alarm (FDA) system is designed based on a robust network technology and tested with simulated network architecture. The simulation process shown an expected performance while working in proposed network tropology to send alarm to the central control room. In this system, it is required to place the fire detection location according to the arrangement of the distributed networking system as building location to analyse its performance. But, for the lab test building localization mechanism is considered with virtual device connection process. Therefore, the virtualization of designed workload makes the system appear more robust then usual. Therefore it requires further tests with more realistic scenario to ensure its fire protection capability and to compare the efficiency over other systems. To improve the fire detection systems of the buildings, we will also focus on control automated solution of predicting the fire behaviour in an efficient manner.

7. Acknowledgement

8. The Authors express their personal appreciation for the effort of Saiyara Shehnaz in proof-reading the paper. The authors gratefully acknowledge the financial support of IIUM Research Management Center (RMC) and Ministry of Higher Education Malaysia under Research Initiative Grant Scheme (RIGS) number RIGS16-084-0248.

8. References

[1] K. C. Lee and h.-h. Lee. (2004) “network-based fire detection system via controller area network for smart home automation”, vol. 50, no. 4, pp. 1093- 1100, November.

[2] J. S. Kim and j. Lyou. (1999) “the communication method at the auto-startup systems using tcp/ip and vxi and expert system (g2)”, vol. 1, no. 2, pp. 141-146, december.
[3] S.-h. Suh and k.-j. Woo. (2003) “serial communication based fault diagnosis of a bldc motor using bayes classifier,” international journal of control, automation, and systems, vol. 9, p.p. 308-314.
[4] Http://www.silentknight.com/pages/default.aspx
[5] Http://www.simplexgrinnell.com
[6] Http://www.notifier.com
[7] Miyagi, p.e.et al. (2002) “petri net approach for modelling system integration in intelligent buildings”. Journal of the brazilian society of mechanical sciences, abcm, rio de janeiro, rj, vol. 23, n.4, pp.341-350.
[8] Ole fanger, p. (2001)"human requirements in future air-conditioned environments."international journal of refrigeration 24.2: 148-153.
[9] Korea fire industry technology institute (2011) “model approval & inspection technical standards for fire detectors”. Notification no. 2011-1 by national emergency management agency in south korea.
[10] Bukowski rw, moore wd (2003) fire alarm signaling systems, 3rd edn. Nfpa inc., pp 31–252.
[11] Wikipedia, http://en.wikipedia.org/wiki/fire_alarm_system, (january 2008).
[12] Wikipedia, http://en.wikipedia.org/wiki/fire_alarm_control_panel, (december 2010).
[13] Liu, z. G., j. M. Makar, and a. K. Kim. (2001) "development of fire detection systems in the intelligent building": p.p. 561-573. Nrcc- 44226.
[14] Yu, liyang, neng wang, and xiaoqiao meng. "real-time forest fire detection with wireless sensor networks." wireless communications, networking and mobile computing, 2005. Proceedings. 2005 international conference on. Vol. 2. Ieee, 2005.
[15] Wang, y., yu, c., tu, r., & zhang, y. (2011). Fire detection model in tibet based on grey-fuzzy neural network algorithm. Expert systems with applications, 38(8), 9580-9586.
[16] Krammer. l., kastner, w., & sauter, t. (2016, may). A generic dependability layer for building automation networks. In factory communication systems (wfcs), 2016 ieee world conference on (pp. 1-4). Ieee.
[17] Steenberg, c. M., & van duijn, n. (2014). U.s. patent no. 8,649,987. Washington, dc: u.s. patent and trademark office.
[18] Bhatt, j., & verma, h. K. (2015). Design and development of wired building automation systems. Energy and buildings, 103, 396-413.
[19] Gangurde, s. S., & tated, k. S. (2016). Zigbee based wireless fire security node design using fpga whose monitoring and controlling through labview. Journal of analog and digital devices, 1(1-2).
[20] Ma, j., yang, x., & wang, p. (2017, february). Design of the intelligent smoke alarm system based on photoelectric smoke. In second international conference on photonics and optical engineering (pp. 1025623-1025623). International society for optics and photonics.
[21] Somov, a., baranov, a., & spirjakin, d. (2014). A wireless sensor–actuator system for hazardous gases detection and control. Sensors and actuators a: physical, 210, 157-164.
[22] Mary, g. I., alex, z. C., & jenkins, l. (2017). Time analysis of wireless controller area network-based fire and gas safety system. International journal of communication networks and distributed systems, 18(1), 1-17.
[23] Pan, z., hariri, s., & al-nashif, y. (2014, november). Anomaly based intrusion detection for building automation and control networks. In computer systems and applications (aiccsa), 2014 ieee/acs 11th international conference on (pp. 72-77). Ieee.
[24] Bharani, j. C., gowtham, m., narmada, k., & prasath, j. R. (2014, march). Fire rescue system in railways using labview. In green computing communication and electrical engineering (icgccee), 2014 international conference on (pp. 1-13). Ieee.
[25] Islam, m. A., & islam, m. M. (2015). A novel signature-based traffic classification engine to reduce false alarms in intrusion detection systems. International journal of computer networks & communications, 7(1), 63.

[26] Mcneil, j. G., & lattimer, b. Y. (2016). Autonomous fire suppression system for use in high and low visibility environments by visual servoing. Fire technology, 52(5), 1343-1368.

[27] Kim, j. H., & lattimer, b. Y. (2015). Real-time probabilistic classification of fire and smoke using thermal imagery for intelligent firefighting robot. Fire safety journal, 72, 40-49.

[28] Djahel, s., doolan, r., muntean, g. M., & murphy, j. (2015). A communications-oriented perspective on traffic management systems for smart cities: challenges and innovative approaches. Ieee communications surveys & tutorials, 17(1), 125-151.

[29] Chen, l., zhou, q., & fang, w. (2016, august). Spatial pyramid pooling in structured sparse representation for flame detection. In proceedings of the international conference on internet multimedia computing and service (pp. 310-313). Acm.

[30] Fang, w., zhen, y., kang, q. S., xi, s. D., & shang, l. Y. (2014). A simulation research on the visual servo based on pseudo-inverse of image jacobian matrix for robot. In applied mechanics and materials (vol. 494, pp. 1212-1215). Trans tech publications.

[31] Deve, k. B., hancke, g. P., & silva, b. J. (2016, october). Design of a smart fire detection system. In industrial electronics society, iecon 2016-42nd annual conference of the ieee (pp. 6205-6210). Ieee.

[32] Di pasquale, f., soto, m. A., & bolognini, g. (2014). Raman based distributed optical fiber temperature sensors: industrial applications and future developments (no. Epfl-chapter-197859, pp. 88-113). World scientific publishing co. Pte. Ltd.

[33] Sun, m., tang, y., yang, s., li, j., sigrist, m. W., & dong, f. (2016). Fire source localization based on distributed temperature sensing by a dual-line optical fiber system. Sensors, 16(6), 829.

[34] Peto, r. (2016). U.s. patent no. 9,285,854. Washington, dc: u.s. patent and trademark office.

[35] Ziemann, o., krauser, j., zamzow, p. E., & daum, w. (2013). Pof-polymer optical fibers for data communication. Springer science & business media.

[36] Decusatis, c. (ed.). (2013). Handbook of fiber optic data communication: a practical guide to optical networking. Academic press.

[37] Mahloo, m., chen, j., & wosinska, l. (2015). Pon versus aon: which is the best solution to offload core network by peer-to-peer traffic localization. Optical switching and networking, 15, 1-9.

[38] Zhang, n., cheng, n., gamage, a. T., zhang, k., mark, j. W., & shen, x. (2015). Cloud assisted hetnets toward 5g wireless networks. Ieee communications magazine, 53(6), 59-65.

IOP Conf. Series: Materials Science and Engineering **260** (2017) 012026 doi:10.1088/1757-899X/260/1/012026

Trajectory tracking in quadrotor platform by using PD controller and LQR control approach

Maidul Islam[1], Mohamed Okasha[2] and Moumen Mohammad Idres[3]

[1, 2, 3] Dept. of Mechanical Engineering, International Islamic University Malaysia (IIUM), Kuala Lumpur 50725, Malaysia

Email: [1]mislam.dipu@gmail.com, [2]mokasha@ium.edu.my and [3]moumen@iium.edu.my

Abstract. The purpose of the paper is to discuss a comparative evaluation of performance of two different controllers i.e. Proportional-Derivative Controller (PD) and Linear Quadratic Regulation (LQR) in Quadrotor dynamic system that is under-actuated with high nonlinearity. As only four states can be controlled at the same time in the Quadrotor, the trajectories are designed on the basis of the four states whereas three dimensional position and rotation along an axis, known as yaw movement are considered. In this work, both the PD controller and LQR control approach are used for Quadrotor nonlinear model to track the trajectories. LQR control approach for nonlinear model is designed on the basis of a linear model of the Quadrotor because the performance of linear model and nonlinear model around certain nominal point is almost similar. Simulink and MATLAB software is used to design the controllers and to evaluate the performance of both the controllers.

1. Introduction

The popularity of quadrotor is increasing because of its versatile applications. It is popular because of not having any runway rather direct vertical take-off and landing capabilities for flight, hovering on a position constantly, agility, easy manoeuvrability and being small in size gives an additional advantage of monitoring an area or a building [1, 2].

Generally controlling a quadrotor is highly challenging because it has to maintain six degrees of freedom by using four motors [3]. In both classical control theory and modern control theory, a good number of controllers have been developed to deal with multiple inputs and multiple outputs (MIMO) system. Several sorts of control techniques are being designed to improve the performance of a quadrotor while these vary from linear control technique to nonlinear control techniques. A linear model for a quadrotor is achieved by the procedure of linearization of the non-linear model around a nominal point while any hover position is chosen in most of the time. Then all the chosen nominal are summed up to generate the trajectory for the quadrotor.

Linear models are mostly chosen because it is simple to design and the implementation of the model is easy on any real platform. But the main drawback of the linear model is using the linearized model when the controller is being designed otherwise the controller will perform extremely bad [4]. PD controller, LQR controller and H_∞ controller are three different well-known linear control systems that are frequently used for quadrotor. Emil Fresk and George Nikolakopoulos [5] presented a solution of attitude control problem of a quadrotor by using PID control. M. Alhaddad et al [6] designed a

IOP Conf. Series: Materials Science and Engineering **260** (2017) 012026 doi:10.1088/1757-899X/260/1/012026

mathematical model and then by applying LQR control technique, they achieved proper stability in attitude control.

Some popular nonlinear control techniques for quadrotors are feedback linearization, model predictive control (MPC), Backstepping controller and Sliding mode controller. R. Bonna and J. F. Camino [7] applied feedback linearization to achieve proper trajectory tracking while small angle approximation was avoided when the controller was designed for quadrotor platform. Bouabdullah and Siegwart [8] applied a modified Backstepping control technique, named as Integral Backstepping control method to achieve a flexible control of the quadrotor.

In this work, a comparative discussion of control techniques with performance indices are presented for both LQR control approach and PD control approach on a quadrotor platform. LQR is an optimal control technique of tuning gains for linear systems by minimizing a cost function in an infinite control horizon. Besides, in PD control technique, two different control parameters are used to minimize the error between desired set points and measured process variables. So, the performance for both the controllers can be improved by proper tuning.

2. Dynamic Model

The quadrotor basically holds a rigid cross-linked structure that has four independent rotors with fixed pitched propellers. Three basic movements are necessary to describe all the movements of a quadrotor. The rotational movement around X axis is described as Roll movement (ϕ) and Y axis is denoted as Pitch movement (θ). Roll movement and pitch movement can be achieved by balancing the speed of motor 2 & 4 and motor 1 & 3 respectively. Lateral and longitudinal acceleration can be achieved respectively by changing ϕ and θ angle.

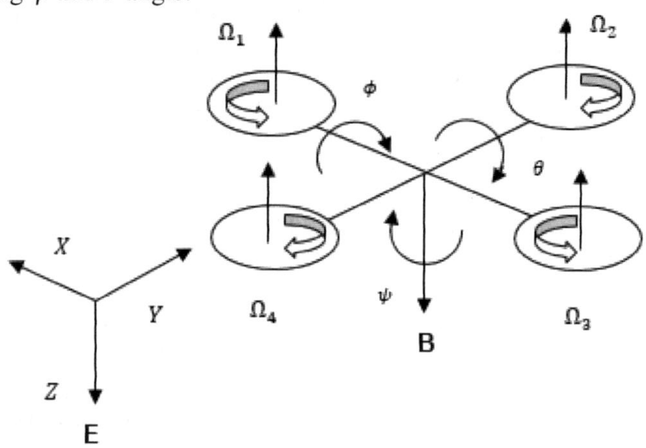

Figure 1. Configuration of quadrotor whereas B and E denotes Body fixed frame and Earth fixed frame respectively

But another rotational movement Yaw (ψ), around Z axis can be achieved by balancing the speed of the motor pair (1, 3) and (2, 4) at the same time.

Every controller input has an effect on a certain movement like U_2 effects on roll movement while U_3 effects on pitch movement and U_4 affects on yaw movement. Besides, U_1has an effect on upward movement along Z axis. So, the control inputs are as followed.

$$U_1 = k_f(\Omega_1^2 + \Omega_2^2 + \Omega_3^2 + \Omega_4^2)$$

$$U_2 = k_f(\Omega_4^2 - \Omega_2^2)$$

$$U_3 = k_f(-\Omega_3^2 + \Omega_1^2)$$

$$U_4 = k_M(\Omega_1^2 - \Omega_2^2 + \Omega_3^2 - \Omega_1^2)$$

IOP Conf. Series: Materials Science and Engineering **260** (2017) 012026 doi:10.1088/1757-899X/260/1/012026

Where,

Ω_i represents the ith motor speed

The initial conditions and nominal parameters for simulation of the quadrotor are shown in table 1 [9].

Table 1. Parameters and initial conditions for simulation.

Symbol	Description	Value	Unit
I	Moment of inertia	$\begin{pmatrix} 7.5e-3 & 0 & 0 \\ 0 & 7.5e-3 & 0 \\ 0 & 0 & 1.3e-2 \end{pmatrix}$	kg.m^2
l	Arm length	0.23	M
I_r	Inertia of motor	6e-5	kg.m^2
k_f	Thrust coefficient	3.13e-5	Ns^2
k_M	Moment coefficient	7.5e-7	Nms^2
m	Mass of quadrotor	0.65	Kg
g	Gravity	9.81	ms^2
k_t	Aerodynamic thrust drag coefficient	$\begin{pmatrix} 0.1 & 0 & 0 \\ 0 & 0.1 & 0 \\ 0 & 0 & 0.1 \end{pmatrix}$	Ns/m
k_r	Aerodynamic moment drag coefficient	$\begin{pmatrix} 0 & 0 & 0 \\ 0 & 0 & 0 \\ 0 & 0 & 0 \end{pmatrix}$	Nm.s

2.1. Transformation Matrix

Since it is required to deal with two different coordinate systems i.e. Earth fixed frame and Body fixed frame simultaneously in order to explain the position and movement of a quadrotor, a transformation matrix is to be derived. Here, R is the required transformation matrix that can describe the position and movement from Earth-inertial frame to Body fixed frame [10].

$$R = \begin{pmatrix} c\theta c\psi & c\psi s\phi s\theta & c\phi s\theta c\psi + s\psi s\phi \\ c\theta s\psi & s\psi s\phi s\theta + c\theta c\psi & c\phi s\theta s\psi - s\theta c\psi \\ -s\theta & s\phi c\theta & c\phi c\theta \end{pmatrix}$$

where, c denotes cos and s denotes sin.

Generally on-board Inertial Measurement Unit (IMU) is used to measure the angular velocity of a quadrotor and it is measured on Body fixed frame. To generate another relationship between Euler rates, $[\dot{\phi}, \dot{\theta}, \dot{\psi}]^{\mathrm{T}}$ that describes the angular rates with respect to time on Earth fixed frame and angular velocity, $[p,\ q, r]^{\mathrm{T}}$ of quadrotor, a new rotational matrix R_r is to be derived again, as follows [11]:

$$\begin{pmatrix} p \\ q \\ r \end{pmatrix} = \begin{pmatrix} 1 & s\phi t\theta & c\phi t\theta \\ 0 & c\phi & -s\phi \\ 0 & \dfrac{s\phi}{c\theta} & \dfrac{c\phi}{c\theta} \end{pmatrix} \begin{pmatrix} \dot{\phi} \\ \dot{\theta} \\ \dot{\psi} \end{pmatrix} \tag{1}$$

IOP Conf. Series: Materials Science and Engineering **260** (2017) 012026 doi:10.1088/1757-899X/260/1/012026

2.2. Translational Motion

Gravitational force (F_g) and aerodynamic drag force (F_a) are to be introduced in order to describe the translational movement of a quadrotor though these forces must be overcome by the motors' thrust (F_{tb}) to achieve any horizontal movement or yaw movement. Here, the translational motion of the quadrotor is described by Newton's second law as follows.

$$m\begin{pmatrix}\ddot{x}\\ \ddot{y}\\ \ddot{z}\end{pmatrix} = \begin{pmatrix}0\\ 0\\ mg\end{pmatrix} + R\begin{pmatrix}0\\ 0\\ -U_1\end{pmatrix} - k_t\begin{pmatrix}\dot{x}\\ \dot{y}\\ \dot{z}\end{pmatrix} \tag{2}$$

2.3. Rotational Motion

Newton-Euler method is used to get rotational equations of motion for the quadrotor [12].

$$I\dot{\omega} = -\begin{pmatrix}p\\ q\\ r\end{pmatrix} \times I\begin{pmatrix}p\\ q\\ r\end{pmatrix} - \begin{pmatrix}p\\ q\\ r\end{pmatrix} \times \begin{pmatrix}0\\ 0\\ I_r\omega_r\end{pmatrix} + \begin{pmatrix}lU_2\\ lU_3\\ U_4\end{pmatrix} - k_r\begin{pmatrix}p\\ q\\ r\end{pmatrix} \tag{3}$$

The dynamic model of a quadrotor can be described by 12 states.

$$X_s = [x\ y\ z\ \dot{x}\ \dot{y}\ \dot{z}\ \phi\ \theta\ \psi\ p\ q\ r]^T \tag{4}$$

A state-space representation can be achieved as follows by considering from equation (2) to (3) for dynamic model.

3. Control Techniques

3.1. LQR (Linear Quadratic Regulator)

LQR is one of the most popular optimal control techniques where a cost function is used that is dependent on the states of any system and control inputs. As LQR needs a linearized model, the dynamic model of the quadrotor is linearized by Jacobian method as follows:

$$\delta\dot{X}_s = A\delta X_s + B\delta U$$

If the model is linearized at an nominal point (X_{ss}, U_{ss}) where $X_{ss} = [x_{ss}, y_{ss}, z_{ss,}\ \psi_{ss}\ ]^T$ and others states are considered as zero with $U_{ss} = [mg\ \ 0\ \ 0\ \ 0]^T$

Where,

$$\delta X = X_d - X_a$$

$$U = U_{ss} + \delta U \tag{5}$$

$$X = X_{ss} + \delta X \tag{6}$$

Now according to the control approach of LQR, a feedback control is to be designed by following equation.

$$U = -K\delta X_s + U_{ss} \tag{7}$$

Where, K is the feedback gain matrix. It has been calculated by minimizing the cost function

$$J = \int_0^\infty (\delta X_s Q\delta X_s + \delta U R\delta U)dt \tag{8}$$

IOP Conf. Series: Materials Science and Engineering **260** (2017) 012026 doi:10.1088/1757-899X/260/1/012026

Where, Q is considered as a 12×12 semi-positive definite matrix and R is a 12×4 positive definite matrix for this quadrotor.

So, the closed loop control system appears as

$$\delta\dot{X}_s = (A - BK)\delta X_s$$

where, K has been calculated by using Q and R matrices.

Then the linear model is replaced by nonlinear model to have proper control of quadrotor with the same gain K.

3.2. PD Controller

PID control approach is one of the most popular control approaches because it is very simple to design. For this model, integrator part of PID controller is avoided because the model contains integrator twice. Therefore, the general model for a PD controller will be as follows:

$$e(t) = X_d - X_a \quad (9)$$

$$u(t) = K_p e(t) + K_d \frac{d}{dt} e(t) \quad (10)$$

Where $e(t)$ symbolizes the error between desired states (X_d) and actual states (X_a), $u(t)$ represents the control output K_p denotes proportional gain and K_d is derivative gain.

In this system, four different control inputs U are achieved by PD controller while U_2, U_3 and U_4 is directly solved by PD control technique and U_1 is derived from the equation of $\ddot{z}_d$ that is also solved by PD control technique as follows.

$$U_1 = m[\frac{g - (\ddot{z}_d + \frac{k_{t_z}}{m}\dot{z})}{c\phi c\theta}] \quad (11)$$

$$U_2 = K_p(\phi_d - \phi_a) + K_d(\dot{\phi}_d - \dot{\phi}_a) \quad (12)$$

$$U_3 = K_p(\theta_d - \theta_a) + K_d(\dot{\theta}_d - \dot{\theta}_a) \quad (13)$$

$$U_4 = K_p(\psi_d - \psi_a) + K_d(\dot{\psi}_d - \dot{\psi}_a) \quad (14)$$

Where, desired roll θ_d and desired pitch ϕ_d are derived from the equation (2) as follows.

$$\theta_d = tan^{-1}(\frac{-c\psi\left(\ddot{x}_d + \frac{k_{t_x}}{m}\dot{x}_d\right) - s\psi(\ddot{y}_d + \frac{k_{t_y}}{m}\dot{y}_d)}{g - (\ddot{z}_d + \frac{k_{t_z}}{m}\dot{z}_d)}) \quad (16)$$

$$\phi_d = tan^{-1}\frac{(-s\psi\left(\ddot{x}_d + \frac{k_{t_x}}{m}\dot{x}_d\right) + c\psi\left(\ddot{y}_d + \frac{k_{t_y}}{m}\dot{y}_d\right))c\theta_d}{g - (\ddot{z}_d + \frac{k_{t_z}}{m}\dot{z}_d)} \quad (17)$$

4. Simulation Results

The two control technique i.e. PD controller and LQR have been implemented on nonlinear model of the quadcopter respectively. The performances for both the controllers are shown in 3D plot for circular and helix trajectory in figure 4 and 5 respectively.

IOP Conf. Series: Materials Science and Engineering **260** (2017) 012026 doi:10.1088/1757-899X/260/1/012026

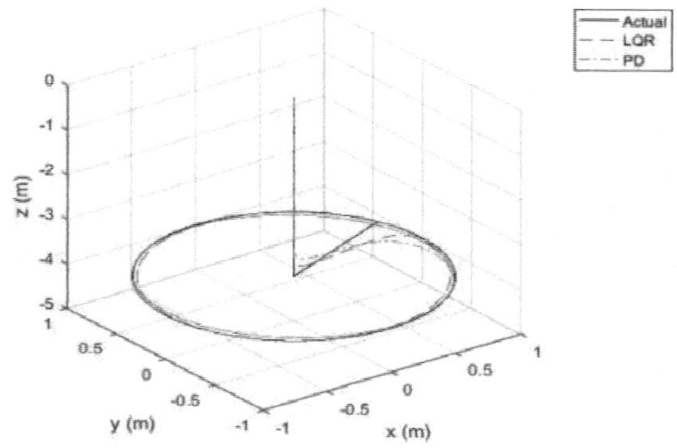

Figure 2. Circular Trajectory.

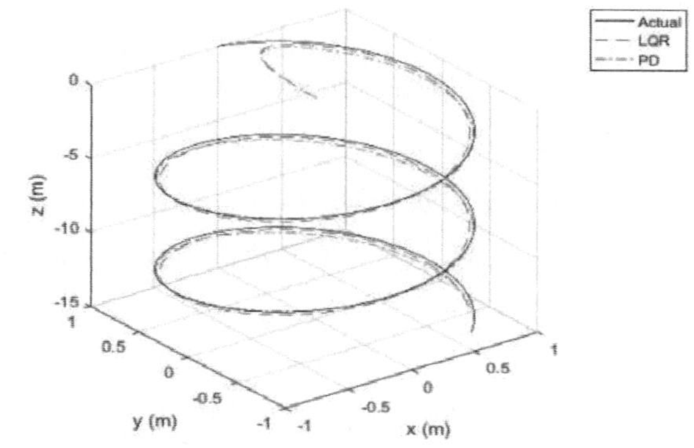

Figure 3. Helix Trajectory.

The suitable gains which are tuned upon several iterations for PD controller are mention in table 2.

Table 2. PD controller gains.

States	K_p	K_d
x	2.2	2.8
y	2.2	2.8
z	7.5	5.0
ϕ	141.0	20.0
θ	5.5	1.5
ψ	141.0	20.0

And the tuned Q and R matrices are given as follows for LQR control techniques in table 3.

Table 3. Q and R matrices for LQR controller

Q	R
$\begin{pmatrix} 21 & 0 & 0 & 0 & 0 & 0 & 0 & 0 & 0 & 0 & 0 & 0 \\ 0 & 14 & 0 & 0 & 0 & 0 & 0 & 0 & 0 & 0 & 0 & 0 \\ 0 & 0 & 81 & 0 & 0 & 0 & 0 & 0 & 0 & 0 & 0 & 0 \\ 0 & 0 & 0 & 6 & 0 & 0 & 0 & 0 & 0 & 0 & 0 & 0 \\ 0 & 0 & 0 & 0 & 6 & 0 & 0 & 0 & 0 & 0 & 0 & 0 \\ 0 & 0 & 0 & 0 & 0 & 6 & 0 & 0 & 0 & 0 & 0 & 0 \\ 0 & 0 & 0 & 0 & 0 & 0 & 11 & 0 & 0 & 0 & 0 & 0 \\ 0 & 0 & 0 & 0 & 0 & 0 & 0 & 16 & 0 & 0 & 0 & 0 \\ 0 & 0 & 0 & 0 & 0 & 0 & 0 & 0 & 56000 & 0 & 0 & 0 \\ 0 & 0 & 0 & 0 & 0 & 0 & 0 & 0 & 0 & 9 & 0 & 0 \\ 0 & 0 & 0 & 0 & 0 & 0 & 0 & 0 & 0 & 0 & 6 & 0 \\ 0 & 0 & 0 & 0 & 0 & 0 & 0 & 0 & 0 & 0 & 0 & 6 \end{pmatrix}$	$\begin{pmatrix} 1 & 0 & 0 & 0 \\ 0 & 1 & 0 & 0 \\ 0 & 0 & 1 & 0 \\ 0 & 0 & 0 & 1 \end{pmatrix}$

Figure 6 and 7, depict the performance of the dynamic model for both the controllers through position tracking (x, y, z) where figure 6 and 7 describe circular trajectory and helix trajectory respectively.

Root-Mean-Square (RMS) error helps to know about the accuracy of actual data with respect to desired data. So, the performance of both the controller can be evaluated by using RMS error while table 2 shows comparison for circular trajectory and table 3 shows for helix trajectory.

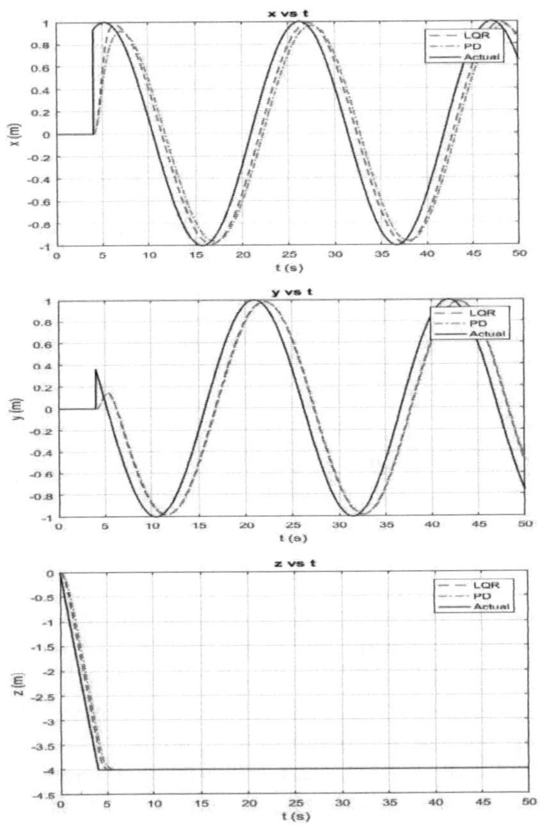

Figure 4. x vs t, y vs t and z vs t in circular trajectory

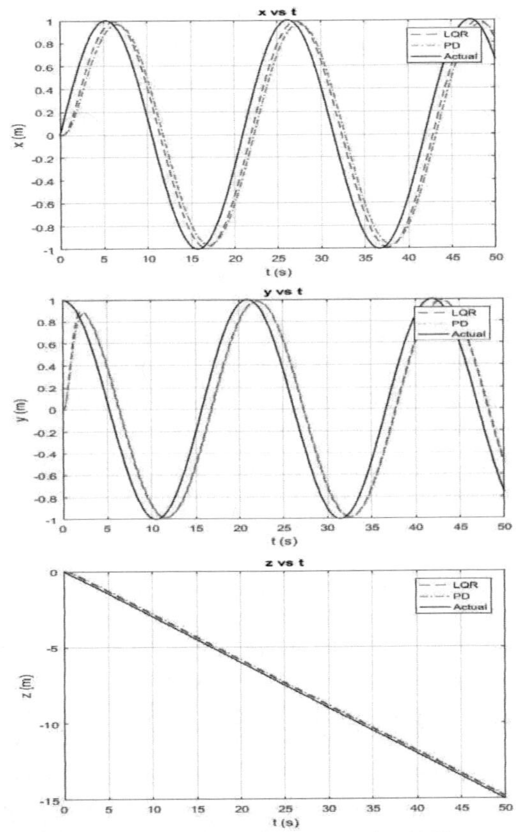

Figure 5. x vs t, y vs t and z vs t in helix trajectory

From table 4, it is found that LQR performs better comparatively PD controller along x and z axes while PD controller shows better performance along y axis than LQR.

Table 4. RMS Error of the controllers for circular trajectory

	x (%)	y (%)	z (%)
RMSE for PD	6.28	3.05	0.21
RMSE for LQR	3.38	3.11	0.21

Table 5. RMS Error of the controllers for helix trajectory

	x (%)	y (%)	z (%)
RMSE for PD	5.15	3.69	2.40
RMSE for LQR	2.40	3.26	2.03

From the result of table 5, between these two controllers along x, y and z axes tracking in helix trajectory, LQR shows better performance comparing to PD control technique.

5. Conclusion

The work introduces with the two different control approaches with their singular characteristics that where these control approaches are used because of wide popularity and design simplicity. In general,

IOP Conf. Series: Materials Science and Engineering **260** (2017) 012026 doi:10.1088/1757-899X/260/1/012026

LQR control approach shows robustness and generates very low steady-state error but keeping update twelve states at the same time may create a transitional delay whereas the fast response is important during flight and the states have no direct access to the plant. However, PD controller can give faster response and overcome steady-state error by adding an additional integral part in controller but it is failed to give robust performance like LQR control approach.

Looking back at the responses, the controllers could not give exact performances as these are supposed to follow the trajectories because of not tuning proper gains while it is the toughest part of designing controllers perfectly.

The gains for PD controllers are tuned by genetic algorithm. As it is difficult to handle ten gains for GA optimization simultaneously, the optimizer tuned the best possible gains as it could. However, if the optimization can be done iteratively, it is possible to tune the perfect gains. On the other hand, Q and R matrices are chosen arbitrarily to calculate gains for LQR control approach and it also can be improved by several processes like Bryson's rule or optimization etc. If the proper gains can be tuned properly, the settling time and rise time can be decreased more.

In the future work, the designed controllers will be implemented on real quadrotor platform considering constraints at control inputs that can help to comprehend and evaluate the most suitable controllers for the quadrotor practically.

6. References

[1] J. Li and Y. Li, "Dynamic analysis and PID control for a quadrotor," in *Mechatronics and Automation (ICMA), 2011 International Conference on*, 2011, pp. 573-578.

[2] S. Bouabdallah and R. Siegwart, "Backstepping and sliding-mode techniques applied to an indoor micro quadrotor," in *Robotics and Automation, 2005. ICRA 2005. Proceedings of the 2005 IEEE International Conference on*, 2005, pp. 2247-2252.

[3] M. Chipofya, D. J. Lee, and K. T. Chong, "Trajectory tracking and stabilization of a quadrotor using model predictive control of Laguerre functions," in *Abstract and Applied Analysis*, 2015.

[4] A. Chovancová, T. Fico, P. Hubinský, and F. Duchoň, "Comparison of various quaternion-based control methods applied to quadrotor with disturbance observer and position estimator," *Robotics and Autonomous Systems,* vol. 79, pp. 87-98, 2016.

[5] E. Fresk and G. Nikolakopoulos, "Full quaternion based attitude control for a quadrotor," in *Control Conference (ECC), 2013 European*, 2013, pp. 3864-3869.

[6] M. Alhaddad, M. Tyupkin, and L. Aeshina, "MODELING AND ATTITUDE CONTROL OF A QUADCOPTER USING LINEAR QUADRATIC REGULATOR," *Молодежь. Общество. Современная наука, техника и инновации,* pp. 173-176, 2016.

[7] R. Bonna and J. Camino, "Trajectory Tracking Control of a Quadrotor Using Feedback Linearization," in *Proc. of the XVII International Symposium on Dynamic Problems of Mechanics DINAME-2015.–2015.*

[8] S. Bouabdallah and R. Siegwart, "Full control of a quadrotor," in *Intelligent robots and systems, 2007. IROS 2007. IEEE/RSJ international conference on*, 2007, pp. 153-158.

[9] S. Bouabdallah, "Design and control of quadrotors with application to autonomous flying," 2007.

[10] A. L. Salih, M. Moghavvemi, H. A. Mohamed, and K. S. Gaeid, "Modelling and PID controller design for a quadrotor unmanned air vehicle," in *Automation Quality and Testing Robotics (AQTR), 2010 IEEE International Conference on*, 2010, pp. 1-5.

[11] H. Bolandi, M. Rezaei, R. Mohsenipour, H. Nemati, and S. M. Smailzadeh, "Attitude control of a quadrotor with optimized PID controller," *Intelligent Control and Automation,* vol. 4, p. 335, 2013.

[12] L. E. Romero, D. F. Pozo, and J. A. Rosales, "Quadcopter stabilization by using PID controllers," *Maskana,* pp. 175-186, 2016.

IOP Conf. Series: Materials Science and Engineering **260** (2017) 012027 doi:10.1088/1757-899X/260/1/012027

Assistive-as-Needed Strategy for Upper-Limb Robotic Systems: *An Initial Survey*

I. M. Khairuddin[1,2], S. N. Sidek[1], H. Md Yusof[1], K. Baarath[2] and A. P. P. A. Majeed[2]

[1]Department of Mechatronics Engineering, International Islamic University Malaysia, Kuala Lumpur, Malaysia
[2]Innovative Manufacturing, Mechatronics and Sports (iMAMS) Laboratory, Faculty of Manufacturing Engineering, Universiti Malaysia Pahang

ismailkhai@ump.edu.my

Abstract. Stroke is amongst the leading causes of deprivation of one's ability in carrying out activities of daily living. It has been reported from literature that, the functional recovery of stroke patients are rather poor, unless frequent rehabilitative therapy is assumed on the affected limb. Recent trends of rehabilitation therapy have also shifted towards allowing more participation of the patient in the therapy session rather than simple passive treatments as it has been demonstrated to be non-trivial in promoting neural plasticity to expedite motor recovery process. Therefore, the employment of rehabilitation robotics is seen as a means of mitigating the limitations of conventional rehabilitation therapy. It enables unique methods for promoting patient engagement by providing patients assistance only as needed basis. This paper attempts on reviewing assist-as-needed control strategy applied on upper-limb robotic rehabilitation devices.

1. Introduction

The life expectancy of the elderly, over the past two decades, particularly those of the age of 60 years and above has increased significantly [1]. The Malaysian Ministry of Health's annual report in 2011 conveyed that the Malaysian population between the age group of 0 to 18 years old records both physical and cerebral palsy disabilities at around 11% and 7%, respectively [2]. In addition, the report also highlighted that there was an increase of stroke patients on an annual basis approximately 300% annually. These statistics reflect the size of the segments in the society that are diagnosed with disabilities that in turn depriving them of performing activities of daily living (ADL) [3].

Rehabilitation is a process by which patients undergo treatment to aid them in regaining and relearning the existing motor skills for ADL [4]. The period of stroke rehabilitation spans very broadly, as the rehabilitation therapy may transpire as soon as in the sub-acute stage (immediately after the incident of stroke) and can reach out into the chronic stage too (six months post stroke) [5]. Direct involvement is predicted to be sufficient on the condition of recovery. Emphasis is often given towards gait rehabilitation whilst the upper extremity is usually overlooked during the initial stages although there is a greater chance for recovery. Therefore, stroke patients often do not reach the maximum ability for recovery in their upper extremity when discharged from inpatient settings [6].

The involvement of robotic devices to grant rehabilitation therapy is a comparatively recent area within the field of robotics in healthcare and arise from the concept of adopting robots to support people with impairments. In turn out, this has contributed to the increase in a myriad of robotic rehabilitation devices. As the rehabilitation device is under the therapist's supervision, the main requirement in the field of robotics rehabilitation is primarily on how therapist's expert knowledge and skills can be embedded into the robotic system with the growing technologies. The system in envisioned to not only being able to provide more assistance but also offer other observations into the efficacy of the treatment.

Rehabilitation robots essentially operate close to the human user, and it should be able to deal with various human joints individually and concurrently imitate human motion. This calls for a robot design that is ergonomic, reliable and user-friendly. Interaction forces between the human user and the robot should also be taken into consideration in controlling the robot in addition to its position. The robot is required to not only be able to carry the user's limb but moreover is able to react to the forces applied by the user's limb as well. Moreover, modulation of the system is desirable as it is able to provide for modifications of the rehabilitation treatment that fits the patient's progress. These conditions bring in some challenges in the control of robots for rehabilitation which are described as follows; 1). Position control: Position control for robots includes either the joint based or space based trajectory tracking that moves as fast as possible or with a uniform velocity. In contrast, the actions of a rehabilitation robot should be smoother and have a velocity profile similar to that of normal human action. 2. Adaptive force control: The robot requires to be able of interacting and reacting to movement caused by the user in real-time. Various kinds of rehabilitation activities will need specific kinds of interaction. Furthermore, modifications to the difficulty or amount of assistance required during therapies should also be made available to provide users with various stages of impairment.

This paper is trying to embark in benchmarking the existing techniques in assist-as-needed strategy for upper limb rehabilitation robot system. The development of the robotic devices is described in section 2.1. Then section 2.2 summarizes the findings from other works that use different assist-as-needed control strategy for the upper-limb robotic rehabilitation.

2. Literature Review

The existing proposed approaches are briefly presented and reported which constitute the dealing with the upper-limb rehabilitation. The existing proposed approaches that are reviewed in this paper covers features that appear significant in control of the upper-limb robotic system.

2.1. Development of Robotic Devices for Upper-Limb Stroke Rehabilitation

Assistive robotics to support people with disabilities including rehabilitation has been an extensive research area for the last few decades to support, strengthen and quantify rehabilitation process. This kind of robot provides autonomous training where patients are involved in the repeated exercise of goal-directed tasks leading to improvements in motor function. It appears that with respect to upper-limb rehabilitation, the earlier versions of robotic manipulators have been used to move the patient's impaired hand and arm to desired position in a 3-D plane enabling the rehabilitation of the elbow and shoulder joints. Controller designs are then incorporated to ensure the safety of such rehabilitative systems to the users by allowing regulated interaction forces [7]. It has been reported that early clinical trials demonstrated that such rehabilitation devices were safe for human use and to augment traditional mode of therapy [8].

Later devices aimed to expand the capabilities of robotic therapeutic devices by targeting the more on the distal segments of the upper-limb, with wrist and hand modules to attach to previously developed arm devices [9], standalone wrist and/or hand devices [10] and also shoulder and elbow devices [11]. Many of these are exoskeleton type devices, which aim to isolate the motion of individual joints. Many of these devices isolate the motion with respect to individual joints and have a limited range of motion (ROM) in comparison with endpoint manipulators. Nonetheless, it is worth to mention that, it rehabilitates the targeted joint(s) efficiently and allow reasonably accurate data collection on the patient's limb motion.

2.2. Assist-as-Needed (AAN) Control Strategy

The main purpose of robotics based rehabilitation is to assist patients to undergo rehabilitation in addition to reduce the dependency on the therapists. However, to complete the prescribed exercise regime demands significant effort from the patient. Therefore, the strategy of assist-as-needed is paramount so as to contribute the minimal amount of robot assistance to aid the patients in rehabilitation. There are various methods that have been developed by researchers and the most basic type of controllers found from the literature and often applied in robotic rehabilitation is impedance controller [12]. The objective of deploying impedance controller is to ensure the impaired upper-limb to move on a specific path during the therapy sessions. The notion behind the aforementioned controller is to compare the kinematics of the robot which is the position and velocity of the desired reference trajectory and to apply force proportional to a weighted sum of the position and velocity errors [13]. Impedance controller is well suited to the robot device for stroke rehabilitation, where the restoring force which is proportional to the perpendicular distance from the path is provided by the robot in the event the patient's hand deviates from the prescribed path [14]. Nonetheless, if the prescribed motion is successfully carried out by the patient, the robot does not apply any force. Another type of controllers requires the determination of both the desired path and the time required for completion. In such a case, the desired trajectory can be defined as minimum jerk trajectory (a specific trajectory which is smooth, and has bell-shaped velocity profile that minimizes the change in acceleration during movement) [15] or an average trajectory that has been pre-recorded from healthy subjects. Based on this definition, then the robot provides an assistive force proportional to the distance the subject's arm that lags the desired trajectory [16]. Although the implementation of this controller is recommended for certain trajectory between the starting point and the target point, there is another study conducted by introducing alternative approach where patients have the option to select the trajectory by their own, while the controller still supports the motion to reach the target [17]. Feasibility study of these systems has been carried out on stroke patients [16]-[18], and it was shown that these algorithms result in a reduction in muscle tone [7], significant functional improvements [19], and better functional improvements than robotic therapy designed to mimic the intensity of traditional therapy [20].

Although this controller has shown positive results in clinical trials, nevertheless, its tendency or "slacking" for the patient to rely heavily on the assistance force have been recorded [21]. In order to compensate this tendency, several studies have been conducted by incorporating the forgetting element in the controller. The goal behind the idea stemmed to reduce the amount of assistance to the patients after each trial if it was successfully implemented, thus further promote more active participation from patients [14], [22], [23]. This type of controllers requires the knowledge of both the estimated inertia and damping of the robotic system as well as the patient's arm in order to compensate the forces arises from these properties. Due to unknowns in the explicit model, researchers have also made an effort to model the patient's effort contribution at various points in the workspace. The contribution of patient's effort is modeled by using radial basis function (RBF) and this estimated function is updated during training [24]. There are also studies on using Bayesian learning techniques to measure the appropriate amount of assistance required to accomplish the task [25]. Pérez-Rodríguez et al. [22], investigated on the path deviations from the desired trajectory that could be anticipated by modelling the patient's ability. The implications of this anticipation lead to the introduction of corrective force that can be applied in prior to the deviations, thus the force supplied by the robot is reduced. In another study, Wolbrecht et al. made a comparison between their adaptive controller with and without a forgetting term in clinical trials on both stroke and healthy subjects [14]. The output obtained from the study indicated that the forgetting term leads to a more active involvement of the patient in the therapeutic process. Following the positive results of this study, a variety of improvements in the modelling of patient's effort such as the directionality of movement were demonstrated. Nonetheless, the ability of the controller is only conducted on subject healthy alone [26]. In order to overcome the problem, a study on the dependency of velocity in the model investigated applied to the stroke patients [26].

Another potential opportunity is to modify the difficulty of the movement task adaptively. In the study, the subjects were asked to track the moving targets in a sinusoidal pattern with the movement of the wrist and the increase of required range of motion after each attempt if successful [27]. In this study, three stroke subjects have been selected and tested and the preliminary results showed the potential of the method to improve a variety of subject's wrist movement. A follow-up study was carried out by [28] where eleven stroke subjects were chosen, and the results showed the improvements in the subjects' active ROM and measurable improvements in motor function. In addition, a different method has been adapted to match the difficulty of the task as being carried out by [29]. In this research, the authors employed machine learning techniques to classify the physiological state of the subject as "relaxed", "medium stressed", or "over-stressed" based on physiological signals (heart rate, respiratory rate, skin temperature, and galvanic skin response) and adjusted the difficulty level of the tasks accordingly. The experimental results suggest the feasibility of the proposed method, nonetheless, it is worth to note that clinical trials on stroke patients has yet been investigated. Table 1 summarizes the literature survey conducted on the topic.

Table 1. Assist-as-needed (AAN) strategies for upper-limb stroke rehabilitation robotic.

Author	Robotic System	Strategies
Pehlivan et al. (2016)	3-DoF wrist exoskeleton	Employ Kalman Filter and Lyapunov method to estimate the patient capability based on joint position and velocity. The assistance is provided to the patient if the patient is unable to perform the motion (position & velocity) while the assistance is decreased if the patient achieves the prescribed motion
Pérez-Rodríguez et al. (2014)	Simulated 8-DoF arm orthosis	A dysfunctional-adapted biomechanical prediction subsystem is developed to provide an anticipatory force-feedback to patient in order to avoid the trajectory deviations. The assistance force is decrease when patient is success to complete the task.
Squeri et al. (2011)	3-Dof wrist exoskeleton	Impedance control scheme is used to generate an assistive force field based on relative positions of the target and the end effector. The assistance is provided based on the error (trajectory tracking) produced by the patient.
Bower et al. (2013)	2-finger grasp robot	State dependence (position and direction) and inertial force are used as input to AAN strategy. The RBF activation function is applied

		to model the effort from patient. The AAN strategy will decrease the assistance to the patient in the event that the patient is able to complete the task by themselves.
Guidali et al. (2011)	7-dof arm exoskeleton	The kinematic error (velocity and time) is used as an input to AAN strategy. The assistance strategy is decrease when the patient is able to achieve desired task.

3. Conclusion

It is apparent from the literature that the assist-as-needed methods encourage active patient participation in movement therapy by providing a bare minimum necessary assistance. Under the assist-as-needed strategy, the robot is no longer needed to contribute full support throughout the path of motion trajectory; whereas the robot can impel subjects to use their muscular strengths while trying to keep up with the predefined motion trajectory. It has also been demonstrated from clinical trials the feasibility of the methods as well as the functional improvements as a consequence of this form of training. However, controlled clinical trials are still needed to compare training methods and quantify possible benefits of such strategy over traditional therapy.

Acknowledgment

The work presented was carried out in the Biomechatronics Research Laboratory of International Islamic University Malaysia. The author wishes to gratefully acknowledge the E-Science grant funding (SF15-015-0065) from the Ministry of Science, Technology and Innovation Malaysia

References

[1] W. H. Organization, "The World Health Report 2013: Research for universal health coverage.," *Sci. Transl. Med.*, vol. 5, no. 199, p. 146, 2013.
[2] MOH, "Annual Report Ministry of Health 2011," p. 351, 2011.
[3] H. S. Lo and S. Q. Xie, "Exoskeleton robots for upper-limb rehabilitation: State of the art and future prospects," *Med. Eng. Phys.*, vol. 34, no. 3, pp. 261–268, 2012.
[4] I. M. Khairuddin *et al.*, "A hybrid joint based controller for an upper extremity exoskeleton," *IOP Conf. Ser. Mater. Sci. Eng.*, vol. 114, p. 12133, 2016.
[5] C. J. Winstein *et al.*, *Guidelines for Adult Stroke Rehabilitation and Recovery: A Guideline for Healthcare Professionals from the American Heart Association/American Stroke Association*, vol. 47, no. 6. 2016.
[6] S. M. Hatem *et al.*, "Rehabilitation of Motor Function after Stroke: A Multiple Systematic Review Focused on Techniques to Stimulate Upper Extremity Recovery.," *Front. Hum. Neurosci.*, vol. 10, no. September, p. 442, 2016.
[7] H. I. Krebs *et al.*, "Rehabilitation robotics: Performance-based progressive robot-assisted therapy," *Auton. Robots*, vol. 15, no. 1, pp. 7–20, 2003.
[8] T. Nef, M. Guidali, and R. Riener, "ARMin III – arm therapy exoskeleton with an ergonomic shoulder actuation," *Appl. Bionics Biomech.*, vol. 6, no. 2, pp. 127–142, 2009.
[9] P. Staubli, T. Nef, V. Klamroth-Marganska, and R. Riener, "Effects of intensive arm training with the rehabilitation robot ARMin II in chronic stroke patients: four single-cases," *J.*

Neuroeng. Rehabil., vol. 6, no. 1, p. 46, 2009.

[10] A. U. Pehlivan, D. P. Losey, and M. K. Omalley, "Minimal Assist-as-Needed Controller for Upper Limb Robotic Rehabilitation," *IEEE Trans. Robot.*, vol. 32, no. 1, pp. 113–124, 2016.

[11] M. H. Rahman, C. Ochoa-Luna, and M. Saad, "EMG Based Control of a Robotic Exoskeleton for Shoulder and Elbow Motion Assist," *J. Autom. Control Eng.*, vol. 3, no. 4, pp. 270–276, 2015.

[12] L. Marchal-Crespo and D. J. Reinkensmeyer, "Review of control strategies for robotic movement training after neurologic injury.," *J. Neuroeng. Rehabil.*, vol. 6, no. 1, p. 20, 2009.

[13] N. Hogan, "Impedance Control: An Approach to Manipulation," *Am. Control Conf. 1984 IS - SN - VO -*, no. March, pp. 304–313, 1985.

[14] E. T. Wolbrecht, V. Chan, D. J. Reinkensmeyer, and J. E. Bobrow, "Optimizing Compliant, Model-Based Robotic Assistance to Promote Neurorehabilitation," *IEEE Trans. Neural Syst. Rehabil. Eng.*, vol. 16, no. 3, pp. 286–297, 2008.

[15] T. Flash and N. Hogan, "The coordination of arm movements: an experimentally confirmed mathematical model.," *J. Neurosci.*, vol. 5, no. 7, pp. 1688–1703, 1985.

[16] M. Guidali, A. Duschau-Wicke, S. Broggi, V. Klamroth-Marganska, T. Nef, and R. Riener, "A robotic system to train activities of daily living in a virtual environment," *Med. Biol. Eng. Comput.*, vol. 49, no. 10, pp. 1213–1223, 2011.

[17] V. Klamroth-Marganska *et al.*, "Three-dimensional, task-specific robot therapy of the arm after stroke: A multicentre, parallel-group randomised trial," *Lancet Neurol.*, vol. 13, no. 2, pp. 159–166, 2014.

[18] M. Mihelj, T. Nef, and R. Riener, "A novel paradigm for patient-cooperative control of upper-limb rehabilitation robots," *Adv. Robot.*, vol. 21, no. 8, pp. 843–867, 2007.

[19] H. I. Krebs *et al.*, "Robot-aided neurorehabilitation: A robot for wrist rehabilitation," *IEEE Trans. Neural Syst. Rehabil. Eng.*, vol. 15, no. 1, pp. 327–335, 2007.

[20] N. Hogan *et al.*, "Motions or muscles? Some behavioral factors underlying robotic assistance of motor recovery," vol. 43, no. 5, pp. 605–618, 2006.

[21] M. Casadio and V. Sanguineti, "Learning, retention, and slacking: A model of the dynamics of recovery in robot therapy," *IEEE Trans. Neural Syst. Rehabil. Eng.*, vol. 20, no. 3, pp. 286–296, 2012.

[22] R. Pérez-Rodríguez *et al.*, "Anticipatory assistance-as-needed control algorithm for a multijoint upper limb robotic orthosis in physical neurorehabilitation," *Expert Syst. Appl.*, vol. 41, no. 8, pp. 3922–3934, 2014.

[23] M. Guidali, U. Keller, V. Klamroth-Marganska, T. Nef, and R. Riener, "Estimating the patient ' s contribution during robot-assisted therapy," vol. 50, no. 3, pp. 379–394, 2013.

[24] A. U. Pehlivan, F. Sergi, and M. K. Omalley, "A subject-adaptive controller for wrist robotic rehabilitation," *IEEE/ASME Trans. Mechatronics*, vol. 20, no. 3, pp. 1338–1350, 2015.

[25] V. Squeri, A. Basteris, and V. Sanguineti, "Adaptive regulation of assistance 'as needed' in robot-assisted motor skill learning and neuro-rehabilitation," *IEEE Int. Conf. Rehabil. Robot.*, 2011.

[26] C. Bower, H. Taheri, and E. Wolbrecht, "Adaptive control with state-dependent modeling of patient impairment for robotic movement therapy," *IEEE Int. Conf. Rehabil. Robot.*, 2013.

[27] L. Masia, M. Casadio, P. Giannoni, G. Sandini, and P. Morasso, "Performance adaptive training control strategy for recovering wrist movements in stroke patients: a preliminary, feasibility study," *J. Neuroeng. Rehabil.*, vol. 6, no. 1, p. 44, 2009.

[28] V. Squeri, L. Masia, P. Giannoni, G. Sandini, and P. Morasso, "Wrist Rehabilitation in Chronic Stroke Patients by Means of Adaptive, Progressive Robot Aided Therapy," *IEEE Trans. Neural Syst. Rehabil. Eng.*, vol. Early Acce, no. 2, pp. 312–325, 2013.

[29] F. J. Badesa, R. Morales, N. Garcia-Aracil, J. M. Sabater, A. Casals, and L. Zollo, "Auto-adaptive robot-aided therapy using machine learning techniques," *Comput. Methods Programs Biomed.*, vol. 116, no. 2, pp. 123–130, 2014.

Nanosatellites constellation as an IoT communication platform for near equatorial countries

A Narayanasamy[1], Y A Ahmad[2*], M Othman[3]

[1] Usaha Prakarsa (M) Sdn Bhd, Shah Alam, Selangor, Malaysia
[2] Electrical and Computer Engineering Department, Kulliyyah of Engineering, International Islamic University Malaysia, Malaysia
[3] Academy of Sciences Malaysia, Malaysia

*yasser@iium.edu.my

Abstract. Anytime, anywhere access for real-time intelligence by Internet of Things (IoT) is changing the way that the whole world will operate as it moves toward data driven technologies. Over the next five years, IoT related devices going to have a dramatic breakthrough in current and new applications, not just on increased efficiency and cost reduction on current system, but it also will make trillion-dollar revenue generation and improve customer satisfaction. IoT communications is the networking of intelligent devices which enables data collection from remote assets. It covers a broad range of technologies and applications which connect to the physical world while allowing key information to be transferred automatically. The current terrestrial wireless communications technologies used to enable this connectivity include GSM, GPRS, 3G, LTE, WIFI, WiMAX and LoRa. These connections occur short to medium range distance however, none of them can cover a whole country or continent and the networks are getting congested with the multiplication of IoT devices. In this study, we discuss a conceptual design of a nanosatellite constellation those can provide a space-based communication platform for IoT devices for near Equatorial countries. The constellation design i.e. the orbital plane and number of satellites and launch deployment concepts are presented.

1. Introduction

The Internet of Things (IoT) application is beginning to multiply significantly. Consumer products, durable goods, cars and trucks, industrial and utility components, sensors and other everyday objects are being attached with Internet connectivity and powerful data analytic capabilities that promise to transform the way we work, live and play. Projections for the impact of IoT on the internet and economy are impressive, with some anticipating as many as 100 billion connected IoT devices by 2025.

The existing wired and wireless network are almost being fully utilized by the current IoT for control and communication of the devices. However, as IoT devices continue to multiply greater data communications and data services are required, these existing networks will become increasingly stressed and congested, especially in remote regions. One of the solutions to overcome the need for greater data communication is investment in space infrastructure. Space-based data communication network have greater impact on humanity by delivering the required additional data communication to every part of the world.

Space-based data communication network is not something new as back in the 1990s there were several large space-based satellite network providers, such as Intelsat, Iridium, GlobalStar, Teledesic, etc. but these providers used bigger satellites which entailed high costs of deployment and operation. However, currently with the greatly advanced of satellite technology, the cost of the development and deployment has reduced significantly. With the current technology, nanosatellites can be mass produced and can be launched by the dozens at a time to Low Earth Orbit (LEO), reducing launch costs, while delivering performance comparable to larger, older satellites at higher orbits [1].

Nanosatellites ranging in mass between 1kg – 10kg are gaining momentum as an additional means to address targeted scientific and engineering programs in a rapid and very much more affordable manner. Within the group of nanosatellites, CubeSats have appeared as a more recognised space platform. The standardization of CubeSat defined in terms of 10cm x 10cm x 10cm in size and approximately 1 kg each called a ‘U’. Originally, CubeSats were deployed as training programme to expose students to the challenges of real world satellite engineering practices and system design. However, CubeSats use has rapidly spread within academia, industry and government agencies globally to achieve various space missions.

A set of satellites distributed over a space to achieve a common mission objective is called constellation. While there are no absolute rules, there are a number of key issues which dominate constellation design trades, namely coverage, number of satellites, launch options, space environment, orbit perturbations for station keeping, collision avoidance, constellation build-up and required replenishment and number of orbital planes [2]. Recently, OneWeb proposed 650 satellites as a LEO constellation around the globe, to provide Internet access to rural and developing areas of the world with significant investments from Virgin Group and Qualcomm. Airbus Defence and Space was recently announced as the manufacturer for these satellites [3].

The good news is, the development of a LEO constellation to provide 365/24/7 coverage for countries near the Equator will be much less complex with only less than twenty-five CubeSats required. The orbit at or close to Equator will be extremely advantageous to countries straddling the Equator, whereby the footprint of the satellites will always cover the Equator and provide frequent revisits. Many existing constellations or geostationary satellites are already available to provide the data communication services, however, these require an extensive infrastructure on the ground for data transmission and most of them are expensive in terms of capital and operational expenditures. These two parameters are the cause for existing services not being applicable for IoT application.

For these reason, we are proposing a CubeSat Equatorial orbit constellation as a communication platform for IoT which will provide continuous coverage for the countries near the Equator at a low cost and low power, requiring smaller sized transmitters on the ground and provide sufficient data rate for the application of IoT devices which will communicate directly to the constellation of CubeSats in space to transfer the data in real time.

2. Space-Based IoT Communication Platform

A space-based IoT communication platform requires an infrastructure that will fit its various applications. The infrastructure requires a network of satellites in the Equatorial plane, ground stations, and IoT terminals that are interconnected. These variations in requirements can be used by different satellite design.

Every satellite has its own size and mass and Table 1 shows the differences between classic, micro, Nano and Pico satellites. Micro, Nano and Pico satellites are classified as small satellites, however, the nano and micro satellites tend to overlap while the definition of Microsatellite could be assumed to range between from 50kg and 100kg.

2.1. Concept of operations

The ground station two functions: telemetry and telecommand of the satellite operations and the data handling of the space IoT data. The terminal is equivalent to a very small device that is able to be placed on strategic devices for applications such as environmental monitoring, oil pipeline monitoring, smart grid, mobile tracking and control of water sources. The IoT terminal should be small but able to communicate with the satellites. The satellites, ground stations, and terminals all have an antenna, transmitter, receiver, and control equipment that enable them communicate with each other.

Table 1. Typical classification of satellites [4]

	Mass (Kg)	*Altitude (Km) / Orbit Period (Hour)*	*Construction Duration (Year)*	*Total Cost (Million, USD)*
Classic	1,000	10,000 5.8	10	150
Micro	10-100	500-2,000 1.6-2	2-5	1-30
Nano	1-10	300-800 1.4-1.7	2-3	0.1-10
Pico	0.1-1	200-600 1.4-1.5	1-2	0.05-2

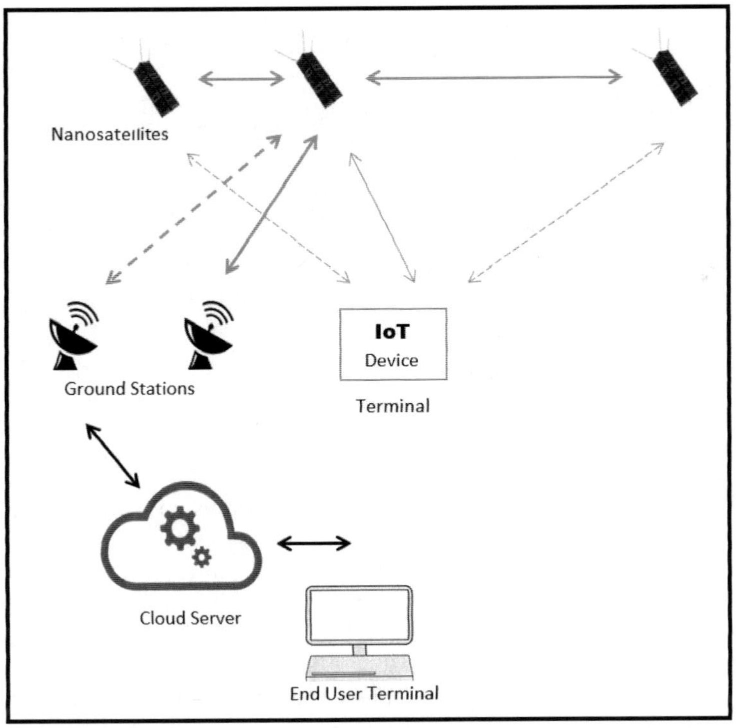

Figure 1. Space based IoT infrastructure concept.

The satellites in the equatorial orbit are in the same orbital plane. They broadcast data to multiple ground stations in different locations through broad or spot beams. The satellite transmitter must provide enough transmission power to Earth surface. The receiver has multiple channels to receive data from multiple IoT devices. The satellites will have the capability of store and forward messages to enable IoT data to be transmitted to the ground stations. An advance mode would enable Inter Satellite Link (ISL) communications to allow IoT data to be passed on to each satellite. However, the satellites must be within line of sight and at the right distance. The IoT devices will directly send message to the satellite and can detect the satellite coming. To save power, the IoT device would be able to implement a sleep mode. The ground stations act as a gateway to the end-user and they would enable the IoT messages to be forwarded to the end-user. The ground station requires a tracking antenna since require to follow the motion of satellite, other functions of the ground station including providing the satellite operation and maintenance.

2.2. Communication system

In developing the communication system, the architecture of the communication architecture itself determines the link requirement for the system. The communication links will allow a satellite system to function by transmitting tracking, telemetry, and command data between the elements [5]. The communication link consists the uplink and downlink. Uplink means sending the signal from Earth station antenna to spacecraft antenna while the downlink means sending the signal from the satellite antenna to the Earth station antenna [6]. Other than the uplink and downlink, the inter-satellite link (ISL) plays an important role in relaying data from and to a satellite in the Equatorial orbit. ISL enable two satellites in orbit to communicate directly without using a ground station on Earth.

For the uplink and the downlink design, the link budget equation is used to determine the communications link information such as antenna parameters, data rate, antenna size, propagation of path length and transmitter power. Using those parameters, we can calculate the E_b/N_0 of the link. The equation of E_b/N_0 is shown as below [7]:

$$\frac{E_b}{N_0} = \frac{P L_l G_t L_s L_a G_r}{k T_s R} \tag{1}$$

- E_b/N_0: Ratio of received energy-per-bit to noise-density
- P: Transmitter power
- L_l: Transmitter-to-antenna line loss
- G_t: Transmit antenna gain
- L_s: Space loss
- L_a: Transmission path loss(function of factors such as rainfall density)
- G_r: Receive antenna gain
- k: Boltzmann's constant
- T_s: System noise temperature
- R: Data rate

For advance communications using ISL, the distance between two satellites must be known. The complexity of calculating the distance between the two satellites is reduced at the Equatorial orbit where all the satellites are in the same orbit plane. Two satellites are required to define the sufficient distance for acceptable space loss based on receiver sensitivity, antenna gain and transmit power. The distance between two satellites can be calculated using orbit plane coordinate system and WGS-84 coordinate system. WGS-84 is a standard that is developed by the United States Department of Defense in order to provide a reference globally mainly for the Global Positioning System (GPS) constellation of satellite. The WGS-84 parameters consist of the Earth semi-major axis, the Earth flattening factor, the Earth Mean Angular velocity and the Earth Geocentric Gravitational Constant. The coordinate of a satellite is based on the Earth Centered Inertial where the x axis is the reference point pointing to the direction of vernal equinox, the z axis points to the Celestial pole or north and y axis is the normal to the x and z axis. The orbit plane coordinate is then transformed into the coordinate (X, Y, Z) of the satellite in WGS-84 coordinate system. The coordinate transformation equation is provided in equation 2 [8] below. Ω and i, denote the ascending nodes and inclination, respectively.

IOP Conf. Series: Materials Science and Engineering **260** (2017) 012028 doi:10.1088/1757-899X/260/1/012028

$$\begin{bmatrix} X \\ Y \\ Z \end{bmatrix} = \begin{bmatrix} x cos\Omega - y cos\Omega \\ x sin\Omega + y\cos\Omega\cos i \\ y\sin i \end{bmatrix} \quad (2)$$

The distance *d*, between the two satellites are denoted by satellite S1 = (X_1, Y_1, Z_1) and satellite S2 = (X_2, Y_2, Z_2) and the distance can be calculated by using equation 3 [8] below:

$$d = \sqrt{(X_2 - X_1)^2 + (Y_2 - Y_1)^2 + (Z_2 - Z_1)^2} \quad (3)$$

The distance *d,* is inserted in the equation for the space loss (L_s) is shown as below [7] where *f* denotes the spectrum of the inter-satellite link.

$$L_s = 147.55 - 20\log d - 20\log f \quad (4)$$

Higher frequency such as Ka and V band frequency are the typical medium to transfer information between satellites. This is because the link is not being affected by rain attenuation which occurs between ground to satellite links [4]. High-throughput optical and Radio Frequency (RF) ISLs have been named as a medium to transfer information between satellites. Most satellites are using RF and microwave link to transfer data. However, the limitations of size on nanosatellites are the key challenges in implementing the ISL.

2.3. Frequency allocations

The frequency allocations are the main criteria in defining a space-based IoT communication platform as it requires approval from the International Telecommunication Union (ITU). The satellite services defined by ITU covers three areas namely mobile-satellite service, fixed-satellite service and space operation service. Mobile satellite services is a radio communication services between a mobile station on earth and one or more satellites, either with or without ISLA fixed-satellite service is between earth stations at fixed positions with one or more satellites are used. A space operation service is radio-communications exclusively used for the operations of a spacecraft, particularly for tracking, telemetry and telecommand. A combination of the three services is required for a fully operational Space Based IoT communication platform.

The driving criteria in defining the communications architecture functions of a space-based IoT communication platform mission are the RF spectra, the orbit, and the operational considerations for implementations, such as the ground network, cloud and IoT technologies, data rate, link availability, and link access time.

2.4. Intersatellite link analysis and results

The position vector of satellite is listed in Table 2 show the parameters of Kepler element that was used.

Table 2. Orbital Elements Simulated

Parameters	*Values*
Angular Momentum, L	53032.8 km^2s^{-1}
Inclination, i	0°
Right Ascension (RA) of the Ascending Node, Ω	61° and 62°
Eccentricity, e	0.01
Argument of Perigee, ω	90°
True Anomaly, v	35°

The link analysis was conducted by using all the parameters that shown in table 3.

Table 3. Intersatellite link analysis parameters

Parameters		*Proposed Value*
Ambient Temperature	T_{amb}	353 K
Centre Frequency	f_c	2.45 GHz
Bandwidth	B	33 MHz
Link Distance	D	70 - 120 km
Transmitter Power	P_T	1 - 4 W
Transmitter Losses	L_T	1 dB
Friis Path Loss	L	139.3 dB
Receiver Losses	L_R	1 dB
Link Margin	LM	2 dB
Signal-to-noise Ratio	SNR	6 dB
Receiver noise Figure	NF	4.6 dB
Noise Floor	N	-105 dB
Antenna System Gain	G_T, G_R	0 - 5 dBi

The position vector was simulated using Matlab.

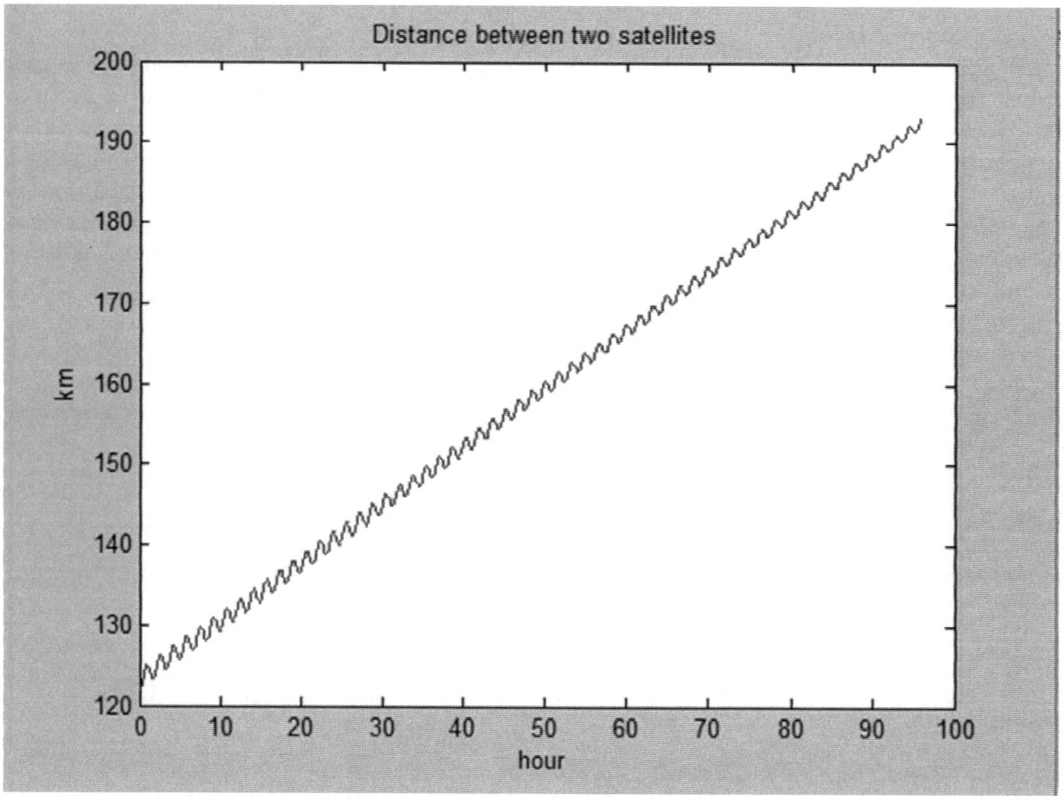

Figure 2. Distance variation between two satellites

Link budget parameter that was list in Table 3 was used and calculated. The main objective was to find the minimum value of Power Received, P_R that was suitable for the ISL to take place. The

IOP Conf. Series: Materials Science and Engineering **260** (2017) 012028 doi:10.1088/1757-899X/260/1/012028

minimum value of P_R was determine to be -140.5 dB. The calculation was made and it shows that there are many time where the value of P_R was higher than -140 dB. Table 4 shows all the value that intersatellite link can take place with variation of distance, transmit power and antenna gain. It is therefore feasible to achieve intersatellite link between two satelltie in orbit.

Table 4. Summary of the result for intersallite link analysis

Distance (km)	P_T *(W)*	*Gain (dB)*	P_R *(dB)*
70	1	1	-139.139
80	1	1	-140.299
90	4	1	-135.302
100	4	1	-136.217
110	1	5	-135.065
120	1	5	-135.821
120	4	1	-137.8
130	1	5	-136.516
140	1	5	-137.10
140	4	1	-139.139

3. Orbit and constellation design

A trajectory is a line traced by a moving body and an orbit is a trajectory that is periodically repeated. While the path followed by the motion of an artificial satellite around the Earth is an orbit, the part followed by a launch vehicle is a trajectory called the launch trajectory [9]. The gravitational force that act on satellite creates a pulled down effect and if satellite did not have any motion of their own they would fall back to Earth, burning up in the reaches of the atmosphere. Instead, the initial energy acting to make satellite motion has a force associated with it pushing the satellite away from the Earth and make satellite rotating around the Earth. For any given altitude, for satellite remains stable in orbit, neither going higher nor lower, there is a specific speed for which gravity and the centrifugal force balance each other. The orbit that is chosen for a satellite depends upon its application, for example, those used for direct broadcast use the Geostationary orbit, those used for satellite phones, are in the Low Earth Orbit satellites while satellites for navigation, like Navstar for Global Positioning System (GPS), occupy the Medium Earth Orbits.

3.1. Constellation design

A Constellation consists of multiple satellites that orbit the Earth in similar, but suitably shifted or arranged orbits. A famous example is the America's Global Positioning System (GPS), whereby a fully operational system comprises a total 24 satellites in six orbital planes at 55° inclination. Four satellites each share the same orbit of 20,200km altitude, but are offset from their neighbours by a 90° longitudinal phase shift. Meanwhile, each orbital plane are separated by 120° and such configuration makes a minimum of 6 satellites always available to be linked to the ground GPS receiver. Due to the orbital period of 12 hours, the configuration of all satellites relative to Earth is exactly repeated twice every sidereal day. GLONASS, the Russian counterpart of the United States' GPS, utilizes a similar constellation of 24 satellites evenly distributed in three planes, with an orbital inclination of 64.8^o [10]. At an altitude of 19, 100km, the orbital period of 11.25 hours.

In the past decade, the high potential of Low Earth Orbit satellite constellations for global mobile communication was realized. The signal paths is shorter for low Earth satellites, signal link can be established from only hand-held devices in comparison requirement of huge user antennas setup for geostationary satellites. By making use of ISL, telephone calls can then be routed around the world to other mobile-phone user or to a suitable ground network terminal. A well-known example is

IOP Conf. Series: Materials Science and Engineering **260** (2017) 012028 doi:10.1088/1757-899X/260/1/012028

IRIDIUM, a 66-satellites constellation at an altitude of 700km, which was put into operation in 1999 [11]. Other constellations have been designed and partly implemented, which include Globalstar with 48 satellites at 1414km altitude, ICO with 10 satellites at 10,400km, ORBCOMM [12] and Teledesic with 288 satellites at 1,350km [13].

There are no absolute rules existing for designing a constellation. However, there are a number of key issues which dominate the constellation design trade. Table 5 shows the dominant parameters for constellation design.

Table 5. Principal issues that dominate the constellations design [2]

Issue	*Why important*	*What determines it*	*Principal issues or options*
Coverage	Principal performance parameter	Altitude, minimum elevation angle, inclination, constellation pattern	Gap times for discontinuous coverage, Number of satellites simultaneously in view for continuous coverage
Number of Satellites	Principal cost driver	Altitude, minimum elevation angle, inclination, constellation pattern	Altitude, minimum elevation angle, inclination, constellation pattern
Launch Options	Major cost driver	Altitude, inclination, spacecraft mass	Low altitude, low inclination cost less
Environment	Radiation level and therefore, lifetime and hardness requirements	Altitude	Option are below, in or above Van Allen radiation belts
Orbit Perturbation (Station keeping)	Causes constellation to disassociate over time	Altitude, inclination and eccentricity	Keep all satellites at common altitude and inclination to avoid drifting apart
Collision Avoidance	Snowball effect can destroy entire constellation	Constellation pattern, orbit control	No option – Must design entire system for collision avoidance
Constellation Build-Up, Replenishment and End-of-Life	Determines level of service over time and impact	Altitude, constellation pattern, build-up and sparing philosophy	Sparing: on-orbit spares vs launch on demand; End-of-life deorbit vs raise to higher orbit
Number of Orbit Planes	Determines performance plateaus	Altitude, inclination	Fewer planes means more growth plateaus and more graceful degradation

A circular Equatorial low Earth orbit constellation will be the simplest constellation as it eliminates many constellation design issues, such as orbit perturbation, number of orbit planes, and at the same time provides continuous coverage for the near Equatorial countries. The current analysis will only focus on the design of constellation with Equatorial orbit for near Equatorial countries.

3.2. Constellation design analysis and results

The main objective of the analysis or the study is to identify the feasibilities and number of satellites for continuous coverage most often requires that the Near Equatorial region of interest must be seen continuously by at least one satellite in the constellation. To narrow down the simulation scenarios, we are required to define few conditions or parameters. The first condition is that the near Equatorial region of interest has been set to ±10° latitudes. Bigger satellites can generate bigger power and can be utilized for communication power system. However, the proposed constellation comprising nanosatellites will have limited power generation, therefore, it will limit the communication distance between the IoT devices and satellites. The trade-off study for this analysis are perform for altitudes between 500km – 800km. The link margin effect from the elevation mask has a strong impact on the communication link availability. A smaller degree of elevation mask means more interference from the atmosphere which interrupts the link availability, while a higher degree of elevation mask will narrow the communication footprint and increase the number of satellites in the constellation. The value for the elevation mask is always dependent on the receiving or transmitting terminals on ground. For this simulation scenario, we have constrained the elevation mask value to be 20°. Based on these parameters we calculate the communication beam angle for altitude 500km – 800km by referring to the Figure 4 and equations (5), (6) and (7). Table 6 shows all the calculated communication beam angle (η) and Earth central angle (λ).

Table 6. Calculated Nadir Angle and Earth Central Angle for altitude 500km – 800km

Altitude (km)	*Orbital Radius, r (km)*	*Nadir Angle, η (deg)*	*Earth Central Angle, (deg)*
500	6878.14	60.62	9.38
550	6928.14	59.89	10.11
600	6978.14	59.19	10.81
650	7028.14	58.52	11.48
700	7078.14	57.86	12.14
750	7128.14	57.23	12.77
800	7178.14	56.61	13.39

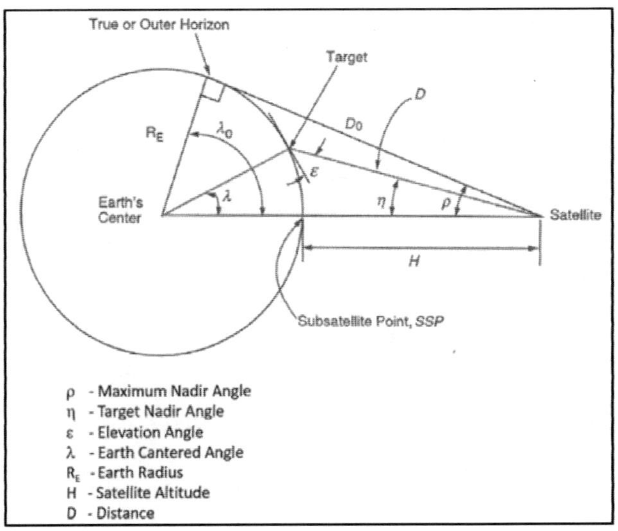

Figure 3. Definition of angular relationship between the satellite and earth's centre. [7]

$$Sin\,\rho = \frac{R_E}{R_E+H} \quad (5)$$

$$Sin\,\eta = Cos\,\varepsilon Sin\,\rho \quad (6)$$

$$\eta + \lambda + \varepsilon = 90 \quad (7)$$

The required number of satellites, *N*, to provide continuous coverage with at least one satellite for the near equatorial region of ±10° latitudes are calculated based on Figure 5 and equation (8) and (9). The results are shown in Table 7.

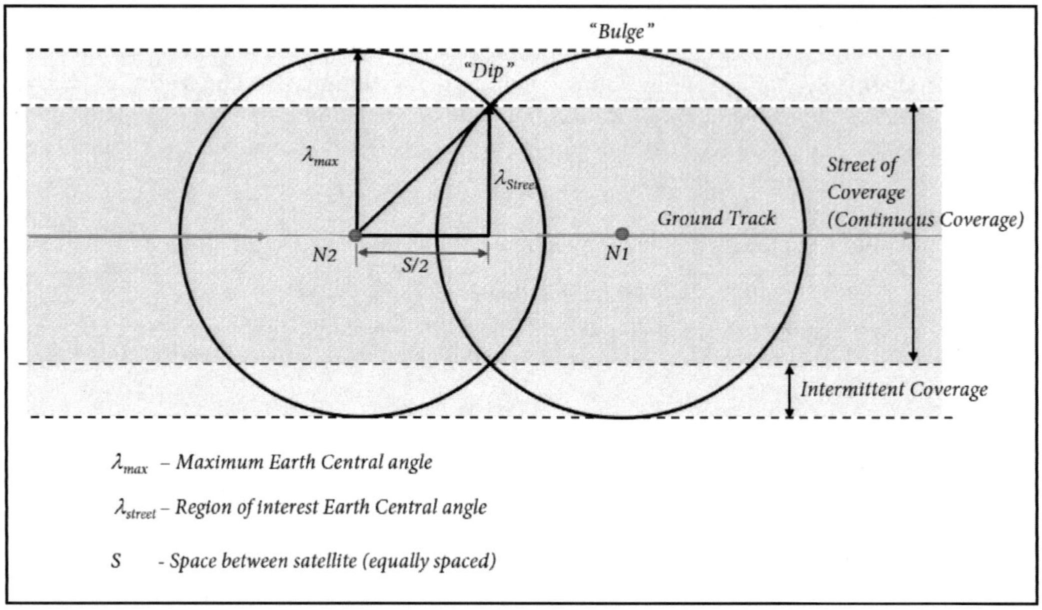

Figure 4. Street of coverage for constellation.

$$S = \frac{360}{N} \tag{8}$$

$$Cos\ \lambda_{Street} = \frac{Cos\ \lambda_{max}}{\cos(\frac{S}{2})} \tag{9}$$

Table 7. Number of Satellites required for continuous coverage of latitude region of ±10

Altitude (km)	*Earth Central Angle, (deg)*	*Number of Satellites, N*
500	9.38	No continuous coverage
550	10.11	121
600	10.81	44
650	11.48	32
700	12.14	27
750	12.77	23
800	13.39	21

The Table 4 shows that at 800km attitude with 20° elevation mask a minimum of 21 satellites will be required to provide continuous coverage for the Equatorial region between ±10° latitudes. Figure 4 shows the satellite coverage on a 2D Earth map. As for coverage redundancy, the constellation can be increase to 25 satellites which will provide four satellites in orbit redundancy.

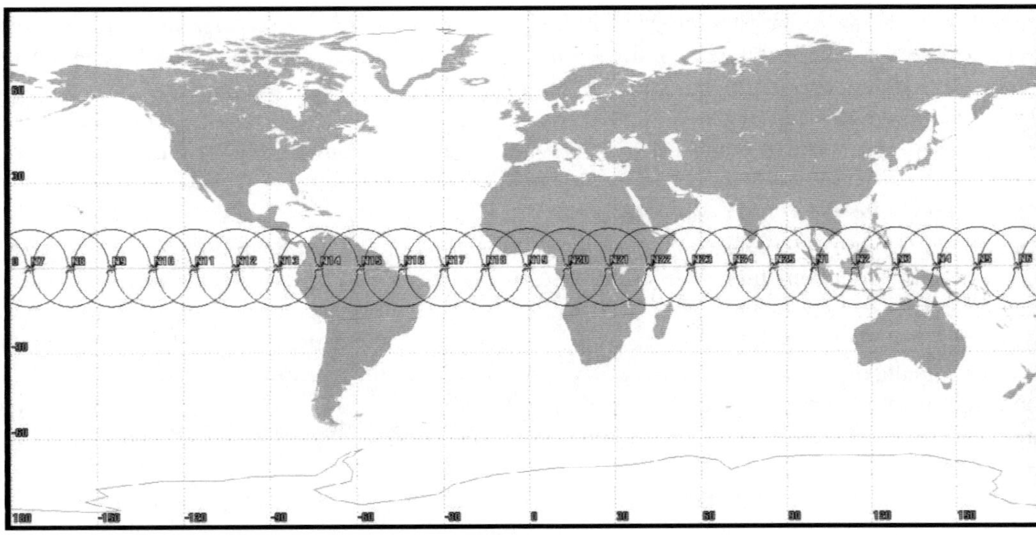

Figure 5. Satellite coverage for equatorial orbit constellation with 25 number of nanosatellites.

4. Launch and constellation deployment

The basic goal of the launch vehicle is to expel hot gases at velocities on the order of 2 to 3 km/s to active a final vehicle velocity of 8 km/s or more, even for low Earth orbit [2]. This means that the mass of any launch vehicle will be dominated by the propellant and that the vehicle itself will need to be designed for minimum excess weight. This in turn implies that the launch will incur the greatest cost and will entail significant risk since each launch vehicle on the launch pad is more than 95% explosive propellant.

4.1. Small satellite launch option

The absence of sufficiently small or inexpensive launch vehicles for the delivery of nanosatellites to orbit presents a significant barrier to the development of nanosatellite missions with smaller budgets. The launch opportunities as a secondary payload for small satellites normally addressed the issue of access to orbit, where normally the main payload will either offer share launch capacity through clustering or share rides or allow to use the excess launch space capacity for a commercial value, which is called piggybacking. Unless arranged through a subsidised launch programme (e.g. NASA CubeSat Launch Initiative and Educational Launch of Nanosatellites), the cost of secondary payload opportunities is generally greater than the specific cost ($/kg) of the launch vehicle itself [14]. However, these opportunities still allow nanosatellites to achieve the orbit at a significantly lower total cost than a dedicated launch. Piggybacking or secondary payload opportunities always constrains by main payload or primary satellite launch schedule and orbit destination, therefore the small satellites need to be flexible in their mission and final orbit destination. Further restrictions on the launch of small satellites by utilising secondary payload opportunities can include the requirement to be compatible with a certain class of deployment mechanism (e.g. P-POD, X-POD, ISIPOD), they reducing the level of certification required by the secondary payloads by isolating them from the launch vehicle and primary payload [15]. All the constrains limit the volume, mass possible provision for deployable structure such as antennas or solar panels

As the growth of the small satellite launch multiplied, a number of new launch services also growing to fill the gap for dedicated launch of microsatellites and nanosatellites via the use of smaller rockets. The goal for the smaller launch vehicle is to provide launch services for payload range from 10 to 300 kg for the same launch cost of current secondry payload. Notable examples include the Virgin Galactic LauncherOne which will be air-launched from the White Knight Two carrier aircraft

IOP Conf. Series: Materials Science and Engineering **260** (2017) 012028 doi:10.1088/1757-899X/260/1/012028

and will have a capacity on the order of 225 kg to LEO [16], a 10kg payload launcher deployed from the XCOR Aerospace Lynx Mk.III suborbital vehicle [17] and DARPA ALASA program involving several companies working towards the launch of a 45kg payload to orbit for less than USD 1 million. These vehicles will support the dedicated launch of microsatellites and nanosatellites, avoiding the potentially mission critical issues related to secondary payload launch opportunities [14].

4.2. Launch and constellation deployment
Traditionally, a constellation of nanosatellites has been populated through many launches. However, due to high mission launch costs in comparison with nanosatellites development cost, launches in this manner are not economically viable for the discussed Equatorial nanosatellite constellation. Since the proposed constellation will only be in single orbital plane in Equatorial orbit, the first possible option will be a dedicated single launch for deployment of all the nanosatellites in the Equatorial orbit constellation by the launcher itself. An earlier planning and mission design of the final stage of the launcher, it should made capable of staying in the final orbit to deploy all the 25 nanosatellites will the required orbital spacing.

Alternately we shall share the launch with other satellites to similar orbit. In the second possible option the use of satellite network deploy adapter is used [18]. The satellite network deployment adapter will be similar to the final stage of the launch vehicle which carries propulsion and will use the advantage of orbit perturbation to transfer the nanosatellite to final orbits with optimum use of fuel. Both the discussed options are considered the use of a single launch for deployment of the full nanosatellite constellation. Thus, option is provided because the launch for any space missions a major contributor of the programme cost and multiple launches will dramatically increase the cost.

5. Conclusion

The main application of IoT is to collect data from sensors that are remotely deployed and the ability to send command remotely to activate the actuators. Satellite communication is a platform to enable IoT in remote applications to be implemented. Satellite communications are enable through a GEO stationary satellite or an established global LEO communications network such as GlobalStar and Iridium. The Conceptual design of a nanosatellite constellation as an IoT communications platform for Near Equatorial countries is feasible as shown in this paper. In realising IoT through satellite, factors such as 1) interoperability between satellite systems, IoT device terminal with sensors and actuators and 2) compatibility of IPv6 on board the satellite [19] has to be thoroughly studied. The study shown that the near Equatorial countries can benefit from the Equatorial orbit constellation by deploying a minimal number of nanosatellites comparing to hundreds of nanosatellites needed for global coverage with Polar orbits. The current antenna technology for nanosatellite application can only produce smaller antenna beam widths. However, the conceptual design study using basic beam width shows the communications platform for the IoT can be achieved and with the advancement of the antenna design in the future for nanosatellite the design of constellation can be improved. The conceptual design study has shown at 800km altitude with a 20° elevation mask, 25 nanosatellites with 4 in-orbit spaces will be sufficient to build the IoT communications platform which will provide continuous coverage for ±10 latitudes near Equatorial countries.

References

[1] *Technology review: why the time seems right for a space based internet service* [Accessed online on November 2016]
[2] Wertz J R 2001 *Mission Geometry: Orbit and Constellation Design and Management* (Dordrecht: Kluwer Academic Publishers)

[3] *One web selects airbus to build 900 internet satellite on spaceflight now* [Accessed online on December 2016]

[4] Lazreg N & Besbes K 2016 *2nd International Conference on Advanced Technologies for Signal and Image Processing (ATSIP)* 601-605

[5] Adaramola M F & Balogun E B 2015 *Computer Engineering and Intelligent Systems* **5** 1

[6] Garg M Franki M & Sembera J 2003 *CubeSat Final report* (Austin: University of Texas)

[7] Wertz J R & Larson W J 1999 *Space Mission Analysis and Design* (Dordrecht: Kluwer Academic)

[8] Huang F Lu X Li Z Zhang J Deng R & Liu G 2015 *J. Software Eng.* **9** 230-241

[9] Maini A K & Agrawal V 2007 *Satellite Technology: Principles and Applications* (Chichester: John Wiley & Sons Ltd)

[10] Ivanov N Salischev V 1992 *The Journal of Navigation* **45** 175-182

[11] Pizzicaroli J C 1998 *Launching and Building the IRIDIUM Constellation; in: van der Ha J. C. (ed.), Mission Design & Implementation of Satellite Constellations* 113-121 (Kluwer Academic Publishers)

[12] Evans M J Maclay T D 1998 *Mission Design of the ORBCOMM Constellation; in: van der Ha J.C.(ed.), Mission Design & Implementation of Satellite Constellations* 103-112 (Kluwer Academic Publishers)

[13] Matossian M G 1998 *A TELEDESIC Space Infrastructure Overview; in: van der Ha J.C. (ed.), Mission Design & Implementation of Satellite Constellations* 153-156 (Kluwer Academic Publishers)

[14] Crisp N H Smith K L & Hollingsworth P M 2014 *Trans. Jpn. Soc. Aeronaut. Space Sci. Aerosp Technol. Jpn.* **12** 39-47

[15] Woellert K Ehrenfreund P Ricco A J Hertzfeld H R 2011 *Cubesats: Cost-effective science and technology platforms for emerging and developing nations*, Adv.SpaceRes.47

[16] Charania A Isakowitz S Pomerantz W Morse B Sagis K 2013 *The 27th Annual AIAA/USU Conference on Small Satellites* vol 626 (Logan: American Institute of Aeronautics and Astronautics (AIAA),

[17] Rodway K Papadopoulos K Greason J 2014 *The 65th International Astronautical Congress* (Toronto: International Astronautical Federation)

[18] King J A & Beidleman N J 1993 *Method and Apparatus for Deploying A Satellite Network*; (Fairfax: Orbital Science Corporation)

[19] Sanctis M D Cianca E Araniti G Bisio I & Prasad R 2016 *IEEE Internet of Things Journal* **3** 113-123

IOP Conf. Series: Materials Science and Engineering **260** (2017) 012029 doi:10.1088/1757-899X/260/1/012029

Wireless Battery Management System of Electric Transport

[1]Ataur Rahman*,[2]Mizanur Rahman, [2]Mahbubur Rashid
[1]Department of Mechanical Engineering ,[2]Department of Mechatronics Engineering,
International Islamic University Malaysia
50728 KL, Malaysia

Corresponding author: arat@iium.edu.my ; ataur7237@gmail.com
Phone:+6-03-61964544

Abstract
Electric vehicles (EVs) are being developed and considered as the future transportation to reduce emission of toxic gas, cost and weight. The battery pack is one of the main crucial parts of the electric vehicle. The power optimization of the battery pack has been maintained by developing a two phase evaporative thermal management system which operation has been controlled by using a wireless battery management system. A large number of individual cells in a battery pack have many wire terminations that are liable for safety failure. To reduce the wiring problem, a wireless battery management system based on ZigBee communication protocol and point-to-point wireless topology has been presented. Microcontrollers and wireless modules are employed to process the information from several sensors (voltage, temperature and SOC) and transmit to the display devices respectively. The WBMS multistage charge balancing system offering more effective and efficient responses for several numbers of series connected battery cells. The concept of double tier switched capacitor converter and resonant switched capacitor converter is used for reducing the charge balancing time of the cells. The balancing result for 2 cells and 16 cells are improved by 15.12% and 25.3% respectively. The balancing results are poised to become better when the battery cells are increased.

1. Background of the Study

The electric transportation is accounted for 35% of the world's vehicles by 2040 due to its potentility to reduce the greenhouse gas (GHG) reduction about 75% [1]. Electric vehicles curb the use of hydrocarbons and ultimately decrease greenhouse gases (GHG). This optimistic outlook is based on lower expected prices for lithium-ion batteries. In addition to being profitable in terms of the environment and the economy, the growth in the use of electric vehicles promotes the emergence of leading-edge technologies, not least energy storage devices and power optimization systems.

Continuing reductions in battery prices could bring the total cost of ownership of electric transport below that for conventional-fuel vehicles by 2025, even with low oil prices. The electric vehicle revolution could turn out to be more dramatic than governments and oil companies have yet realized. New research [2] suggests that further, big reductions in battery prices lie ahead, and that during the 2020s EVs will become a more economic option than gasoline or diesel cars in most countries. It is forecasted that sales of electric vehicles will hit 41 million by 2040, representing 35% of new light duty vehicle sales. This would be almost 90 times the equivalent figure for 2015, when EV sales are estimated to have been 462,000, some 60% up on 2014. This projected change between now and 2040 will have implications beyond the car market. The research estimates that the growth of EVs will mean they represent a quarter of the cars on the road by that date, displacing 13 million barrels per day of crude oil but using 2,700 TWh of electricity. This would be equivalent to 11% of global electricity demand [3].

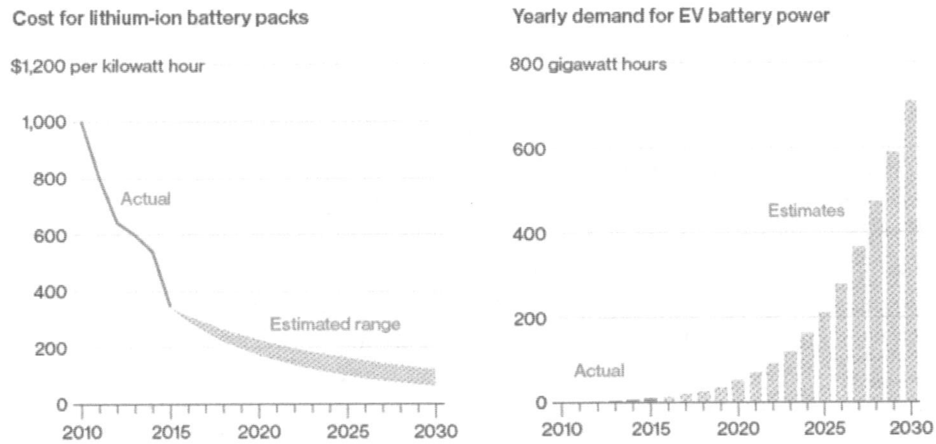

Fig. 1: Battery price forcasting [2].

Lithium-ion battery costs have already dropped by 65% since 2010, reaching $350 per kWh last year. We expect EV battery costs to be well below $120 per kWh by 2030 [1]. The crude oil price recovering to $50, and forcasted trending back up to $70-a-barrel or higher by 2040 which might increase the sales of electric transport.[3]. Electric transport in two categories – battery electric vehicles, or BEVs, that rely entirely on their batteries to provide power; and plug-in hybrid electric vehicles, or PHEVs, that have batteries that can be recharged but have conventional engines as back-up. The best-selling BEV over the last six years has been the Nissan Leaf, and the best-selling PHEV the Chevrolet Volt. Battery prices fell 35% last year and are on a trajectory to make unsubsidized electric vehicles as affordable as their gasoline counterparts in the next six years, according to a new analysis of the electric-vehicle market by Bloomberg New Energy Finance (BNEF). That will be the start of a real mass-market lift off for electric cars. The electric cars on the road will rise as its price will be equal to the price of IC engine power conventional car by 2022.

Rechargeable lithium batteries batteries are utilized extensively (Kim et al., 2014]. This technology has attracted more attention, especially for an electric vehicle, due to a clean alternative, domestic energy independence, cost and savings [5]. There are some important issues for li-ion batteries such as durability, reliability, and safety that need to be addressed in all sectors of portable application [6]. Li-ion batteries have not only the advantage of high energy density and power capacity, but also have the special features of low self-discharge rate (6%- 10%), high terminal voltage, high discharge current, non-memory effect and longer life cycle [7]. In addition, if Li-ion batteries are charged and discharged partly, there is no effect on the available capacity due to free of memory effect. Since EV requires high voltage and high current, Li-ion battery can be used which has the higher voltage than other rechargeable batteries. The nominal-cell voltages of Li-ion battery, lead-acid battery, Ni-Cd and Ni-MH

IOP Conf. Series: Materials Science and Engineering **260** (2017) 012029 doi:10.1088/1757-899X/260/1/012029

battery are 3.3–3.7V, 2V, 1.2V and 1.2 V respectively [8]. To supply high voltage and current to the load in the application of plug-in hybrid electric vehicle (PHEVs), electric vehicles (EVs), a large number of battery cells are connected in series (S) and parallel (P) connection [9]. The cell combination of a battery pack configuration can be 1S2P,1S3P and 2S2P [10,11]. A battery pack of 13 modules with 2S2P is being used in IIUM electric vehicle [12, 13].

Battery pack is a non-linear system, the available capacity affected by the charging-discharging cycles, over charging, under discharging, and excessive temperature. The battery pack should be supervised by battery management system (BMS) to offer safe operation and optimum performance [14]. In other words, BMS provides appropriate battery utilization, guaranteed maximum performance, monitoring present battery pack's condition and diagnosis [15]. A BMS acts as a medium among the battery and other modules (charging source and load) in improving vehicle performance and optimizing operation safely [16]. A battery management system (BMS) with hardware and software controlling should consist several parts: charging, discharging, thermal management, charge balancing, SOC estimation and communication. In practice, a BMS collects the all information from the sensors (voltage, current, temperature, and SOC) connected to a battery pack and take a specific action by activating the precise circuitry during charging and discharging operation.

The objective of this article is to develop wireless battery management system to control the operation of evaporatve thermal management system to maintain the battery temperature ≤ 40°C and balance the SOC of the cells with the maximum variation of (0.75 ± 5%) Ah.

2. DEVELOPMENT OF ELECTRIC VAN

The electric van is developed from this study by using a 14 kWh $LiFePO_4$ battery of nominl voltage 148 V and capacity of 86 Ah. The 14 kWh battery pack is made with 88 cells. Each cell has nominal voltage of 7.5 V and capacity of 86 Ah. The van is powered by an 48 kW AC motor and a 550 A controller. A two phase evaporative thermal management system has been developed to control the battery temperature in the range of 30-40°C with the wireless battery management system. A wireless battery management system has been used to maintain the cell's SOC variation of (0.75 ± 5%) Ah.

2.1 Two-phase thermal Management System

The IIUM has developed a battery thermal management system (Fig.2) to cool the battery pack by a direct refrigerant-based evaporative cooling. Heat generated from the battery is absorbed by the evaporating refrigerant inside the cooling duct and then dissipated heat to the surrounding air at the condenser. Thus the proposed system was able to maintain the temperature of the battery in the range of 30-40^{0}C. The combined effect of our battery cell's charge balancing system and battery thermal management system could enhance the battery life from 8 years to 10.5 years (projection based on field experimental result both in IIUM campus and Sepang F1 circuit). In this project, we have done the required modification of our final evaporative thermal management system by extensively testing the performance with emplying wireless battery management system to maintain the battery temperature in the range of 25 -35^{0}C.

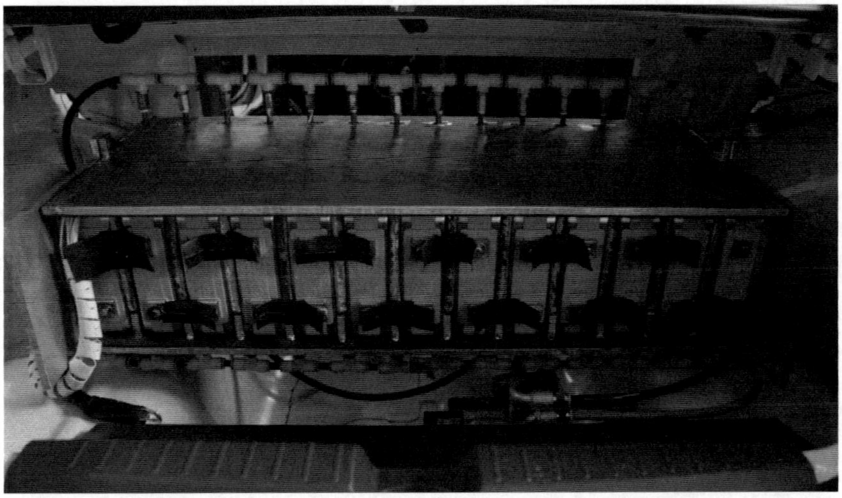

Fig. 2: Battery thermal management system [11]

2.2 Wireless Battery Management System

The central control system of BMS communicates with its sub-systems through the Controller Area Network (CAN) bus or serial bus to provide safe and efficient performance [17]. BMS collects the information from the sensors (voltage, current, SOC, impedance, internal and ambient temperature) relate to battery cells or pack, process, controls its sub-systems and send that information to the user end to monitor through communication interface. Two types of communication system are being employed with BMS, which are wired and wireless communication. For wired communication, each battery cell connected to BMS. Since, battery pack consists of a large number of cells, the complex connection causes the correction of the electric contact, which may lead to whole system failure [11]. Instead of wired, wireless communication is more preferable because of low cost, reliable, easy implementation. This study presence the wireless battery management system which has been built with sensors and ZigBee.

Wireless network deployed in BMS to monitor instantaneous current, voltage, SOC, capacity (AH) and temperature of battery cell or module. It provides wireless bidirectional communication between BMS and monitoring (user end control) units where BMS can be controlled from the user end. Fig.3 shows battery management system with ZigBee wireless communication. The design model is divided into two parts: battery management system (BMS) and monitoring/user end controlling unit. The BMS mainly consists of a thermal management system, a charging system and a balancing system, a battery pack, different kinds of sensors (SOC, voltage, current,

IOP Conf. Series: Materials Science and Engineering **260** (2017) 012029 doi:10.1088/1757-899X/260/1/012029

and temperature), two master controllers, and four slave controllers. The monitoring/user end controlling unit is made of a display unit, a user controlled unit. To communicate between two parts, two ZigBee wireless communication devices are used because of its special feature such as, low power consumption, low cost, high reliability and low data rates [19,20, 21].

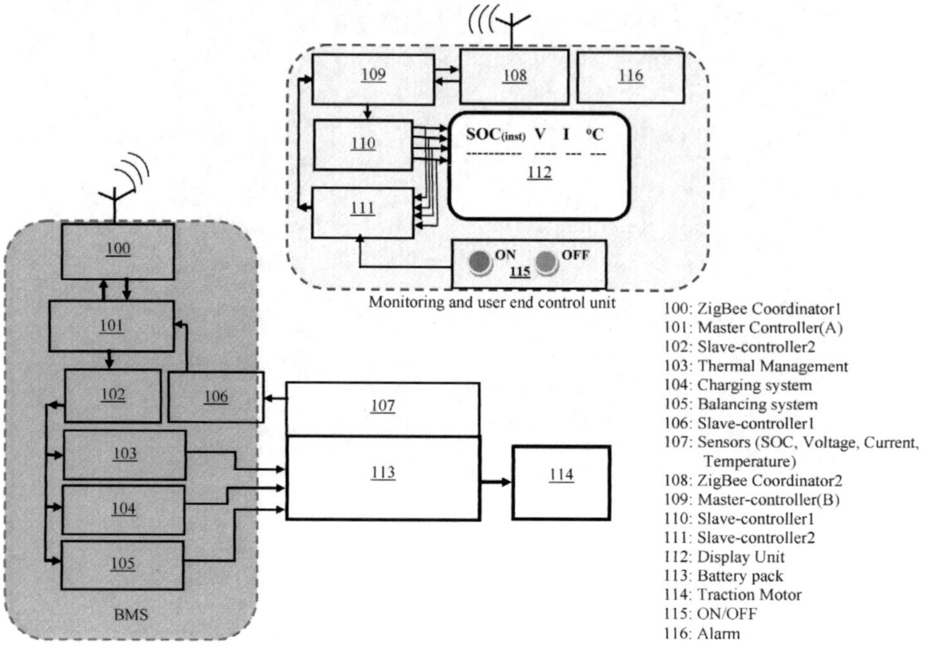

Fig.3. BMS with wireless communication.

Charging and discharging operation are very common phenomena of battery monitoring and controlling in order to provide safety and better performance. In the BMS part, slave controller 1 (106) collects the information regarding a battery pack from the employed analog sensors (SOC 107-1, voltage 107-2, current and temperature 107-3) and mapped into its specific ranges. After that, slave controller (106) sends the processed information to the master controller (A) (101) as shown in Fig.4. On basis of the receiving information, the master controller (A) (101) takes two decisions: communication and control signal generation. To communicate with the monitoring device, master controller (A) (101) activates ZigBee coordinator 1 (100) to send the information. Besides, it sends a control signal to the slave controller 2 (102) to turn ON/ OFF switches for the thermal management, charging operation, charge balancing system after checking the allowable range of voltage (2.7-4.2V), temperature ($<46^0$C) and SOC (30%-85%). On the monitoring and user controlled side, Master controller (B) 109 receives the sending information via ZigBee coordinator 2 108. Then Master controller (B) (109) processed the information and sent signal to slave controller 1 106 to display information or show a warning message. Slave controller 2 111 receives two information either copy of information from Slave controller 1 (110) or user command (ON/OFF) (115) and sends to the

IOP Conf. Series: Materials Science and Engineering **260** (2017) 012029 doi:10.1088/1757-899X/260/1/012029

master controller (B) (109). Then master controller (B) 109 receives the request and sends to master controller (A) (101) to take necessary actions on maintaining the battery temperature.

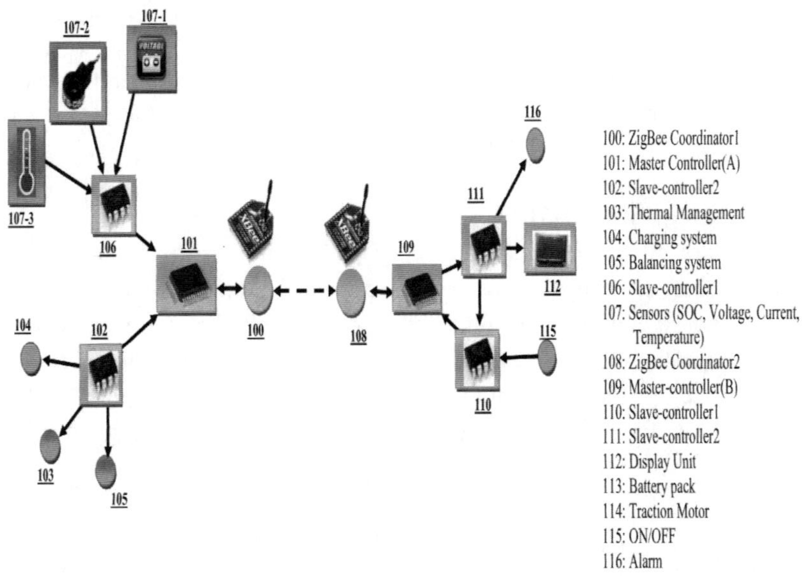

Fig.4. Point to point topology base network model for BMS.

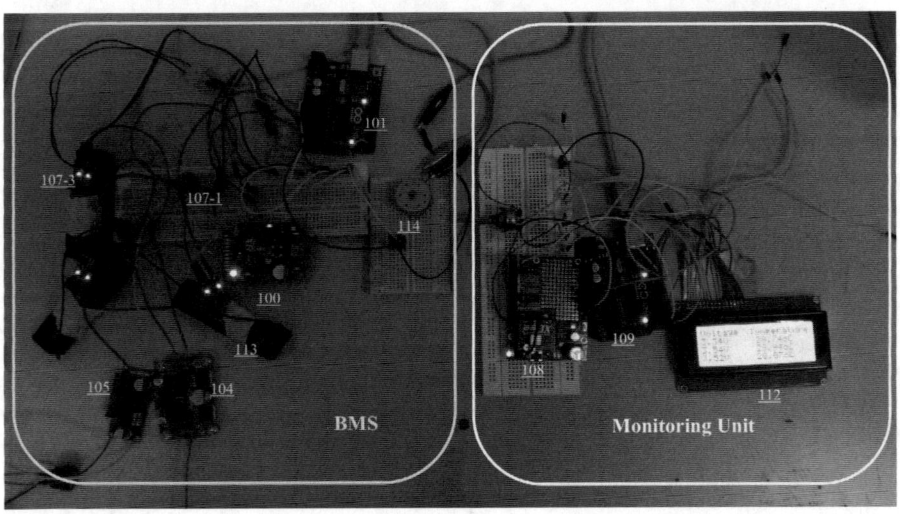

Fig.5. Wireless battery management system (for Single battery module: 2 Cells in series and 2 in parallel)

3. RESULTS AND DISCUSSION

The wire less battery management system (WBMS) has been developed with developing an algorithm based on the battery electro chemistry ($LiFePO_4$) for maintaining the accurate state-of-charge (SoC) and state-of-discharge (SoD) of the

IOP Conf. Series: Materials Science and Engineering **260** (2017) 012029 doi:10.1088/1757-899X/260/1/012029

individual cells. This allows the vehicle to utilize the maximum energy available from the battery for a given drive cycle whilst maintaining pack balance within the range of optimal functionality. This system is capable of preventing overcharging and over-discharging of $LiFePO_4$ cell. The battery was made with 22 modules (M) each of nominal voltage (NV) of 7.5V and capacity of 86Ah connected in series. Based on the topology, the experimental work has been setup and shown in Fig 5. The performance of developed WBMS was conducted in laboratory and measured the voltage of the 1st module and 22nd module. The difference was found from 1st M to 22nd M: 0.66% at full charge and 0.95% at full discharge.

Charge Balancing with WBMS

The simulation results are carried by using PSIM 9.3 on multistage LC networks for a four series connected battery cells with and without individual resistance (1.5Ω). Considering the simulation time limits, the battery cells are modeled with 10F of electrolytic capacitor. The values of LC networks are 5uH of inductor and 10uF of capacitor. The cell voltage is intentionally made different that are 3.2V, 3.5V, 3.4V and 3.6V respectively. The switching frequency 22 kHz has been chosen for all balancing process.

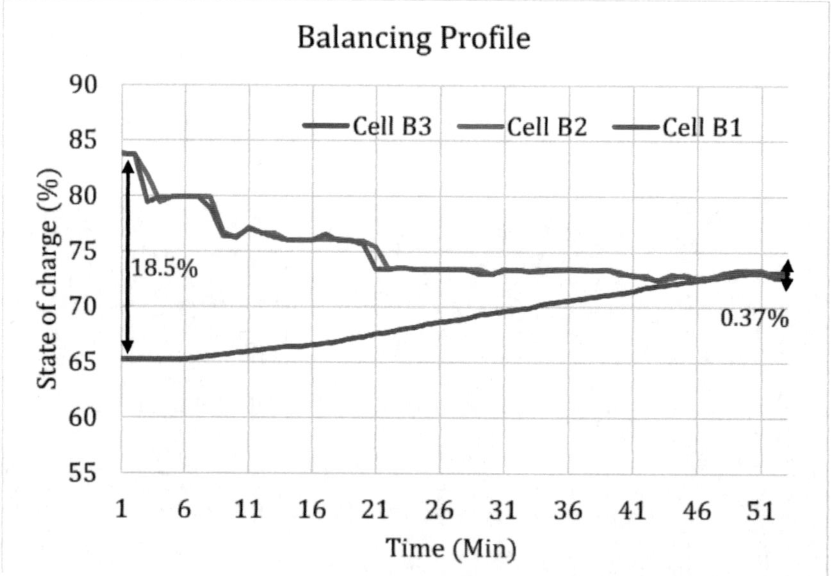

Fig.6: Module SOC during discharging.

Experiment has been conducted for a module, which SOC was recorded before using the modules to power the motor as 83.82%, 83.8%, 83.78% and 65.25% of the cell B1, B2, B3, and B4 respectively. During discharging the modules with the other modules about 55 minutes in laboratory to power the motor, the cell's SOC was recorded and presented in Fig. 6. All cells were in balance in 53 minutes during discharging. The result indicates that the potentiality of the wireless cell balancing is not highly effective. However, it has positive effect on the battery lifespan by preventing the cells from damage or even fire.

IOP Conf. Series: Materials Science and Engineering **260** (2017) 012029 doi:10.1088/1757-899X/260/1/012029

Battery Temperature with WBMS

Battery temperature has been measured based on the sensors employed evaporative control battery thermal management system (ECBThMS) during discharging current for the 60-70 A and 220-240 A by two different mode: with and without WBMS. The results are presented in Table 1 and 2. The positive difference shows that the modules SOC were in the range of 83-84%, which were uniformly discharge current and temperature goes up uniformly while the negative module shows that the module has SOC below 70%, which contribute more temperature. In overall, the temperature difference was not significant.

Table 1: Battery temperature for the discharge current 60-70A

Battery Discharging Time (min)	Battery Temperature Control by ECBThMS		
	With WBMS	Without WBMS	Difference
0	26.1	26.16	0.06
10	26.3	26.41	0.11
20	26.9	27.15	0.25
30	27.6	27.59	-0.01
40	28.4	28.45	0.05
50	29.3	29.33	0.03
60	30.1	30.13	0.03
70	31.2	31.20	0.00
80	32.8	32.74	-0.06
90	34.4	34.51	0.11
95	35.0	35.14	0.14

Table 2: Battery temperature current for the discharge current 220-240A

Battery Discharging Time (min)	Battery Temperature Control by ECBThMS		
	With WBMS	Without WBMS	Difference
0	26.4	26.27	0.1
5	34.5	34.31	0.2
10	39.1	38.87	0.2
15	44.3	44.11	0.2
18	46.4	46.42	0.0

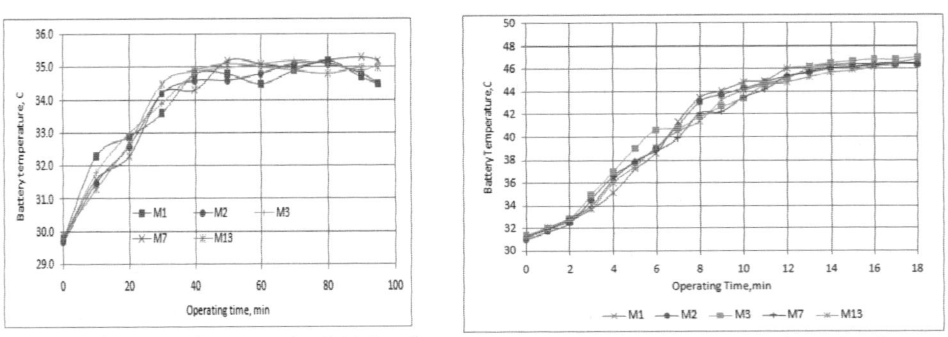

(a) Vehicle operating speed of 60 km/h (b) Vehicle operating speed of 120 km/h

Figure 7: Battery temperature of the electric vehicle with WBMS equipped EC-BThMS

Figure 7 shows the temperature of the modules has been recorded experiment by using the WBMS for the vehicle speed of 60 km/h and 120 km/h. The result shows that the temperature spike of the battery is proportional with the discharge current (i.e, $T \propto I$, where I is current and T is temperature), which is mainly for the ohmic affect (i^2R). It is found that the 48 kW motor has drawn 200 A current to speedy the e-coaster 120 km/h and the battery temperature spike to 49°C in 20 minutes. While the battery temperature is recorded maximum of 35°C for the vehicle speed 60 km/h due to the current drawn by the motor from the battery about 70 A.

The energy efficient of the developed EC-BThMS has been compared with two other EVs using air cooling battery thermal management system (AC-BThMS 1 and AC-BThMS 2). For these AC-BThMS, the all fans for air blowing were kept running during operation of the vehicles. The result of the cooling systems has been shown in Table 3. Table 3 indicates that the EV with EC-BThMS can save 17.69% more energy than the EV with AC-BThM 1 and 23% than the EV with AC-BThMS 2. This is because the potentiality of the EC-BThMS which is able to maintain the temperature of the battery pack in the range of 20ºC – 40ºC. The result of Table 5.2 shows that the EV with EC-BThMS can save 25% more time than the EV with AC-BThM 1 and 23% than the EV with AC-BThMS 2 for the quarter mile acceleration; 20.21% more time than the EV with AC-BThM 1 and 21.28% than the EV with AC-BThMS 2 for the two laps travelling.

Table 3: Performance analysis of EC-BThMS over two others AC-BThMS I

Type of BCThMS	Performance Measurement Mode				
	1/4 mile accelerat ion (sec)	Time for 11.2 km traveling (Sec)	Max. velocity (within fastest time in 11.2 km) (km/h)	Farthest distance (km)	Energy saving with EC-BThMS (%)
EV with EC-BThMS	22	17:15	119	68.62	-
EV with AC-BThMS 1	51	26:51	119	59.94	17.69
EV with AC-BThMS 2	25	26:19	91	57.75	23

The correlations between measured (experimental) and predicted (simulation) values of temperature profile has been illustrated in Figure 5.19. The correlation coefficient of temperature profile at discharge current 80 A or the goodness of fit is found as 0.9730. It indicates that the predicted data over the measured data have a closed agreement and thus, validity of the simulation result. However, the mean relative error of measured and predicted values from the mathematical model on temperature profile is found as 9.027%. The relative error of simulation values over the experimental values is due to the effect of outdoor environment.

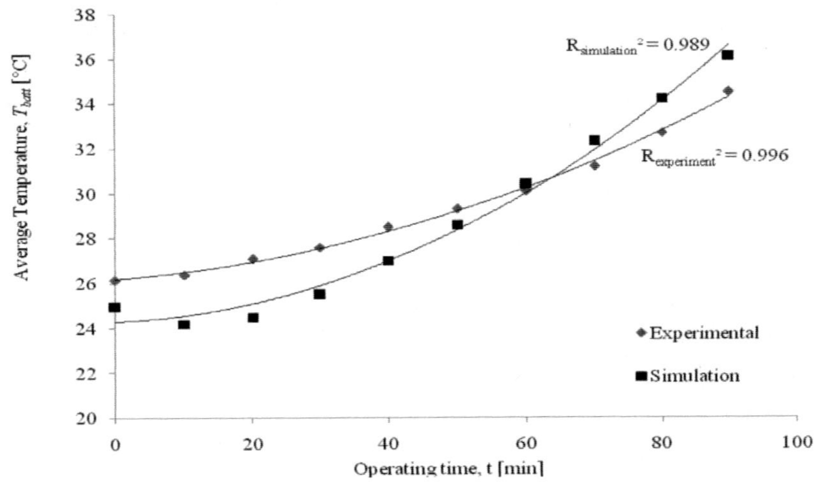

Fig.8: Validation of battery temperature rises between experimental and simulation.

4. Conclusion

The Evaporative cooling battery thermal management system is found to be an efficient for the enhancement of battery ifespan. The enhancement of the battery ifesapn would be more by using the wireless battery management system which balance the battery cells in shorter time with keeping the cells variation about 2.5 to 5%. The WBMS is important for the electric vehicle battery mainly to balance the charge while for controlling the battery temperature it is not so efficient or otherwise.

Reference

1. Colin McKerracher (2017). https://about.bnef.com/blog/electric-vehicles-to-be-35-of-global-new-car-sales-by-2040/#_ftnref1.
2. https://about.bnef.com/blog/electric-vehicles-to-be-35-of-global-new-car-sales-by-2040/#_ftn2.
3. https://www.bloomberg.com/features/2016-ev-oil-crisis/img/ev-sales.jpg.
4. Kim, M.-Y., Kim, C.-H., Kim, J.-H., & Moon, G.-W. (2014). A chain structure of switched capacitor for improved cell balancing speed of lithium-ion batteries. IEEE Transactions on Industrial Electronics, 61(8), 3989-3999.
5. Guo, X., Kang, L., Huang, Z., Yao, Y., & Yang, H. (2015). Research on a Novel Power Inductor-Based Bidirectional Lossless Equalization Circuit for Series-Connected Battery Packs. Energies, 8(6), 5555-5576.
6. Hua, C.-C., & Fang, Y.-H. (2016). Battery charge equalization circuit with a multi-winding transformer. Journal of the Chinese Institute of Engineers, 1-9.
7. Gering, K. L., Sazhin, S. V., Jamison, D. K., Michelbacher, C. J., Liaw, B. Y., Dubarry, M., & Cugnet, M. (2011). Investigation of path dependence in commercial lithium-ion cells chosen for plug-in hybrid vehicle duty cycle protocols. Journal of Power Sources, 196(7), 3395-3403.
8. Hua, C., & Fang, Y.-H. (2016). A charge equalizer with a combination of APWM and PFM control based on a modified half-bridge converter. IEEE Transactions on Power Electronics, 31(4), 2970-2979.

9. Uno, M., & Tanaka, K. (2011). Single-switch cell voltage equalizer using multistacked buck-boost converters operating in discontinuous conduction mode for series-connected energy storage cells. IEEE Transactions on Vehicular Technology, 60(8), 3635-3645.
10. Dubarry, M., Devie, A., & Liaw, B. Y. (2016). Cell-balancing currents in parallel strings of a battery system. Journal of Power Sources, 321, 36-46.
11. Rahman, Ataur., Helmi, Ahmed., Hawlader, MNA (2017) 15. Two-Phase Evaporative Battery Thermal Management Technology for EVs/HEVs. International Journal of Automotive Technology, Vol.18 (5), 75–88.
12. Narayanaswamy, S., Kauer, M., Steinhorst, S., Lukasiewycz, M., & Chakraborty, S. (2016). Modular Active Charge Balancing for [-2pt] Scalable Battery Packs. IEEE Transactions on Very Large Scale Integration (VLSI) Systems.
13. Rahman, A., Nur, F., Hawlader, M., & Afroz, R. (2015). Fuzzy controlled evaporative battery thermal management system for EV/HEV. International Journal of Electric and Hybrid Vehicles, 7(1), 22-39.
14. Stuart, T. A., & Zhu, W. (2011). Modularized battery management for large lithium ion cells. Journal of Power Sources, 196(1), 458-464.
15. Lee, Y.-J., Khaligh, A., & Emadi, A. (2009). Advanced integrated bidirectional AC/DC and DC/DC converter for plug-in hybrid electric vehicles. IEEE Transactions on Vehicular Technology, 58(8), 3970-3980.
16. Xing, Y., Ma, E. W., Tsui, K. L., & Pecht, M. (2011). Battery management systems in electric and hybrid vehicles. Energies, 4(11), 1840-1857.
17. Wu, Y., Liao, X., Chen, W., & Chen, D. (2011). A Battery Management System for electric vehicle based on Zigbee and CAN. Paper presented at the Image and Signal Processing (CISP), 2011 4th International Congress on.
18. Yu, C., Cui, Y., Zhang, L., & Yang, S. (2009). Zigbee wireless sensor network in environmental monitoring applications. Paper presented at the 2009 5th International Conference on Wireless Communications, Networking and Mobile Computing.
19. Chen, D., & Wang, M. (2006). A home security zigbee network for remote monitoring application. Paper presented at the Wireless, Mobile and Multimedia Networks, 2006 IET.
20. Han, D.-M., & Lim, J.-H. (2010). Design and implementation of smart home energy management systems based on zigbee. IEEE Transactions on Consumer Electronics, 56(3), 1417-1425.
21. Mathew, S. A., Prakash, R., & John, P. C. (2012). A smart wireless battery monitoring system for Electric Vehicles. Paper presented at the 2012 12th International Conference on Intelligent Systems Design and Applications (ISDA).

IOP Conf. Series: Materials Science and Engineering **260** (2017) 012030 doi:10.1088/1757-899X/260/1/012030

Time response for sensor sensed to actuator response for mobile robotic system

N S Amir and A A Shafie

Department of Mechatronics, Kulliyyah of Engineering, International Islamic University Malaysia, Jalan Gombak, 53100 Kuala Lumpur, Malaysia

E-mail:syamimi.amir25@gmail.com, aashafie@iium.edu.my

Abstract. Time and performance of a mobile robot are very important in completing the tasks given to achieve its ultimate goal. Tasks may need to be done within a time constraint to ensure smooth operation of a mobile robot and can result in better performance. The main purpose of this research was to improve the performance of a mobile robot so that it can complete the tasks given within time constraint. The problem that is needed to be solved is to minimize the time interval between sensor detection and actuator response. The research objective is to analyse the real time operating system performance of sensors and actuators on one microcontroller and on two microcontroller for a mobile robot. The task for a mobile robot for this research is line following with an obstacle avoidance. Three runs will be carried out for the task and the time between the sensors senses to the actuator responses were recorded. Overall, the results show that two microcontroller system have better response time compared to the one microcontroller system. For this research, the average difference of response time is very important to improve the internal performance between the occurrence of a task, sensors detection, decision making and actuator response of a mobile robot. This research helped to develop a mobile robot with a better performance and can complete task within the time constraint.

1. Introduction

Developing a mobile robot is a complex process. Robots have to simultaneously avoid an obstacles while performing the tasks to achieve goal. Tasks may need to be done within a time constraint. Since time and performance matters, robot controllers should be able to send command immediately to its devices. Time constraint is an important aspect to ensure good operation of mobile robot and can result in better performance [8]. With the rise in research and development of autonomous robots in the past decade, there has been an increased focus on control strategies for the robots to achieve robust and optimal performance [12].

The response time of an autonomous mobile robot is presented as an internal performance that characterize a robot regardless of a particular task. The response time is defined as the time between the occurrence of a task, its detection by a robot, the decision making, and the response [1]. This system response is important for mobile robot, especially when it needs to turn left or right immediately after its sensor detect an obstacle, for instance [2].

Real-time systems are computer systems that monitor, respond, or control to an external environment and it is connected to the computer system through sensors, actuators, and other input-output interfaces [3]. Since actuators perform actions in response to the sensed tasks, real-time communications and fast reaction are necessary [4]. Real time system have two types, hard real time

IOP Conf. Series: Materials Science and Engineering **260** (2017) 012030 doi:10.1088/1757-899X/260/1/012030

system and soft real time system. Hard real time system is the system that should complete the task within the time constraint while soft real time system is a system that require performance assurances from the operating system [5]. In this research, the system analyse the time response starting from the detection of an obstacle by the sensor, until the actuator responds; thus it falls within the type of soft real time system.

Buttazzo et al. designed and implemented a software architecture for designing robots with real time constraint in their research. The architecture of the programming consists of four parts, namely action layer, control layer, communication layer and herd real-time kernel (HARTIK) which controls the other three layers. They concluded that their architecture provide a flexible framework for the development of robots [6].

Brega et al. conducted a case study that discussed the needs and requirements for real time robots in research, education and real world applications. They noted that the increasing complexity of mobile robots set higher requirements for the hardware of the mobile robots, software and the operating system [7].

Navigation of mobile robots is solved with microcontrollers with different memory capacities and operating speeds that can handle incoming and outgoing signals. The tasks of the robots may vary significantly, from simple tasks to the execution of complicated tasks. This fact determines the architecture and complexity used for the robot's navigation in maze solving [8].

2. Methodology

The experiment of this research was carried out with line following and obstacle avoidance tasks. Two identical mobile robot were used; both equipped with four sensors and two actuators. They differ in the number of microcontroller used because this research was carried out to compare the time response between sensor detection and actuators response for one and two microcontrollers used in a mobile robot.

2.1 Task

The tasks of this experiment is that the robot have to follow the line and make a correct decision at the junction until it reaches the finishing line. The mobile robot also have to avoid any obstacle in front of it. The time were recorded starting from the line detection until the actuator responds.

2.2 Measurement technique

The proposed system design is implemented on Arduino which consists of Atmel as processor running real time application, where the ultrasonic, sharp IR sensor and line sensor are the input which send the acquired data to the main microcontroller. The control module will make the decision when conditions are met.

One of the sensor used is ultrasonic sensor. It transmits ultrasonic wave towards external environment from the tranceiver, then the wave is reflected by the object and received by the sensor as shown in Figure 1. The time taken for the signal to travel is then calculated by the sensor and converted into a value that represents the distance between the sensor and the object in front of it [9].

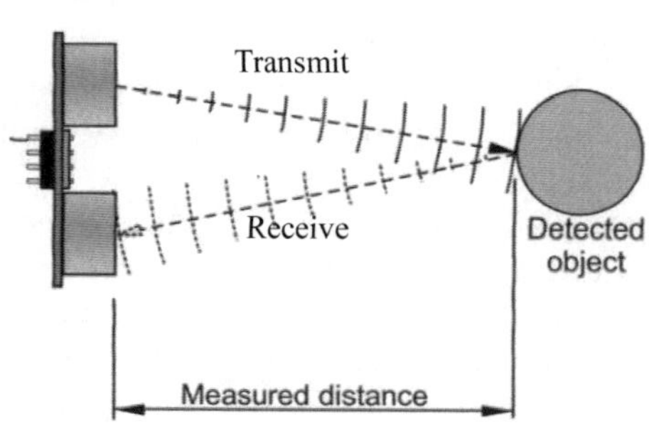

IOP Conf. Series: Materials Science and Engineering **260** (2017) 012030 doi:10.1088/1757-899X/260/1/012030

Figure 1. Principle of ultrasonic sensor [10]

Line sensor is also being used in this mobile robot. Basically, line sensor is similar with IR sensor which is made up of two components: a light transmitter and a light receiver. Different colours reflect different intensity of light from emitter to the receiver. Figure 2 shows the principle of IR sensor.

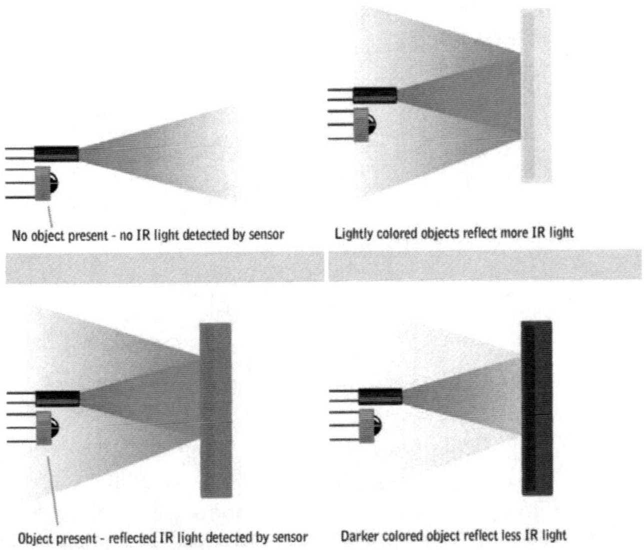

Figure 2. Principle of IR sensor [11]

The distance between the obstacle and mobile robot is calculated using Equation 1 for IR sensor and Equation 2 for ultrasonic sensor in order to get the distance in centimetre.

Distance = 6050/ sensor value (1)

Distance = sensor value / 58 (2)

2.3 Hardware Setup

The proposed system is shown in Figure 3 and Figure 4. Figure 3 shows single microcontroller system whereas Figure 4 shows multiple microcontroller system. In both cases same application was executed on both platforms and response time is calculated.

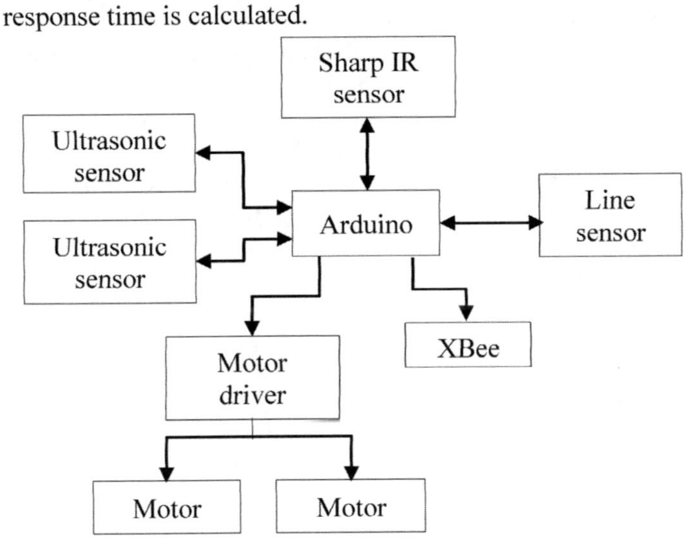

Figure 3. Block diagram of the overall single microcontroller designed system.

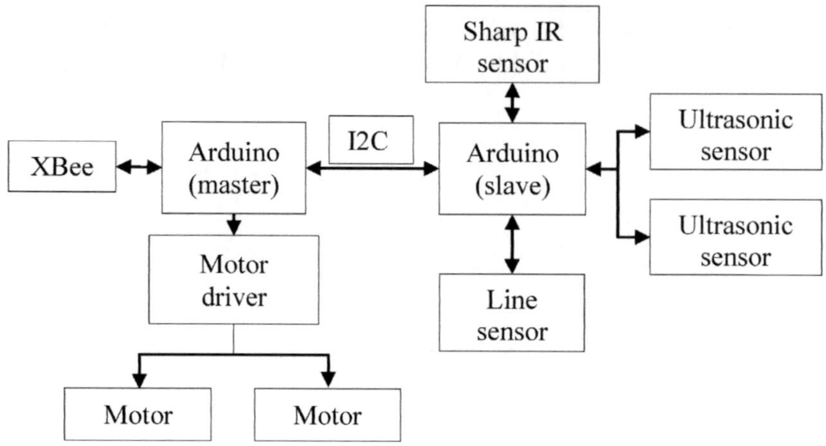

Figure 4. Block diagram of the overall multiple microcontroller designed system.

A microcontroller is a small computer on a single integrated circuit containing processor core, memory and programmable input/output peripherals. Microcontrollers are used in automatically controlled devices such as automobile engine control systems, remote controls and other embedded systems. Arduino microcontroller are used in this research because it consists of a standard programming language compiler and a boot loader that executes on the microcontroller. Furthermore, Arduino has a serial monitor that can show the data transmission when using serial communication. Thus, makes it easier to test the sensor since it show results or sensor values.

There are 3 sensors used in this research which are ultrasonic sensor, sharp IR sensor and line sensor. Two ultrasonic sensor are used to detect an obstacles at the right side and left side of the mobile robot while the sharp IR sensor is used to detect an obstacle in front of the mobile robot. The other sensor is line sensor that is used to detect line for line following task.

For this research, the data from the microcontroller at the mobile robot will be sending wirelessly from the mobile robot to the PC. Transmitter wireless module will send the time response between sensors sensed to the actuator response at the mobile robot to the PC and the receiver module will receive the data for a record. Wireless module that will be used in this research is XBee wireless module. Two XBee are needed in this research because one of the XBee act as a transmitter and the other one is a receiver. One XBee will be connected to the mobile robot and the other XBee will be connected to the PC.

Two DC motors at the mobile robot will act with the help of motor driver circuit. The motor driver is used to receive the signals from the control module in the form of binary logic and it will be supplied to the DC motors with sufficient voltage and current supply. The motor driver will also protect the control module from DC motors if it gets damaged.

2.4 Software Setup / Algorithm Used For The Functioning Of The Robot

Arduino has general purpose input and output pins where these pins can be used by the ultrasonic sensor and DC motor. In this research, the ultrasonic sensor is triggered and once an echo is generated, the time will start. A signal is generated based on the threshold value to the DC motor driver then the timer is stopped. The time value between the sensors sense to the output response gives the response time of mobile robot. The structure of the propose architecture is shown in Figure 5.

IOP Conf. Series: Materials Science and Engineering **260** (2017) 012030 doi:10.1088/1757-899X/260/1/012030

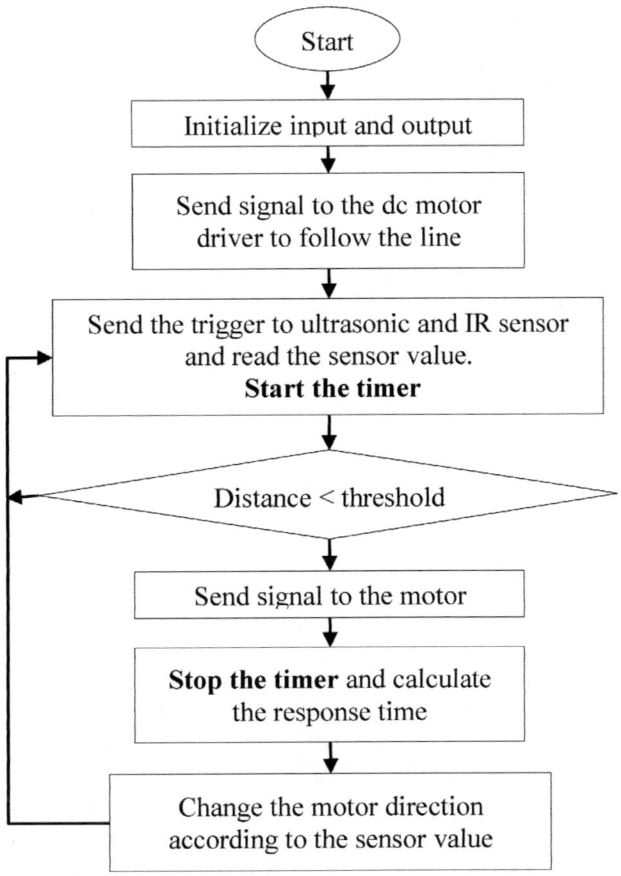

Figure 5. Flow chart of the system.

Experiments were performed to measure the time response from sensor sense to the actuator respond. These date can be collected via the command line in the programming. An obstacle avoidance was used to measure the time response and task performance. The program structure has five standards in the obstacle avoidance behaviour. The number of standards are determined by the possibility of sensors detection in the environment. The standards are used to control the speed and steering angle of the motor as shown in Table 1. However, if all sensors detect the obstacle, the mobile robot will turn back. In this research, the number of rule are five based on the combination of three sensors as an input with a steering angle manipulation.

Table 1. Obstacle avoidance behaviour standard base.

Sensor 1 (left)	Sensor 2 (front)	Sensor 3 (right)	Steering angle
Not detect	Detect	Not detect	Straight
Not detect	Detect	Detect	Turn left
Detect	Detect	Not detect	Turn right
Detect	Detect	Detect	Turn back right
Not detect	Not detect	Not detect	Straight

Two different experiments were conducted: those using 1 microcontroller and using two microcontroller in a robot. Each set of experiments had total of three runs and all trials have same obstacle and same tasks.

IOP Conf. Series: Materials Science and Engineering **260** (2017) 012030 doi:10.1088/1757-899X/260/1/012030

3. Results

In this paper, evaluation is conducted to analyse the mobile robot performance based on number of microcontroller. The proposed system is provided with an algorithm that helps in obstacle avoidance in a new environment. The data that are needed to be pre-defined are the sensor distance, speed and steering angle. Since these information are used for the obstacle avoidance to check whether the system generates correct output depending on the value it received from the ultrasonic and sharp IR sensor.

This section presents the results of the research about the calculation of the response time between sensors detection and output response in real environment. When using one microcontroller, the response time of the mobile robot is shown in Table 2. The response of the system was measured, the result shows that the response time from ultrasonic sensor or sharp IR sensor to DC motor are 50 ms to 53 ms.

Table 2. Time response for 1 microcontroller.

Obstacle	Test 1 (ms)	Test 2 (ms)	Test 3 (ms)
1	50.92	51.33	51.05
2	50.54	51	50.74
3	51	50.94	50.65
4	51.5	52.29	50.25
5	51.73	51.2	52.25
6	52.2	52.27	52.25
7	52.19	52.25	52.25
8	51.71	51.19	51.2

The same application was carried out for two microcontrollers and the response time of the system is shown in Table 3. The response of the system was measured, the result shows that 25 ms to 29 ms as the response time from ultrasonic sensor or sharp IR sensor to DC motor.

Table 3. Time response for 2 microcontroller.

Obstacle	Test 1 (ms)	Test 2 (ms)	Test 3 (ms)
1	26.5	26	26.62
2	25.94	26.76	26.59
3	26.19	27.56	26.65
4	26.63	26.65	26.64
5	27.29	28.07	25.94
6	27.2	28.67	27.53
7	27.56	28.2	28.07
8	27.53	27.67	26.69

The response time for different number of microcontroller from one microcontroller to two microcontroller was calculated and tabulated in graph shown in Figure 6. The graph shows the difference response between one microcontroller and two microcontroller, there was an average difference of 24.4 milliseconds. The timer should start and end accurately to calculate the distance between the obstacle detection and response time.

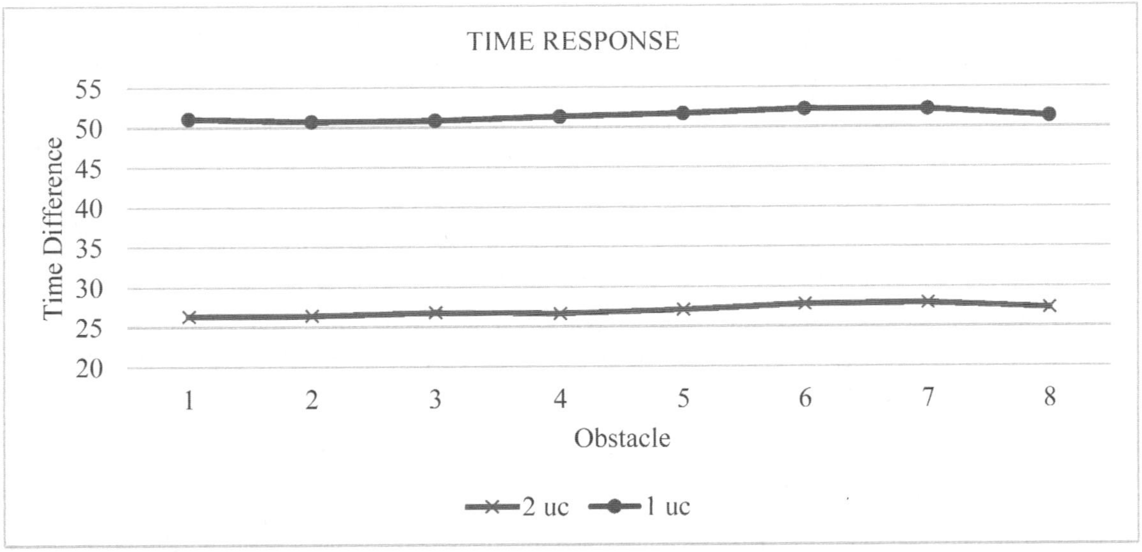

Figure 6. Graph of the difference response between one microcontroller and two microcontroller

T-test was applied to analyse the data collected. From the calculation, the u1 = 51.45351, δ1 = 0.569261, u2 = 27.0472, δ2 = 0.597558 and the T value is 83.6433.

As k = 8+8-2 = 14, $t\alpha$ = 2.145 (from critical value (t_{crit}) table [12]). The inequality 83.6433 > 2.145 holds, the null hypothesis H_0 (i.e. u1 = u2) is rejected, which means that the second experiment (2 microcontroller) performs significantly better than the first one, the probability of this statement to be erroneous is 0.05.

4. Conclusion

The overall research shows that ultrasonic sensor can detect the obstacle and communicate with control module to avoid the obstacle in 50 ms to 53 ms for one microcontroller and 25 ms to 29 ms for two microcontroller. The results show that two microcontroller system have better response time compared to the one microcontroller system and the probability of this statement to be erroneous is 0.05. The average difference between the response time of one and two microcontroller is 24.4 milliseconds. This response time is important in many real time applications especially maze solving.

References

[1] Baums A and Gordyushin A 2013 Response Time of a Mobile Robot *Automatic Control and Computer Sciences,* pp 352-358

[2] Atmadja W, Christian B and Kristofel L 2014 Real Time Operating System on Embedded Linux with Ultrasonic Sensor for Mobile Robot *International Conference on Industrial Automation, Information and Communications Technology (IAICT)*, Bali

[3] Shaw A 2001 Real-Time Systems and Software *Software Focus*, Middlesex, UK, John Wiley & Sons, Ltd., pp. vol **2**, Issue 2, p 81

[4] Ngai E C H, Lyu M R and Liu J 2006 A Real-Time Communication Framework for Wireless Sensor-Actuator Networks *IEEEAC*

IOP Conf. Series: Materials Science and Engineering **260** (2017) 012030 doi:10.1088/1757-899X/260/1/012030

[5] Murikipudi A, Prakash V and Vigneswaran T 2015 Performance Analysis of Real Time Operating System with General Purpose Operating System for Mobile Robotic System *Indian Journal of Science and Technology,* p 6

[6] McKenzie A, Dawson S, Hu F and Anderson M 2011 Using Sun's Java Real-Time SystemtoManage Behavior-Based Mobile Robot Controllers *Journal of Robotics,* p. 11

[7] Brega R, Tomatis N and Arras K O 2000 The Need for Autonomy and Real-Time in Mobile Robotics: A Case Study of X0/2 and Pygmalion *International Conference on Intelligent Robots and Systems*

[8] Matijevics I 2006 Microcontrollers, Actuators and Sensors in Mobile Robots *4th Serbian-Hungarian Joint Symposium on Intelligent Systems*, Subotica, Serbia

[9] Srinivasan K and Gu J 2007 Multiple Sensor Fusion in Mobile Robot Localization *Electrical and Computer Engineering Conference*, Canada

[10] Kelemen M, Virgala I, Kelemenová T, Miková Ľ, Frankovský P, Lipták T and Lörinc M 2015 Distance Measurement via Using of Ultrasonic Sensor *Journal of Automation and Control,* pp. 71-74

[11] Ismail R, Omar Z and Suaibun S 2016 Obstacle-avoiding robot with IR and PIR motion sensors *AEROTECH VI - Innovation in Aerospace Engineering and Technology*, Kuala Lumpur

[12] Nehmzow u 2006 Scientific Methods in Mobile Robotics (Quantitative Analysis of Agent Behaviour), united Kingdom: Springer

[13] Pandey A, Jha S and Chakravarty D 2016 Modeling and Control of an Autonomous Three Wheeled Mobile Robot with Front Steer *IEEE International Conference on Robotic Computing*

[14] Ahmad A, Rathore M M, Paul A, Hong W H and Seo H 2016 Context-Aware Mobile Sensors for Sensing Discrete Events in Smart Environment," *Journal of Sensors,* p. 10

IOP Conf. Series: Materials Science and Engineering **260** (2017) 012031 doi:10.1088/1757-899X/260/1/012031

Improvement of Dimensional Accuracy of 3-D Printed Parts using an Additive/Subtractive Based Hybrid Prototyping Approach

A N M Amanullah Tomal[1], Tanveer Saleh[1], Md. Raisuddin Khan[1]

[1] Smart Structures, System and Control Research Laboratoty, Department of Mechatronics Engineering, International Islamic University Malaysia, Kuala Lumpur 50728, Malaysia

E-mail: tanveers@iium.edu.my

Abstract. At present, two important processes, namely CNC machining and rapid prototyping (RP) are being used to create prototypes and functional products. Combining both additive and subtractive processes into a single platform would be advantageous. However, there are two important aspects need to be taken into consideration for this process hybridization. First is the integration of two different control systems for two processes and secondly maximizing workpiece alignment accuracy during the changeover step. Recently we have developed a new hybrid system which incorporates Fused Deposition Modelling (FDM) as RP Process and CNC grinding operation as subtractive manufacturing process into a single setup. Several objects were produced with different layer thickness for example 0.1 mm, 0.15 mm and 0.2 mm. It was observed that pure FDM method is unable to attain desired dimensional accuracy and can be improved by a considerable margin about 66% to 80%, if finishing operation by grinding is carried out. It was also observed layer thickness plays a role on the dimensional accuracy and best accuracy is achieved with the minimum layer thickness (0.1 mm).

1. Introduction

Rapid Prototyping (RP) technology, also known as 'Additive Manufacturing (AM)' or 'Layered Manufacturing (LM)' is a key method to manufacture components without the involvement of the expert model-manufacturers' services (1). RP technology is able to construct the complete object directly from CAD (Computer Aided Design) model without help of any types of heavy and complex tooling. While with traditional methods the prototype needed to be constructed and finished manually (2). This technology has gained significant attention due to its capability to overcome many drawbacks of old-fashioned manufacturing techniques. Its ability to form almost any geometric feature or shape is a great advantage of AM (3) . On the other hand, CNC (Computer Numerical Control) machining process is a subtractive manufacturing process with high-precision, high-flexibility, high-accuracy and high-speed However CNC machining is not cost effective to use it for producing prototypes. Among the vast area of RP, a number of methods are used: Fused Deposition Modelling (FDM), Stereolithography (SLA), Selective Laser Sintering (SLS), Selective Laser Cladding (SLC) Laminated Object Manufacturing (LOM), Multi-jet Modelling or Solid Imaging (SI), Laser Engineering Net Shaping (LENS), Selective Binding (SB)(4, 5). Due to inexpensive machinery, ease of operation, and durability of built objects, FDM is considered as the most appropriate process for RP due to its ease of operation, inexpensive

IOP Conf. Series: Materials Science and Engineering **260** (2017) 012031 doi:10.1088/1757-899X/260/1/012031

machinery and durability of built parts (6, 7). Final parts from FDM have good mechanical properties compared to other RP technologies (8). In this process, a plastic material (ABS or PLA) is melted and extruded through a nozzle layer-by-layer by means of a feed motor to build the required object. Generally, this plastic material is supplied in the form of a filament. Resistive heaters are contained by the nozzle that help to keep the temperature of the plastic just above its melting point. So, the plastic can easily flow through the nozzle and create the layer. After that, it hardens instantly and bonds to the previous layer. This procedure is continued till the shape of the prototype is being created is shown in figure 1.

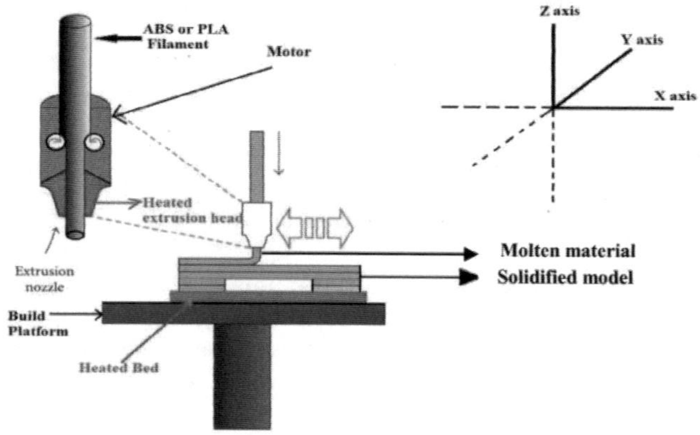

Figure 1. FDM Process(9)

High surface roughness is one of the most noteworthy disadvantages of FDM process, specially if compared to other processes. The FDM can yield parts with complex features: nevertheless, for the reason of having residual stresses and shrinkage which occurred during running the RP process, the precision and accuracy of the produced product is usually low. Staircase effect causes poor surface quality that is another limitation of FDM (10). The layer thickness was found out as the most important aspect for the finished surface in FDM process (11). If we want to reduce the layer thickness, then we have to increase build time that will create another issue. So, for the solution of the above problems we are integrating the CNC grinding operation for final finishing operation on the FDMed parts. The finishing of additively manufactured (AM) components may be categorized into three mechanisms, namely: (i) mechanical conversion and machining e.g. grinding, milling and shot-peening; (ii) thermal processes including electron beam and laser melting; (iii) electrochemical and chemical processes, such as electro polishing and etching. Several subtractive processes have been vastly used in near-net shaping processes, such as die-casting, casting and moulding. This has now been extended to additive manufacturing, allowing feature geometries to be realized with greater accuracy and surface quality via subtractive processing. CNC turning was used to finish AM parts (12). Taminger et al. (13) used high speed milling to finish aluminium AM parts. Grinding has also been used to finish MAM parts (14, 15)Complex and intricate geometries pose a challenge for conventional grinding. In an attempt to alleviate this, Beauchamp et al. (16) used shape-adaptive grinding to finish AM parts. Above mentioned finishing of the AM parts were performed in different workstations. Some other researches have been done to finish the AM parts in a single workstation combining additive and subtractive manufacturing processes to achieve short build time and high accuracy. A hybrid process was developed with the combination of three-axis milling and SLC in which the laser head and the CNC milling spindle were attached on two vertically isolated axes (17).Similar to this another process was developed where CO2 laser welding was used as the RP process (18). Some other researches have been done where the only change in the implementing of RP process for example gas metal arc welding (GMAW) (19) and arc welding (20). Another type of hybrid process was developed by using laser cladding (LC) technology and five-axis machining as RP process and machining process respectively. But the limitation of that machine was its complicated mechanism (21). A different approach was adopted for the combination of conventional machining process and RP process (22). Any type of machinable material could be used for the fabrication of the product with that method but may not be simply valid to products with a complex shape. This drawback was solved by using multi-axis (5-axis) milling machine to construct the undercut feature, lastly screws and pins were used to make the assembly (23). A more robust assembly was reported in which ultrasonic welding was used to connect sheet material (24). A hybrid RP system has been developed with the combination of five-axis machining and FDM. The FDM extruder was installed in a relation to the cutter spindle so that two modes could be interchangeable by 180° rotation of the axis which prevented the interference between the cutter spindle and FDM extruder (25). But at the same time, this design was bulky and the system needs more motors and actuators. Recently we have developed a new hybrid machining system which incorporates Fused Deposition Modelling (FDM) as RP Process and CNC grinding operation as subtractive manufacturing process in a single platform (9). It is figured out that a little research has been performed on a hybrid system for thermoplastic materials

2. Experimental Setup and Procedure:

2.1. Newly Developed Hybrid Machine:

Incorporating the CNC spindle and the FDM extruder together on a single workstation without unnecessary bulky mechanism was the major issue with respect to designing this hybrid machining system. Maintaining alignment of CNC Spindle and FDM heat extruder during changeover process was the main issue for this one. In our newly developed hybrid machine we have used a rotary stage along Z-axis to incorporate the spindle rectangular shaped aluminium attachment on which the heat extruder and spindle head is attached respectively just parallel to the rotary stage. The rotary stage can be operated both manually and automatically. Another rectangular shaped aluminium attachment was attached perpendicularly to the rotary stage on which IR sensors were placed in a significant distance. These sensors were used to omit the misalignment problem during exchanging the equipment of CNC machining and FDM process.The rotary stage is able to be moved accordingly the difference of analog signals coming from the IR sensors. When the the equipment (spindle or heat extruder) is at the correct position, two reflected IR signals come from at the same distance that means there is no difference between the signals as well as no voltage difference. Similarly, if the reflected IR signals come from different distances then there will be voltage difference which indicates that the object is not at the aligned position.

IOP Conf. Series: Materials Science and Engineering **260** (2017) 012031 doi:10.1088/1757-899X/260/1/012031

Figure 2(a) and 2(b) shows the 3D design of the newly developed Hybrid machine with CNC spindle and FDM extruder respectively.

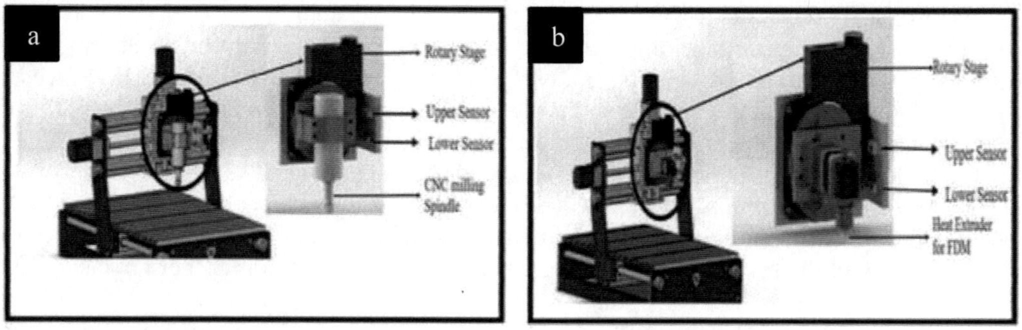

Figure 2. 3D design of the Hybrid Machine with (a) CNC spindle and (b) FDM heat extruder(9)

2.2. Control System of the Hybrid Machine

The total control system for the newly developed hybrid machine is PC-based. The whole control unit for the machine was developed using separate controller, circuitry to handle the CNC machining and FDM process. LabVIEW programming was used to control the whole hybrid machining system from a user friendly GUI. We have used NI PCI-7344, UMI-7764 and NI DAQ USB 6211 for movement of the machine. We have used three servo motors, one stepper motor, one rotary stage and two IR sensors which are desired to be controlled: three motors for the X, Y and Z axis, and the other one is for the FDM heat extruder. NI PCI 7344 motion controller was used for controlling the four motors. This controller is able to achieve simultaneous three-axis motion trajectories through linear, spherical, circular or helical interpolation, which will placate all our requirements for concurrent three-axis machining and also for the FDM operation.

2.3. Procedure

To design and Slice the CAD model of the required parts we have used SolidWorks and Slice3R respectively. After getting the G codes of the CAD model, firstly we produced the FDM part on our hybrid machining system and after that without removing the part from the position we did the grinding on the outer surface of the part. This was done by implementing CNC Spindle along the Z axis. We have done the same procedures several times for the objects with different layer thickness.

3. Results and Discussion:

The aim of this research was to identify the differences in dimensions between the 3D design and the completed object made by newly developed hybrid machine. To get a better dimensional accuracy and to identify the effect of layer thickness we produced objects with 3 different layers such as 0.1 mm, 0.15 mm and 0.2 mm. At first FDM was used to build several cubes layer by layer. After that we did CNC grinding operation to finish the produced FDM objects. For making the FDM part we used PLA filament, and diameter of the filament was 1.75 mm. The design dimensions of the cube were 15mm×15mm×10mm. For different layers we got different dimensions of the completed object from the hybrid machining system. Figure 3(a) shows that when the layer thickness is 0.1mm the layers were uniformly overlapped whereas figure 3(b) shows the layers were not uniformly overlapped when the layer thickness is 0.15 mm. Objects with layers thickness 0.2 mm have several gaps between two layers and causes the worst dimensional accuracy which is shown in figure 3(c).

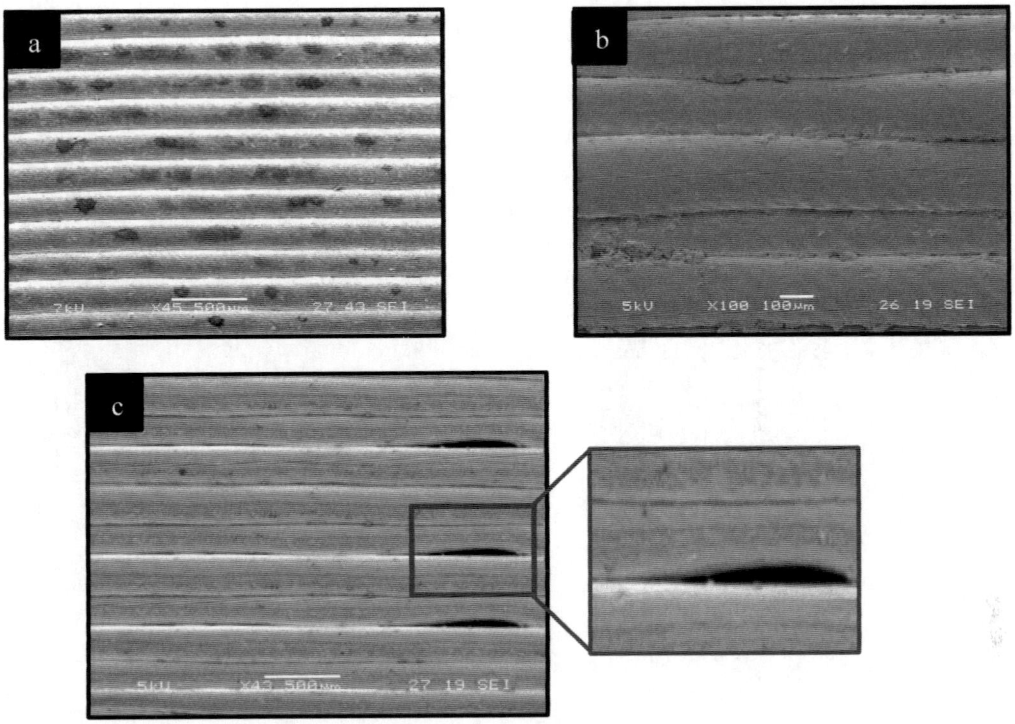

Figure 3. Surface of the FDM part (without machining) with layer thickness (a) 0.1mm (b) 0.15 mm and (c) 0.2 mm (with gap showing insight)

Table 1 shows the dimensions of the produced objects from newly developed hybrid machining system.

Table 1. Measured Dimensions (mm) for different layer thickness

Layer Thickness(mm)	Measured Dimensions (mm)	
	Without Machining	With Machining
0.1	15.25mm ×15.28mm ×10.07mm	15.05mm ×15.10mm ×10.07mm
0.15	15.57mm ×15.65mm ×10.13mm	15.25mm ×15.22mm ×10.13mm
0.2	16.05mm ×16.10mm ×10.35mm	15.55mm ×15.35mm ×10.35mm

The differences were 1.67%, 1.87% and 0.7% for each dimension , respectively while the part is without machining and the layer thickness was 0.1mm. As the targeted dimension for both length and width were same 15mm, that's why the differences were almost same. After completing the FDM operation we did the grinding operation along the outer surface (length and width) of the objects. Then we got an improvement upto 80% accurate dimension in both length and width compared to the targeted dimension. The results of the research suggest that Hybrid process lies in ± 0.05 mm, ± 0.25mm and ± 0.55mm limit in regard to dimensional accuracy for layer thickness 0.1mm, 0.15 mm and 0.2 mm respectively of 3D printed parts. The comparison of each dimensions for different layer thickness are shown in the following figure 4(a) and 4(b). Standard error for both cases without and with machining is also shown in the figures. We have figured out that there was a significant change in the dimension of height with the increment of layer thickness. The possible causes for the difference in dimensions between the C and the completed component are described in the following. Firstly, the diameters of the fused filament produced by the FDM process were inconsistent. The variation of the fused filament diameter caused the variation of the final FDM part that consisted of the filaments. Secondly, as the

IOP Conf. Series: Materials Science and Engineering **260** (2017) 012031 doi:10.1088/1757-899X/260/1/012031

FDM extruder changed the moving direction, it was not possible to maintain its speed at a constant value and the FDM extruded more and less PLA material at the locations where the speed slowed down and the speed was up respectively, thus causing the PLA filament to become thicker and create the dimensional variation

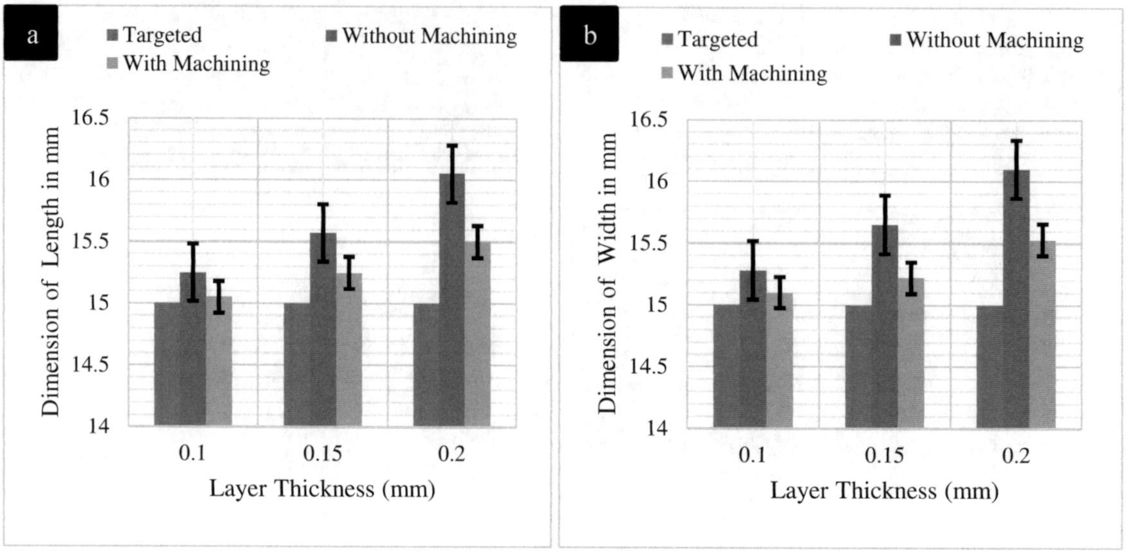

Figure 4. Comparison of dimensions in (a) Length and (b) width between Targeted and final parts

which is specially happened when the layer thickness was 0.15 mm and 0.2mm.Thirdly, when the layer thickness is kept (0.2mm) half of the nozzle diameter (0.4mm) the fused filaments did not stick with each layer properly; gaps existed between the melted filaments which caused difference in dimensions of the final parts. The reason of the gap was firmly correlated to the layer thickness and also to non-uniform diameter of the melted filament. Another cause for the gap was the environment. As our hybrid machining system is not covered that's why environmental air affected the melted filament to stick properly.

4. Conclusion:

In this paper, it has been shown that better dimensional accuracy is achievable with small layer thickness. When it is kept ¼ th of the nozzle diameter then there are small difference in dimensions compared to the targeted dimensions having almost uniformly fused filament. On the other hand when the layer thickness is ½ of the nozzle diameter then there are gaps between the layers. With our newly developed hybrid system we were able to improve the dimensional accuracy by a significant margin (upto ~80%). Better control of the speed of the nozzle and the temperature of both heater and heat bed might solve the issue of gap between layers which would help to achieve more accurate dimensions.

References:

[1] Upcraft S, Fletcher R 2003 The rapid prototyping technologies. *Assembly Automation* **23(4)** pp 318-30.

[2] Novak-Marcincin J, Novakova-Marcincinova L, Barna J, Janak M 2012 Application of FDM rapid prototyping technology in experimental gearbox development process. *Tehnički vjesnik.* **19(3)** pp 689-94.

[3] Hanser Gebhardt A 2003 *Rapid Prototyping* (Gardner Publications Inc.), Cincinnati.

[4] Jacobs PF 1992 *Rapid prototyping & manufacturing: fundamentals of stereolithography* Society of Manufacturing Engineers.

[5] Grimm T. 2004 *User's guide to rapid prototyping*, Society of Manufacturing Engineers.

[6] Levy GN, Schindel R, Kruth J-P 2003 Rapid manufacturing and rapid tooling with layer manufacturing (LM) technologies, state of the art and future perspectives *CIRP Annals-Manufacturing Technology* **52(2)** pp 589-609.

[7] Rosochowski A, Matuszak A 2000 Rapid tooling: the state of the art. *Journal of Materials Processing Technology* **106(1)** pp 191-8.

[8] Percoco G, Lavecchia F, Galantucci LM. 2012 Compressive properties of FDM rapid prototypes treated with a low cost chemical finishing *Research Journal of Applied Sciences, Engineering and Technology* **4(19)** pp 3838-42.

[9] Amanullah A, Saleh T, Khan R 2017Design and Development of a Hybrid Machine Combining Rapid Prototyping and CNC Milling Operation. Procedia Engineering **184** pp 163-70.

[10] Pandey PM, Reddy NV, Dhande SG 2003 Improvement of surface finish by staircase machining in fused deposition modeling *Journal of Materials Processing Technology* **132(1)** pp 323- 31.

[11] Anitha R, Arunachalam S, Radhakrishnan P 2001 Critical parameters influencing the quality of prototypes in fused deposition modelling *Journal of Materials Processing Technology* **118(1)** pp 385-8.

[12] Spierings A, Starr T, Wegener K 2013 Fatigue performance of additive manufactured metallic parts *Rapid Prototyping Journal* **19(2)** 88-94.

[13] Taminger KM, Hafley RA, Fahringer DT, Martin RE 2004 *Effect of Surface Treatments on Electron Beam Freeform Fabricated Aluminum Structures* (DTIC Document)

[14] Löber L, Flache C, Petters R, Kühn U and Eckert J 2013 Comparison of different post processing technologies for SLM generated 316l steel parts *Rapid Prototyping Journal* 19(3) pp 173-9.

[15] Rossi S, Deflorian F, Venturini F 2004 Improvement of surface finishing and corrosion resistance of prototypes produced by direct metal laser sintering. *Journal of Materials Processing Technology* **148(3)** pp 301-9.

[16] Beaucamp AT, Namba Y, Charlton P, Jain S, Graziano AA 2015 Finishing of additively manufactured titanium alloy by shape adaptive grinding (SAG). *Surface Topography: Metrology and Properties* **3(2)** p 024001.

[17] Jeng J-Y, Lin M-C. 2001 Mold fabrication and modification using hybrid processes of selective laser cladding and milling *Journal of Materials Processing Technology* 110(1) pp 98-103.

[18] Choi D-S, Lee S, Shin B, Whang K, Song Y, Park S, et al. 2001 Development of a direct metal freeform fabrication technique using CO 2 laser welding and milling technology *Journal of Materials Processing Technology* **113(1)** pp 273-9.

[19] Song Y-A, Park S. 2006 Experimental investigations into rapid prototyping of composites by novel hybrid deposition process *Journal of Materials Processing Technology* **171(1)** pp 35-40.

[20] Karunakaran K, Suryakumar S, Pushpa V, Akula S 2010 Low cost integration of additive and subtractive processes for hybrid layered manufacturing *Robotics and Computer-Integrated*

Manufacturing **26(5)** pp 490-9.
[21] Kerschbaumer M, Ernst G 2004 Hybrid manufacturing process for rapid high performance tooling combining high speed milling and laser cladding. Proceedings of the 23rd International Congress on Applications of Lasers and Electro-Optics (ICALEO), San Francisco, CA.
[22] Hur J, Lee K, Kim J. 2002 Hybrid rapid prototyping system using machining and deposition. *Computer-Aided Design* **34(10)** pp 741-54.
[23] Salloum T, Anselmetti B, Mawussi K. 2009 Design and manufacturing of parts for functional prototypes on five-axis milling machines *The International Journal of Advanced Manufacturing Technology* **45(7)** pp 666-78.
[24] Karunakaran K, Bernard A, Suryakumar S, Dembinski L and Taillandier G. 2012 Rapid manufacturing of metallic objects *Rapid Prototyping Journal* **18(4)** pp 264-80.
[25] Lee W-c, Wei C-c, Chung S-C. 2014 Development of a hybrid rapid prototyping system using low-cost fused deposition modeling and five-axis machining. *Journal of Materials Processing Technology* **214(11)** 2366-74.

Acknowledgments
Authors would like to thank IIUM for providing all experimental facilities to carry out the research work.

IOP Conf. Series: Materials Science and Engineering **260** (2017) 012032 doi:10.1088/1757-899X/260/1/012032

The effect of electrohydrodynamic force on the lift coefficient of a NACA 0015 airfoil

Y Yusof[1], A Hossain, A H Abdullah[2], Rizal M E Nasir[1], A Hamid[1] and N Muthmainnah M N[1]

[1] Faculty of Mechanical Engineering, Universiti Teknologi MARA (UiTM), Shah Alam, Selangor
[2] Universiti Kuala Lumpur, Malaysian Institute of Aviation Technology (UniKL MIAT), Dengkil, Selangor

Email: yaakob2241@salam.uitm.edu.my

Abstract. Lift, the force component that is perpendicular to the line of flight, is generated when a small aircraft moves through the air. With the help of the sets of flaps and slats on its wing, the pilot controls his aircraft manoeuvring in the air. In this study, we preferred to cut the drawbacks of the flaps system by introducing the electrohydrodynamic actuator. Widely known as plasma actuator, it is able to improve the induced lift force as well as the efficiency of a small aircraft system. A dielectric-barrier-discharge actuator using a 6 kV AC power supply was developed and tested on a NACA 0015 airfoil using copper as the electrodes and kapton as its dielectric component. The experimental results showed that it was successful in presenting a positive effect of the plasma actuator on the lift coefficient of the airfoil at smaller angle of attack, where enhancements ranged between 0.7% and 1.8%. However, at a higher angle, the results were not as swayed as it was desired since the energy exerted by the plasma actuator on the lift performance of the airfoil was inadequate. Further tests are needed using higher rated voltage supply and other equipment to improve the capability of the actuator in refining the aerodynamic performance of the airfoil.

1. Introduction

This is considered to be an introductory study to understand the improvement an electrohydrodynamic (EHD) actuator or widely known as the plasma actuator brings to the aerodynamic performance of a small aircraft system. Aircraft is lifted by its special design of wing, termed as airfoil [1]. The airfoil's lift and drag forces are induced by its motion when cutting through the air. Angle of attack (AoA) and aircraft's speed play important roles as they determine the performance and stalling point of the aircraft [2].

To reduce the effect of turbulence and to help small aircraft maintains its lift at a certain limit, conventional flaps and slats on the wing of the small aircraft are the workable solution for now. Though they are capable, they have some drawbacks, which include putting more weight to the aircraft, complexity of the linkages system, its maintenance cost and also higher fuel consumptions due to the weight needed to be lifted [2] & [3].

As the old system is fully mechanical where the pilot of the small aircraft needs to control the flaps and ailerons using pedal and lever as shown in Figure 1, plasma actuator is totally an electronic system which enables it to be controlled by modern control system. According to [4], plasma actuator can be used to modify the lift and drag forces, just as what the flaps did before.

IOP Conf. Series: Materials Science and Engineering **260** (2017) 012032 doi:10.1088/1757-899X/260/1/012032

Plasma actuator can also be used for longitudinal control of the leading edge of an airfoil without employing the conventional hinged flap used on normal airplane. These small size actuators reduce kilograms of weight, depending on the size of the airplane and its conventional flap system. It just needs to be laminated at the designated area at the airfoil and does not require slots or any mechanical change to the wing [5]. In addition, it requires less maintenance routine and cost as compared to the old system [6]. Thus, introducing a plasma actuator will improve these weaknesses, other than giving additional benefits such as increasing and reducing the lift and drag force, respectively, as well as delaying the stalling point [7].

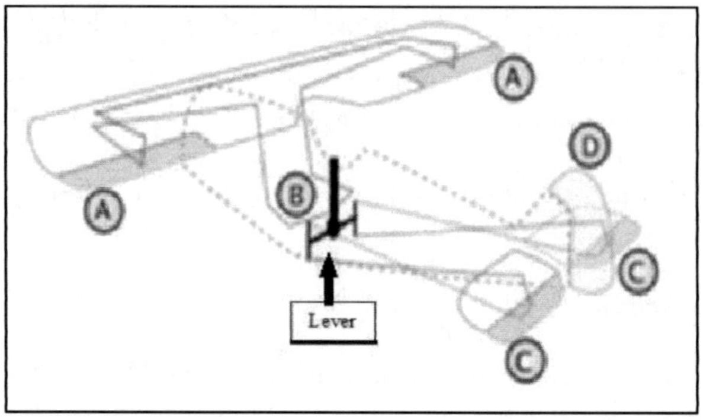

Figure 1. The conventional lever and flaps system. The notations of "A", "C", and "D" were the flaps of the small aircraft, controlled by the lever at "B". (image taken from en.wikipedia.org)

Plasma actuator is a type of electrical actuator that is capable of ionizing the flowing air and add localized momentum to the flow through collision process of the migrating charged particles due to the large potential difference with the natural surrounding gasses [2]. This creates the EHD forces that modify the boundary layer of airflow, and also affect the moving parts in that region. The induced flow should be laminar and parallel to the surface to realize flow re-attachment and minimize boundary layer turbulence [7]. Plasma actuator needs high voltage power supply to operate [8] & [9].

In terms of electrical power, a plasma actuator offers an efficient system in transferring the electrical energy to the mechanical form, because with minimum power input, it can generate high energy density and produce great effect on the airfoil or propeller's blade [5]. It is also fully electronic, which makes it suitable for feedback control [10]. It is a long way to go though, because a lot of studies are required in understanding the effects of the environment, where temperature, humidity and pressure play important roles.

There are different types of plasma actuator, mainly in the research area. Among others include the corona-based discharge, dielectric-barrier-discharge (DBD), sliding discharge actuator, local arc filament and sparkjet [11] & [2] & [10]. The one being used in this study is the DBD type plasma actuator. A DBD places dielectric material in between two electrodes that were connected to a high voltage source. In generating UV emission and low temperature plasma, the discharge performed at the surface of a dielectric layer is usually asymmetric configuration [12].

The parameters involved in studying aerodynamic environment are, not restricted to, airstream velocity induced from the thrusters, temperature and the pressure of the surrounding air as well as the air density and the angle of attack. This study focused on only two parameters which were the AoA and airstream velocity. Speed of the aircraft and the AoA are two of the main parameters that determine the action of an aircraft – the lift force induced. Thus, by engaging the open circuit wind tunnel, these two parameters were made the main control factors of the experiment. By varying these two, the effect of the plasma actuator on the lift performance of the NACA 0015 airfoil was studied. Besides providing significant contributions to the lift performance, these two parameters were made easily controllable in the wind tunnel. This experiment dealt only with the lift force to study the plasma actuator's effectiveness.

The objective of this study involves the development of a plasma actuator system and a study on how effective it is in improving the lift performance of a NACA0015 airfoil. In analysing this, the performance of the airfoil with and without the plasma actuator was compared and examined.

1.1 Design Parameters for DBD Plasma Actuator
There are several important parameters for the actuator that determine the power requirement of the system, as well as the output drive the plasma actuator is capable of. Firstly, the dielectric. Dielectric is an insulating material with special characteristics. When voltage difference is applied through the dielectric, electric polarization occurs. This makes the electricity passes through the dielectric, and this phenomenon is called dielectric breakdown. In the case of a DBD, it generates the plasma. In terms of the dielectric choices of a plasma actuator, [13] discussed the details of their experimental data, in which they compared the types of dielectric including kapton – our choice for this system because of its availability in terms of the required thickness and size; teflon, plexiglass and plexi-kapton material, using copper as the electrode.

Secondly, the length of the electrode. The length of the electrode is equally important on the plasma generated. According to the experiment by [14], the length of the electrode significantly affected the thrust. They found that when the length of the plate decreased, the thrust generated was increased. The next parameter is the electrode's gap. The gap between the top and bottom electrode gives some significance too. According to the experiment by [15], it was found that a good gap size was between -2mm to 1mm, where the negative sign indicated that the bottom electrode was overlapped by the top. In this experiment, a 0.9mm-thick dielectric was being used to test the effectiveness of the power generation with gap size ranging from -2mm to 5mm. It was also found that at 3mm gap and above, the supply voltage must be above 6 kV in order for the plasma actuator to discharge power.

Another consideration is the position of the actuator on the wing. The best location of the plasma actuator needs to be determined to have the maximum output out of it. [4] suggested that the best location to put the actuator is chordwise at the leading edge of the airfoil. The last factor is the supplied electrical potential energy. This turn-on-voltage is very important in supplying enough energy for the electrons to cross the gap between the electrodes and the dielectric to create the plasma.

2. Experimental setup and procedure

The experimental setup consisted of three main parts. They were the airfoil as the base of the plasma actuator, wind tunnel as the main testing apparatus, and the plasma actuator itself. The airfoil selected was NACA 0015 of the NACA 4-digit series, where each digit in each series gave geometrical definition of the dimension and shape of the airfoil. This selection was made due to the availability of the model in the laboratory. Besides, NACA 0015 was one of the common airfoils being used in researches; thus, more references and analysis on NACA 0015 were available in the publications for references. The airfoil used had a chord length of 0.125m, and the reference area was 0.0159 m^2. Reference area is the area of cross section of the airfoil at the chord line. The wind tunnel used in this experiment was an open circuit wind tunnel, as shown in Figure 2.

The dielectric barrier discharge plasma actuator used in this experiment consisted of two copper tapes, each with thickness of 0.1mm. The copper tapes was separated by 0.2mm thick kapton tape as dielectric. The 0.2mm dielectric was built by stacking three layers of kapton tape, which was about 60μm thick each, on top of each other. The width of the top electrode was 2mm, while its length was 65mm. For the bottom electrode, its width was 4mm and its length was 75mm. The power to the PA was supplied by an inverter as shown in Figure 3. The inverter's rating was 6.0 kilovolt (kV) AC peak-to-peak.

This actuator was attached on the top surface of the NACA 0015. The location of the actuator on NACA 0015 was where the separation was expected to happen, which was slightly off the leading edge of the airfoil. When the plasma actuator was set up on the airfoil, the airfoil was mounted at the designated place inside the test section, as shown below in Figure 4. To operate the plasma actuator, the switch was turned on and as the voltage supply went into the plasma actuator, it turned on and discharge happened. To stop the action of the plasma actuator, the power button was switched off. The performance of the system's lift force with actuator turn-on and turn-off was analyzed as well as other observations during the experiment.

Figure 2. Open circuit wind tunnel used in this experiment.

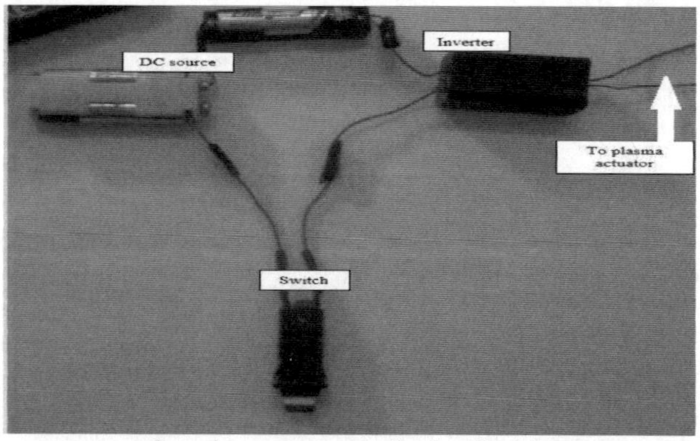

Figure 3. The inverter setup used as the power supply for the plasma actuator in this experiment.

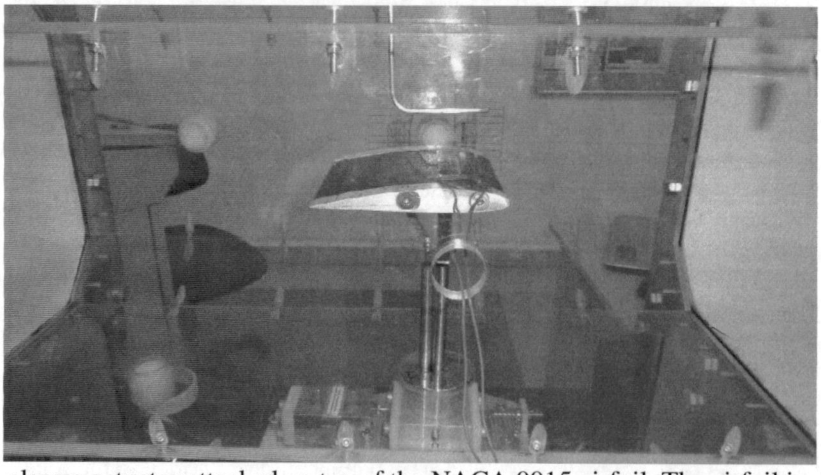

Figure 4. The plasma actuator attached on top of the NACA 0015 airfoil. The airfoil is mounted at the adjustable hook inside the test section of the wind tunnel.

During the experiment, the speeds of the air and the AoA of the airfoil were varied.. The lift force induced was recorded. For each airstream velocity and its pair of AoA, two sets of lift force data were recorded. One set was the value of the lift forces where the plasma actuator was turned off, while another set of data was when the plasma actuator was turned on.

The data of the lift forces collected were then converted to the correspondent lift coefficient, C_L, using equation 1 below,

$$C_L = \frac{L}{\frac{1}{2}\rho v^2 S} \quad (1)$$

where L is the lift force in N, ρ is the air density in kg/m^3, v is the stream velocity in m/s and S is the reference area in m^2. The C_L data was plotted into a C_L vs. AoA plot and analyzed. The data of the plasma-off and plasma-on were compared and analyzed to perceive what contribution a plasma actuator brought to the aerodynamic characteristics of an airfoil.

One thing that is important in studying fluid flow is the characteristics of the flow. In order to compare or analyse different situations or specimens, there must be characteristics of the air that need to be set in a controlled situation. Reynold's Number (Re) can be used as the reference tool. The speeds of airflow inside the wind tunnel used in this testing were 40 m/s and 45 m/s. These speeds were selected in order to see the difference of flow for different Re. The Re is defined based on the chord length shown below;

$$Re = \frac{\rho v c}{\mu} \quad (2)$$

where ρ is the air density in kg/m^3(taken as 1.1839), v is the stream velocity in m/s, c is the chord length in m, and μ is the dynamic viscosity of the air in kg/(m)(s) (1.84×10^{-5}).

3. Experimental outcome

To study the effectiveness of the plasma actuator in enhancing the lift performance, the results from the experiment were compared between the plasma OFF and plasma ON. Plasma OFF is when no plasma actuator is initiated in the system, whereas plasma ON is when the plasma actuator is activated. Theoretically, to see the positive effects of the plasma actuator on the NACA 0015 airfoil's aerodynamic performance, the values of C_L for plasma ON must be of higher values compared to the values of C_L during plasma OFF [16]. Higher values of C_L for plasma ON as compared to plasma OFF indicated that there were improvements in the airfoil's lift performance induced by the plasma actuator itself.

The Reynolds number selected for the analysis were 3.2×10^5 and 3.6×10^5 in accordance to the respective airstream velocity, while the AoAs were 12°, 14°, 16° and 18°. Table 1 presents the values of C_L for plasma OFF and plasma ON at each Reynolds numbers and AoA. From the tabulated data, we can see whether the C_L for plasma ON is higher from the C_L for plasma OFF. Negative values in the "Diff" column of Table 1 indicated that C_L of plasma ON was lower than the C_L of plasma OFF, while positive values indicated otherwise. The differences shown at the extreme right column of the tables are based on the plasma OFF since it is the reference value. From Table 1 (a) and (b), we can see that the percentage differences were sometimes negative and sometimes positive.

At lower to medium AoA of 12 to 14 degrees, we can undoubtedly tell that the percentage differences were of positive values, ranging from 1.0% to about 1.8% of improvements from the OFF conditions for Reynold's number 3.2×10^5, and from 0.7% to 1.6% for Reynold's number 3.6×10^5. This might indicate the positive effect of the EHD forces that we are expecting it to provide to the lift performance of the airfoil. However, at higher AoA of 16 and 18 degrees, the value became inconsistent and more negative. This means that the values of C_L for plasma ON were not always

IOP Conf. Series: Materials Science and Engineering **260** (2017) 012032 doi:10.1088/1757-899X/260/1/012032

consistently higher than the plasma OFF, which showed that the actuators were hardly making its account. This might be because, at higher degrees, the flow is reaching turbulence, thus making the power produced by the actuator hardly seen affecting the flow characteristics. Figure 5 (a) and (b) illustrate the results graphically.

Table 1. Differences of C_L values between plasma OFF and plasma ON for Reynolds number (a) $3.2x10^5$ and (b) $3.6x10^5$.

(a)

No	AoA (°)	Lift coefficient, C_L		
		Plasma OFF	**Plasma ON**	**Diff. (%)**
1	**12**	1.256	1.278	1.787
2	**14**	1.385	1.399	1.015
3	**16**	1.568	1.571	0.172
4	**18**	1.682	1.673	-0.509

(b)

No	AoA (°)	Lift coefficient, C_L		
		Plasma OFF	**Plasma ON**	**Diff. (%)**
1	**12**	1.290	1.300	0.747
2	**14**	1.354	1.376	1.620
3	**16**	1.526	1.512	-0.906
4	**18**	1.624	1.619	-0.304

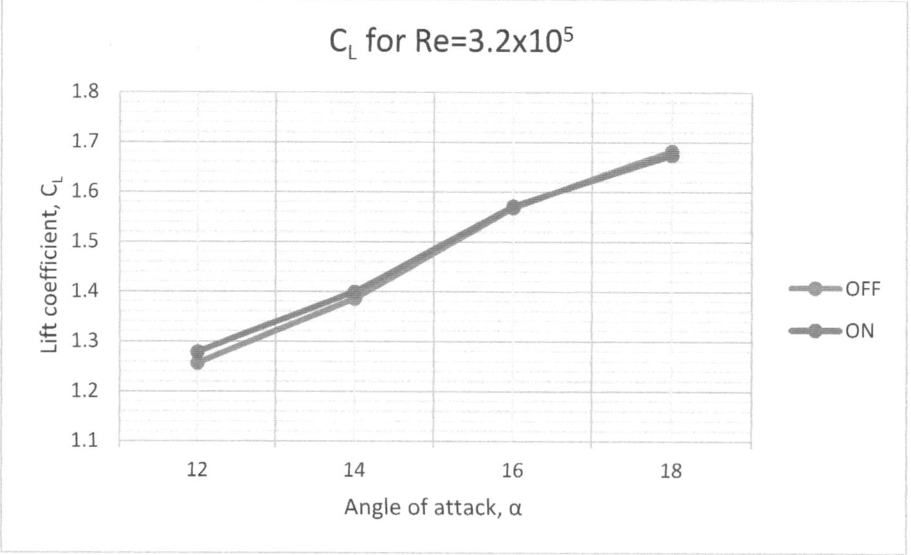

(a)

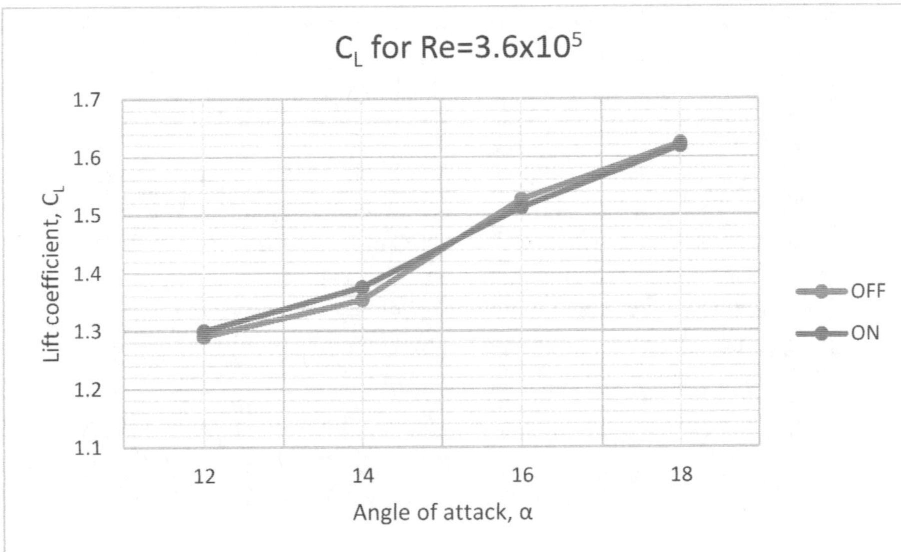

(b)

Figure 5. C_L plots of plasma OFF and plasma ON for Reynolds number (a) $3.2x10^5$ and (b) $3.6x10^5$.

The results of this experiment also informed us on the importance of a suitable power generator for plasma generation. Referring to the literatures, this experiment realized a few reasons based on the inconsistency of the data recorded. The main differences between this study and other literatures showed that power supply might not be sufficient enough in producing adequate plasma thrust to affect the airflow, especially at above medium AoA. [17] explained that a voltage source of 20 kV was needed to effectively reattached the separated flow, while [18] used a power supply of 13 kV to increase the lift performance of more than 30% of its original value. Furthermore, [3] supplied their plasma actuator with 16kV and achieved 50% improvement in its lift performance. Although these experiments setup were different with ours, they gave indications on the suitability of power needed to supply the plasma actuator in order for it to give good results in improving the aerodynamic performances of the airfoil. In addition, although we can see that the plasma was generated due to the existence of the bluish discharge, this showed that the plasma generated did not sufficiently affect the airflow characteristics that surrounded the airfoil.

Although the results of the experiment did not give us convincing conclusions, there were also other observations to note from the experiment. Firstly was the effect of the gap between the top and bottom electrodes, as briefly explained in its design parameters. By using a relatively low 6 kV-rated voltage source, gap of the electrodes can give significant remarks. The plasma generated was not uniform and continuous when there was a gap of 1mm between the two electrodes, whereas when the gap was 0mm, the discharge was much better in its form as shown in Figure 6. This showed that when the gap was increased, the plasma generated was less and as a result less effect can be produced by the plasma.

Another observation was that after long hours of experiment, a clear white line was formed at the top of the covered electrode as shown in Figure 7 below. This might due to the degradation effect [12]. This degradation effect was caused by the ion bombardment from the top to the bottom electrode. The degradation of the dielectric material was related to the high voltage applied, and it could be lessened by increasing the electrode gap.

IOP Conf. Series: Materials Science and Engineering **260** (2017) 012032 doi:10.1088/1757-899X/260/1/012032

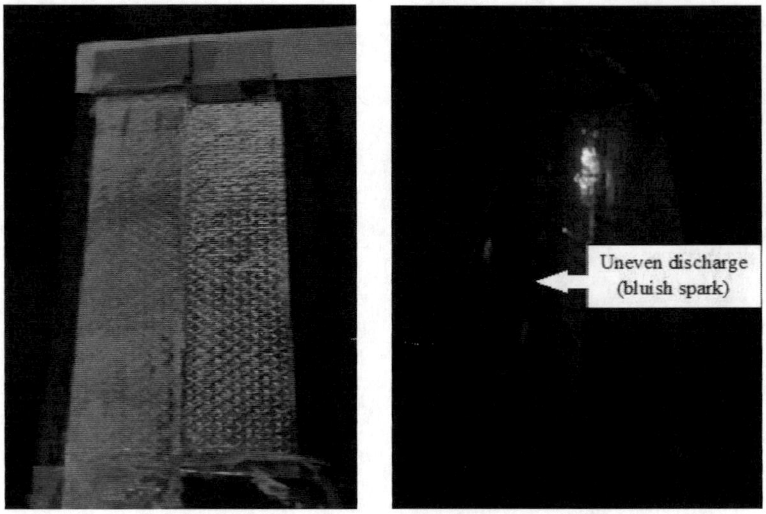

Figure 6. The bluish plasma generated when the electrode gap is a) 0mm and b) 1mm.

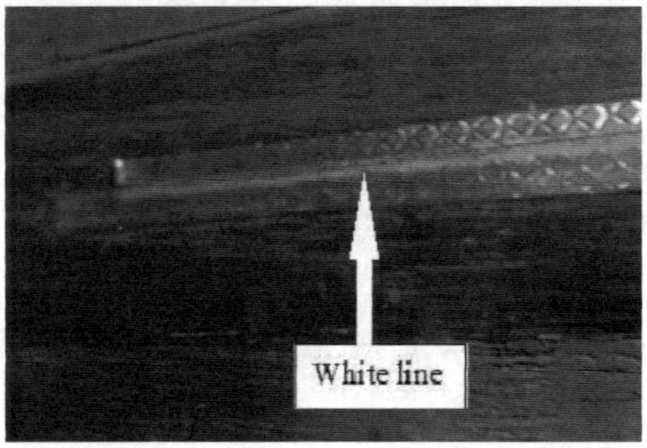

Figure 7. Thick white line formed on top of the covered electrode after long hours of experiment.

4. Conclusion

It is known from various findings that a plasma actuator is able to affect the aerodynamic performance of airfoils. Among others, it can increase lift performance, reduce the drag force and delay the separation of flow. The intent of this work was to study the effect of plasma actuator on the lift performance of NACA 0015 airfoil.

Throughout this report, the procedures of the experiment as well as the complete results were detailed out. The experiments were successful in showing the effect of the plasma actuator on the lift performance of a NACA 0015, even though the results were not as swayed as it was expected. The improvement on the lift performance of the airfoil was found to be positive at lower to medium AoA, with improvements ranging between 0.7% and 1.8%, as compared to the original plasma OFF value. However, at higher degrees, the thrust generated was not adequate. This, as we can summarize, might be mainly due to the insufficient power supplied to the actuator. In addition to that, we studied the traits of the plasma actuator, especially on the effects of other parameter such as electrode gap, on the plasma generated.

There are many things that can be done in order to improve the research of plasma actuator effect in the future. Firstly is by using a higher and adjustable power supply. Our fixed 6 kV-rated inverter was proven insufficient to generate enough thrust for the whole experiment. Increasing the voltage to the actuator is the easiest way to increase actuator thrust; thus, increasing the actuator effectiveness [19] in influencing the aerodynamic performance. Improvement in the aspect of the equipment and apparatus used in the experiment can also benefit this study. Besides power supply, employing a high speed and resolution camera could examine more aspect and traits of the plasma, such as the uniformity of the plasma discharge, the form of the surrounding air flow, and the fluid flow characteristics ambiances the airfoil. This will aid in further developing a mathematical model for the DBD plasma actuator and its computational fluid study.

5. Acknowledgements

The author would like to thank UiTM and LESTARI Fund (Grant No: 600-IRMI/MYRA 5/3/LESTARI (0417/2016)) for the funding of research.

References

[1] I. H. Abbott and A. E. D. Von, *Theory of Wing Sections - Including a Summary of Airfoil Data*. New York: Dover Publications, Inc., 1959.

[2] R. Sosa and G. Artana, "Steady control of laminar separation over airfoils with plasma sheet actuators," *J. Electrostat.*, vol. 64, no. 7–9, pp. 604–610, 2006.

[3] N. Benard, J. Jolibois, and E. Moreau, "Lift and drag performances of an axisymmetric airfoil controlled by plasma actuator," *J. Electrostat.*, vol. 67, no. 2–3, pp. 133–139, 2009.

[4] M. P. Patel, T. T. Ng, S. Vasudevan, T. C. Corke, and C. He, "Plasma Actuators for Hingeless Aerodynamic Control of an Unmanned Air Vehicle," *J. Aircr.*, vol. 44, no. 4, pp. 1264–1274, 2007.

[5] M. L. Post, "Plasma Actuators for Separation Control on Stationary and Oscillating Airfoils," Notre Dame, Indiana, 2004.

[6] J.-S. Yoon and J.-H. Han, "One-equation modeling and validation of dielectric barrier discharge plasma actuator thrust," *J. Phys. D. Appl. Phys.*, vol. 47, no. 40, p. 405202, 2014.

[7] J. R. Roth and X. Dai, "Optimization of the Aerodynamic Plasma Actuator as an Electrohydrodynamic (EHD) Electrical Device," in *44th AIAA Aerospace Sciences Meeting and Exhibit*, 2006, no. January, pp. 1–28.

[8] V. Boucinha, R. Weber, and A. Leroy, "Study of A DBD Plasma Actuator Dedicated to Airflow Separation Control," in *IEEE 35th International Conference on Plasma Science*, 2008, vol. 40.

[9] K. Shimizu, Y. Mizuno, and M. Blajan, "Basic study on flow control by using plasma actuator," *2013 IEEE Ind. Appl. Soc. Annu. Meet.*, vol. 0, no. 3, pp. 1–6, 2013.

[10] L. N. Cattafesta and M. Sheplak, "Actuators for Active Flow Control," *Annu. Rev. Fluid Mech.*, vol. 43, no. 1, pp. 247–272, 2011.

[11] E. Moreau, C. Louste, and G. Touchard, "Electric wind induced by sliding discharge in air at atmospheric pressure," *J. Electrostat.*, vol. 66, no. 1–2, pp. 107–114, 2008.

[12] A. Ragai, H. Rigit, K. C. Lai, D. Boon, and L. Bong, "Degradation of a Dielectric Barrier Discharge Plasma Actuator," *Int. Conf. Prop. Appl. Dielectr. Mater.*, pp. 569–572, 2009.

[13] G. Neretti, A. Cristofolini, C. a. Borghi, A. Gurioli, and R. Pertile, "Experimental Results in DBD Plasma Actuators for Air Flow Control," *IEEE Trans. Plasma Sci.*, vol. 40, no. 6, pp. 1678–1687, Jun. 2012.

[14] R. Durscher and S. Roy, "Evaluation of thrust measurement techniques for dielectric barrier discharge actuators," *Exp. Fluids*, vol. 53, no. 4, pp. 1165–1176, 2012.

[15] A. R. H. Rigit and I. Dakek, "Performance of a Dielectric Barrier Discharger Plasma Actuator," pp. 1–5, 2010.

[16] P. Zhang, B. Yan, A. B. Liu, and J. J. Wang, "Numerical Simulation on Plasma Circulation Control Airfoil," *AIAA J.*, vol. 48, no. 10, pp. 2213–2226, 2010.

[17] G. Touchard, "Plasma actuators for aeronautics applications - State of art review-," *Int. J. Plasma Environ. Sci. Technol.*, vol. 2, pp. 1–25, 2008.

[18] H. Liang, S. Cao, S. Zhang, and J. Bai, "Experimental investigation of airfoil suction side flow separation control by spanwise nanosecond actuation," *Proc. - 3rd Int. Conf. Meas. Technol. Mechatronics Autom. ICMTMA 2011*, vol. 2, pp. 108–111, 2011.

IOP Conf. Series: Materials Science and Engineering **260** (2017) 012032 doi:10.1088/1757-899X/260/1/012032

[19] J. W. Ferry, "Thrust Measurement of Dielectric Barrier Discharge Plasma Actuators And Power Requirements for Aerodynamic Control," Missouri University of Science and Technology, 2010.

IOP Conf. Series: Materials Science and Engineering **260** (2017) 012033 doi:10.1088/1757-899X/260/1/012033

Network Function Virtualization (NFV) based architecture to address connectivity, interoperability and manageability challenges in Internet of Things (IoT)

Shariq Haseeb, Aisha Hassan A. Hashim, Othman O. Khalifa and Ahmad Faris Ismail

1 Kulliyyah of Engineering, International Islamic University Malaysia, Jalan Gombak, 53100 Kuala Lumpur, Selangor, Malaysia

shariqkhan1@yahoo.com

Abstract. IoT aims to interconnect sensors and actuators built into devices (also known as Things) in order for them to share data and control each other to improve existing processes for making people's life better. IoT aims to connect between all physical devices like fridges, cars, utilities, buildings and cities so that they can take advantage of small pieces of information collected by each one of these devices and derive more complex decisions. However, these devices are heterogeneous in nature because of various vendor support, connectivity options and protocol suit. Heterogeneity of such devices makes it difficult for them to leverage on each other's capabilities in the traditional IoT architecture. This paper highlights the effects of heterogeneity challenges on connectivity, interoperability, management in greater details. It also surveys some of the existing solutions adopted in the core network to solve the challenges of massive IoT deployments. Finally, the paper proposes a new architecture based on NFV to address the problems.

1. Introduction

IoT is an extension of traditional computer based Internet model to a geographically distributed, heterogeneous and constrained model of connected things. The "things" in IoT can be computers, sensors, actuators and software systems that possess communication ability. The aim of IoT is to eliminate as much as possible, erroneous human roles by automating much of the communication and actuation process [1]. Most useful IoT applications aim to provide value to people by utilizing and combining data received from various IoT systems connected to the Internet rather than single isolated sensor network.

Typical IoT architecture consists of complex systems that perform various functions ranging from sensors that collect information about the environment, tracking people, tracking things, collecting physiological measurements, collecting machine data and stores them in IoT backend with or without the help of a gateway. IoT backend is responsible for fusing these data with the help of automated business logic found in process automation to provide feedback to actuators or application. The actuators and applications automate tasks for the benefit of people and processes. The aim of IoT is to eliminate as much as possible, erroneous human roles by automating much of the communication and processes [1].

The typical IoT architecture assumes that all devices are able to interoperate and communicate with the backend which is typically cloud based. However, this is not always true because there are too many device vendors with proprietary communication stack [2], too many communication layer protocols [2], constrained nature [3] of IoT device and too many application layer interfaces [2,3]. This heterogeneity leads to isolated vertical solutions that never communicate with other systems for exchanging valuable information thus, not allowing the true potential of IoT. Moreover, it is virtually impossible to manage a large number of heterogeneous IoT devices.

There is a dire need to manage these unprecedented challenges that were not anticipated when Internet protocol was first adopted globally. The sheer amount of predicted IoT devices entering the digital world, with vast array of technologies incorporated into them, are going to be a challenge the existing architecture at every layer from connectivity to network protocols to session protocols to application protocols and network management protocols. Moreover, successful IoT deployment would also need to resolve the traditional network security challenges like End-to-End Security, Data Security, Device Identity, Personal Data Protection, Access Control and Distributed Denial of Service attack. On top of the traditional challenges, IoT devices would also face Authorization, Authentication, Integrity and Confidentiality challenges because of the constrained computation and power on the devices [4].

This paper highlights some of the existing challenges and the efforts being made by the research community to resolve these challenges. It further introduces the concepts of NFV and proposes a new architecture based on NFV to tackle these challenges. The remainder of this paper is organized into section for existing IoT architecture, connectivity challenges, interoperability challenges, management challenges, NFV concepts and NFV inspired IoT architecture.

2. Typical IoT architecture

A typical IoT architecture is shown in figure 1 [5] which, consists of two types of IoT devices with either sensing capability, actuating capability or both capabilities built into them. These IoT devices could also connect directly to the IoT backend if they are IP capable or could connect to the backend with the help of an IoT gateway.

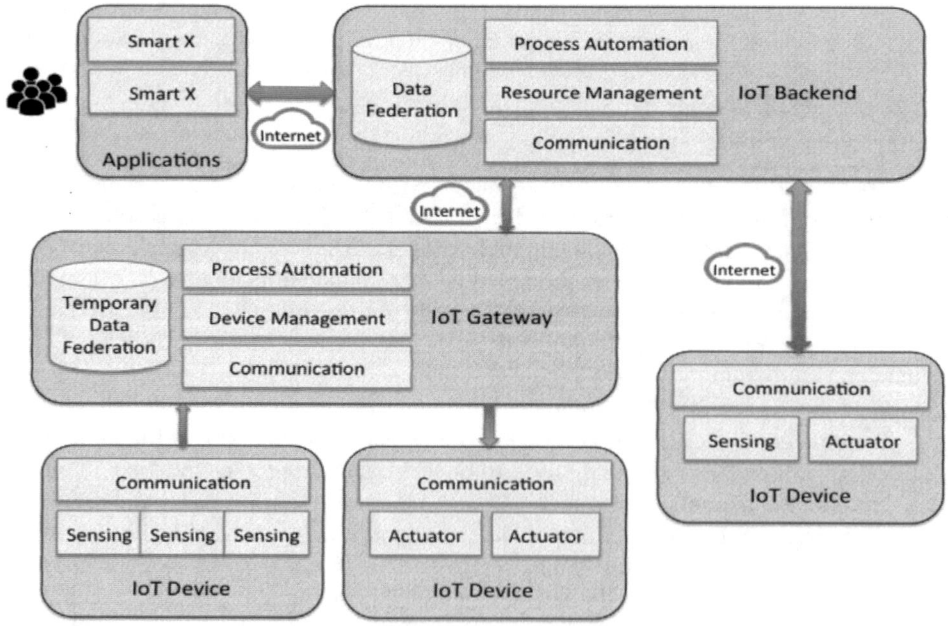

Figure 1. Typical IoT architecture

A typical gateway is meant to store, translate and forward data to the IoT backend system. Some of the more complex gateways perform proprietary device management for the IoT devices that they are serving. Furthermore, some fog [5] enabled IoT gateways also perform process automation before forwarding the processed data to the cloud backend for storage and further processing.

The cloud based IoT backend basically has the function of a large data store for analytics and process automation for actuation. Applications then communicate with the cloud backend to consume data for providing value to the users.

The entire IoT architecture, is designed to aid data flow from sensors to applications. However, the fundamental flaw in the architecture is that it offers a set of interfaces for IoT devices to push data to the backend. If the IoT device is not capable of communicating with the back end then they are unable to take advantage of the existing data on the backend. Furthermore, the IoT backend does not facilitate device to device communication, device discovery, device reachability and network management. The IoT backend is also limited by the capabilities of the physical IoT device and has no mechanism to overcome that. Further limitations are highlighted in the next sections of this paper.

3. Connectivity challenges in IoT

Connectivity is the fundamental component of IoT because transport of data from one IoT device or system to another depends very much on where they are able to connect with each other or not. Difference in connectivity protocols can be studied at various layers of the communication stack.

3.1. Physical layer

Deployment if IoT solutions depends largely on the availability and cost of the physical hardware that operates on the allowed frequency band in the country of deployment. Since the cost of IoT deployment depends largely on the mass production ability of a particular sensor/actuator, it is necessary to have some form of agreement on the allocated license frequency band for IoT. Different worldwide regulatory bodies define the frequency spectrum in different countries. For example, Federal Communications Commission (FCC) regulates radio transmissions for US and Conference of Postal and Telecommunications Administration (CEFT) for Europe. It is not an easy task to come up with a common licensed band for IoT since government auctions are used worldwide to sell spectrum bands to operators. [3] This leads to the first interoperability problem in IoT where devices manufactured for one country may not work in another.

Table 1. Worldwide ISM bands [3]

Region	Bands
America	315/433/915MHz and 2.4GHz
Europe	433/868MHz and 2.4GHz
Africa	433MHz and 2.4GHz
East Asia	315/426/950MHz and 2.4GHz
South Asia	315/433/470/780MHz and 2.4GHz
South East Asia	433/915MHz and 2.4GHz
Australia	433/915MHz and 2.4GHz

Many of the vendors have already started manufacturing IoT products on the ISM bands to work around the license spectrum issues. However, ISM bands are also not the same around the world. Table 1 [3] summarizes the available ISM bands in different parts of the world. From the table, it can be observed that the popular bands for deploying IoT solutions are 433MHz, 868MHz, 915MHz and 2.4GHz.

IOP Conf. Series: Materials Science and Engineering **260** (2017) 012033 doi:10.1088/1757-899X/260/1/012033

3.2. MAC layer

Presence of multiple vendor solutions and low-power communication technologies are the main cause of concern at the MAC layer. Different vendors have their preference on the MAC layer technologies for various reasons such as team capability, cost, ease of use etc. Various technology options at the data link layer include wired LAN, Wi-Fi, Bluetooth, ZigBee, 6LoWPAN, proprietary Sub-1GHz protocols, Power Line Communication (PLC), Fieldbus and many others. Moreover, many of the MAC protocols have already been enhanced to cater for better battery life. For example, Bluetooth has been modified to Bluetooth Low Energy (BLE) and Wi-Fi has been modified to Low-power Wi-Fi to suite the IoT power requirement [4].

Most standardization bodies are spending excessive amount of time trying to address heterogeneity at the MAC layer [4]. However, the implementation of some of these technology spans over different layers of the OSI model and makes it difficult to come to a consensus. Furthermore, standards are simply recommendations and not certification so the vendors have a choice to comply, partially comply or not to comply with the standards.

The research community is exploring cross layer protocols such has 6LoWPAN to unify the MAC and IP layer for communication [4]. However, 6LoWPAN requires the IoT devices to host computation and storage capability and that increase the cost of IoT device.

3.3. Network layer

In order to exchange information with each other, IoT devices need to be connected to the Internet or have a mechanism to push data to the Internet. Hence, they need to support the TCP/IP protocol suite that is too bulky and not cost effective for the IoT device. The Internet Engineering Task Force (IETF) has several recommendations for low powered IoT stacks that will be discussed next.

One of the biggest challenges of using reduced TCP/IP in IoT is Maximum Transmission Unit (MTU) size. IoT devices operate with much smaller MTU of about 127 bytes compared to computers, which typically assumes a minimum MTU of 1500 bytes. IPv6 specifications further complicate the situation because they include two design decisions that cause problems for small-MTU links. Firstly, fixed 40-byte header length adds too much overheads for IoT devices that produce small packets with little data. Secondly, IPv6 specifies the network to support a minimum MTU of 1280 bytes and this is unrealistic for constrained devices. The rationale behind fixed-header length is to improve protocol-processing speed and minimum MTU of 1280 bytes is to avoid in-network fragmentation. The proposed solution to this is to introduce padding and then header compression, however, padding introduces unnecessary overheads and header compression requires the nodes to have high computation and hence, higher cost of manufacturing [4].

Typical IP protocol is designed for Firstly, multi-access link, where multiple nodes are connected to the same access network and secondly, point-to-point links where there are only two nodes on a single link. However, in most IoT deployments, there is a need for mesh network where a collection of layer-2 links is joined without a presence of a layer-3 routing device. This poses a problem for legacy protocols such as Address Resolution Protocol (ARP), Dynamic Host Configuration Protocol (DHCP), Neighbor Discovery and many routing protocols as they depend on link scoped multicast, which does not work over mesh networks. Researchers are developing new intra-subnet routing protocol but these protocols are energy hungry and affect the lifetime of the IoT devices [4].

Multicast protocol is at the core of IP protocol. However, multicasting in IoT network requires the constrained IoT devices to forward packets for other nodes, thus reducing the lifespan of the battery. Furthermore, many IoT devices are designed to enter into sleeping state after transmission and are unable to process multicast packets. The solution to this problem is to redesign the IP protocol to operate only on unicast. However, devices that operate on full-IP protocol will note be able to communicate on unicast based IP protocol [4].

Traditional mesh routing protocols also need to be modified for use in IoT networks because they are designed to flood the network while IoT networks don't perform well under those scenarios. There

IOP Conf. Series: Materials Science and Engineering **260** (2017) 012033 doi:10.1088/1757-899X/260/1/012033

are already some proposals at IETF 802.15.5 standards to support link-layer routing for mesh networks [4].

The current IP mobility protocols are also not suitable for IoT use cases because firstly, they impose too many message exchanges for establishing mobility that affects the battery life of the IoT devices and secondly, and IoT device may not have the full mobile IP stack deployed on them as they are memory and CPU constrained [6]. The research community is considering many variations of the network-based mobility schemes. One such scheme is Sensor Proxy Mobile IPv6 (SPMIPv6) which introduces few components such as Sensor Local Mobility Anchor (SLMA) and Sensor Mobility Access Gateway (SMAG). SLMA could be a high-end computer as it resides in traditional network while SMAG could be one of the IoT nodes that have higher resource compared to the other nodes in the Network [7,8]. Inter-MARIO and LoWMob are yet other protocols for mobility management in IoT. They are designed with the make-before-break concept in mind where several static nodes in the IoT network known as partner nodes serve as access points for the MN and preconfigure future partner nodes before the MN moves to those networks [7]. However, To-date mobility in IoT is still a research focus area as the existing schemes are a mere adoption of existing mobility concepts in IPv4 and IPv6 which require the nodes that are involved in the mobility process to have high computational and power resources.

3.4. Transport layer

Transport layer in the Internet provides congestion control, guaranteed and in-order delivery of packets. These provisions are effectively used by the TCP protocol. However, TCP is not suited for IoT applications as IoT devices offer a varying traffic pattern due to their limitations. For example, firstly, due to the energy conservation requirements, IoT devices usually go into sleep mode after transmission or receiving the packet, making it difficult to maintain the communication and acknowledgement channels. Secondly, IoT devices send/receive only small amounts of data and TCP handshakes impose too much overheads for the devices. Thirdly, IoT applications have low latency requirements and TCP has too much delay in setting up communication. Finally, IoT applications operate in lossy domain, so in-order delivery and retransmission would block the communication until all previous packets have been successfully delivered. [4]

Some of the IoT protocols like ZigBee still maintain support for TCP, while the rest of the protocols like BACnet and CoAP have proposed to build TCP like transport functionality in the application layer and use UDP as the transport protocol.

4. Interoperability challenges in IoT

Heterogeneity in IoT solutions leads to Interoperability problems between IoT systems. Without interoperability, IoT deployments will be isolate to their use case and the data generated by one system can be consumed by other systems. Interoperability issues happen due to multi-vendor solutions and presence of legacy sensor network based solutions that have been deployed before the concepts of IoT was introduced.

4.1. Multi-Vendor problem

Standardization bodies only offer best practices for IoT. These recommendations only serve as guidelines for bodies that offer certification for IoT products. The certification bodes often adopt just a portion of the standards. Four well known bodies that manage certification programs today to ensure heterogeneous interoperability between IoT devices are Wi-Fi Alliance for wireless LAN technology, Bluetooth Special Interest Group (SIG) for Bluetooth enabled devices, ZigBee Alliance for ZigBee compatible devices and LoRa alliance for wide area IoT devices [3].

However, the problem lies in the fact that these certifications drive up the cost of IoT deployment and many vendors choose not to conform to the standards so that they can keep the cost down and offer value added services that can give them edge over their competitors.

4.2. Legacy solutions

Many vendors and solution providers have been offering closed loop vertical solutions like building management, home automation, vehicle tracking, personnel tracking, etc. The problem with these solutions is that they were never designed to connect to the Internet so their data is contained within their domain. The challenge faced today by IoT because of these solutions is that firstly, it is too expensive to upgrade these systems to take advantage of IoT and secondly, the data models deployed in these systems are based on old database technology like relational databases, which may not conform to the concepts of non-relational Big Databases. This gives opportunity to the research community to develop IoT middleware that can translate between legacy systems and newer IoT system. The middleware employs the concept of protocol translation and offers node resource discovery so that IoT systems can take advantage of the legacy systems [8].

5. Management challenges in IoT

Network management is the biggest challenge in large-scale network deployments. In IoT networks, this is an even bigger problem due to the sheer number of IoT devices. Traditional management protocols for remote control, monitoring and maintenance such as Simple Network Management Protocol (SNMP) will not work with IoT devices because they are too demanding in terms of computation, storage and networking resources. Moreover, they need the IoT devices to support the IP protocol stack [8,10].

Table 2. Management issues

Management issue	Description
Configuration Management	• How to setup IoT device one at a time and in batches • IoT device ownership • IoT device to single and multiple application relationship • Network connectivity • Asynchronous Transaction support • Network re-configurability
IoT Device Control	• Turning IoT device on and off • Disconnecting IoT device from network • Waking up and sending IoT devices to sleep
Monitoring	• Determining the status of IoT devices. To know if they are running, listening, down, sleeping etc. • Network status monitoring • Network topology discovery • Notification • IoT device logs

IoT Device maintenance	• Detecting network failure • Detecting Device failure • Over the air software update • Patch updates • Protocol version update
IoT Device Performance	• Monitoring network performance • Monitoring IoT device QoS
Security and Privacy	• Authorization of IoT devices • Authentication of IoT devices • Access control of IoT devices • Security bootstrapping mechanisms • IoT device ownership • IoT device to application authorization
IoT Device Energy Management	• Management of energy resource • Energy level management • Estimated lifespan before battery change

Table 2 summaries all the management issues in IoT that are being considered by researchers in this field. Each area in the first column is further described in the second column. Only when all of these issues are resolved, IoT would become manageable and would require lesser human intervention [11,12,13].

6. Concepts of NFV

NFV has been popularized by the telecommunication companies (Telcos), who have been suffering from the interoperability problems in their core networks. In 2012, more than 20 worlds' largest Telcos formed the Industry Specification Group (ISG) within the European Telecommunications Standard (ETSI) device a mechanism to virtualize their propriety core network hardware. It was aimed at addressing the operational challenges, high costs of managing closed and proprietary appliances and heterogeneity of device.

Heterogeneity is telco networks results in following issues [14]:

- Fixed Configuration: the telco hardware is configured with fixed IP locations that remain unchanged for years resulting in very rigid resource allocation.
- Manual management: configuration and management of telco equipment requires movement of physical staff and can be done one at a time. Hence, it's difficult to implement a common centralized policy.
- Rapid growth of IP end points: Telcos offer Internet access to large number of users and this is expanding as the consumer base grows. Hence its difficult if no central provisioning is established.
- Network endpoint mobility: networks are fixed and too rigid to move around so requirements for mobility takes long time to reconfigure the network for different scenarios.
- Elasticity: there is no mechanism to upgrade and downgrade a physical hardware based on demand so the networks have to be over provisioned.
- Multi-tenancy: usually one end-point equipment is configured for a single tenant and dynamic or multiple tenant is impossible on vendor define hardware as vendors always wants Telcos to buy more hardware.

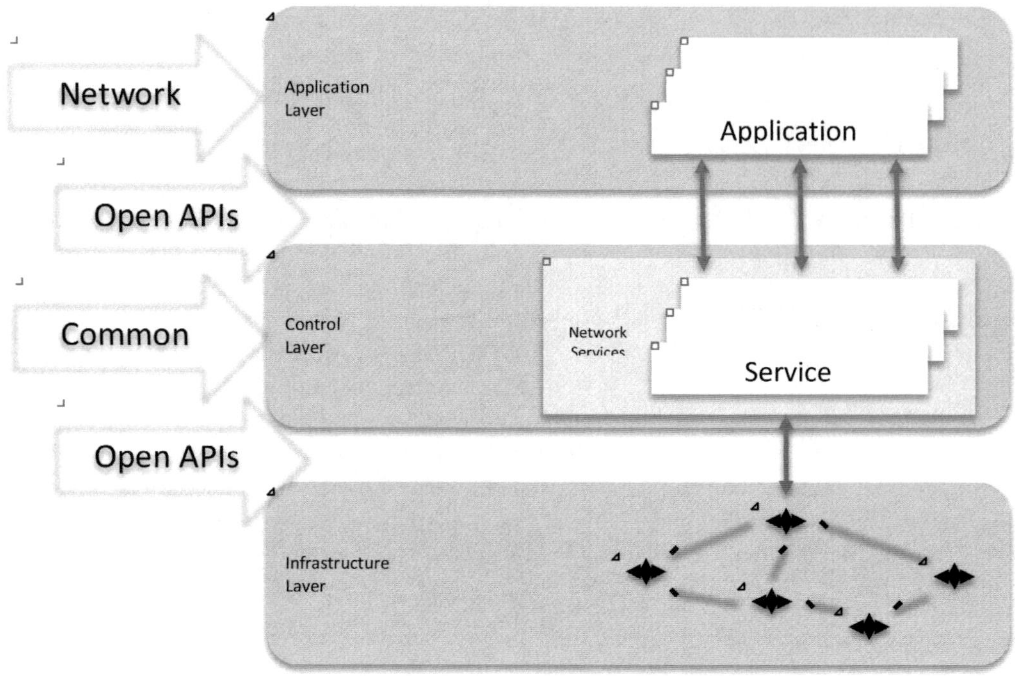

Figure 2. Typical NFV Architecture

In order to solve all these interoperability and management issues, ETSI has proposed a new architecture shown in figure 2 [15]. In this new architecture, Telco equipment (infrastructure layer) has been changed to dumb switches with only packet forwarding capabilities. Control layer has been separated from the hardware in order to enable centralized policy and dynamic resource allocation based on the need. Finally, application layer has been moved to the cloud to have virtually unlimited resources. This application layer could function as a firewall, DNS, load balance etc. and then communicates via software APIs to the control and infrastructure layer. Resulting in the entire core network to operate as a software that is independent of the limitations of the hardware and vendor specific protocols [16].

IoT today faces similar problems as Telcos did with their core network equipment and the NFV route seems like the best option for them. We have been inspired by this approach to solve the problem cause by heterogeneous devices in IoT. Our proposed architecture for virtualizing IoT devices is discussed in the next section.

7. Proposed IoT Architecture

Our proposed IoT architecture leverages on the existing architecture and the lessons learned from the NFV architecture. The new architecture is shown in figure 3 where, the IoT device could simply be a dumb device with purely a communication interface coupled with sensing or actuating capabilities. They could either directly connect to the IoT backend of connect with the help of an IoT Gateway. The IoT gateways would also be lightweight because they only need to maintain a device registry to know the IoT devices connected to it. The gateways also need to host two communication stacks for connecting to the IoT devices and the IoT backend.

The real change has to be made on the IoT backend that has to host more capabilities than the traditional IoT backend. The IoT backend would reside on the cloud to take advantage of the virtually unlimited resources available in the cloud infrastructure. The IoT backed would need to allow device virtualization where each dumb IoT device would be represented by a virtual IoT device. The virtual

IoT device will emulate the physical device capabilities but would be represented as a software code in the IoT backed. The important parameters of an IoT device that need to be emulated are as follows:

- Device ID: a unique device ID to identify it on the IoT backend. This is could follow the standard MAC address or an incremental email address format.
- IoT Device IP (Optional): this is the IP address of the IoT device in its current network.
- Gateway ID (Optional): a unique gateway ID to identify it on the IoT backend. This is could follow the standard MAC address or an incremental email address format.
- Gateway IP (Optional): this is the IP address of the gateway in its current network.
- CPU Utilization: the CPU utilization of the IoT device as a data stream to the IoT backend.
- Memory Utilization: the memory utilization of the IoT device as a data stream to the IoT backend.
- Battery Level: the battery utilization of the IoT device as a data stream to the IoT backend.
- Data Stream: type of data that the IoT device is generating.
- Service Stream: types of services offered by the IoT device.
- Status Stream: current status of the IoT device that could be online, offline, asleep etc.
- Location (optional): longitude and latitude of the IoT device.

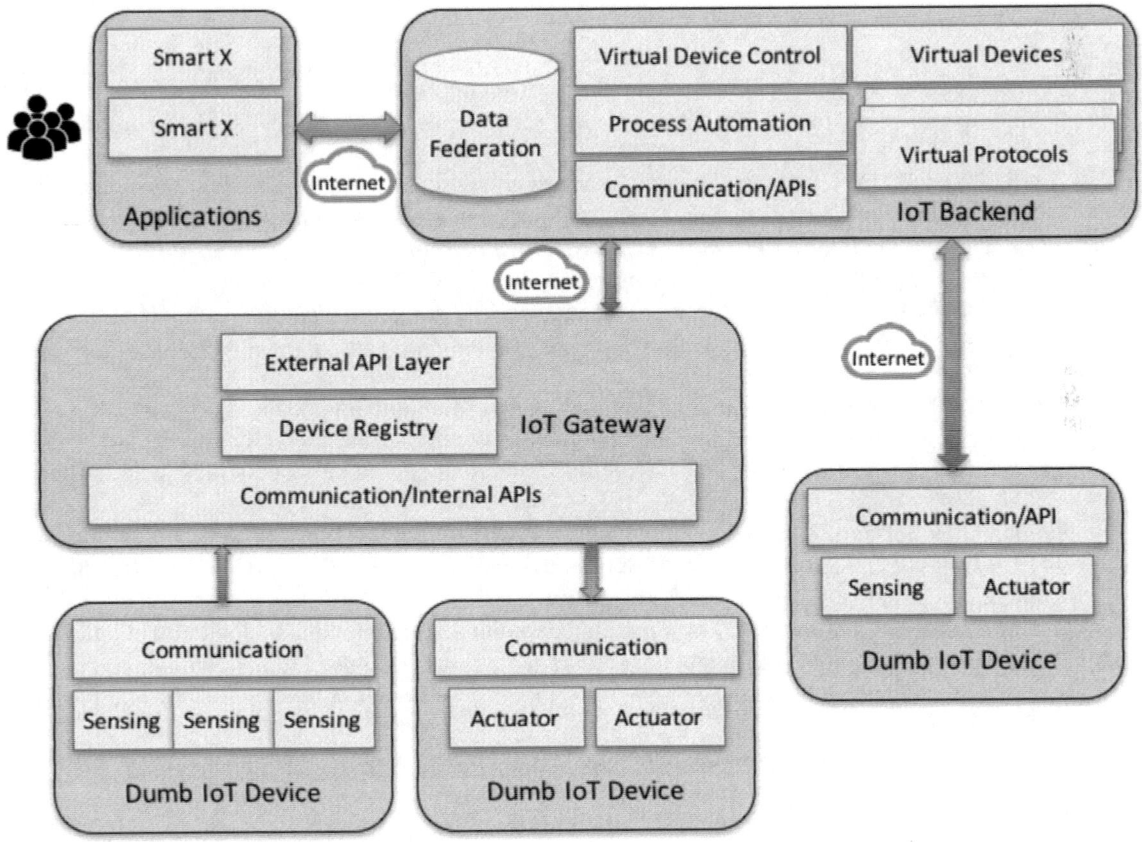

Figure 3. Proposed IoT Architecture

With the help of these parameters, a virtual IoT device can be created in the IoT backend. The IoT backend would need the status updates from the IoT devices either periodically or in real-time to maintain the device.

Once a device has been virtualized, it is no longer constrained to the physical limitations and vendor specific restrictions. The IoT backend can group the virtual devices into virtual networks by assigning IP protocols suite to the virtual devices. An IP capable virtual device can than take advantage of more complex protocols like Domain Name Service based Service Discovery, Constrained Application Protocol (COAP) for application layer protocol and Reachability Protocol (REAP) for device status discover in order to achieve IoT device interoperability. Simple Network Management Protocol Version 3 (SNMPv3) can also be used to manage large number for virtual IoT devices.

8. Conclusion

Internet designed for unconstrained devices with large amount of computation power. It was never designed for IoT, however the demand for IoT has forced the research community to retrofit the existing protocols into IoT and develop the IoT architecture. The existing IoT architecture suffers from connectivity, interoperability, manageability challenges due to the heterogeneous nature of the devices that are controlled by vendors. The Telcos faced similar heterogeneity problem in their core networks and are resorting to NFV for solving their problems. In this paper, we proposed a new architecture for IoT that has been inspired by concepts of NFV in order to tackle the problems in IoT. Future work for this architecture is to simulate the operations of the architecture and demonstrate it with a testbed.

References

[1] 811106755049Elkhodr M, Shahrestani S and Cheung H 2016 The Internet of Things : New Interoperability, Management and Security Challenges *Int. J. Netw. Secur. Its Appl.* 8 85–102

[2] Sutaria R and Govindachari R 2013 Making sense of interoperability: Protocols and Standardization initiatives in IOT *2nd International Workshop on Computing and Networking for Internet of Things (CoMNet-IoT) held in conjunction with 14th International Conference on Distributed Computing and Networking (ICDCN 2013)* pp 2–5

[3] Reiter G 2014 *Wireless connectivity for the Internet of Things. Europe, 433, 868MHz*

[4] Shang W, Yu Y, Zhang L and Droms R 2016 *Challenges in IoT Networking via TCP/IP Architecture* vol 8

[5] 5' Constant N, Borthakur D, Abtahi M, Dubey H and Mankodiya K 2017 Fog-Assisted wIoT: A Smart Fog Gateway for End-to-End Analytics in Wearable Internet of Things

[6] Silva R, Silva J and Boavida F 2014 Mobility in wireless sensor networks–Survey and proposal *Comput. Commun.*

[7] Sinniah G R, Suryady Z, Sarwar U and Hoey T K 2013 Node Mobility Scheme for IP and Non-IP Wireless Personal Area Network Nodes using 6LoWPAN *SENSORCOMM 2013 Seventh Int. Conf. Sens. Technol. Appl.* 83–9

[8] Richards T and Srivastav V 2014 Enabling communication of non-IP device in an IP-based infrastructure *US Pat. 8,837,485*

[9] Silva R, Silva J and Boavida F 2014 Mobility in wireless sensor networks–Survey and proposal *Comput. Commun.*

[10] Sheng Z, Yang S, Yu Y and Vasilakos A 2013 A survey on the ietf protocol suite for the internet of things: Standards, challenges, and opportunities *IEEE Wirel.*

[11] Marotta M A, Both C B, Rochol J, Granville L Z and Tarouco L M R 2015 Evaluating management architectures for Internet of Things devices *IFIP Wirel. Days* 2015–Janua

[12] Vasilomanolakis E, Daubert J, Luthra M, Gazis V, Wiesmaier A and Kikiras P 2016 On the Security and Privacy of Internet of Things Architectures and Systems *Proceedings - 2015 International Workshop on Secure Internet of Things, SIoT 2015* pp 49–57

[13] Sutaria R and Govindachari R 2013 Making sense of interoperability: Protocols and Standardization initiatives in IOT *2nd International Workshop on Computing and Networking for Internet of Things (CoMNet-IoT) held in conjunction with 14th International*

Conference on Distributed Computing and Networking (ICDCN 2013) pp 2–5
[14] Bernardos C J, Rahman A, Zuniga J C, Contreras L M and Aranda P 2016 *Network Virtualization Research Challenges (Research Report No.03)*
[15] Malcolm B, Nigel D, Rob D, Paul D, Steve F, Dave H, Mandar J and Kam L 2014 *SDN Architecture*
[16] Shin M K, Nam K, Pack S, Lee S and Krishnan R 2016 *Verification of NFV Services : Problem Statement and Challenges (Research Report No.02)*

IOP Conf. Series: Materials Science and Engineering **260** (2017) 012034 doi:10.1088/1757-899X/260/1/012034

Determining the Time of Flight and Speed of Sound on Different types of Edible Oil

[1]N. A. Azman and [2]S. B. Abd Hamid.

Department of Mechatronics Engineering, International Islamic University Malaysia, Jalan Gombak, 53100 Kuala Lumpur, Selangor, Malaysia.
[1]ajeera_won@yahoo.com and [2]syamsul_bahrin@iium.edu.my

Abstract. Edible oil is most often plant-based oils that have been extracted from various seeds. There are cases where the fully virgin edible oil was found to be a fraud. The adulterated edible oil indicates the intentional, fraudulent addition of extraneous, improper or cheaper ingredients puts into the oil or the dilution or removal of some valuable ingredient of the oil in order to increase profits. Hence, decrease the reliability of the Malaysian food product quality. This research was done by using the method of time of flight obtained using the Texas Instrument board, TDC1000-TDC7200 EVM connected to an ultrasonic transducer with 1 MHz frequency. The authors measured the time of flight and temperatures controlled from 20^0C to 40^0C of five vegetable oils (olive oil, sunflower oil, corn oil, coconut oil, and mustard oil). The value is observed and compared with other research from the literature review. From the study, time of flight values decreases exponentially while speed of sound value increases. This relationship will be useful in spectrum unfolding method to investigate the adulteration in different type of edible oil.This research outcome is to investigate the quality value of the different type of edible oil while eliminates the issues where the quality of Malaysian food product is not reliable.

1. Introduction

Malaysian product has been exported worldwide centuries ago and have received much admiration and respect from importing countries due to its high quality and reliability. However, the economy is growing faster and there must be many ways to increase the value and trust of Malaysian Product. One of the ways is by reducing the fraud from the processed food. While, processed food means foods that require little processing or production and packaged to stay fresh and save time for consumers. A few examples of processed foods include canned and frozen fruits and vegetables, packaged foods labelled "natural" or "organic," such as cereals, fresh meat and poultry, and jarred baby foods. Also, the foods with health and nutrition claims on the label, such as "may reduce risk of heart disease," "low in fat" or "high in calcium" and foods fortified with nutrients such as fiber, vitamin D and omega-3 fatty acids [1].

Hence, one way to reduce the fraud in pure liquid material is by being able to determine the type of liquid before been sent to consumers. When compared to chromatographic techniques, infra-red spectroscopy is an attractive option for detection of adulterations in oils and fats due to the speed of analysis, and minimal sample preparation [2]-[3]-[8]. Instrumental techniques based on chromatography and spectroscopy have received particular attention from researchers worldwide since they are efficient and able to control adulteration practices in food lipids [3].

IOP Conf. Series: Materials Science and Engineering **260** (2017) 012034 doi:10.1088/1757-899X/260/1/012034

Ultrasonic Spectroscopy is the characterization of the ultrasonic response of a material to the propagation of a low energy ultrasonic wave which can measure the absolute frequency dependent and attenuation of liquids. The ultrasonic spectroscopy characteristics which is Non-destructive, Non-Intrusive and in-situ is one major contribution [4]. The method of ultrasound spectroscopy is suitable for determining the properties of liquids.

A standard development organization that serves as an open forum for the development of international standard control value called ASTM International (American Society for Testing and Materials) is useful for the characterization of materials depends on the materials speed of sound, density, velocity, melting points and boiling points. Differential scanning calorimetry technique for detecting lard adulteration in butter required sample preparations kept in glass vials under refrigerated condition with temperature control and tends to be laborious since the sample need to undergo the blend process according to percentage of lard in buffer [4].

The frequency dependence of liquids is proved where the frequency used in the related research controlled from 1MHz, 2MHz, 3MHz, and 5MHz to study the different acoustic impedance value in different liquid. The result showed for the acoustic impedance is dependent on the frequency of the ultrasound [5]. Attenuation occurs with any type of signal, whether digital or analog. Attenuation is a natural consequence of signal transmission over long distances [6].

Determination of the real time density to monitor the ultrasound velocity are further classified techniques into Multiple Reflection Method (MRM), Reference-Reflection Method (RRM), Transmission Reflection Method (TRM), and Angular Reflection Method (ARM). ARM and RRM are proven suitable for highly sound absorbing liquids, however requires calibration measurements [7]. Sound absorption refers to the process by which a material, structure, or object takes in sound energy when sound waves are encountered, as opposed to reflecting the energy [6].

The changes in the quality of liquid are premeditated from the changes in the parameters such as viscosity, density, ultrasonic velocity, acoustic impedance, and intermolecular free length [9]-[10]. Each oil type have their own physical properties divided in density, viscosity, and saturated acid. Pressure-viscosity coefficient increases exponentially with the adiabatic bulk modulus [11]. The relationship demonstrates that the pressure-viscosity coefficient can be predicted from the adiabatic bulk modulus [11]. Based on the relationship on the adiabatic bulk modulus, the value for the speed of sound for different type of oil can be calculated.

2. Experimental

2.1 Experimental Set Up

The TDC1000-TDC7200EVM evaluation module is an ultrasonic for Time-of-Flight (TOF) applications operating with ultrasound frequencies up to 4MHz. The TDC1000-TDC7200EVM is then connected to the computer through USB. For the research, the 1MHz transducer is used.

2.2 Preparation of sample

Olive oil, Mustard oil, coconut oil, Sunflower oil and Corn oil were tested. All samples were obtained from commercial sources and used as supplied without further purification. The list of tested oils and their physical properties are given in Table 1.

Table 1. Physical Properties of vegetables oil

Oil name	Density ρ[g/cm3] 15 ^{0}C	Viscosity ν[mm2/s] 40 ^{0}C
Olive Oil	0.917	38.7
Mustard Oil	0.920	44.1
Coconut Oil	0.926	26
Sunflower Oil	0.919	35
Corn Oil	0.942	31

IOP Conf. Series: Materials Science and Engineering **260** (2017) 012034 doi:10.1088/1757-899X/260/1/012034

2.3 Time of flight measurement

Time of flight was measured by the TDC1000-TDC7200EVM board. The board is connected to an ultrasonic transducer. The 1MHz transducer is attached to a beaker. The board combines with the ultrasonic transducer and measures the ultrasound wave spread in a sample liquid. The basic principle of this device is the ultrasonic wave pulse echo launched from the transmission transducer propagates in the sample and its echo is received to the same transducer. It is converted as start and stop pulse by the TDC1000-TDC7200EVM board, while the distance from the start to the stop pulse is known as time of flight. Time of flight is the duration it takes for the ultrasound to travel through the media and return back to the receiver.

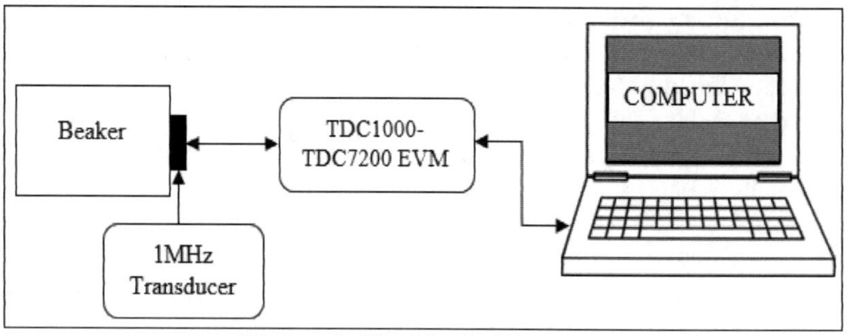

Figure 1. The block diagram.

3. Results and Discussion

Time of flight (TOF) describes a variety of methods that measure the time that it takes for an object, particle or acoustic, electromagnetic or other wave to travel a distance through a medium. This measurement can be used for a time standard, as a way to measure velocity or path length through a given medium, or as a way to learn about the particle or medium such as composition or flow rate. The advantages of using the techniques is the traveling object may be detected directly. The changes in the composition of the liquid can be detected directly when there is absolute changes in the time of flight. For the research, the value for time of flight proves the changes in speed of sound values which change proportionally with the temperatures since the molecular bonding is loosened due to the increase in temperature. Refer to table 2 showed below, the value of time of flight is increasing when the temperature is increasing. Based on the formulae calculating the speed of sound, where C = Speed of sound of a medium, D = distance.

$$C = \frac{2D}{Time\ of\ flight} \quad (1)$$

The speed of sound is inversely proportional to the value of time of flight on a different medium. Based on the molecular structure of an oil, they are composed of molecules known as triglycerides, which are esters composed of three fatty acid units linked to glycerol. An increase in the percentage of shorter-chain fatty acids and/or unsaturated fatty acids lowers the melting point of an oil.

The different type of edible oil, which are olive oil, mustard oil, corn oil, sunflower oil, and coconut oil was chosen based on their popular availability on the market. It proves that the consumer choses to buy these type of edible oil on their daily cooking basis rather than any other vegetables oil which is hard to come by. Oils recognised as essential nutrients for consumers. They provide the most concentrated source of energy of any foodstuff, supply essential fatty acids and contribute greatly to the cooking satisfaction. The results from this research showed the different types of edible oil is investigated in three different controlled temperature to observed the effect of temperature on the values of speed of sound.

To achieve the different in temperature, the oil is kept in refrigerated condition to obtain 20 °C, while the oil is heated to obtain the 30 °C and 40 °C. Then, the oil is put inside a closed beaker to avoid any disturbance from outside air or any unwanted noise. The time of flight measurement is obtain and the changes in speed of sound is observed. From the result, the time of flight values is increasing and compared to the speed of sound value is decreasing. Obviously, the changes in temperature also affect the value in both the time of flight and speed of sound on different type of edible oil.

Table 2. Comparison the experimental and calculated results for different type of edible oil. T: temperature, U: Time of Flight, S: Experimental speed of sound, K: Referred speed of sound value based on literature.

Type of Edible oil	T (°C)	U (ns)	S (m/s)	K (m/s)	Error (%)
Olive Oil	20	120077	1432	1431	-0.07
	30	121969	1410	1430	1.39
	40	124500	1381	1397	1.14
Mustard Oil	20	118642	1449	1490	2.75
	30	121356	1417	1426	0.63
	40	124205	1384	1362	-1.61
Corn Oil	20	119404	1440	1460	1.37
	30	121636	1414	-	-
	40	123912	1388	-	-
Sunflower Oil	20	121826	1418	1423	0.35
	30	121783	1418	-	-
	40	123784	1389	-	-
Coconut Oil	20	-	-	-	-
	30	125488	1370	1378	0.58
	40	127948	1344	1361	1.25

The transducer was mounted on acrylic beaker to maximize the sensor sensibility and flexibility. The sensor accuracy were determined based on the percentage error. The speed of sound from the literature is compared with the speed of sound obtained from the experiment. The percentage error is in the range of -5% up to +5%. The accuracy of the transducer is reliable and the value obtained can be used for further examining the quality of the different types of edible oils. The transducer is compatible with the TDC1000-TDC7200EVM in determining the fraud in the pure quality edible oil. The difference value in speed of sound and time of flight can be used to determine the type of liquid tested and using the technique spectrum unfolding approach, supposed the quality either the edible oil is pure or fraud can be investigated.

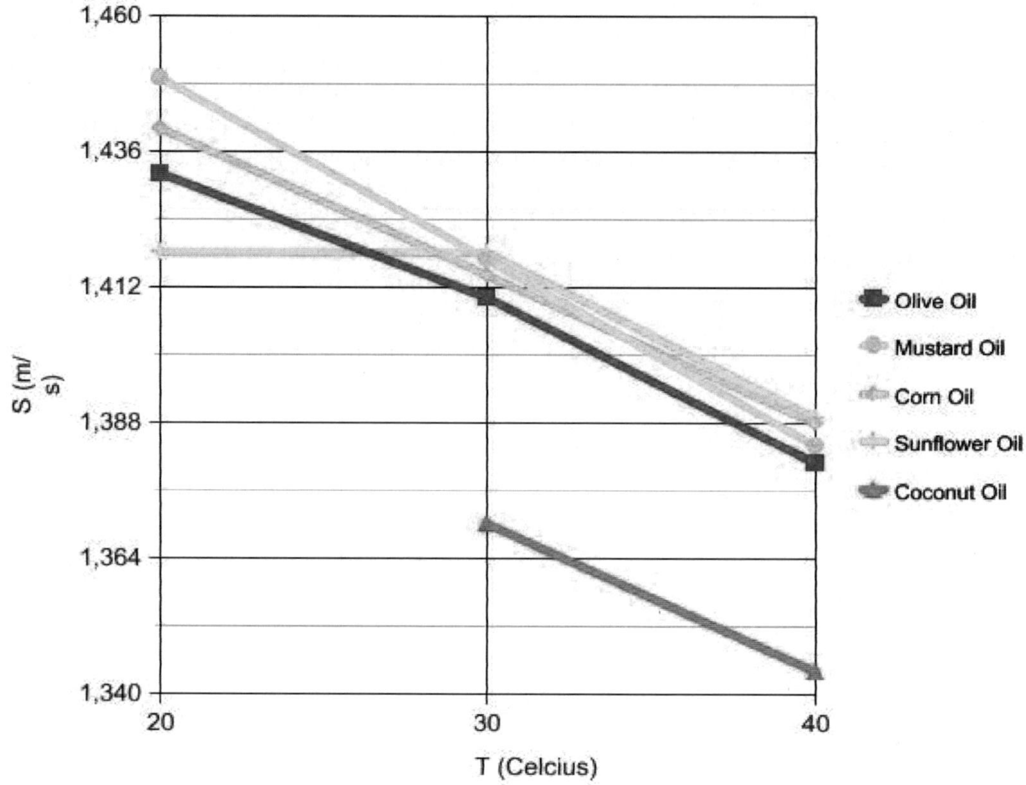

Figure 2. Relationship between the temperature and speed of sound for different type of edible oils.

4. Conclusion

The ultrasonic studies in liquids are of great use in understanding the nature and strength of molecular interactions in oil. This research works on demonstrating the usefulness of the ultrasonic technique for on-line process monitoring the composition of different type of edible oil. The simplicity, portability and low cost of ultrasound devices make them essential elements in research laboratories, pilot plants and processed food industries.

The relationship between the time of flight and the speed of sound is investigated in this study, being compared with corn oil, olive oil, sunflower oil, coconut oil, and mustard oil. The following conclusion can be drawn from the study: From the study, it is found that time of flight values decreases exponentially with the increase of speed of sound. This relationship will be useful in spectrum unfolding method to investigate the adulteration in different type of edible oil. The spectrum unfolding is a new method to unfold the time of flight and the speed of sound to actually gain the critical values of the oil so that the system is able to demonstrate the type of oil and what is inside the combination of few oils mixed together. This will actually help increase the Malaysian Food Qualities.

References

[1] International Food Information Council Foundation. "W*hat is a Processed Food? You Might Be Surprised"*. September, 2010.

[2] J. M. N. Marikkar, M.E.S Mirghani and I. Jaswir. "*Application of Chromatographic and Infra-red Spectroscopic Techniques for Detection of Adulteration in Food Lipid"*. International Institute for Halal Research and Training, International Islamic University Malaysia, Malaysia, 2016.

[3] R. Tweedie. "*Process Ultrasound Spectroscopy for In-Situ Characterization*", Industrial Tomography System. Accessed 12 April, 2016.
[4] Nurrul hidayah, A.F. Arieff, S.R. Rohman, A. Amin, I. Shuhaimi, and M. Khatib. "*Detection of butter adulteration with lard using differential scanning calorimetry*", International Islamic University Malaysia, Malaysia, 2014.
[5] SK Mahammad Ali and Basharath Ali. "*Acoustics Impedance Studies in Some Commonly Used Edible Oils*", IJISET, International Journal of Innovative Science, Engineering & Technology, *India*, 2014.
[6] James A. Zagzebski. "*Essential of Ultrasound Physics*", Mosby Inc, 1996.
[7] S. Hoche, M.A. Hussein, and T. Becker. "*Ultrasound-based density determination via buffer rod technique: a review*", Journal of Sensors and Sensor System, 2, Pages 103-125. 2013.
[8] Rubalya Valantina, S.Chandiramouli, and R.Neelamegam. "*Detection of adulteration in olive oil using rheological and ultrasonic parameters*", International Food Research Journal 20(6): 3197-3202, 2013.
[9] O.O. Fasina and Z. Colley. "*Viscosity and specific heat of vegetables oils as a function of temperature: 35°C TO 180°C*", International Journal of Food Properties, 11: 738–746, Department of Biosystems Engineering, Auburn University, Auburn, AL, USA, 2008.
[10] Formo, M. W. "*Physical properties of fats and fatty acids*." Bailey's industrial oil and fat products 1: 177-232, 1979.
[11] Nobuyoshi Ohno, Hirohito Mori, Toshifumi Mawatari, BoZhang, Bunji Ono, and Sobahan Mia. "*Prediction of Pressure-Viscosity Coefficient of Environmentaly Friendly Vegetable Oils from Adiabatic Bulk Modulus Based on Sound Velocity under Atmospheric Pressure*", Department of Mechanical Engineering, Saga University, Saga, Japan. 2014.

Acknowledgments
This research is supported by the Fundamental Research Grant Scheme (FRGS) by Ministry of Higher Education Malaysia (MOHE) to the Department of Mechatronics Engineering at the International Islamic University Malaysia (FRGS16-065-0564)

IOP Conf. Series: Materials Science and Engineering **260** (2017) 012035 doi:10.1088/1757-899X/260/1/012035

Closed-loop model identification of cooperative manipulators holding deformable objects

A A Alkathiri[1], R Akmeliawati[2] and N Z Azlan[3]

[1,2,3] Department of Mechatronics Engineering, International Islamic University Malaysia, Kuala Lumpur, Malaysia

E-mail: rakmelia@iium.edu.my

Abstract. This paper presents system identification to obtain the closed-loop models of a couple of cooperative manipulators in a system, which function to hold deformable objects. The system works using the master-slave principle. In other words, one of the manipulators is position-controlled through encoder feedback, while a force sensor gives feedback to the other force-controlled manipulator. Using the closed-loop input and output data, the closed-loop models, which are useful for model-based control design, are estimated. The criteria for model validation are a 95% fit between the measured and simulated output of the estimated models and residual analysis. The results show that for both position and force control respectively, the fits are 95.73% and 95.88%.

1. Introduction

System Identification is a methodology for constructing mathematical models of a system from its measured inputs and outputs. This method uses grey box or black box models on the model spectrum, which are less reliant on first principles like white box models [1]. System Identification has been used in many applications, and is mainly for obtaining the mathematical model of a system of interest for control design purpose. There are many different techniques in system identification, however, in general, one can classify them into two main approaches based on the domain; time-domain approaches, which use the time series data of the input-output of the system and frequency-domain approaches, which use the frequency response of the system and/or the corresponding spectral plots.

The typical procedure in system identification involves determination of model structure, parameter estimation and validation. In such case, statistical approach or artificial intelligence techniques can be used for obtaining the mathematical model based on the experimental data, such as ARX, ARMAX, artificial neural networks (ANN), Box-Jenkins, and many other techniques.

In terms of applications, system identification has been used widely. In this paper, as the system of interest has some similarity to link manipulators, one can find many applications of system identification, for examples, in modelling of single-link and two-link flexible manipulators in [2]-[5], the actuation subsystems of a heavy-duty electrohydraulic harvester manipulator in [6] and of a two-link pneumatic artificial muscle manipulator using genetic algorithm in [7].

In this paper, the system of interest is a pair of cooperative manipulators used to hold a deformable object despite the unknown stiffness and position of the deformable object without causing permanent visual deformation. Deformable objects are objects whose shape changes due to an applied external force, for examples, paper cups, tomatoes, and a bag of sand. The difficulty of manipulating deformable object is mainly due to the its nonlinear elasticity, friction and parameter variations [8]. Ability to hold deformable objects is important in robotics technology. The applications can be found in in industrial fields such as the food industry and recycling industry [9], as well as the medical field [10].

This paper presents closed-loop system identification of the cooperative manipulators for holding deformable objects that is proposed in [11]. The system consists of a master and slave system. The paper is organized as follows. Section 2 presents the methodology used including the description of the designed system. Section 3 presents the mathematical models of the master and slave manipulators as the results of the system identification described in Section 2. Finally, Section 4 concludes the paper.

2. Methodology

The system identification is performed on a closed-loop system, i.e. the manipulators together with the controllers, in order to obtain the input-output data through experiments. Tomatoes were used as the deformable objects. The mathematical models will be presented in form of transfer functions for the master and slave systems as will be explained in more details in the end of this section.

2.1. System Description

Figure 1 shows the prototype of the system. It is made of two 1-degree of freedom (DOF) cooperative manipulators; each is powered by a DC motor. One of the manipulators ('master') is position-controlled with encoder feedback where the input is the desired position and the output is the actual position, while the other ('slave') is force-controlled which tracks the desired force and uses a force sensor as the feedback [11].

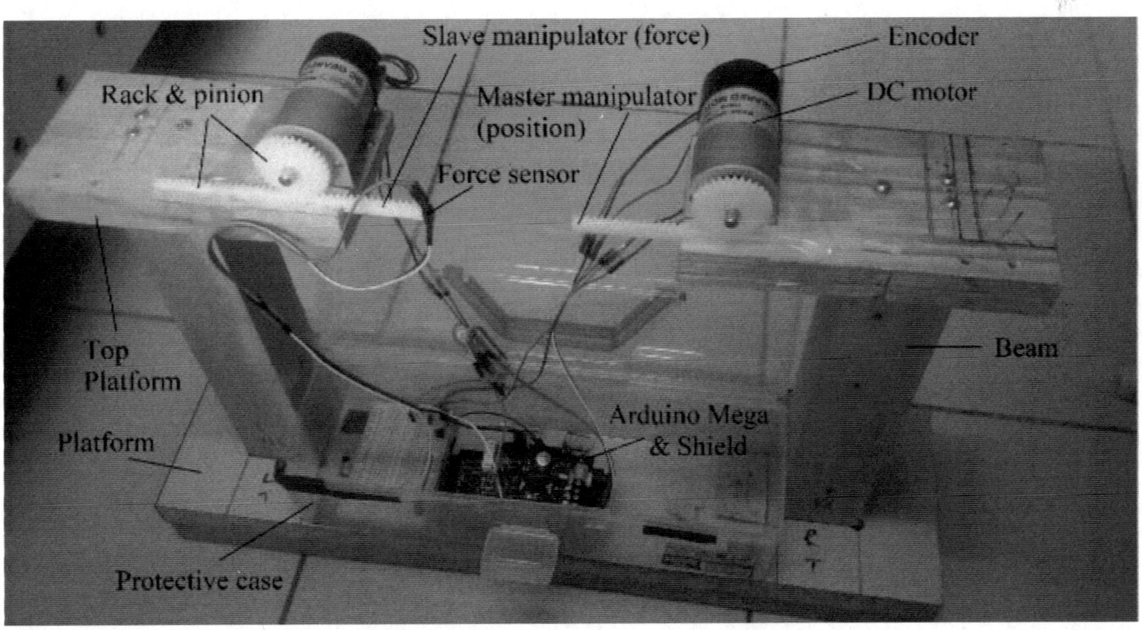

Figure 1. Manipulators prototype [11].

Proportional-Integral-Derivative (PID) controllers have been designed for the purpose of holding a deformable object in place by bringing the master manipulator to the desired position and exerting desirable amount of force at the slave manipulator. Thus, two sets of input and output data can be acquired. Figure 2 shows the block diagram of the system.

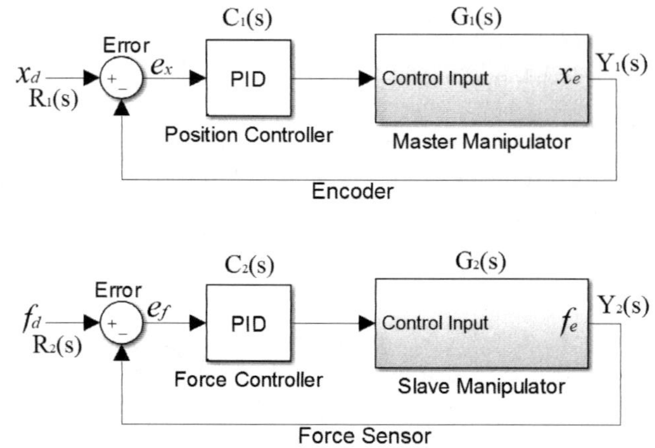

Figure 2. Block diagram of the system [11].

In [11], PID controllers are tuned to acquire the desired output as shown in Figure 3 and Figure 4, and while the controllers' parameters are known, the models of the master and slave manipulators are not mathematically identified. The PID gains for each controller are as follows. For the master (position) manipulator, $Kp = 3100$, $Ki = 20$ and $Kd = 150$, and for the slave (force), $Kp = 800$, $Ki = 200$ and $Kd = 500$. The tuning was performed through trial-and-error with the initial gains were obtained using the Ziegler-Nichols method.

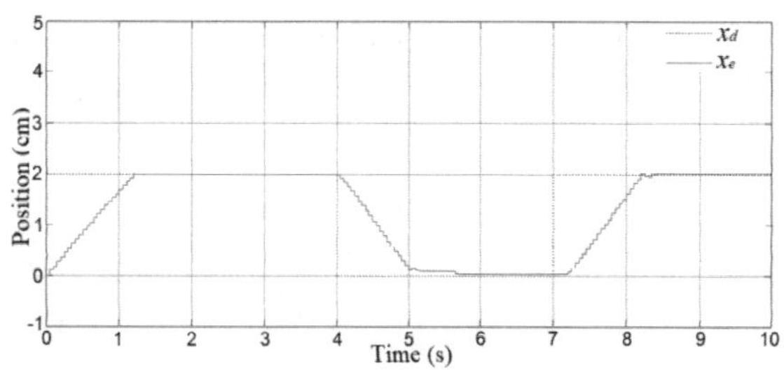

Figure 3. Position tracking of the master manipulation, where x_d and x_e are desired and actual positions, respectively [11].

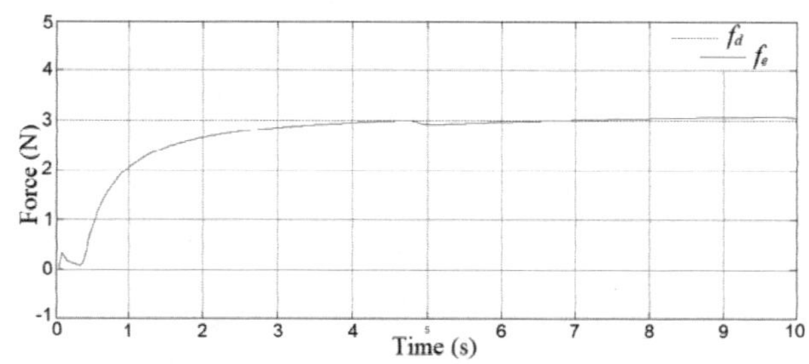

Figure 4. Force tracking of the slave manipulator, holding a tomato, where f_d and f_e are desired and actual forces, respectively [11].

2.2. Structure of Mathematical Models

This section discusses the models and parameters of the closed-loop transfer functions of the manipulators.

After acquiring data from the experiments, the model set is chosen (which in this case is transfer functions). Then, the criterion of fit is chosen; in this paper, a model with 95% fit against the validation data is considered a "good" model. Residual analysis test is also carried out to ascertain that the cross correlation function between the residuals and input is mainly within the confidence region, which also translates to the "goodness" of the model [12]. This process is repeated until the optimal tradeoff between model fidelity and model complexity is achieved, i.e. the least complex model that achieves at least a 95% fit. Performing identification in closed loop is useful for model-based control design [13].

MATLAB'S System Identification Toolbox [14], which provides convenient GUI menu, is used in the system identification process.

For the master manipulator, a sixth order transfer function with six poles and five zeros is chosen (equation (1)). The order was chosen by trial and error while observing the output response of the system.

$$\frac{Y_1(s)}{R_1(s)} = \frac{b_5s^5 + b_4s^4 + b_3s^3 + b_2s^2 + b_1s^1 + b_o}{a_6s^6 + a_5s^5 + a_4s^4 + a_3s^3 + a_2s^2 + a_1s^1 + a_0} \quad (1)$$

where $R_1(s)$ and $Y_1(s)$ are the input and output of the system (desired and actual position), respectively, a_6 through a_0 are the denominator coefficients, and b_5 through b_0 are the numerator coefficients.

Similarly, for the slave manipulator, a fourth order transfer function with four poles and three zeros is selected as in equation (2).

$$\frac{Y_2(s)}{R_2(s)} = \frac{d_3s^3 + d_2s^2 + d_1s^1 + d_o}{c_4s^4 + c_3s^3 + c_2s^2 + c_1s^1 + c_0} \quad (2)$$

where $R_2(s)$ and $Y_2(s)$ are the input and output of the system (desired and actual force), respectively, c_4 through c_0 are the denominator coefficients, and d_3 through d_0 are the numerator coefficients.

3. Results and Analysis

In this section, the validation of the closed-loop transfer function models obtained from model estimation is discussed.

Figure 5 shows the comparison between the measured output of the master manipulator and the simulated output of the model in (1). The fit between the measured and simulated output is 95.73%, which is above 95%, and therefore is considered satisfactory. The coefficients of (1) have been obtained from the estimation. Therefore, (1) now becomes

$$\frac{Y_1(s)}{R_1(s)} = \frac{0.2369s^5 + 6.302s^4 + 9.741s^3 + 264.3s^2 + 57.06s^1 + 942.4}{s^6 + 5.395s^5 + 56.18s^4 + 169.3s^3 + 471.8s^2 + 634.7s^1 + 957.5} \quad (3)$$

Likewise, Figure 6 compares the measured output of the slave manipulator to the model simulated output and the fit between them is 95.88%, well above 95%. From the estimation, the coefficients of (2) have been obtained. Therefore, (2) now becomes

$$\frac{Y_2(s)}{R_2(s)} = \frac{2.192s^3 - 17.06s^2 + 116.6s^1 + 33.33}{s^4 + 9.501s^3 + 103.1s^2 + 172.6s^1 + 31.91} \quad (4)$$

The higher order of the master manipulator model can be attributed to the complexity of its feedback mechanism that requires more pre-processing filters, which is the rotary encoder as opposed to the force sensor of the slave manipulator. Consequently, the master manipulator model will be more complex and burdensome for running simulations.

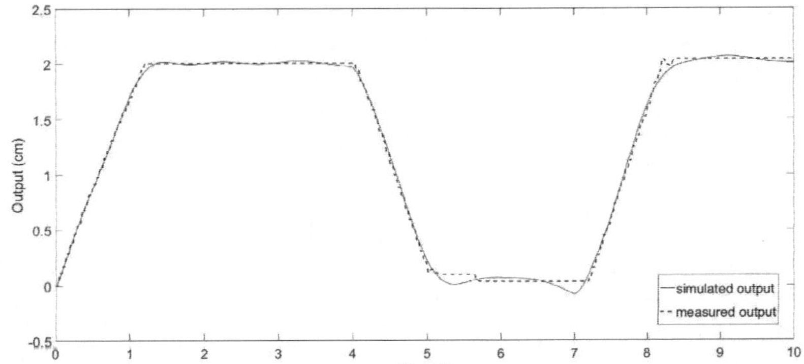

Figure 5. Measured and simulated model output of the master manipulator.

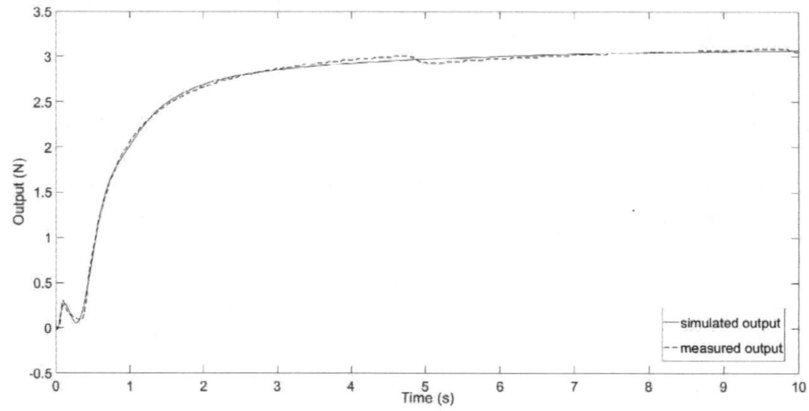

Figure 6. Measured and simulated model output of the slave manipulator.

Figures 7 and 8 show the cross-correlation between the input and the residuals for each input-output pair for the master and slave manipulators, respectively. In both cases the cross correlation function is between the confidence region (dashed lines). This shows that the residuals of the models are not correlated with and are independent from past inputs. Therefore, validating the "goodness" of the models.

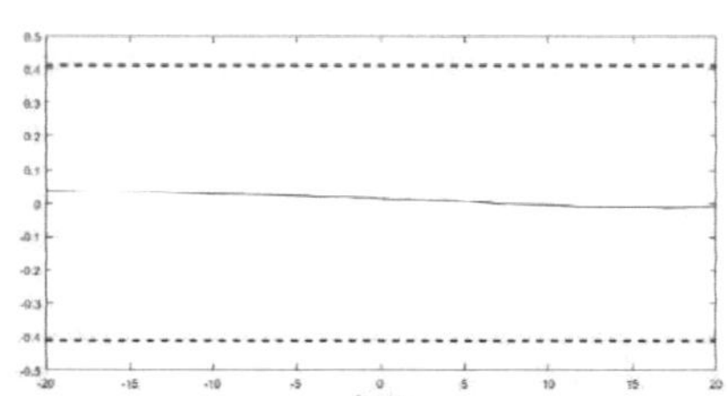

Figure 7. Cross correlation for $Y_1(s)$ and $R_1(s)$ residuals.

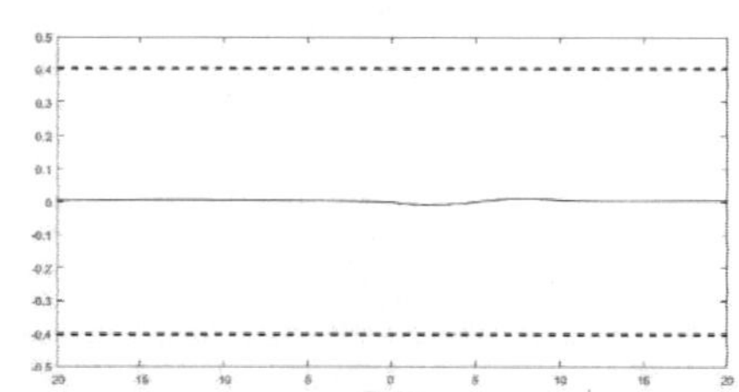

Figure 8. Cross correlation for $Y_2(s)$ and $R_2(s)$ residuals.

4. Conclusion

The closed-loop transfer function models of cooperative manipulators holding deformable objects have been estimated and validated using the system's closed-loop input and output data. The fits between the measured and simulated output of the estimated models are 95.73% and 95.88% for the master and slave manipulators, respectively, above the 95% criterion for a "good" model. The obtained models are useful for model-based control design. The residual analysis also shows the cross correlation functions being within the confidence region for both cases.

References

[1] Ljung L 2010 Perspectives on system identification *Annual Reviews in Control* **34** 1-12

[2] Yazdizadeh A, Khorasani K and Patel R 2000 Identification of a two-link flexible manipulator using adaptive time delay neural networks *IEEE Transactions on Systems, Man and Cybernetics* B **30** 165-172

[3] Liu K and Sun X 2001 system identification and model reduction for a single-link flexible manipulator *Journal of Sound and Vibration* **242** 867-891

[4] Mohd Yatim H, Mat Darus I and Mohamad M 2012 Parametric identification and dynamic characterisation of flexible manipulator system *2012 IEEE Conference on Control, Systems & Industrial Informatics* 16-21

[5] Deshmukh S, Zambare H, Mate K, Shewale M and Khan Z 2015 System identification and PID implementation on double flexural manipulator *2015 International Conference on Nascent Technologies in the Engineering Field (ICNTE)* 1-5

[6] Papadopoulos E, Bin Mu and Frenette R 2003 On modeling, identification, and control of a heavy-duty electrohydraulic harvester manipulator *IEEE/ASME Transactions on Mechatronics* **8** 178-187

[7] Aun K and Anh H 2007 System modeling identification and control of the two-link pneumatic artificial muscle manipulator optimized with genetic algorithms *2007 IEEE International Conference on Control and Automation* 501-506

[8] Wada T, Hirai, S, Kawamura S and Kamiji N 2001 Robust manipulation of deformable objects by a simple PID feedback *IEEE International Conference on Robotics and Automation* **1** 85-90

[9] Hirai S, Tsuboi T and Wada T 2001 Robust grasping manipulation of deformable objects *IEEE International Symposium on Assembly and Task Planning* 411-416

[10] Cleary K and Nguyen C 2001 State of the art in surgical robotics: clinical applications and technology challenges *Computer Aided Surgery* **6** 312-328

[11] Alkathiri A A 2017 *Control of Cooperative Manipulators in Holding Deformable Objects* undergraduate (International Islamic University Malaysia)

[12] Ljung L 1999 *System Identification: Theory for The User* (New Jersey: Prentice Hall)

[13] Hjalmarsson H, Gevers M and de Bruyne F 1996 For model-based control design, closed-loop identification gives better performance *Automatica* **32** 1659-73

[14] Ljung L 2015 *System Identification Toolbox™ User's Guide* (Natick: The MathWorks, Inc.) Online: https://fenix.tecnico.ulisboa.pt/downloadFile/845043405443234/ident_r2015a.pdf

IOP Conf. Series: Materials Science and Engineering **260** (2017) 012036 doi:10.1088/1757-899X/260/1/012036

Effect of Temperature on Ultrasonic Signal Propagation for Extra Virgin Olive Oil Adulteration

N.A. Alias[1], S.B. Abdul Hamid[2], A.Sophian[3]

Department of Mechatronics Engineering, Kulliyyah of Engineering, International Islamic University Malaysia, 53100 Gombak, Selangor

ain_alias@yahoo.com[1], syamsul_bahrin@iium.edu.my[2]

Abstract. Fraud cases involving adulteration of extra virgin olive oil has become significant nowadays due to increasing in cost of supply and highlight given the benefit of extra virgin olive oil for human consumption. This paper presents the effects of temperature variation on spectral formed utilising pulse-echo technique of ultrasound signal. Several methods had been introduced to characterize the adulteration of extra virgin olive oil with other fluid sample such as mass chromatography, standard method by ASTM (density test, distillation test and evaporation test) and mass spectrometer. Pulse-echo method of ultrasound being a non-destructive method to be used to analyse the sound wave signal captured by oscilloscope. In this paper, a non-destructive technique utilizing ultrasound to characterize extra virgin olive oil adulteration level will be presented. It can be observed that frequency spectrum of sample with different ratio and variation temperature shows significant percentages different from 30% up to 70% according to temperature variation thus possible to be used for sample characterization.

1. Introduction

Olive oil is one of the health beneficial vegetables oil to humans to consume in their daily life with the proper balanced in diet [10]. This oil is commonly used as significant ingredient in cooking, salad dressing or in food products formulation. It is also stated in the Quran that olive tree is considered as a blessed tree in Islam and a source of a good nutrition to life [9]. As of April 2017, current price of virgin olive oil in Ringgit Malaysia is about RM 19,343.41 per metric ton or about RM 19.34 per litre. In this case, olive oil could be considered as one of the edible oils that can be classified as a high value due to its expensive price in the market.

High value products are of the subjected to adulteration involving the replacement of high-cost ingredient with lower grade and cheaper substitutes and this sometimes can be difficult to detect [2]-[7]. Adulteration of olive oil with various seed oils is common problem affecting the quality and commercial value of the products. This practice will cause a major loss in economic value and commercial exploitation of olive oil [3]. Olive oil is a commonly adulterated food product of the world due to its relatively low production and expensive prices compared to other vegetables and seed oils as previously expressed. Adulteration of olive oil is a serious problem for regulatory agencies, oil suppliers and could also be a threat to the health of consumers. Hence, there are several techniques that has been introduced specifically to detect adulteration of olive. Current common methods are using cylindrical capacitive sensor as dielectric-based system, controlled cooling of differential scanning calorimetry (DSC), mass spectrometry, standard method by The American Society for Testing and Materials International (ASTM International) such as Density test, Evaporation test, Distillation test, Chemical Marker test and Gas Chromatography [4]. These methods may be used to determine the adulteration but tend to be fixed with laboratory apparatus and can cause delay in validation. Ultrasound is one of in-situ approach available to detect adulteration but prone to error due to temperature variance and mass chromatography requires a complicated sample preparation.

Ultrasound technology is an environmentally friendly, externally applied, non-invasive and well proven in medicine, industry and sciences. Common application of ultrasound in industry is non-destructing test (NDT), thickness measurement, flow speed measurement, characterization of dispersions, manipulation of particles, enhancement of flow in microchannel (ultrasonic streaming) and fish detection (ultrasound real-time identification)

There are two basic ultrasonic testing that is either the reflected waves (pulse-echo) or the transmitted waves (through transmission). Pulse echo system are more useful compared to through transmission in terms of application as they required one-sided access to the object being inspected or tested. Ultrasound wave travel at a speed of sound that is related to compressibility (K) and density of a material combined with the laws of conservation of mass and momentum, directly imply the existence acoustic waves [6]. Other than that, viscosity of fluid that undergoes propagation of ultrasonic wave will change the waveforms due to attenuation and the dispersion of the propagation medium. In the ultrasound system used for experimental purposes, common methods involve are based either on the measurement of reflection coefficients and as well as by the determination of the characteristics of propagation (velocity and attenuation). Measurement of velocity propagation and attenuation when propagate ultrasonic waves allows to determine the physical quantities of the medium. In this case, it shows that is possible to characterize the adulteration of extra virgin olive oil using ultrasound measurements.

According to Pascal Laugier and Guillaume Ha¨ıat, temperature of the medium also effect the speed of sound (c) as temperature dependence of c. Previous research has proved that speed of sound in the water is 1482 m/s at 20 °C and varies between 20 °C and 37 °C with a temperature coefficient about 2.5m/s/°C [5]. Moreover, when heat is subjected to fluid, the particles will speed up and spread out immediately and vice versa. This will result on fluid to expand as the volume increase as well as causes the density to increases. (D=M/V if V gets smaller M/V gets bigger). This shows that temperature of sample that undergone propagation of ultrasound waves should be taken into consideration during experiment as it will affect viscosity of fluids. This paper is organized as follows, as in next section is materials and methodology used followed by results and discussion in section 3 and section 4 is a conclusion of the paper.

2. Materials and Methodology

Sample used in determination of virgin olive oil being mixed with other filler material of palm oil was poured into glass with a volume of 300 ml. Water is used as reference sample and sample used for the analysis contained mixture of extra virgin olive oil with palm oil bought from local supermarket. Sample used are 100% olive oil and ratio of 50:50 and 30:70 of olive oil and palm oil. Thus, the same concentrations are used which related to adulteration of olive oil with other filler materials. Temperature range are taken based on standard ambient and storage temperature of extra virgin olive oil which are 27 ^{0}C, 30 ^{0}C and 33 ^{0}C. Temperature of the sample is controlled using digital thermometer and heater was used to increase the temperature of the sample.

In order to generate a time domain signal of the transducer, a 5Vp-p continuous sine wave with a frequency of 40 kHz was excited using function generator model INSTEK GFG-8217A (1MHz). Received signal was digitized by oscilloscope model Tektronix TDS 1002B (60MHz). Signal data are taken manually using USB drive features of oscilloscope. Single ultrasonic transducer model JSN SR04 with acoustic emission frequency of 40 kHz is used for transmitting and receiving ultrasonic signal. Figure 1 will illustrate the measurements of ultrasonic system setup and figure 2 shows the example of transmitting signal of continuous wave.

IOP Conf. Series: Materials Science and Engineering **260** (2017) 012036 doi:10.1088/1757-899X/260/1/012036

Ultrasound signal generation

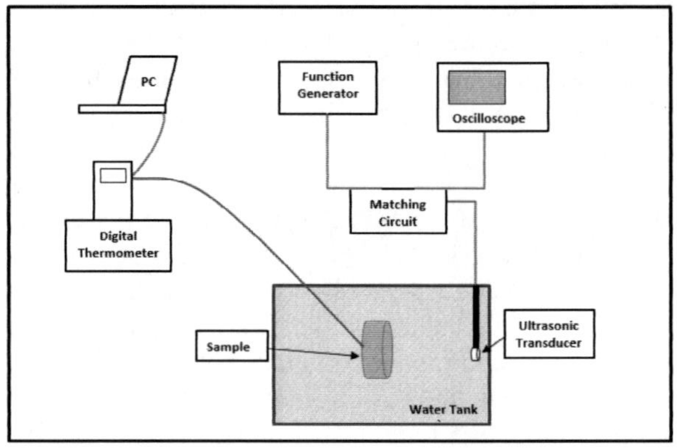

Figure 1: Experimental setup of sound wave propagation and signal acquisition

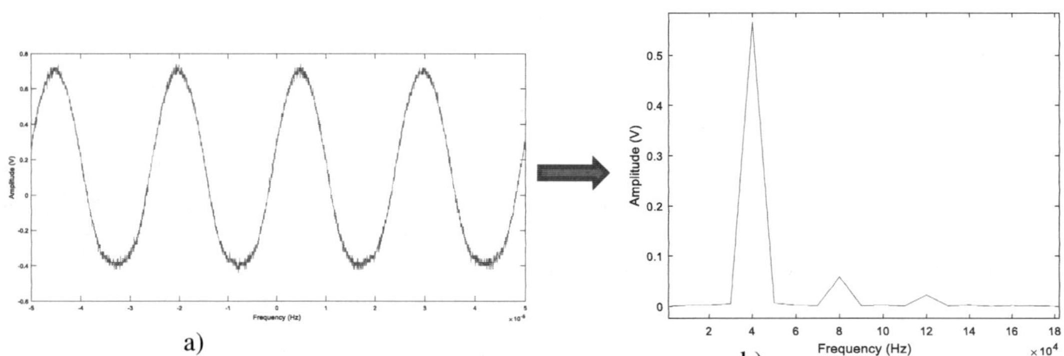

Figure 2: a) Continuous sine wave transmitted signal b)Frequency spectrum of this wave

IOP Conf. Series: Materials Science and Engineering **260** (2017) 012036 doi:10.1088/1757-899X/260/1/012036

Results

Table 1 : Different sample ratio and variation temperature spectra

<table>
<tr><th rowspan="2">Sample</th><th colspan="3">Temperature (^0C)</th></tr>
<tr><th>27^0C</th><th>30^0C</th><th>33^0C</th></tr>
<tr><td>100% Olive Oil</td>
<td>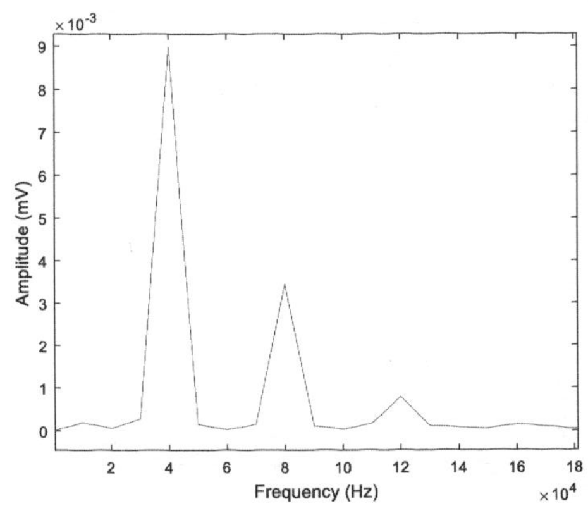
</td>
<td>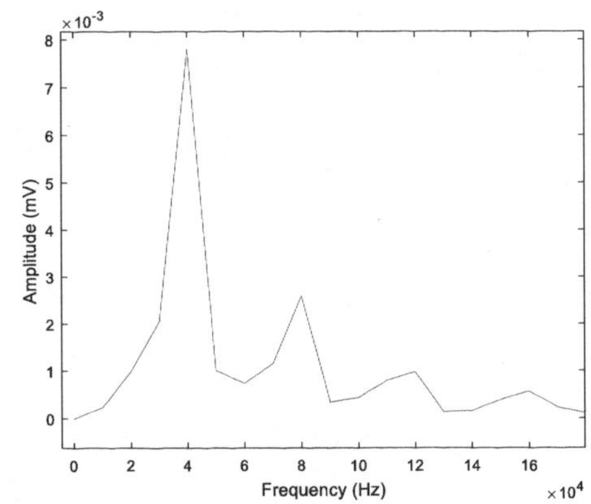
</td>
<td>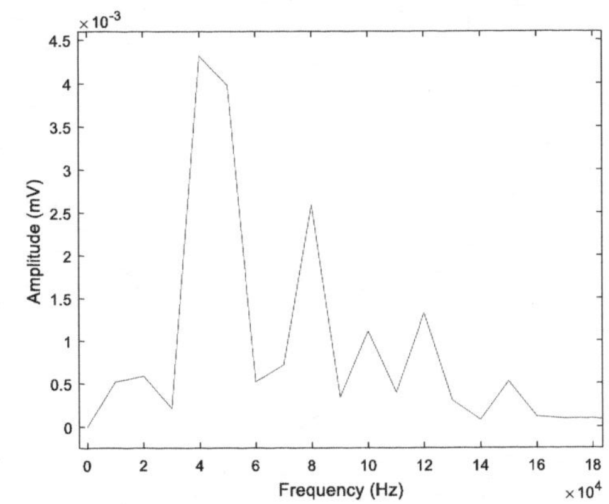
</td></tr>
<tr><td>50% Olive Oil + 50% Cheap Oil</td>
<td>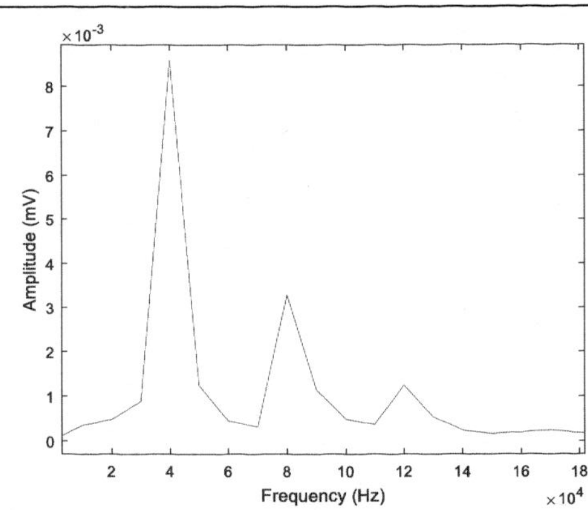
</td>
<td>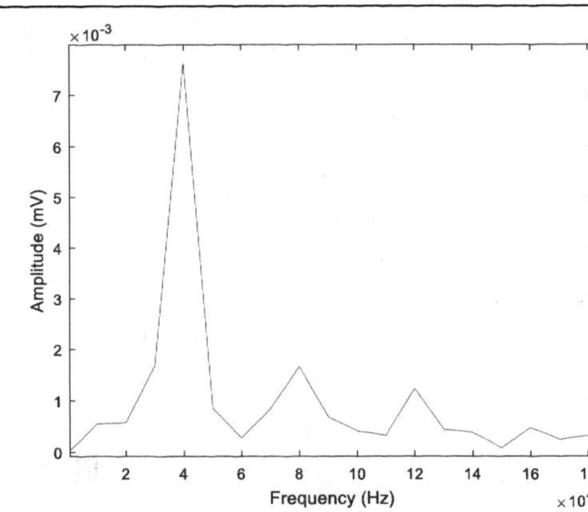
</td>
<td>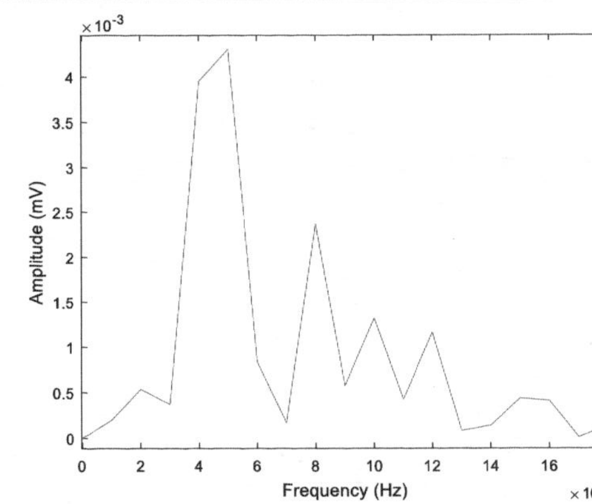
</td></tr>
</table>

IOP Conf. Series: Materials Science and Engineering **260** (2017) 012036 doi:10.1088/1757-899X/260/1/012036

3. Results and Discussion

Results in figure 3 shows that 100% olive oil sample possessed different frequency spectrum of receiving signal and it is possible to characterize adulteration of olive oil. Temperature also plays important role in the formation of frequency spectrum. Variation of temperature effect the received signal as speed of sound, viscosity and density change, thus effect the frequency spectrum as well. Figure 3 shows received signal of olive oil with different sample temperature. First peak of the spectrum shows that higher temperature resulted in lower amplitude of the spectrum and vice versa. This has proved that temperature of sample does affect the received signal of ultrasound and first peak of the signal can be used for further analysis of the sample used.

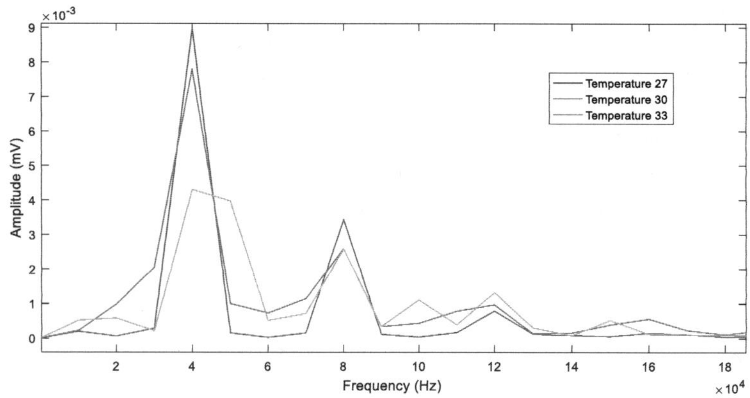

Figure 3: Frequency spectrum of 100% olive oil sample with different temperature

Table 2: Data obtained from frequency spectrum of different temperature and sample used

Sample	Temperature (^{0}C)	Amplitude (V)		
		First peak	Second peak	Third peak
100% Olive Oil	27	0.00906	0.00342	0.00133
	30	0.00793	0.00261	0.00094
	33	0.00451	0.00259	0.000809
50:50 ratio	27	0.00806	0.00331	0.00125
	30	0.00772	0.00241	0.00124
	33	0.00398	0.00168	0.00117
30:70 ratio	27	0.00735	0.00327	0.00176
	30	0.00695	0.00331	0.00101
	33	0.00571	0.00249	0.000806

Table 3: Percentage different of frequency spectrum amplitude

Sample	Temperature (^{0}C)	Percentage different	
		First peak to second peak	Second peak to third peak
100% Olive Oil	27	62%	61%
	30	67%	64%
	33	42%	68%
50:50 ratio	27	58%	62%
	30	68%	48%
	33	57%	30%
30:70 ratio	27	55%	46%
	30	52%	69%
	33	62%	61%

Table 1 shows the frequency spectral of received signal that goes with different initial condition. All sample that undergoes propagation of sound wave showed different frequency spectrum under the same condition of temperature and pressure and possessed different spectral with the variation of temperature. Percentage different of one peak to another peak is calculated using values obtained from the graph as shown in table 2. Calculations shows that percentage are varies from 30% to 70%. This shows that changes of the amplitude are quite significant. Hence, further analysis on frequency spectrum which is in frequency domain can be done to characterize the analysed materials from different aspects as of density and viscosity compare to time domain analysis. Thus, this present work shows that characterization of the materials can be done using ultrasound signal based on the different values of amplitude obtained from the spectrum.

4. Conclusion

Based on the observation of the frequency spectrum of the sample undergoes continuous sine wave of 40kHz frequency digitized by the oscilloscope, it is possible to use frequency spectrum for further analysis of sample characteristics based on results that shows shift of peak to peak values of the spectrum amplitude which is quite significant. Temperature of sample should be taken into considerations as the spectra of propagation of ultrasound signal also changed.

5. Acknowledgement

This research is supported by the Fundamental Research Grant Scheme (FRGS) by Ministry of Higher Education Malaysia (MOHE) to the Department of Mechatronics Engineering at the International Islamic University Malaysia (FRGS16-065-0564).

6. References

[1] Baroni, D. B., & Bittencourt, M. S. Q. (2011). Ultrasonic Spectral Analysis For Nuclear Fuel Characterization.

[2] Calorimetry, D. S. (n.d.). Detecting the Adulteration of Extra Virgin Olive Oil by Controlled Cooling DSC, 3–5.

[3] Dourtoglou, V. G., Dourtoglou, T., Antonopoulos, A., Stefanou, E., Lalas, S., & Poulos, C. (2003). Detection of olive oil adulteration using principal component analysis applied on total and regio FA content. Journal of the American Oil Chemists' Society, 80(3), 203–208. https://doi.org/10.1007/s11746-003-0677-1

[4] Gupta A. K. and SharmaR.K. (2002). A new method for estimation of automobile fuel adulteration, 357–371.

[5] Oliveira, P. A., Silva, R. M. B., Morais, G. C., Alvarenga, A. V, & Félix, R. P. B. C.-. (2016). Speed of sound as a function of temperature for ultrasonic propagation in soybean oil. Journal of Physics: Conference Series, 733, 12040. https://doi.org/10.1088/1742-6596/733/1/012040

[6] Sudhan, M., & Gudur, R. (2013). Spectral Ultrasound Characterization of Tissues and Tissue Engineered Constructs.

[7] Tay, a, Singh, R. K., Krishnan, S. S., & Gore, J. P. (2002). Authentication of Olive Oil Adulterated with Vegetable Oils Using Fourier Transform Infrared Spectroscopy. LWT - Food Science and Technology, 35(1), 99–103. https://doi.org/10.1006/fstl.2001.0864

[8] Azadmard-Damirchi, S., & Torbati, M. (2015). Adulterations in Some Edible Oils and Fats and Their Detection Methods. Journal of Food Quality and Hazrds Control, 2, 38–44.

[9] Qur'an, (24:35)(23:20)

[10] Maria L.Palmetti, 2011,Palm Oil : Nutrition, uses and impact, Nova Science Publishers, Inc. New York

IOP Conf. Series: Materials Science and Engineering **260** (2017) 012037 doi:10.1088/1757-899X/260/1/012037

Design of Dual Band Microstrip Patch Antenna using Metamaterial

Islam Md Rafiqul, Adel A.A Alsaleh, Mimi Aminah W. N., M. Sarah Yasmin, Norun Farihah A.M.

Department of Electrical & Computer Engineering, Faculty of Engineering International Islamic University Malaysia, Jalan Gombak, 53100 Kuala Lumpur, Malaysia

rafiq@iium.edu.my, deul89@yahoo.com, mimie@iium.edu.my, norun@iium.edu.my smohamad@iium.edu.my

Abstract. Metamaterial has received great attention due to their novel electromagnetic properties. It consists of artificial metallic structures with negative permittivity (ε) and permeability (μ). The average cell size of metamaterial must be less than a quarter of wavelength, hence, size reduction for the metamaterial antenna is possible. In addition, metamaterial can be used to enhance the low gain and efficiency in conventional patch antenna, which is important in wireless communication. In this paper, dual band microstrip patch antenna design using metamaterial for mobile GSM and WiMax application is introduced. The antenna structure consists of microstrip feed line connected to a rectangular patch. An array of five split ring resonators (SRRs) unit cells is inserted under the patch. The presented antenna resonates at 1.8 GHz for mobile GSM and 2.4 GHz for WIMAX applications. The return loss in the FR4 antenna at 1.8 GHz is -22.5 dB. Using metamaterial the return loss has improved to -25 dB at 2.4 GHz and -23.5 dB at 1.8 GHz. A conventional microstrip patch antenna using pair of slots is also designed which resonates at 1.8 GHz and 2.4 GHz. The return loss at 1.8 GHz and 2.4 GHz were -12.1 dB and -21.8 dB respectively. The metamaterial antenna achieved results with major size reduction of 45%, better bandwidth and better returns loss if it is compared to the pair of slots antenna. The software used to design, simulate and optimize is CST microwave studio.

1. Introduction

The prefix Meta is an ancient Greek means beyond which is used to refer to materials that do not exist in nature [1]. A metamaterial is an artificial compound material that has a precise structure, which offers properties not found in nature [2]. In other words, metamaterial is designed to obtain special electromagnetic properties such as negative permittivity or permeability, enormous handedness and zero refractive index [3]. The same like naturals materials deciding the properties of metamaterials can be done by their arrangement and components. Hence, the component must be designed with precise pattern to achieve particular properties and produce resonant structures. Sometimes Metamolecules or meta-atoms are the name given to these kinds of components. They are arranged in a periodic way in one, two, or three dimensions. Back in history, the first idea of metamaterial consist of double negative materials and negative refraction and it was proposed by Veselago in 1967 and Mandelstam in 1945[4] [5]. However, in 2000, Smith and Pendry were the first to do an experiment to verify the feasibility of negative refraction [6]. Even though a major progress was seen in the new century. Nevertheless, clear and precise description to metamaterial is still debatable. However, the most important concept which

is clear and essential in metamaterial is that the dimensions of unit cells and periodicity must be much less than the operating wavelength. This is to obtain quasi-homogeneous material [2] [3].
The usage of metamaterial is wide in antennas and microwave devices. An enormous number of antennas and microwave devices have been designed and fabricated with unique properties. In addition, flexible portable devices like laptops, mobile phones, wearable device, antennas with tuneable functions based on variable structures are still crucially important and demanded.

In this paper, dual band microstrip patch antenna design using metamaterial for mobile GSM and WiMax application is introduced. The antenna structure consists of microstrip feed line connected to a rectangular patch. An array of five split ring resonators (SRRs) unit cells is inserted under the patch. The presented antenna resonates at 1.8 GHz for mobile GSM and 2.4 GHz for WIMAX applications. The performance and size of metamaterial dual band antenna are compared with that of conventional dual band antenna.

2. Metamaterial Design

The metamaterial unit cell is designed and simulated by using the Eigenmode solver in CST. The result investigates the dispersion characteristics. For the boundary conditions, they are set to be in state of perfect magnetic boundary condition (PMC) and this in Z-direction while in Y and X-direction they are set to be in periodic condition. The LHM unit cell structure consists of two rectangular split ring resonator (SRR) and thin wire (TW). The rings are printed on the substrate. The substrate is a FR4 substrate. The length of the outer ring is L1 = 7.4mm, and the inner ring is L2 = 5.4mm. The gap between the rings W1 = W2 = 0.5mm and a small brick is inserted in the centre of the inner ring. The cut in both rings is G1 = 1.6mm. The permittivity of the substrate is 4.4. The unit cell with it dispersion diagram are shown in Figures 1 and 2.

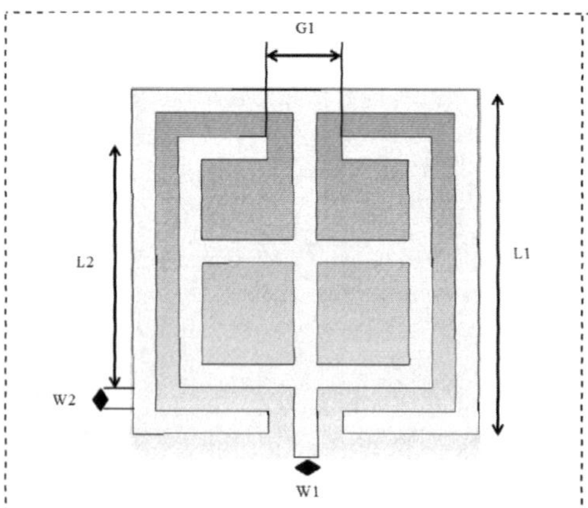

Figure 1. Left-handed unit cell.

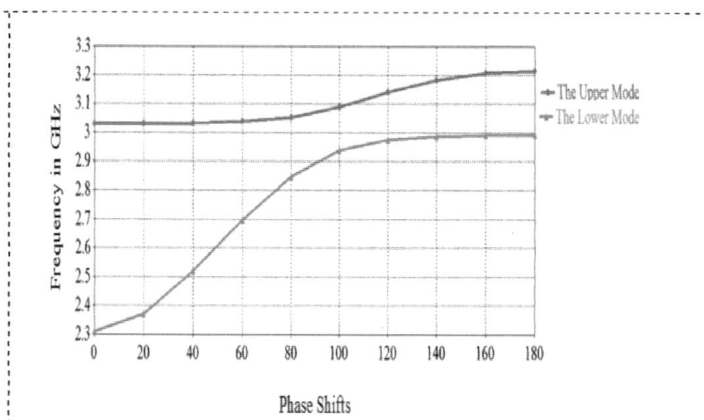

Figure 2. Dispersion diagram.

To prove that this unit cell is useful for wireless application, calculation to get the resonant points must be done. Four cells will be considered. Number of resonant points are given by [5]

$$\text{Resonant points} = 2N - 1 \quad (1)$$

These frequencies can be found by sampling the dispersion diagram as [5]:

$$\beta l = m\pi \quad (2)$$

Where:
m = resonance index: 0, ±1, ±2, ...
l = physical length of the unit cell
β = phase shift in degrees

By increasing the number of unit cells, resonance points can be further sampled from the dispersion diagram of the unit cell, which results additional resonance frequencies.

From the equation (1), the number of resonance points is 7. Moreover, their corresponding mode indices are m=0, ±1, ±2, ±3 by using L=7.4 mm. Therefore, the sampled dispersion diagram from equation (2) is as follow, m = 0, β = 0, m = ±1, β = +24.3° and β = -24.3°, m = ±2, β = +48.6° and β = -48.6°, m = ±3, β = +72.9° and β = -72.9°. The expected resonant frequencies are shown in the Table 1. As it is presented in the table, these frequencies occurred in wireless band. Therefore, this unit cell can be used for wireless application.

Table1. Expected resonance frequencies

Mode of indices	M = -3	M = -2	M = -1	M = 0	M = +1	M = +2	M = +3
The Resonant Frequencies (GHz)	2.79	2.59	2.4	2.31	3.03	3.03	3.05

3. Metamaterial Antenna Design

Firstly, the microstrip patch antenna has been modelled. The FR4 substrate permittivity used is 4.5 and the thickness of the substrate is 1.6 mm. the patch thickness is 0.1 mm. A microstrip feed line is used to feed the antenna as shown in Figure 3. The Table 2 presented all antenna parameters. The conventional patch antenna resonates at 1.8 GHz. Secondly, the dual band metamaterial antenna is designed. An array of split ring resonators is placed under the patch and in the substrate as shown in Figure 3. The split ring resonators array (SRRs) are located in the near filed of the patch.

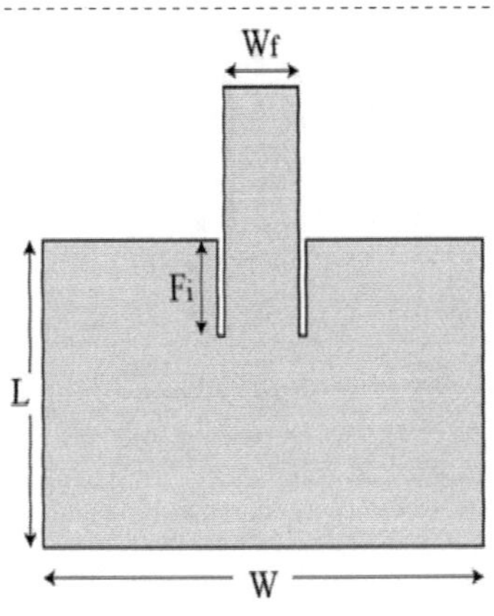

Figure 3. The conventional antenna

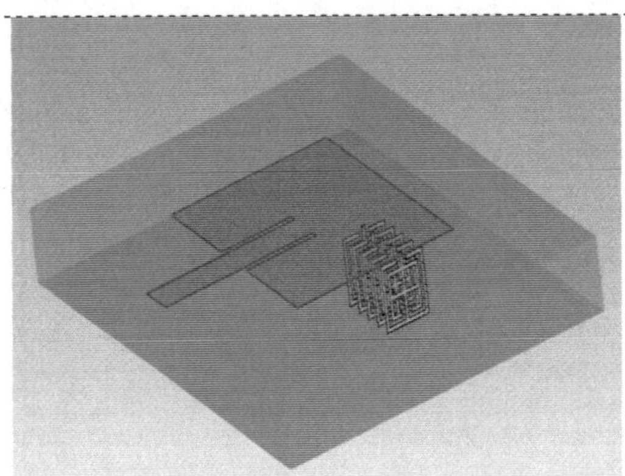

Figure 4. Metamaterial antenna showing unit cells.

Table 2. Comparision between microstrip patch antenna and metamaterial antenna parameters.

Parameter	Microstrip patch antenna	Metamaterial patch antenna
Substrate length	76	56
Substrate width	102	82
Patch length	38	28
Patch width	51	41
Feed slot cut, Fi	12.5	12.5
Feed line length	31.5	26.5
Feed line width, Wf	8.7	6
Substrate thickness, h	1.6	8.3

The resonance characteristics of the antenna will be affected by the permittivity and permeability of the manufactured substrate. A mutual capacitance will be enlarged between inclusions when the SRRs are placed under the patch. The spacing between SRRs will be optimized to achieve dual band frequency at 1.8 GHz and 2.4 GHz. Table 2 shows the parameters in (mm) of the microstrip patch antenna, the metamaterial patch antenna and the difference between both of them. From Table 2, it is seen that a major miniaturization happened to the microstrip patch antenna when the metamaterial was used. The metamaterial antenna achieves a 40% size reduction in the area of the patch and the substrate if it is compared to the conventional patch antenna where no metamaterial has been used. In addition, a miniaturization occurred in the feed line length and width.

4. Characteristics Comparison Between Conventional and Metamaterial Patch Antenna

Dual band frequency was successfully achieved at 1.8 GHz and 2.4 GHz by metamaterial antenna. The return loss at 1.8 GHz is -25 dB and at 2.4 GHz is -23.5 dB. Figure 5 is showing a comparison between the conventional antenna with single band frequency and metmaterial antenna which is showing dual band frequency.

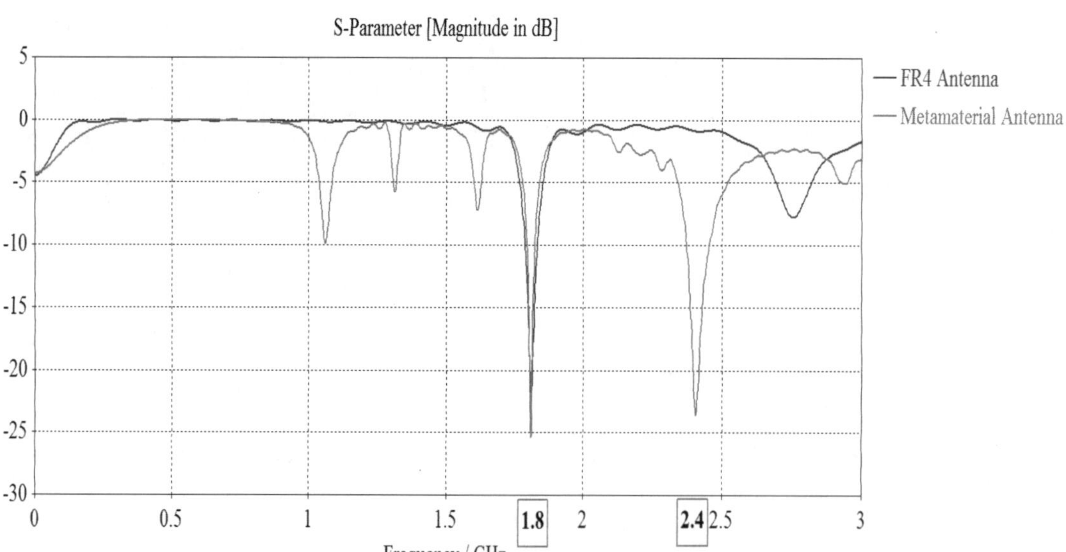

Figure 5. Comparision between FR4 antenna and metamaterial antenna.

IOP Conf. Series: Materials Science and Engineering **260** (2017) 012037 doi:10.1088/1757-899X/260/1/012037

In addition, the Figure shows improvement in the return loss at 1.8 GHz. In conventional antenna the return loss at 1.8 GHz was -22.5 when at metamaterial it increased to be -25. The Fairfield directivity for both resonant frequencies is shown at Figures 6-9. In Figure 6, the angular width and main lobe direction are shown having the value of 91.8 deg and 8.0 deg respectively. Figure 7 shows that the maximum directivity at 1.8 GHz in metamaterial antenna is 5.459 dBi while the total efficiency is -4.595 dB. In Figure 8, the angular width and main lobe direction are displayed having the value of 76.6 deg and 6.0 deg respectively. The side lobe level is -14.1 dB. In addition, Figure 9 shows that the maximum directivity at 2.4 GHz in metamaterial antenna is 7.501 dBi while the total efficiency is -4.396 dB.

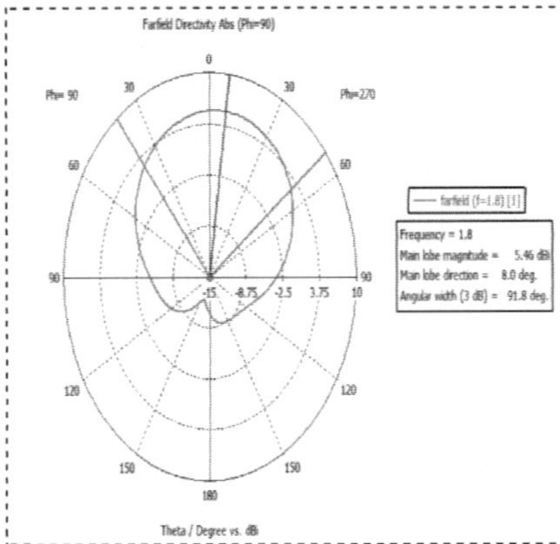

Figure 6. Fairfield directivity in polar form at 1.8 GHz.

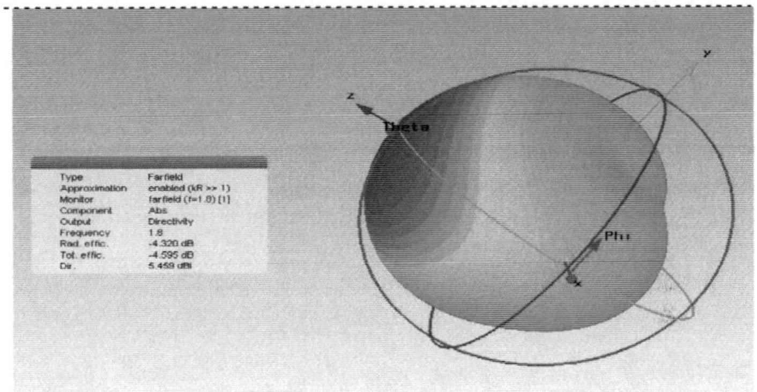

Figure 7. Fairfield directivity in 3D form at 1.8 GHz.

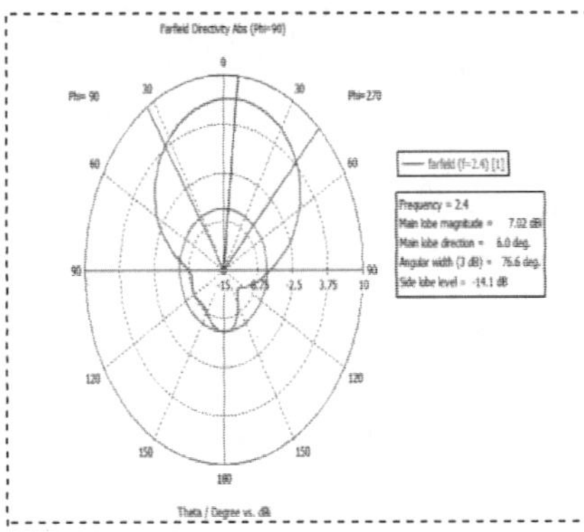

Figure 8. Fairfield directivity in polar form at 2.4 GHz.

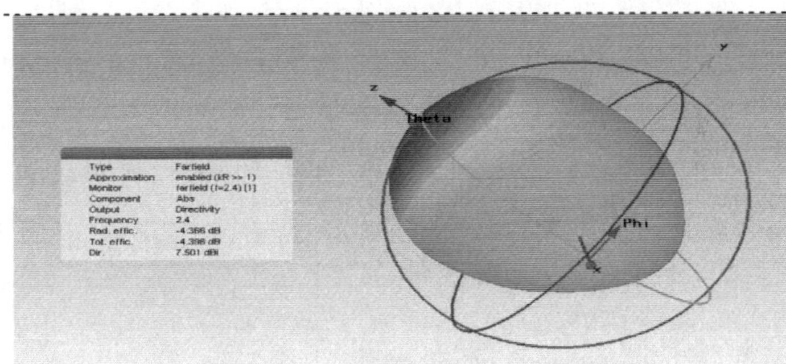

Figure 9. Fairfield directivity in 3D form at 2.4 GHz.

5. Comparision between Pair of Slots Antenna and Metamaterial Antenna

In a previous work [7-8], microstrip patch antenna with pair of slots was designed. In this section, a comparison is done between microstrip patch antenna with pair of slots and metamaterial antenna. Figure 10 shows that the metamaterial antenna is superior to slotted antenna. By using pair of slots, it was possible to obtain 2.4 GHz resonant frequency accurately when the resonant frequency is nearest for 1.8 GHz. However, by using metamaterial , dual band resonant frequencies are found accurate at 1.8 GHz and 2.4 GHz.

In term of the return loss, the metamaterial shows it is superiority again. In the antenna with pair of slots, the return loss at 1.8 GHz and 2.4 GHz were -12.1 dB and -21.8 dB respectively, yet in when the metamaterial is used the return loss at 1.8 GHz and 2.4 GHZ were -25 dB and -23.5 dB respectively. Hence, the return loss improved significantly at 1.8 GHz and fairly at 2.4 GH. In addition, it is noticed that the bandwidth is improved at 2.4 GHz and a better directivity is received at the same resonant frequency with 7.501 dBi. Moreover, by using metamaterial, size miniaturization is achieved. Table 3 displays dimensions difference between pair of slots antenna and metamaterial antenna.

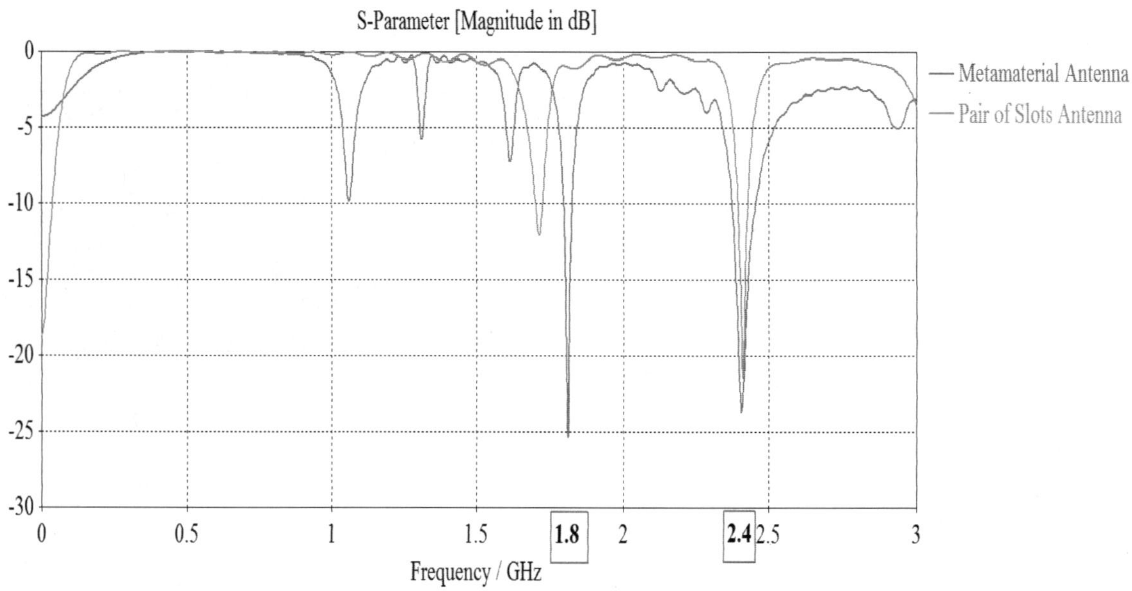

Figure 10. Comparision between pair of slots antenna and metamaterial antenna.

Table 3. Comparision between pair of slots antenna and metamaterial antenna parameters.

Parameter	Pair of slots antenna	Metamaterial patch antenna
Patch length	40	28
Patch width	53	41
Feed slot cut, Fi	12.5	12.5
Feed line length	31.5	26.5
Feed line width, Wf	8.7	6
Substrate thickness, h	1.6	8.3

Table 3 shows clearly size reduction in the patch length and width in metamaterial antenna compared to pair of slots antenna. This reduction corresponds with total area reduction of 45%. A noticeable size reduction is also seen in the feed line length and width while the substrate thickness is increased to form the magneto dielectric substrate.

6. Conclusions

A dual band microstrip patch antenna using metamaterial for mobile GSM and WiMax application is introduced. The antenna structure consists of microstrip feed line connected to a rectangular patch. An array of five split ring resonators (SRRs) unit cells is inserted under the patch. The designed antenna resonates at 1.8 GHz for mobile GSM and 2.4 GHz for WIMAX applications.. The return loss at 1.8 GHz is -25 dB and at 2.4 GHz is -23.5 dB. A conventional microstrip patch antenna using pair of slots is also designed. The pair of slots antenna also resonates at 1.8 GHz and 2.4 GHz. The return loss at 1.8 GHz and 2.4 GHz were -12.1 dB and -21.8 dB respectively. By comparing pair of slots antenna with metamaterial antenna, it is found that the return loss at 1.8 GHz improved significantly while it is slightly improved at 2.4 GHz. Bandwidth has improved and maximum directivity increased at 2.4 GHz. Moreover, a major miniaturization occurs to the microstrip patch antenna when the metamaterial was used. The metamaterial antenna achieves a 45% size reduction in the area of the patch if it is compared to the pair of slots antenna. In addition, a miniaturization occurred in the feed line length and width.

Acknowledgement
Authors are grateful to Research Management Centre, International Islamic University Malaysia to support this research through research grant RIGS16-065-0229.

Reference
[1] Rahmat-Samii, Y. (2006). Metamaterials in Antenna Applications : 2006 IEEE International Workshop, 2(c), 1–4.
[2] Christophe Caloz. (2005). Electromagnetic Metamaterials : Transmission Line Theory And Microwave The Engineering Approach.
[3] Sihvola, A. (2007). Metamaterials in electromagnetics. Metamaterials, 1(1), 2–11. Https://doi.org/10.1016/j.metmat.2007.02.003
[4] Veselago, V. G., & P. (1968). The Electrodynamics Of Substances With Simulataneously Negative Values Of E And μ, 10(4).
[5] Nur, Abdirashid Mohamed and Wan Nordin, Mimi Aminah and Islam, Md Rafiqul and Habaebi, Mohamed Hadi (2015). Investigation of metamaterial unit cells based on dispersion characteristics, ARPN Journal of Engineering and Applied Sciences, 10 (23). pp. 17380-17385. ISSN 1819-6608.
[6] Shelby, R. A., Smith, D. R., & Schultz, S. (2001). Experimental verification of a negative index of refraction. Science (New York, N.Y.), 292(5514), 77–79. https://doi.org/10.1126/science. 1058847
[7] Adel A.A Alsaleh , Md. Rafiqul Islam , Mimi Aminah Wan Nordin , Shadi Al-Askari (2016). Design and Optimization of Dual Band Microstip Patch Antenna Using Slots Pair, Proceedings of 6th International Conference on Computer and Communication Engineering, DOI 10.1109/ICCCE.2016.98, pp. 439-442, July 25-27, 2016.
[8] Mohammad Shawkat Habib, Islam Md Rafiqul, Khaizuran Abdullah and M Jamil Jakpar (2014). U-Slot Rectangular Patch Antenna for Dual Band Application, Proceedings of the 1st International Conference on Communication and Computer Engineering ICOCOE, pp 111-120, DOI: 10.1007/ 978-3-319-07674-4_12, May 20-21, 2014.

IOP Conf. Series: Materials Science and Engineering **260** (2017) 012038 doi:10.1088/1757-899X/260/1/012038

Optimum concentric circular array antenna with high gain and side lobe reduction at 5.8 GHz

Mohammed Zaid, Md Rafiqul Islam[1], Mohamed H Habaebi, AHM Zahirul Alam, Khaizuran Abdullah

Department of Electrical & Computer Engineering, International Islamic University Malaysia, Jalan Gombak, 53100 Kuala Lumpur, Selangor, Malaysia

mziium@gmail.com, rafiq@iium.edu.my, habaebi@iium.edu.my, zahirulalam@iium.edu.my, khaizuran@iium.edu.my

Abstract. The significance of high gain directional antennas stems from the need to cope up with the everyday progressing wireless communication systems. Due to low gain of the widely used microstrip antenna, combining multiple antennas in proper geometry increases the gain with good directive property. Over other array forms, this paper uses concentric circular array configuration for its compact structure and inherent symmetry in azimuth. This proposed array is composed of 9 elements on FR-4 substrate, which is designed for WLAN applications at 5.8GHz. Antenna Magus software is used for synthesis, while CST software is used for optimization. The proposed array is designed with optimum inter-element spacing and number of elements achieving a high directional gain of 15.7 dB compared to 14.2 dB of available literature, with a high reduction in side lobe level of -17.6 dB.

1. Introduction

With the developing of wireless communications, it is nowadays required to design directional antennas with high gain and well matched impedance [1,2]. Microstrip patch antenna has been widely used owing to its small size, low profile, lightweight, low cost and easy to conform [3,4]. However, a main drawback of the microstrip patch antenna is the low gain it exhibits [5,6]. The increased gain is achieved by means of simultaneous usage of several antennas [7]. With the arrangement of array elements in a specific configuration, radiation pattern can be scanned electronically in a given direction [8]. In this paper, a resonant frequency of 5.8 GHz is used for WLAN as the 5 GHz frequency band suffers from less interference, supports higher data rate and can carry up to 23 non-overlapping 20 MHz channels and 12 non-overlapping 40 MHz channels compared to a maximum of 3 non-overlapping 20 MHz channels of the 2.4 GHz [9,10]. Among the commonly used arrays, concentric circular array is intended in this paper for its compact structure and inherent symmetry in azimuth, thus, being a highly desirable choice for full peripheral coverage [11].

Many researches have tried to increase the gain with the use of arrays. However, such studies still lack to achieve an optimum array that provides high gain with minimal number of array elements for low cost and small size design. The optimum design proposed in this paper is attained through many stages, starting from the careful selection of single element using an expert synthesis software (Antenna Magus), with low cost substrate (FR-4), selecting proper array geometry, and forming

[1] Corresponding author.

minimal number of elements that achieve high gain with reduced side lobe level through optimized inter-element spacing.

2. Design and simulation of single array element

It is often ambiguous to find the best approach for an antenna design. Published literature may be unclear or contradictory and thus provide insufficient information for confident design choices. Therefore, careful selection and design are performed using an expert design system (Antenna Magus) that is thoroughly tested for a wide range of design criteria. Then, a full-wave simulation using Computer Simulation Technology (CST) is performed to optimize the selected element.

2.1. Selection of single element

Modern wireless systems require high gain, low profile, lightweight, and simple structure antennas to ensure reliability, mobility, and high efficiency [12]. Other characteristics are also considered for preferred design, such as patch, planar, inset-fed, integrated, and low cost antennas. To fulfill such characteristics, the expansive antenna database of Antenna Magus is used.

After exploring the antenna database taking into account the above characteristics, from over 300 antennas two candidate antennas are obtained, from which one is then selected. These two antennas are rectangular inset-fed microstrip patch antenna and circular inset-fed linearly polarized patch antenna. The two candidate antennas are then designed using Antenna Magus and simulated using CST, which shows the circular antenna outperforming the rectangular one, as demonstrated in the figures below.

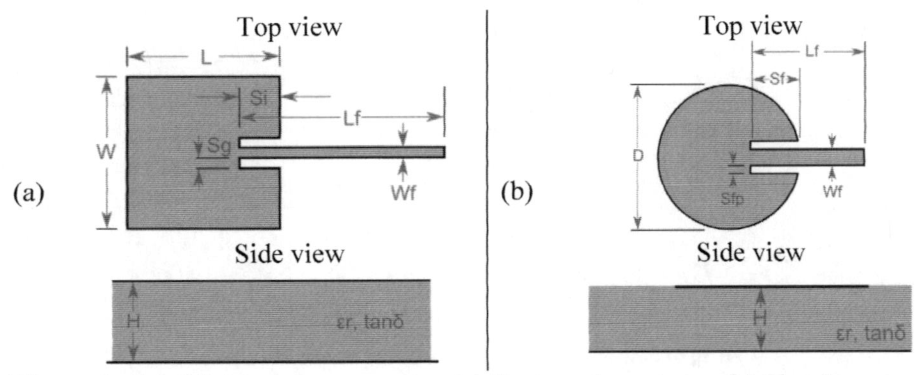

Figure 1. Candidate antenna geometry (a) Rectangular antenna (b) Circular antenna.

Table 1. Antenna physical parameters (a) Rectangular antenna (b) Circular antenna.

	(a)			(b)		
	Description	Value		Description	Value (50Ω)	Value (75Ω)
L	Patch length	11.94 mm	D	Patch diameter	14.57 mm	14.2667 mm
W	Patch width	15.88 mm	H	Substrate height	1.2 mm	1.2 mm
W_f	Feed line width	2.859 mm	ε_r	Relative permittivity	4.3	4.3
L_f	Feed line length	14.30 mm	W_f	Feed line width	2.334 mm	1.097 mm
H	Substrate height	1.47 mm	L_f	Feed line length	14.30 mm	14.70 mm
ε_r	Relative permittivity	4.3	S_f	Feed inset	6.216 mm	6.057 mm
tanδ	Loss tangent of the substrate medium.	0	S_{fp}	Spacing between patch and feed line.	249.3 μm	249.3 μm
S_i	Feed inset from edge of patch.	4.151 mm	tanδ	Loss tangent of the substrate medium.	0	0
S_g	Spacing between feed line and patch.	299.1 μm				

IOP Conf. Series: Materials Science and Engineering **260** (2017) 012038 doi:10.1088/1757-899X/260/1/012038

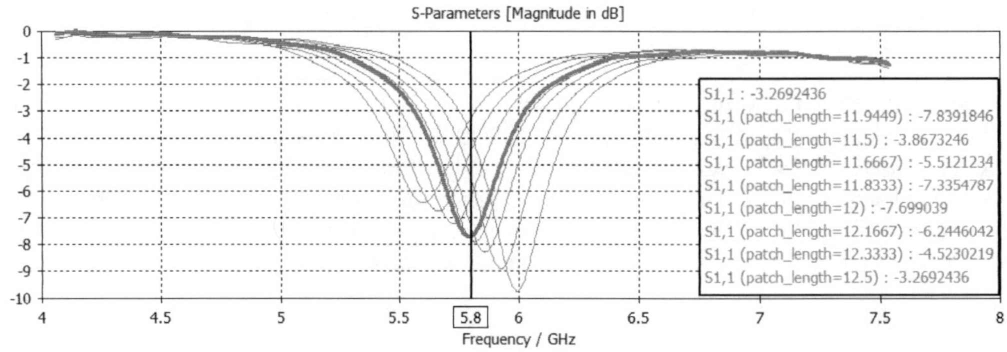

Figure 2. Return loss of rectangular patch antenna.

As Figure 2 shows, the return loss for rectangular antenna is about -7.7 dB, indicating low performance, while for circular antenna it is about -10.3 dB as shown in Figure 3, which indicates a better performance than the rectangular one, since return loss should not exceed -10 dB. Besides, the voltage standing wave ratio (VSWR) for rectangular antenna is about 2.4 that is above the good match level, while it is 1.9 for the circular antenna, which indicates good antenna match. Thus, among the two candidate antennas, the circular antenna is the one selected.

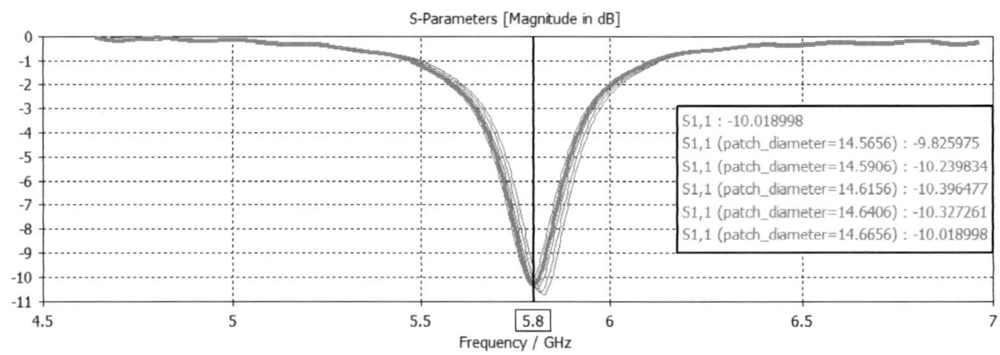

Figure 3. Return loss of circular patch antenna.

The two candidate antennas are simulated using the default standard input impedance of 50 Ω. It is then found that the circular antenna with the other standard impedance of 75 Ω is outperforming, where the return loss is about -13.3 dB compared to about -10.3 dB of that with 50 Ω, as illustrated in Figure 4. In addition, the VSWR is decreased to about 1.6 with 75-ohm circular antenna.

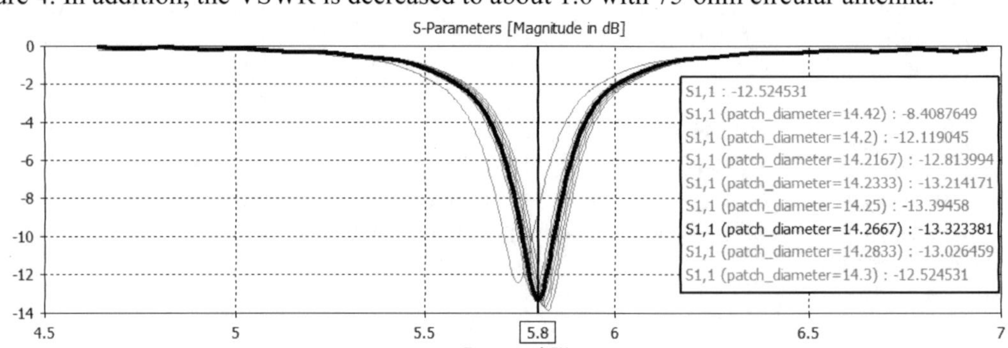

Figure 4. Return loss of circular patch antenna with 75 Ω.

2.1.1. Selected antenna element. The 75-ohm circular inset-fed linearly polarized patch antenna is finally the one selected to be used for circular array, as this selection shows reduced return loss of -13.3 dB and VSWR of 1.6, indicating better performance over the rectangular antenna and the circular one with 50 Ω. The selected antenna is then optimized to resonate at 5.8 GHz.

The geometry of the selected antenna element is as shown in Figure 1(b), where the patch shape is circular that is placed on a rectangular FR-4 substrate with input impedance 75 Ω operating at 5.8 GHz. The FR-4 substrate is used for its low cost in order to achieve a low cost array antenna.

3. Design and simulation of antenna array

Usually the radiation pattern of a single antenna element is relatively wide, and each element provides low directivity (gain). In many applications, it is necessary to design antennas with very directive characteristics (very high gains) to meet the demands of long distance communication. This can be achieved by forming an array of radiating elements in an electrical and geometrical configuration. Five controlling parameters shape the overall pattern of an antenna array that is composed of identical elements. These elements include the geometrical configuration of the overall array, spacing between elements, excitation amplitude and phase of each element, and relative pattern of the individual elements [8].

3.1. Optimum number of array elements

In this paper, it is aimed to achieve a high gain that is beyond 15 dB while maintaining a minimum number of array elements to fulfill a low cost design.

As Table 2 depicts, circular array is designed with different number of elements to reach the desired gain, which is achieved by a number of eight elements. A central element is added to increase the steering capability of the array.

Table 2. Comparison between arrays.

No. of Elements	Directivity (dBi)	Gain (dB)
1	6.8	5.7
2	9.4	8
4	13.3	11.8
(4 + 1)[a]	13.7	12.6
8[b]	16.2	14.8
8[c]	16.3	15.2
(8 + 1)[a]	16.7	15.4
(8 + 1)[d]	17.2	15.7

[a] With central element.
[b] Partially uniform.
[c] Fully uniform.
[d] Non-uniform.

From Table 2, it can be seen that the non-uniformly spaced eight-element array with central element is achieving the highest value of gain with good directive property, as the higher the number of array elements the higher the gain, and thus the array directivity. Thus, this number of elements is deemed to be the optimum number for a low cost design fulfilling the intended gain.

3.2. Optimum inter-element spacing

Inter-element spacing between array elements should be designed carefully, as a spacing below half wavelength of the operating frequency will increase mutual coupling, while a spacing greater than one wavelength will exhibit grating lobes which degrade the array performance [13].

Figure 5 portrays the partially uniform eight-element array with different inter-element spacing, where the array with a spacing of 0.6λ is shown to be more directional with minimised side lobes.

As the inter-element spacing increases the array starts to exihibt greater side lobes. Similarly, as the inter-element spacing decreases below 0.5λ, the array starts to face high mutual coupling between array elements, since higher proximity between array elements causes higher signal interaction.

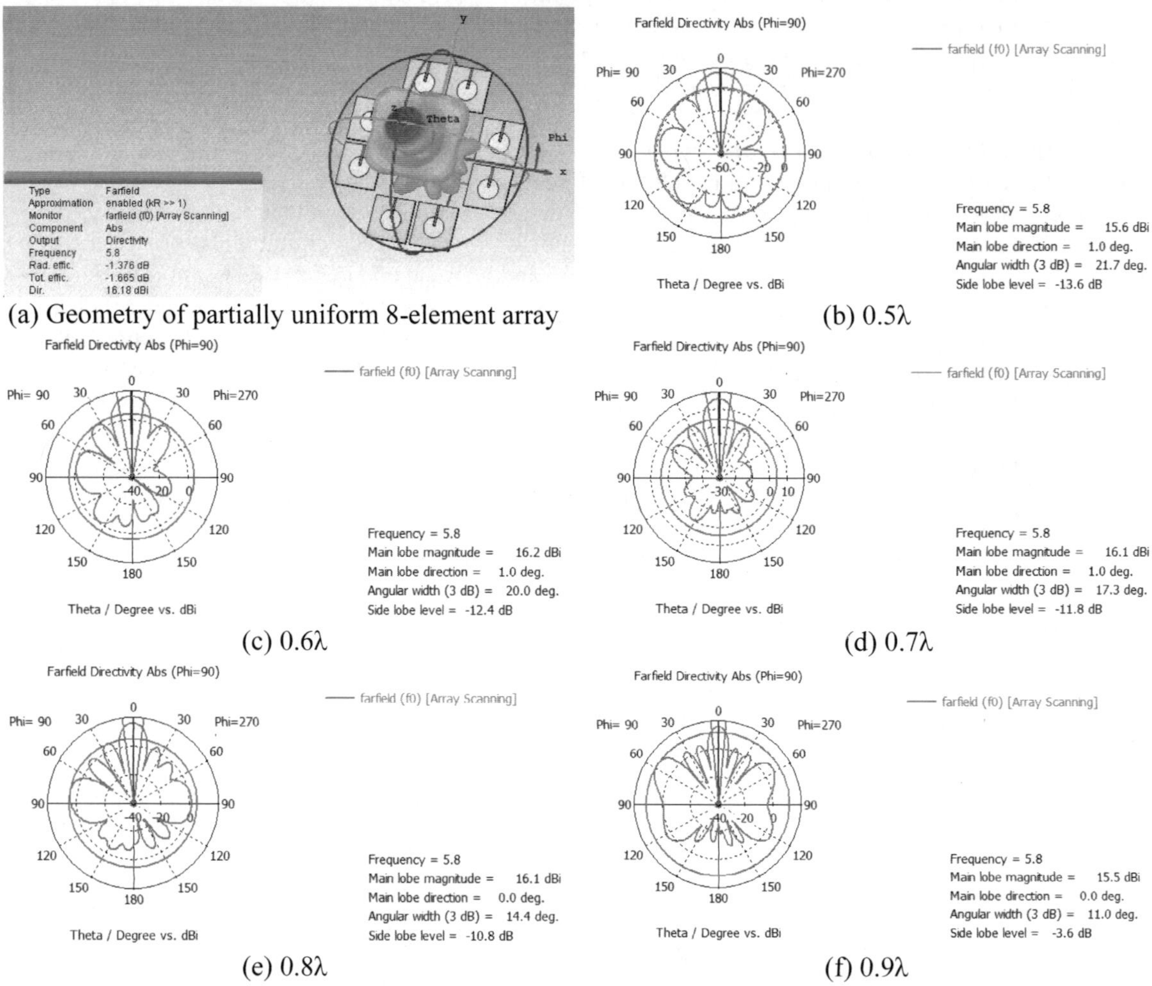

Figure 5. Polar plot of eight-element circular array with different inter-element spacing.

Figure 6 illustrates a nine-element array, where (a) is fully uniform with an optimum uniform spacing of 0.7λ, while (b) is non-uniformly spaced showing the optimum non-uniform configuration. The non-unform array is outperforming, as the directivity is 17.2 dBi compared to 16.7 dBi of the uniform one, and the side lobe level is -17.6 dB compared to -11 dB, respectively.

As compared to the nine-element rectangular array, the circular array outperforms the rectangular array as demonstrated in Figure 6 (c), where the directivity is 17.2 dBi for the circular compared to 15.6 dBi for the rectangular array. In addition, the side lobe level for the circular is -17.6 dB while it is -13.5 dB for the rectangular array.

The optimum inter-element spacing for non-uniform eight-element circular array is about 0.6λ, while it is 0.7λ for uniform eight-element circular array with central element. For non-uniform eight-element circular array with central element, the average inter-element spaing is about 0.71λ. Since geometrical configuration is a controlling parameter that shapes the overall pattern of an antenna array, good directional pattern and gain are achieved through non-uniform geometry manipulation.

IOP Conf. Series: Materials Science and Engineering **260** (2017) 012038 doi:10.1088/1757-899X/260/1/012038

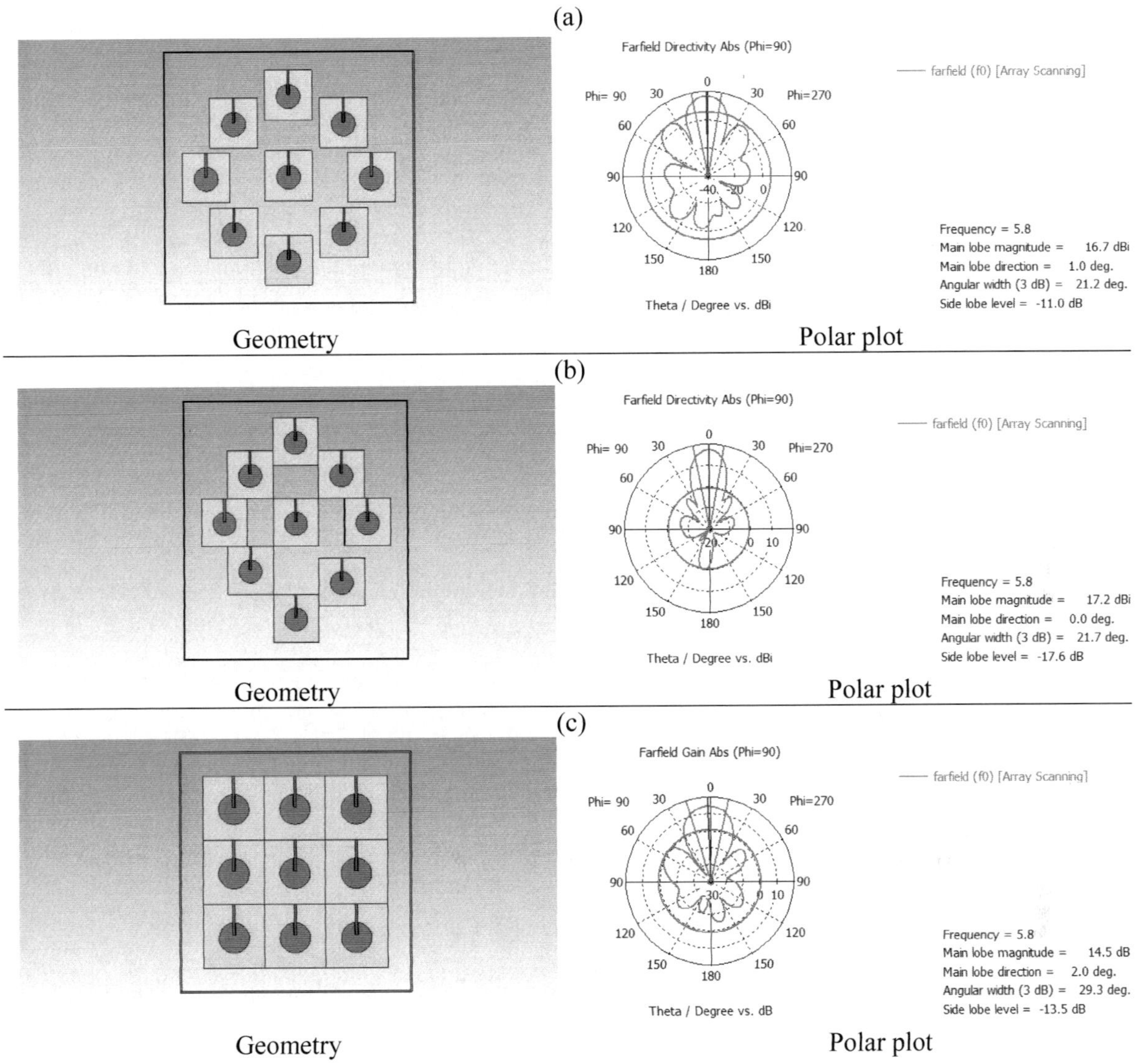

Figure 6. 9-element array (a) Uniform 8-element circular with central element (b) Non-uniform 8-element circular with central element (c) 9-element rectangular array.

4. Proposed circular array design

From the discussion above, the proposed circular array is the eight-element circular array with central element that is non-uniformly spaced, since this design achieves reduced return loss, provides higher gain, and produces uniform directional radiation pattern as a result of optimum non-uniform arrangement.

As demonstrated by Figure 7, the proposed design shows a return loss of -16 dB, while the mutual coupling is minimized to -27 dB, indicating good performance as mutual coupling is below than -20 dB. The design is optimized to resonate at the frequency of 5.8 GHz.

Compared to available literature [14], the proposed design demonstrates to be outperforming, where the gain is 15.7 dB compared to 14.2 dB of the available literature, considering the same number of elements.

(a)

(b)

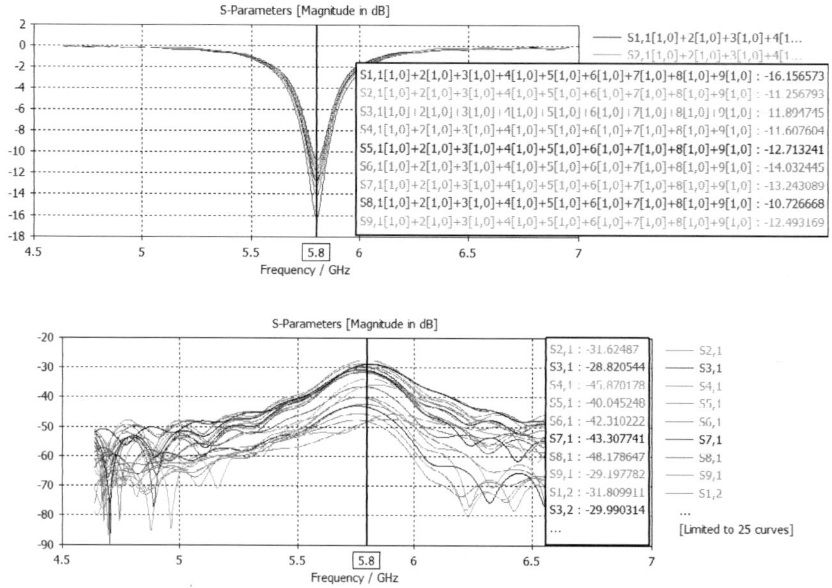

Figure 7. Eight-element circular array with central element (a) Return loss (b) Mutual coupling.

5. Conclusion

In this paper, an eight-element circular array with central element is designed for WLAN applications at 5.8 GHz. The design is optimized to have a minimum number of array elements for low cost design with optimized inter-element spacing in such a way to avoid the exhibition of grating lobes. The proposed design realizes high gain of 15.7 dB with high reduction in side lobe level of -17.6 dB.

Acknowledgement

Authors are grateful to Research Management Centre, International Islamic University Malaysia to support this research through research grant RIGS16-065-0229.

References

[1] Zhang Z, Xu G, Zheng Z, Chen X. A high gain array antenna using Giuseppe Peano Fractal Configuration. InAntennas and Propagation (APCAP), 2014 3rd Asia-Pacific Conference on 2014 Jul 26 (p 540). IEEE.

[2] Habaebi MH, Janat M, Islam MR, Hamida BA. Phased array antenna metamaterial based design operating in millimeter wave for 5G mobile networks. InResearch and Development (SCOReD), 2016 IEEE Student Conference on 2016 Dec 13 (pp. 1-4). IEEE.

[3] George RR, Castro AT, Sharma SK. Comparison of a four stage sequentially rotated wideband circularly polarized high gain microstrip patch array antennas at Ku-band. InAntennas and Propagation (EUCAP), 2017 11th European Conference on 2017 Mar 19 (p 2307). IEEE.

[4] Osman AS, Islam MR, Habaebi MH. Modeling of multiband/wideband stack series array antenna configuration for 5G application. InComputer and Communication Engineering (ICCCE), 2016 *Int. Conf.* on 2016 Jul 26 (pp. 349-354). IEEE.

[5] Nayna TF, Haque E, Ahmed F. Design of a X band defected ground circular patch antenna with diamond shaped slot and ring in patch for UWB applications. InSignal Processing, Communication, Power and Embedded System (SCOPES), 2016 *Int. Conf.* on 2016 Oct 3 (p 559). IEEE.

[6] Khan HA, Ullah S, Afridi MA, Saleem S. Patch antenna using EBG structure for ISM band wearable applications. InIntelligent Systems Engineering (ICISE), 2016 *Int. Conf.* on 2016 Jan 15 (p 157). IEEE.

[7] Myakochin YO. Usage of the antenna array for radio communication in locomotive engines in Russian Railways. InJournal of Physics: Conference Series 2017 Jan (Vol. 803, No. 1, p. 012104). IOP Publishing.
[8] Constantine AB. Antenna theory: analysis and design, fourth edition, John wiley & sons. 2016.
[9] Karmakar R, Chakraborty S, Chattopadhyay S. Impact of IEEE 802.11 n/ac PHY/MAC high throughput enhancements over transport/application layer protocols-a survey. arXiv preprint arXiv:1702.03257. 2017 Feb 10 (pp. 1-39).
[10] Adnan NH, Rafiqul IM, Alam AZ. Massive MIMO for Fifth Generation (5G): Opportunities and Challenges. InComputer and Communication Engineering (ICCCE), 2016 *Int. Conf.* on 2016 Jul 26 (pp. 47-52). IEEE.
[11] Ouyang T, Qian L, Ding L, Yang F. Mutual coupling calibration for uniform circular array using a single source. InWireless Communications and Signal Processing (WCSP), 2014 Sixth *Int. Conf.* on 2014 Oct 23 (p 1). IEEE.
[12] Dalli A, Zenkouar L, Bri S. Comparison of circular sector and rectangular patch antenna arrays in c-band. Journal of Electromagnetic Analysis and Applications. 2012 Nov 14;4(11): p 457.
[13] Affandi A, Husain M, Kasim N. Optimization for spherical phased array antenna. 2013;2(2): p 58.
[14] Ghayoula E, Bouallegue A, Chouinard JY, Ghayoula R, Smida A. Circular array synthesis using Taguchi algorithm for reliable IEEE 802.11 MIMO applications. InWireless Communications and Mobile Computing Conference (IWCMC), 2017 *13th Int.* 2017 Jun 26 (p 854). IEEE.

IOP Conf. Series: Materials Science and Engineering **260** (2017) 012039 doi:10.1088/1757-899X/260/1/012039

Applying Mechatronics to Improve the Safety of Children in Vehicles - What Can Be Done?

Khairul Hazziq Zufar and Ahmad Jazlan

Department of Mechatronics Engineering
Kulliyah of Engineering,
International Islamic University Malaysia,
Kuala Lumpur, Malaysia

Abstract. Nowadays, the media have reported an increasing number of cases where children are accidentally being trapped in vehicles while they parents and guardians are away attending to other matters. In this paper we discuss the feasibility of applying Mechatronics to improve the safety of children in vehicles with the ultimate goal of developing a means for parents,guardians and authorities to be informed if ever there is a child trapped in a vehicle and in need of urgent assistance. We have also presented some preliminary experiments we have carried out for a safety alert system which is currently being developed in our lab.

1. Introduction

Nowadays, cases regarding children trapped in vehicles can be observed through media frequently. Recently, Nurul Hidayah Bahaudin reported that a child as young as 6 years old had faced a deadly fate when his transporter forgot to send him to his evening school[1]. It was stated that the young boy was trapped inside the van for as long as 3 hours. The driver sent the boy to a nearby clinic but he was announced dead while receiving treatment. The doctor declared that the boy suffered heatstroke due to excessive exposure to heat.

Ishimine defined heatstroke as a situation where the core temperature of the body reaches 40°C and usually comes with the failure of the nervous system as the side effect due to excessive high temperature exposure [2]. The heat management mechanism of a growing young child's body is still under development. Due to the weak heat management, the body temperature of a child has a higher rate of temperature increment compared to an adult. It can be observed that the public is still taking vehicular heat stroke lightly. Most of them see the mishap as something that will never happen to them. As reported by McLaren et al., leaving the window car down did not reduce the rate of the rising temperature and the maximum achievable temperature [3].

Due to the ignorance of parents and the negligence of responsibility, the child is doomed to a deadly fate. Parents and caregivers should be well aware of the deadly consequences. Mechatronics has a promising potential to be applied in the development of a safety alert system to warn parents and authorities about the possibility of a child being trapped inside a vehicle. Such an alert system would need to be comprehensive, robust and consists of multiple features such as presence sensing, temperature sensing and real time alert mechanisms to drivers and caregivers to alert them of the presence of their young ones. It is hoped that with the development and eventual usage of such a

IOP Conf. Series: Materials Science and Engineering **260** (2017) 012039 doi:10.1088/1757-899X/260/1/012039

safety alert system the number of child deaths due to heatstroke in vehicles can be reduced or prevented all together.

2. Cases of Children Trapped in Vehicles

A number of research had been conducted to rule out some of the main causes of these accidents. Some of the main causes are as follows:

A) Parents negligence in leaving their child unintentionally
B) Dramatic change of routine
C) Stress or any form of mental fatigue

A hectic lifestyle contributes significantly to this problem. Looking at such cases that happened either in Malaysia or abroad, there are some common factors that causes parents to forget. Majority of these cases happen because the drivers claimed that they forgot due to the sudden change in their routine on a daily basis. Busy drivers usually rush to work every day but suddenly they have to break their routine to send their child to the nursery or school. The brain may not be able to comprehend such sudden changes to routine immediately. The driver may also place the child on the rear seat, causing the lack of interaction between the driver and the passenger, consequently resulting in the driver forgetting that there is indeed a child in the rear seat.

Considering a psychological aspect of the case, psychiatrists suggested that the element of stress was also as a common factor in the equation. Stress causes an unstable state of mind, thus making the driver less alert to surroundings. The stressful experience may occurred before or during the driving. In addition due to other problems such as lack of sleep, the brain may not able to process information such as detecting a sleeping child's presence inside the vehicle.

3. Hot Car Phenomenon

A study had been conducted by Jan Null from Department of Meteorology & Climate Science, San Jose State University to further understand the dynamics of a vehicle heating up due to surrounding conditions. According to the report, the interior and the windows of the car are said to be transparent to the sun's wave of radiation. Therefore, the wave heats up the object within the vehicle. With the increasing temperature of objects inside the vehicle such as dashboards, steering wheel and seats, these objects are capable to heat up the ambient air inside the car to conduction. These objects simultaneously release longwave radiation, making the interior of the car the most optimal "oven" due to lack of ventilation[4].

Some drivers deliberately leave their children inside the car with the intention to quickly perform a task and get back thinking that their children will indeed be safe during such a short duration. This however is a dangerous practice even if the windows of the vehicle are left open for ventilation. According to McLaren et al., winding down the windows of the car slightly do not decrease the rate of temperature increase inside the car or even decrease the limit of the temperature possibly achieved [3].

IOP Conf. Series: Materials Science and Engineering **260** (2017) 012039 doi:10.1088/1757-899X/260/1/012039

4. Existing Safety Empowering Technologies

Currently, the number of cases regarding child being trapped inside vehicles is still high. So far, there is no system that is being used by car manufacturers in Malaysia and abroad specifically to prevent such cases. There are however third-party products in the form of car accessories which are currently being marketed and commercialized especially in the United States. Augenbraun in his survey reported that a baby products manufacturer known as TOMY developed a product that may be one of the solutions to this predicament [5].

Produced in 2013, TOMY's Smart Car Seat was designed to communicate with the driver of the vehicle through a dedicated communication technology named iAlert. TOMY stated that the product features temperature, angle and motion sensor that will notify the driver if the condition of the car is not suitable or conducive to accommodate a child inside the car. Even though the system may sufficient to address the problem, users reported that the system failed to achieve the 0% percent failure rate.

Another currently commercialized product is the Baby Alert's ChildMinder SoftClip System Digital Wireless Technology Monitor [6]. This product was designed based on proximity sensing technology. The product comes in the form of a pair of sensors. One is supposed to be carried by the driver or the caregiver and the other pair will be attached to the baby car seat. As stated by Baby Alert, the pair carried by the caregiver will gives out an alert if the pairs are separated by a distance of more than 15 feet for approximately 6 minutes. Therefore, the caregiver will be reminded of the child's presence. Baby Alert also developed the alternate version of the product called ChildMinder Infant-Toddler ElitePad System. This product uses a pressure pad to detect the presence of the child instead of the proximity sensing principle. However, Augenbraun highlighted "…most deaths happen when there is a change in routine." He also stated the user is prone to forget the device because of its flaw in the design where the device needs to be handed over to the other caregiver if there is a change in role. Thus, the process will open up some room for errors [5].

Bologna reported in The Huffington Post that two fathers from Florida invented a device that can also be one of the gadgets that are capable to address this problem. Friedman and Shamma developed the idea and named it, Sense-A-Life. According to the article, the device was designed as a wireless and Bluetooth-capable and able to send inputs to the user through a smartphone app. With a pressure sensor installed inside the baby seat, the system is capable to read out the presence of a baby with a minimal weight of 0.9 kilograms. The system is also capable to sense if a driver is stepping out of the car while leaving the child inside the car without any supervision. If the situation occurs, the system will send an alert to the forgetful driver. A second driver can also be alerted if no response is being detected from the first driver [7].

The products described above are some options that were already commercialized. However, there are still room for improvements in the design of these products, especially those that may improve the reliability and the failure rate for long term use. There are still individuals or research teams working hard to innovate and invent foolproof systems to ensure that one day, the number of cases children trapped in vehicle can be reduced to 0%.

In 2014, Oshman Engineering Design Kitchen from Rice University developed an infant safety system called Infant SOS. The team of students had developed the system with the idea to keep the core temperature of the child's body below the life-threatening level. The fundamental aspect of this project that the system was developed to be attached or to be integrated with a normal baby seat. The system also produce alerts sufficient enough to tell the driver about the condition of the child. The system also features a relive system namely the cooling system where the system will ensure the temperature is maintained at a suitable level [8].

IOP Conf. Series: Materials Science and Engineering **260** (2017) 012039 doi:10.1088/1757-899X/260/1/012039

In 2003, a patent was granted to Bosch Robert Corp, Dulin and Seip in which they had developed a system to ensure prevention of passengers being trapped inside a vehicle [9].

In these existing systems, the parameters that are being measured and sensed inside the interior of a vehicle are a) if the temperature rise to a predetermined value, b) if the vehicle is in motion, and c) if the passenger is present. With these parameters, the mechanisms were to be executed based on the value of the measured parameters. Another feature that can be noted is that the system provides a "sleep mode" when there are no passengers that are being left behind the moment the driver leave the car. This feature was beneficial to reduce overall power consumption and longevity of the device.

Sasidharan and Kanagarajan reported that they had developed a safety alert system where their system monitors parameters such as oxygen concentration, temperature and humidity regularly. As mentioned by them, they took account the oxygen depletion rate with the accumulation percentage of poisonous gas such as carbon monoxide as gas suffocation is also considered as one of the causes of children's fatalities inside a car. Another aspect of their project that can be noted is that the system provides a form of a relief to the cabin of the vehicle. Basically, when the predetermined value of the temperature from the sensor achieved, the system will roll down one of the window of the car to a predetermined height of the window [10].

Another notable research is the work of Soto, Munoz-Sandoval, Ortiz, Nunez-Gozalez and Aguilar from Mexico. Their report entitled " A Multi-agente System to Avoid Heatstroke in Young Children Left in Baby Car Seats Inside Vehicles" divided the whole system into 2 subsystems namely User Agency and Monitoring Agency. The monitoring agency consists of a coordinator agent, a temperature sensor, a GPS module, a microphone and a motion sensor. As for the user agency, it works as the interface between the on-board system and the user. The user agency allows the system to directly communicate with the user and vice versa. In this particular work, Soto & et.al used the Arduino and Raspberry Pi to execute the written programs. With this configuration, the reported system uses Arduino to obtain value of designated parameters from sensors, initially connected to Arduino. The data then would be processed by the computing power of Raspberry Pi. Therefore, the Raspberry Pi acted as the brain of the system [11].

Lastly, a local team of researchers from Faculty of Computer Science and Information Technology, University Tun Hussein Onn, Malaysia developed a system to minimize heatstroke incidents among children. The proposed system utilized Bluetooth communication to transfer data to users and at the same time, the system used an Estimote beacon to approximate the distance of the driver from the car. As described in their publication, Hairulnizam et. al proposed an Android OS as their platform as because majority of OS users worldwide use Android OS [12].

5. Building a Safety Alert System Using a Mechatronics Approach

The foundation of the safety alert system which we are currently developing is made up of three core components. Firstly, the system will detect the voltage produced from the vibration from the engine of the vehicle. Any vibration which is detected indicates that the engine is still running, presumably the car is still in motion or the driver is still present in the car. Secondly, the system will also measure the temperature of the interior of the vehicle up to 40°C. Exceeding this temperature will put the child in a very dangerous situation. Lastly, the system will use a PIR motion sensor to sense any kind of presence in the car by detecting any motion inside the vehicle. Therefore, the circuit assembly of inputs will consist the circuit of vibration sensor, temperature sensor and PIR motion sensor. As for the programming of the software that will be uploaded into ATMega328p, the codes will be programmed in C language. When the codes had been written, with the developed algorithm, the system will process the inputs and produces the output in a form of text in the LCD display, LED to alert the presence of the child.

IOP Conf. Series: Materials Science and Engineering **260** (2017) 012039 doi:10.1088/1757-899X/260/1/012039

The monitoring unit for this proposed design will be consists of piezovibration sensor, temperature sensor, PIR sensor, reed switch and SKM53 GPS Module. As for the coordinating unit, the system will be solely be control by Arduino Uno running on the ATMega328p microprocessor. Lastly, the real-time output unit will be consisting on LCD display, LEDs and GPRS Shield V3.0.

The following section of this article describes the components which will be used in our proposed safety alert system. Figure 1 is a our system block diagram which describes our proposed method on how these components will work together to function as a complete safety alert system.

A) Vehicular Vibration Sensing

For this project, the vibration sensor will be one of the fundamental component in this system. During the initial phase of the execution of the program, the sensor will measure the vibration produced by the engine of the car. In the design, the sequential task of the system will be executed when the vibration sensor detects zero vibration from the engine. This indicates that the driver is ready to leave the vehicle. Due to the designated working environment for the vibration sensor, type of vibration that will be used is a piezo vibration sensor. Piezoelectric sensors can easily be adopted into the system due to its high temperature tolerance and flexibility in working with any type of force. This sensor can also be programmed to interface with any microcontroller such as Arduino Uno. Piezoelectric based sensors generate energy when they are put into pressured situations such as when the sensors are being pushed, pulled or pressed. This type of sensors also have the capability to produce AC voltage as high as 50V if it being configured correctly. Zener diodes are commonly used as a tool to scale down the voltage to a acceptable and safe levels. Piezoelectric sensors have the potential capability to produce high range of AC voltage when force are being applied.

B) Temperature Sensing

The second sensor that acts as a monitoring agent for this system is the DHT11 temperature and humidity sensor. For this sensor, the temperature measuring mechanism work with the NTC temperature sensor mounted inside. This sensor produces digital signal with auto calibration to ensure lower failure rate and prolonged performance for long term use. This sensor is highly compatible with Arduino Uno thus making it very flexible to be programmed for designated applications. Within the proposed system, this sensor will continuously measure the ambient temperature within the interior of the car. When the temperature reaches the maximum threshold value, the next sequential task will be executed.

C) Motion Sensing

Another sensor or input that forms the core function of the system is the PIR motion sensor. A PIR sensor can be affected positively and negatively by many uncertainties or variables. This motion sensor works with the principal of differential changes. This is because the output of this motion sensor is triggered by the change of pulses back and forth between positive and negative differential changes. The body of the sensor is made up of metal that can positively affect its performance. The detection range of this type of motion sensor can be widened by changing the refraction range of the lens. The commonly used lens are the Fresnel lenses that gives out large refraction of IR wave.

D) Magnetic Switching

To complement the system, one of the mechanism developed for the this project will be using the reed switch or also known as a magnetic switch. The reed switch is common among daily appliances such as mobile phones and laptops. The basic principle of this switch is that the switch is N/O or Normally Open type of switch within the circuit. Therefore, with this principle, the user can determine the time to make the circuit closed by ensuring that the pair of the switch come into contact with each other,

IOP Conf. Series: Materials Science and Engineering **260** (2017) 012039 doi:10.1088/1757-899X/260/1/012039

thus making the circuit closed. This switch is commonly made from ferromagnetic material to ensure a long shelf life.

E) Global Positioning Systems

In a life-threatening situation such as a child being trapped in a vehicle and the parents are not responding to initial alerts and notifications, the authorities need to dispatched to rescue the child as fast as possible. The authorities such as fire departments and police need to have the information regarding the exact location where the child is trapped. Therefore, they need any specific information about the location as fast as they can. Because of this reason, the development of this project will be including location tracking so that the location can be relayed to the authorities beforehand enabling efficient routes for the authorities to arrive.

For location tracking, a GPS module can be integrated together with the Arduino Uno, allowing the system to send specific locations (in term of latitude and longitude). GPS module also features real time clock tracking where the latitude and longitude of the location will be send together with real time stamp so that the authority can approximate their Estimated Time of Arrival (ETA) efficiently.

F) Coordinating Unit

In order to achieve the intended objective for this project, the "brain" of the system has been selected based on some of the features that must be taken into consideration such as size, number of GPIO pins, CPU clock speed, power consumption rate and mode of voltage input and also the programming flexibility. After an extensive research conducted among the common used microcontrollers and development boards, Arduino Uno has been chosen as the main controller of the system.
The Arduino Uno's main processor unit is the ATmega328p microprocessor. With the combination of Arduino's development board, the Arduino Uno features 14 digital GPIO pins. Within these pins, there are 6 of these pins can be used as PWM out pins. Analog sensors such as temperature sensors and light intensity sensors can be attached at one of the 6 analog pins on the board. The board also includes a 16 MHz crystal that can be used as clock for serial data transmission. With this, Pulse Width Modulation can be used to allow DC simulating AC with preferred parameters such as signal amplitude, frequency and duty cycles.

G) GSM Module

Developed in 1970, GSM became a widely used medium of communication worldwide. GSM provides a channel for digital cellular transmission such as voice messages and text messages using fixed frequency specially for GSM communications. Through TDMA method, GSM network convert the analog data into digital while compressing its size. Next, the digitized data is send to client data with rate from 64kbps to 120 Mbps. To use the GSM module with Arduino, the module needs to interfaced with a level shifter IC Max232. With a SIM card, the AT command can be transmit to the recipient of the command.

IOP Conf. Series: Materials Science and Engineering **260** (2017) 012039 doi:10.1088/1757-899X/260/1/012039

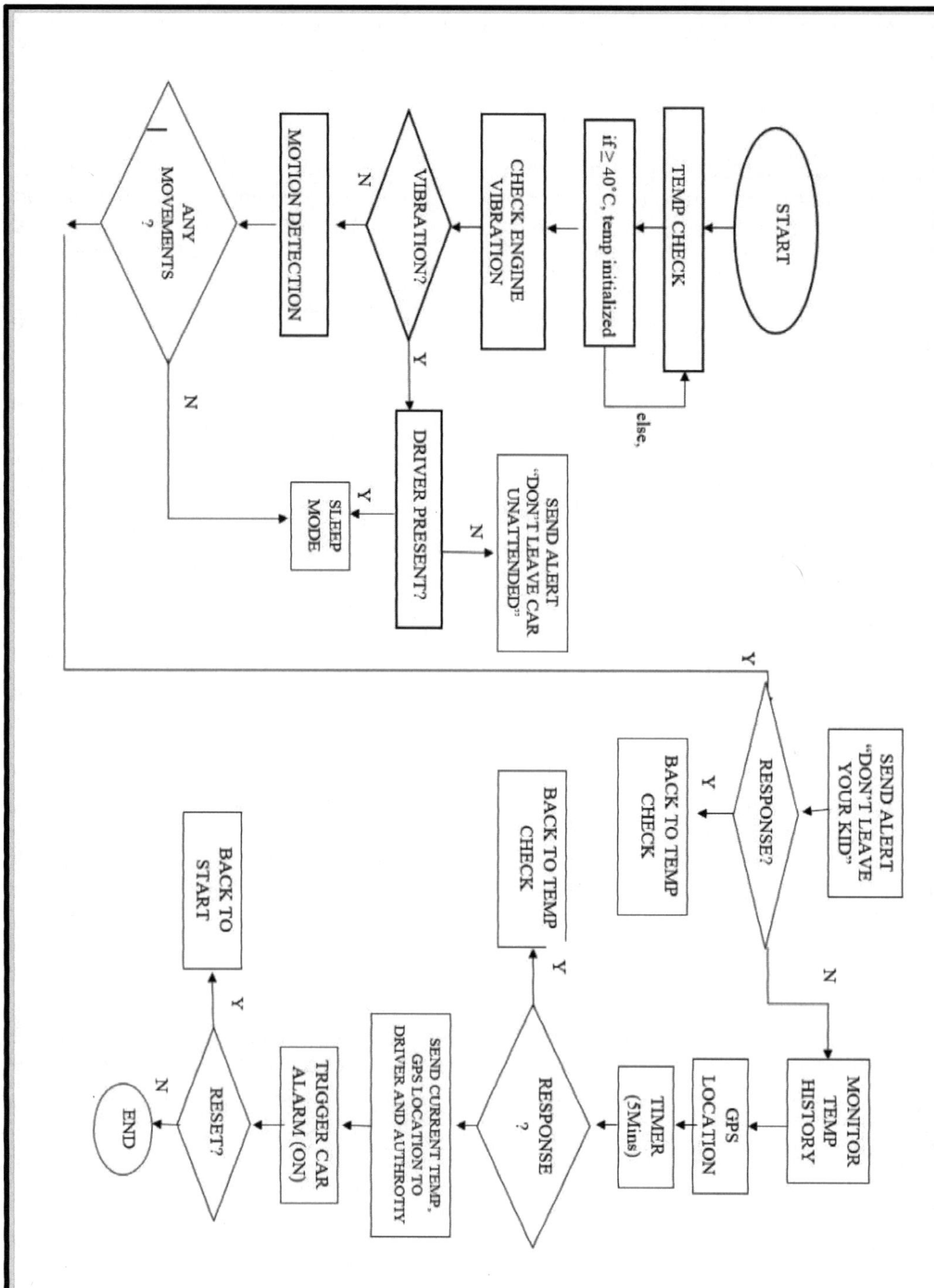

Figure 1: Proposed System Block Diagram.

6. Preliminary Experiments

To gain an insight on how temperature gradually increases inside the cabin of a vehicle, a preliminary experiment was conducted with the objective to observe the rate of temperature increase. With the collected data, further analysis can be done and crucial information will be noted to ensure that the developed system can effectively and efficiently handle the high temperature without facing any equipment failure while being continuously subjected to heat.

Another objective of this experimentation is that previous research by McLaren showed that the interior temperature of the vehicle quickly escalated 56°C to 67°C when the atmospheric temperature rises beyond 30°C[3]. This drastic increase in temperature occurs even though the surrounding temperature is constant. This result can be related closely with Malaysia's climate. According to the monthly report of temperature by Malaysian Meteorological Department, their latest report of March 2017 showed average morning temperature range of 22.1°C to 34.2°C [14]. This temperature range may result in drastic temperature escalation inside the cabin of the vehicle.

This experiment was conducted based on McLaren's work [3]. As mentioned, the objective of this experiment was to study the variation in temperature in the interior of the vehicle and the effect of the surrounding temperature to the variation of temperature inside the vehicle. At 1.00 PM, 6th May 2017, the experiment site was at IIUM Convest Hill, parking area. The vehicle, a Perodua Myvi, black in colour, was parked under full sunlight oriented at 45 degrees from the sunlight as shown in Figure 3to reduce the effect of direct sunlight through the windshield of the vehicle that may cause slight deviation from the actual value of temperature data. The duration for this experiment was 15 minutes with the interval data logging of 1 seconds per data. Before the apparatus were placed inside the vehicle, the door of the vehicle was opened to ensure that the initial temperature of the interior achieved the atmospheric temperature. The temperature data recorded shows continuously increasing temperature inside the vehicle as shown in Figure 4, although the atmospheric pressure at that time was constant. This is consistent with the fact stated by McLaren [3].

As for data logging, the configuration of the sensors and Arduino is showed in Figure 2. The configuration was integrated with PLX DAQ, a data logging software that can be integrated together with Microsoft Excel with real time data logging features. The setup was then placed inside the vehicle.

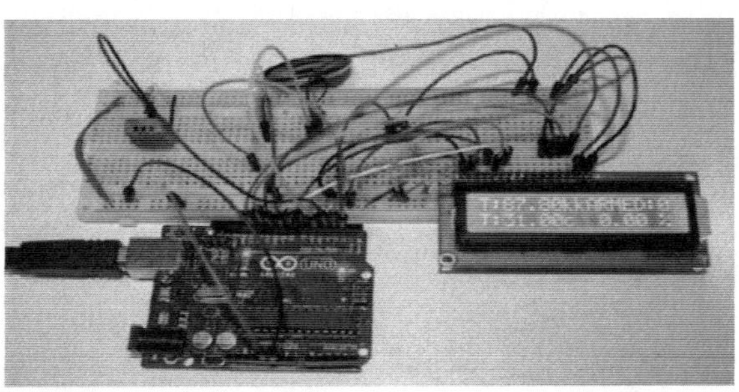

Figure 2: Configuration for preliminary experiment

Figure 3: Vehicle and Experiment Site.

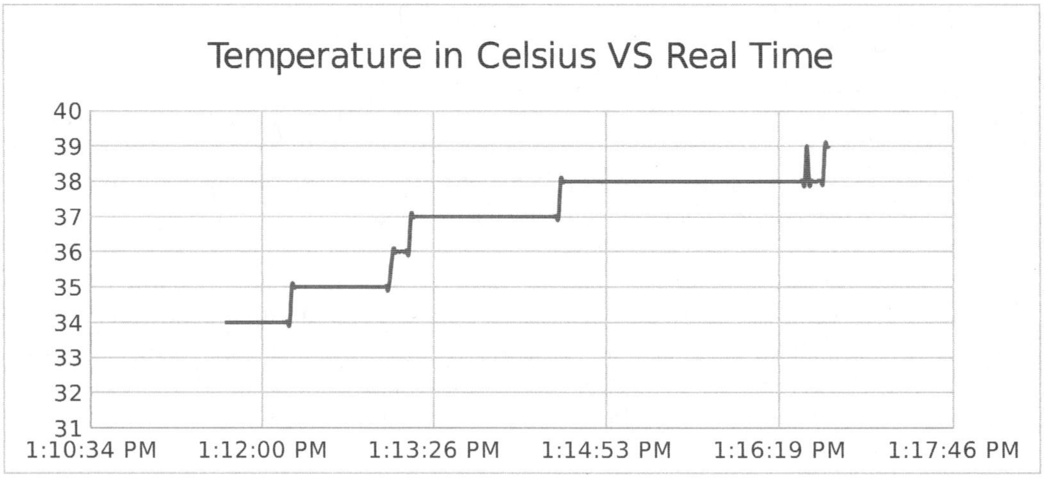

Figure 4: Variation of Temperature inside Vehicle with Respect to Time.

7. Conclusion

This paper presents a review of existing technology available to prevent children from being accidentally trapped inside vehicles. We have also presented a framework for developing our own safety alert system to address drawbacks in existing technologies. We have presented some preliminary results regarding temperature variations inside a vehicle. Further work is underway to develop a fully functional safety alert system embedded with communication abilities.

8. References

[1] Bahaudin, N. (2017), 'Saya maafkan dia'. HM Online. Retrieved 1 May 2017, from http://www.hmetro.com.my/node/223840.

[2] Ishimine, P. (2016). Heat stroke in children. Uptodate.com. Retrieved 2 May 2017, from https://www.uptodate.com/contents/heat-stroke-in-children#H26

[3] McLaren, C. (2005). Heat Stress From Enclosed Vehicles: Moderate Ambient Temperatures Cause Significant Temperature Rise in Enclosed Vehicles, *Pediatrics*, Vol 116, No 1., pp.e109 - e112.

[4] Null, J. (2016). Fact Sheet - Heatstroke Deaths of Children in Vehicles. Retrieved 7 May 2017, from http://www.noheatstroke.org/original/index.html

[5] Augenbraun, E. (2014). How technology could help prevent kids' deaths in hot cars | KidsandCars.org. Kidsandcars.org. Retrieved 7 May 2017, from http://www.kidsandcars.org/2014/07/10/how-technology-could-help-prevent-kids-deaths-in-hot-cars/

[6] Baby Alert International. ChildMinder® Elite Pad Single System. Retrieved from https://www.babyalert.info/childminder-elite-pad-single-system.html

[7] Balogna, C. (2016). Dads Invent Simple Device To Prevent Hot Car Fatalities. The Huffington Post. Retrieved 30 April 2017, from http://www.huffingtonpost.com/entry/dads-invent-simple-device-to-prevent-hot-car-fatalities_us_56f53ed7e4b0a3721819b041

[8] OEDK - Rice University - Member public profile Infant SOS. (2014). Oedk.rice.edu. Retrieved 30 April 2017, from http://oedk.rice.edu/Sys/PublicProfile/25607033

[9] Robert Bosch Corporation. (2002). Hot Vehicle Safety System And Methods Preventing Passengers Entrapment and Heat Suffocation. U.S Patent, WO2002087910 A3, issued May 22, 2002.

IOP Conf. Series: Materials Science and Engineering **260** (2017) 012039 doi:10.1088/1757-899X/260/1/012039

[10] Sasidharan, S, and Kanagarajan, V. (2015). Vehicle cabin safety alert system. In Proc: 2015 International Conference on Computer Communication and Informatics (ICCCI 2015), Coimbatore, pp.8-11.

[11] Barrera, P.S. Sandoval, G. M., Ortiz, G.C., González R. N. and Aguilar, E. R. (2014). A Multi-agent System to Avoid Heatstroke in Young Children Left in Baby Car Seats inside Vehicles. In Proc: 22014 International Conference on Computational Science and Computational Intelligence, Las Vegas, NV, 2014, pp. 245-248.

[12] Mahdin, H., Omar, A., Yaacob, S., Kasim, S., & Fudzee, M. (2016). Minimizing Heatstroke Incidents for Young Children Left inside Vehicle. *IOP Conference Series: Materials Science And Engineering*, *160(2016)*, 012094.

[13] Gabay, J. (2011). Fundamentals of Piezoelectric Shock and Vibration | DigiKey. Digikey.com. Retrieved 27 April 2017, from https://www.digikey.com/en/articles/techzone/2011/dec/fundamentals-of-piezoelectric-shock-and-vibration-sensors

[14] Monthly Weather - Malaysian Meteorological Department. (2017). Met.gov.my. Retrieved 4 May 2017, from http://www.met.gov.my/in/web/metmalaysia/publications/bulletinpreview/monthlyweather

Object Detection Technique for Small Unmanned Aerial Vehicle

M. Faiz Bin Ramli**, **Ari Legowo***, **Syariful Syafiq Shamsudin****

*Department of Mechanical Engineering, Kulliyyah of Engineering, International Islamic University Malaysia
**Faculty of Mechanical and Manufacturing Engineering, Universiti Tun Hussein Onn Malaysia

legowo@iium.edu.my

Abstract. Obstacle detection and avoidance is desirable for UAVs especially lightweight micro aerial vehicles and is challenging problem since it has payload constraints, therefore only limited sensor can be attached the vehicle. Usually the sensors incorporated in the system is either type vision based (monocular or stereo camera) or Laser based. However, each of the sensor has its own advantage and disadvantage, thus we built the obstacle detection and avoidance system based multi sensor (monocular sensor and LIDAR) integration. On top of that, we also combine SURF algorithm with Harris corner detector to determine the approximate size of the obstacles. In the initial experiment conducted, we successfully detect and determine the size of the obstacles with 3 different obstacles. The differences of length between real obstacles and our algorithm are considered acceptable which is about -0.4 to 3.6.

1. Introduction

An Unmanned Aerial Vehicle (UAV) as the name implies, it is an aircraft without a pilot on-board or attached to it. It can fly autonomously by pre-programmed the UAV's system or controlled manually by the pilot on the ground. Hypothetically, UAVs have a great potential to replace manned aircraft in many tasks or missions (e.g. surveillance, search and rescue mission, attack mission, collecting data, and etc.). For example, in 1946 to 1948, United States Air Force and Navy used unmanned B-17 and F6Fs to fly into nuclear clouds within minute after bomb detonation to collect radioactive samples [1]. The concern to operate UAVs in a low altitude either indoor or outdoor are indeed very demanding due to the growing number of small and micro air vehicles.However, to achieve robust level on self-governing system is very challenging and it is important needs for modern-day UAVs.

Obstacle detection and avoidance system highly depend on sensors that being employed on the UAVs, type of the most research and considered popular choice for the system will be a Vision based and Laser based. The selection of detection and avoidance systems are very well connected to the UAVs itself. Besides, selecting a proper sensor to be placed on board of the UAVs plays a critical role for the system, and each of the techniques has its own advantages and disadvantages.

For example, vision based method has rich information regarding bearing angle of the detected obstacles, while distance from the vehicle to obstacles are poorly recognized. On the other hand, Laser based method is able to find accurately the distance of the obstacles to the UAVs, but weak in determine the bearing angle. In order to encounter these disadvantages, different type sensors can be fused or combined together to make the system operate into a reliable detection and avoidance system.

Integration of different sensors (e.g Laser and Optical) allows the extraction of information which cannot be acquired by a single sensor [2]. It will upgrade and increase the capability of the system to estimate the attitude of the obstacles.The concepts of multisensor fusion on one system can be traced back to the humans and animals, they naturally perfomed the sensors fusion to do lots of things (e.g identication of threats, assessment of surrounding enviroments,etc). For example, precise evaluation of some substances by animals can be done by combining different senses such as sight, touch, smell and taste [3].

Nowadays, commercial UAVs are already equiped with single camera (monocular vision) on-board and can be pre-programed to use features based detection techniques. Hence, to take advantage of this matter, the research project are focusing on monocular vision features based taken from computer vision technique. Plus, considering the size, weight, power consumption and capability to extract useful information, camera is the most competitive tool for UAV [4].Therefore, this dissertation mainly aims at developing a real-time obstacles detection and avoidance system based on multi-sensor integration (combination of optical and laser sensors) to achieve autonomous operation for UAVs. Plus, the aims is also to estimates or approximates the size congfiguration of the obstacles for safe avoidance purposes.

2. Related works

Obstacles detection and avoidance system is desirable for small and lightweight UAV and is challenging problem since the payload constraints, it only allows monocular cameras to act as a sensor to detect the obstacles. Eventhough, monocular cameras cannot measure the depth of scenes in an image, various cues can be computed to help with the hitch. In [5,6]uses expansion cues to detect a frontal obstacles by using only single camera. He used Speeded Up Robust Features (SURF) features matches in combination with template matching image to compare relative obstacles sizes. Abdulla Al-Kaff et al [7] estimates the size ratios of the approaching obstacles from the sequences image frames. He used SIFT [8] featuresdetected points and contructs a convex shape from these points, after that, he observed size changes of the convex shape when obstacles are approaching.In [9], a MOPS are used to detect the outline or edges of the obstacles. Then, they apply SIFT features to the image of the obstacles to detect the internal outline. Cooper Bills et al [10], proposed a method to fly the UAV in indoor enviroments by using perspective cues.However their experiments, are strictly applied to corridor and stairs regions. Detection of obstacles based on texture variatons has been demonstrated by G.C.H.E de Croon [11]. By using the facts that when observers approach an object, there are two effects happen which are the image size of the object increases in the image and the detailed texture in view becomes more and more visible.

On the other hand, there is also method used by monocular vision known as optical flow. It is a natural solutions to navigation and obstacle avoidance problem, as motivated by insect and birdflights [12]. Green and Oh [13] has built MAV platform than can enabled a quick transition from cruise flight into hovering mode to avoid collision by using this method. However, they stated that optical flow has limitations when it comes to approaching an obstacles directly (frontal obstacle), because the detection very poor. Simon zigg et all [14] use the method to navigate the UAV through indoor corridors. Zufferey et all [15], use optical flow on their microflyer weighing only 30 g to be fly in small indoor environment.

In [16], sensor fusion of ultrasonic (US) and infrared (IR) sensors is developed to obtain reliable range data for obstacle detection purposes. Results from the project showed that sensor fusion provided accurate range estimation by reducing noises and errors that were present in individual sensors measurements. In [17], they used a combination of Ka-band radar and optical cameras as their detection system. Tomic et all [18] use full capability of laser based sensor and vision based sensor. He integrate a wide LIDAR sensor and stereo vision camera in it UAV.Shadib et all [19] employed optical sensors and U.S sensor for their autonomous mobile robot. They determine the distance to obstacles by measuring the time difference between the transmission of ultrasonic wave and the reception of reflected wave.

3. Methodology

An obstacle detection technique is described as based on the combination of features based detection by monocular camera and laser sensors.Features based detection is one concept in computer vision field to extract the interesting or distinct features on an image such as edges, corners and blobs. It is generally used for finding an object (e.g matching features in one image to another) and tracking.

IOP Conf. Series: Materials Science and Engineering **260** (2017) 012040 doi:10.1088/1757-899X/260/1/012040

Two of widely know features based detection technique which are SURF created by Herbet Bay [20,21,22] and Harris corner by Chris Harris [23] will be employed in this project.

3.1. Obstacle detection

The presented obstacle detection method are based on combination output from different sensors which is camera and Laser sensor. It allows the measurement of length of the obstacles from far for the purpose of earlier decision making in avoidance and also safe avoidance. It is desired for obstacle detection and avoidance system to have good approximation on the length of the obstacles rather than using a tolerance.This is true when a UAV has to fly through tight spaces and enviroment that filled with obstacles. In pursuance of that matter, we will describe the process in the following chapter.

3.1.1 Laser sensor detection

Laser sensors can be used to measure a distance to obstacles with very high accuracy. On top of that, the range of detection very high as compared to U.S sensor and I.R sensor. However, most of worthy distance laser sensors available also known as LIDAR (e.g Hokuyo,SICK,etc) is heavy and big in size. Thus, it is not suitable for application in small UAVs. In this project, we adopted small and compat LIDAR for the system, which use only single ray of beam to find the distance. This sensor performs initial detection by first capture the appropriate distance to obstacles and instruct the camera to capture the reference image frame. Secondly, inform the camera sensors to capture target image frame after avoidance distance has been detected (see figure 1).

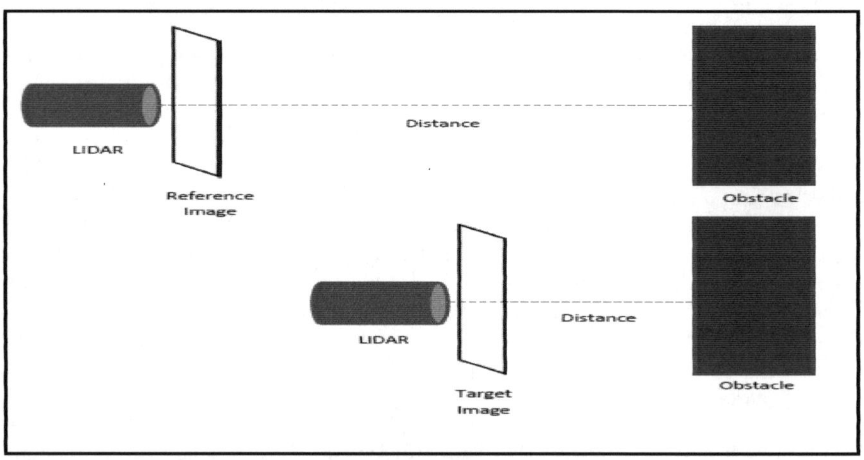

Figure 1:Image frame illustration

3.1.2 SURF detection

Since the system equipped with LIDAR sensor to measure the distance to obstacles and instead of detecting the incoming obstacles by using size expansion, we used the SURF algorithm to help recognize and identify the obstacle size, particularly on length of obstacles.SURF algorithm also can be considered as a Blob detector because the scale produce by this algorithm will direcly propotional to the size of detected blob or in another words size of particular object in an image. Therefore, we took advantage of this characteristic to distinguish the detected obstacles from background of the image. This is neccesary to allow for the calculation of obstacle's length. However, we must determine the rightful scale changes between obstacles that are being captured in reference image frame (RIF) and the one in target image frame (TIF) as shown in figure 2.

Figure 2: SURF keypoints features

We have conducted several experiments to determine the appropriate scale changes by varying the distance between RIF and TIF. For this experiment, TIF distance has been fixed to 4ft.This is due to the fact that, if the TIF distance keep increasing, then obstacle size or blob will be small in the image as compared to the surrounding. As a result, the SURF algorithm will produce a lot of keypoints to the background rather than to the obstacles itself. On the other hand, if the TIF distance is small enough, the avoidance planning by the UAV towards the obstacles will be a problem. These scale changes are the scales that are able to separate or distinguish the obstacles from background of an image.

Table 1: Scale changes

RIF (ft)	RIF-TIF (ft)	Scale changes
5	1	0.8
6	2	0.9333
7	3	1.2
8	4	1.8667
9	5	2
10	6	2.8
11	7	3.3333
12	8	3.4667
13	9	3.6

In adjusting the distance of RIF, amount of matching feature keypoints between frames should be take into account. The higher amount of matching feature keypoints detected, the higher chances of length of obstacles approximation will be near to true value.

3.1.3 Location maxima keypoints

SURF algorithm generate feature keypoints with sets of 2 data which are scale of the features and location of the features with regard to an image. After the obstacle features keypoints have been subtracted from the background of an image, we need to find the maxima (minimum/maximum) location in X-direction of the features keypoints located on the obstacle. Any features keypoints that lies within the boundary of maxima location will be removed.

3.1.4 Binary image conversion

The image frame in TIF is converted to binary from greyscale image. Binary image contain only 2 pixels value which are 0= white and 1=black, while greyscale pixels value is from 0 to 255. To perfomed this conversion we use luminace level of 0.5, which means any greyscale image pixels higher than 127.5 will replace into white otherwise black. Having performed this conversion, edge of the obstacle can be localized or transformed into many steps corner (see figure 3). We take this advantage to know the edge coordinate in the image by applying Harris corner detector.

Figure 3: Corner step edge

3.1.5 Finding discontinuities

If Harris corner detector is apply to the binary image, it produce a lot of corner points in the background as well as in the obstacle. Thus, some method to extract corner points only at the edge of the obstacle are needed because corner point coordinates can be used latter to find pixels and real dimension relation. Small window or patch is created to find the discontinuities of the Harris corner points in the image. The patch is of size 30 x 40 pixels and it search the discontinuities at step size of 1 pixels on both side (left and right) of the image. If the discontinuities is found, then we assume the point of discontinuities in the image is edge coordinates of the obstacle (see figure 4).

Figure 4: Moveable patch window

3.1.6 Finding the length

As previously stated, pixels to real dimension relation is used to find the real dimension (length) of the obstacles. Besides, a simple triogonometry function is also incorporated in the calculation.

$$PD = RCL - LCL \quad (1)$$

$$IL = (TIF \times tan31°) \times 2 \quad (2)$$

$$\frac{P}{ft} = \frac{IL}{1280} \quad (3)$$

$$TL = PD \times P/ft \quad (4)$$

3.2 System configuration

The platform used in the experimentation is the A.r drone 2.0 elite edition, a low-cost MAV built by the company Parrot. This vehicle was selected because it is low-cost commercial UAV available in market, stable flying and proper size configuration. The cameraattached to A.r drone has 1280x720 resolution and horizontal field of view is 62 degree. On the other hand, LIDAR-Lite from Garmin will act as laser sensor distance in this project. It has low weight of 22 g, hence, making it suitable as the additional sensor for UAV application. In addition, the range that the LIDAR-Lite can cover is very impressive which is about 40 m. In this research, the main tool or software that we used for analysis and calculation is Matlab 2016a. The software contain a computer vision system toolbox which makes it possible to use as tool development for our obstacle detection and avoidance system. All our algorithm processed on ground laptop which is quad core Intel i7 running windows operating system.

4. Experiment results and discussion

Experiments are perfomed by moving the UAV from RIF position to TIF position and record the image frame captured. There are 3 obstacles used to be detect by the algorithm in the experiment and each has different sizes. However, for our initial experiment and algorithm validation, we only carry out the experiment with UAV at RIF position equal to 10 ft and stop at TIF position equal to 4ft. In our future experiment, we will test the algoritm at different RIF position to simulate the scale changes as indicated in previous table 1. The algorithm process which produce the approximation length dimension of the obstacles are to be validated with real length of each of the obstacles. From the experiment, the algorithm is able to detect and determine the length of 3 obstacles used in the experiment. It shown that the accuracy of algorithm is exceptional because the length generated by the algorithm (True length) as compared to real length is only by within -0.4 to 3.6 cm. In obstacles 3, the difference is slightly greater than the rest because during the experiment, the captured image frame by the camera at TIF position may got tilted a little to the left or right which is not exactly perpendicular to obstacle. As a consequence, egde coordinate of the obstacle 3 detected by Harris corner detector will be underestimated and so, the length approximated is much less than the real length.

Table 2: Results

Obstacles	Real length (cm)	True length (cm)	Difference
1	58	59.8	1.8
2	39	38.6	-0.4
3	55	51.4	3.6

IOP Conf. Series: Materials Science and Engineering **260** (2017) 012040 doi:10.1088/1757-899X/260/1/012040

5. Result and Discussion

In this research project, multi sensor integration for obstacle detection and avoidance system has been proposed. We have combine monocular sensor with LIDAR sensor to help increase the robustness and accuracy of the system. By, using this system with combination of SURF algorithm and Harris corner detector, we are able to determine the approximate size of the obstacles. Experiment have been performed with 3 different sizes of obstacles and the results shows that our algorithm closely approximate the length of the obstacles.

In our future work, the algorithm will be implement in real time environment including the development of avoidance algorithm once we have the output from detection algorithm. We also eager to investigate the detection and avoidance decision mechanism when multiple obstacles are introduced. In the next experiment, all of RIF position and variety of obstacles will be tested to futher validate the algorithm.

References

[1]. Stephen A. Cambone (2005),"*Unmanned aircraft system roadmap 2005-2030*", Office of The Secretary of Defense.
[2]. David L. Hall, James Llinas (2001),"Introduction to multi sensor data fusion", Proceedings of the IEEE, vol 85.
[3]. David L. Hall, Amulya K. Garga (1999)," , Pitfalls in data fusion (and how to avoid them)", Proc. Eurofusion 99 Stratford on Avon UK.
[4]. Farid Kendoul (2011)," Survey of advances in guidance , navigation, and control of unmanned Rotorcraft system", Journal of Field Robotics 29(2).
[5]. Tomoyuki Mori, Sebastian Scherer (2013)," First result in detecting and avoiding frontal obstacles from a monocular camera for micro unmanned aeril vehicles", IEEE International Conference on Robotics and Automation (ICRA).
[6]. Wilbert G. Aguilar, Veronica P. Casaliglla, Jose L. Polit (2017)," Obstacle avoidance for Low-cost ", IEEE 11th International Conference on Semantic Computing.
[7]. Abdulla Al-kaff,Qinggang Meng, David Martin, Arturo de la Escalera, Jos´e Mar´ıa Armingol (2016),"Monocular Vision-Based Obstacle Detection/Avoidance for Unmanned Aerial Vehicles", IEEE Intelligent Vehicles Symposium (IV).
[8]. David G. Lowe (2004),"Distinctive Image Features from Scale-Invariant Keypoints", International Journal of Computer Vision.
[9]. Jeong-Oog Lee, Keun-Hwan Lee, Sang-Heon Park, Sung-Gyu Im, Jungkeun Park (2010),"Obstacle avoidance for small UAVs using monocular vision", Aircraft Engineering and Aerospace Technology: An International Journal 83/6 (2011) 397–406.
[10]. Cooper Bills, Joyce Chen, Ashutosh Saxena (2011),"Autonomous MAV flight in indoor environments using single image perspective cues", IEEE International Conference on Robotics and Automation (ICRA).
[11]. G. de Croon, E. de Weerdt, C. De Wagter, B. Remes (2010), " The appearance variation cue for obstacle avoidance",IEEE International Conference on Robotics and Biomimetics (ROBIO).
[12]. Haiyang Chao, Yu Gu, Marcello Napolitano (2014)," A Survey of Optical Flow Techniques for Robotics Navigation Applications", J Intell Robot Syst.
[13]. William E. Green, Paul Y. Oh (2008),"Optic-Flow-Based Collision Avoidance",IEEE Robotics & Automation Magazine.
[14]. Simon Zingg, Davide Scaramuzza, Stephan Weiss, Roland Siegwart (2010),"MAV navigation through indoor corridors using optical flow", IEEE International Conference on Robotics and Automation (ICRA).
[15]. Jean-Christophe Zufferey, Dario Floreano (2006),"Fly-Inspired Visual Steering of an Ultralight Indoor Aircraft", IEEE transactions on robotics, vol 22.
[16]. RethNaraj Rambabu, Muhammad Rijaluddin Bahiki, Syaril Azrad,(2015),"*Multisensor Fusion Based UAV Collision Avoidance System*", Asia Pacific International Symposium On Aerospace Technology (APISAT). Department of Aerospace Engineering Jurnal Teknologi.
[17]. GiancarmineFasanoa, DomenicoAccardoa, AnnaElenaTirria, AntonioMocciaa, EttoreDeLellisb (2015),"Radar/electro-optical data fusion for non-cooperative UAS sense and avoid",Elsevier Aerospace Science and Technology.
[18]. Teodor Tomic, Korbinian Schmid, Philipp Lutz, Andreas Domel, Michael Kassecker,Elmar Mair, Iris Lynne Grixa, Felix Ruess, Michael Suppa, Darius Burschka (2012),"Towards a fully autonomous UAV", IEEE ROBOTICS & AUTOMATION MAGAZINE.

IOP Conf. Series: Materials Science and Engineering **260** (2017) 012040 doi:10.1088/1757-899X/260/1/012040

[19]. Fayaz Shahdib, Md. Wali Ullah Bhuiyan,Md. Kamrul Hasan, Hasan Mahmud (2013)," Obstacle Detection and Object Size Measurement for Autonomous Mobile Robot using Sensor", International Journal of Computer Applications (0975 – 8887) Volume 66– No.9.

[20]. Herbert Bay, Tinne Tuytelaars, and Luc Van Gool (2006)," SURF: Speeded Up Robust Features", Proc. European Conf. Computer Vision.

[21]. Edouard Oyallon, Julien Rabin (2015),"An Analysis of the SURF Method", IPOL Image processing online.

[22]. Jacob Toft Pedersen (2011)," SURF: Feature detection & description".

[23]. Chris Harris, Mike Stephens (1988)," A combined corner and edge detector", In Proceedings of the Alvey Vision Conference.

IOP Conf. Series: Materials Science and Engineering **260** (2017) 012041 doi:10.1088/1757-899X/260/1/012041

Performance analysis of rain attenuation on earth-to-satellite microwave links design in Libya

Islam Md Rafiqul, Asma Ali Hussein Budalal, Mohamed H Habaebi, Khairayu Badron, Ahmad Fadzil Ismail

Department of Electrical & Computer Engineering, Faculty of Engineering International Islamic University Malaysia, Jalan Gombak, 53100 Kuala Lumpur, Malaysia

rafiq@iium.edu.my, asma.budalal@ymail.com, habaebi@iium.edu.my, khairayu@iium.edu.my, af_ismail@iium.edu.my

Abstract. Performances of earth-to-satellite microwave links operating in Ku, Ka, and V-bands are degraded by the environment and strongly attenuated by rain. Rain attenuation is the most significant consideration and challenge to design a reliable earth-to-satellite microwave links for these frequency bands. Hence, it is essential for satellite link designer to take into account rain fade margin accurately before system implementation. Rain rate is the main measured parameter to predict of rain attenuation. Rainfall statistical data measured and recorded in Libya for the period of 30 years are collected from 5 locations. The prediction methods require one minute integration time rain intensity. Therefore, collected data were analyzed and processed to convert into one–minute rain rate cumulative distribution in Libya. The model proposed by ITU-R is used to predict and investigate rain fade based on converted 1-minute rain rate data. Rain fade predicted at two locations are used for performance analysis in terms of link spectral efficiency and throughput. V-band downlink shows that 99.99% availability is possible in all the Southern part stations in Libya at 0.29 bps/Hz spectral efficiency and 20.74 Mbps throughput when 72 MHz transponder band width is used which is not feasible in Northern part. Results of this paper will be a very useful resource to design highly reliable earth-to-satellite communication links in Libya.

1. Introduction

Future satellite communication systems are demanding towards the use of higher frequency bands in order to achieve larger bandwidth[1]. These bands offer many advantages to satellite communication services and applications, in terms of spectrum availability, reduced interference potential and minimized equipment size[2]. However, rain attenuation is an important factor that limits the use of frequency above C-band in tropical and sub-tropical regions. Atmospheric hydrometer affects the characteristic of the satellite signals[3]. It causes the reduction in the power density of an electromagnetic wave as it propagates through the rainy path[4]. It is directly responsible for link outages and causes link performance degradation. Rain attenuates the signal, disturbs the polarization and increases system temperature. Thus, accurate prediction of rain attenuation is very significant to design reliable microwave link at higher frequencies. The prediction methods require rain intensity at one- minute integration time. According to Owolawi [6-7] the high integration time may not respond to rapid changes in the rainfall rate CD.

In this paper, one-minute integration time rain rate cumulative distribution (CD) is derived from rainfall measured at 5 meteorological stations in Libya for 30 years in order to investigate the link performance in terms of spectral efficiency and, throughput. The Rainfall rate R (mm/h is the most significant parameter to predict rain attenuation in radio wave propagation. Rain fade is directly responsible for link outages at Ku, Ka and future V-bands.

In Şen & Eljadid [5] represented Mediterranean climate region such as Libya, whereas the temperate rain is mostly stratiform and widespread. Therefore, the conditions of the electromagnetic wave propagation absolutely different in Libya because of rain intensities vary significantly over large areas in the country. The highest and the lowest rain rate in the study of interested locations are considered and analyzed in terms of link spectral efficiency and throughput. The model proposed by International Telecommunication Union-Radio wave Propagation (ITU-R) is used to predict and investigate rain fade based on 1-minute rain rate data. Rain fade predicted at five locations are used for performance analysis in terms of link spectral efficiency and throughput.

2. Collection and Processing of Data

Libya lies approximately within the latitude 34°~33°N and longitude 10°-23°E. Libya's coastline is the longest of any African country bordering the Mediterranean. Therefore, Libya influenced directly and strongly by the Mediterranean climate, which is characterized by hot, dry, sunny summers and a winter rainy season [13]. A long- term cumulative rainfall data measured and recorded at the Libyan National Meteorological Center (LNMC) were collected. This study utilizes the data from a climatic period of (1981-2010). Readings of rain gauge are recorded 3 hourly or 8 readings per day at five stations. Hence, 2880 of rainfall readings per year have been considered in rain rate estimation. In this work rain rate in mm/h for each year have been calculated in these stations. The values of R obtained by averaging the data of rain rates calculated from 1981 to 2010. The average value of rain rate $R_{0.01\%}$ expressed in mm/h was obtained [6,13] and shown in Table 1.

Table 1. Measured and converted rainfall data at 5 selected sites

Station Name	Maximum rainfall in a day (mm)	Average number of rainy days/year	Maximum rainfall in a month (mm)	Average $R_{0.01\%}$ (mm/h)	ITU-R $R_{0.01\%}$ (mm/h)
Tripoli city	120.5	52.8	330	45	30
Shahat	86.5	70	434.8	51.7	50
Tripoliairport	130.2	46	173.4	35.8	30
Sebha	29	3.63	29	4.18	10
Benini	53.4	57.72	234.7	21.3	25

3. Rain Attenuation Prediction

The estimation of the long-term statistics of attenuation due to rain for maximum and minimum $R_{0.01}$ in Libya has been calculated by ITU-R procedure. Based on rainfall rate exceeded in 0.01% of the time of the year, the rain rate has been derived from long-term measured data [8-10]. Table.3 presents some of the local geometry parameters used for ITU-R 2012 methodology on the 30.5-E ∘ARABSAT - 5A links at different elevation angles to each of the stations at C - Ku- Ka- V -band downlink with horizontal (H), vertical (V) ,and circular (C) polarization. For the different percentage of time of the year (P %). For estimating of rain fade at a given location, the following two components need to be predicted [11-12]

- Specific attenuation – attenuation due to the rain per unit length of signal path
- Effective path length.

IOP Conf. Series: Materials Science and Engineering **260** (2017) 012041 doi:10.1088/1757-899X/260/1/012041

Table 2. Specifications which are considered to predict rain attenuations at 2 locations

Point of rainfall rate for the location R0.01% mm/h	Shahat	Sebha
Point of rainfall rate for the location R0.01% mm/h	51.7	4.2
Height of ES above MSL in m	621	432
Latitude of ES	32.48°	27.01°
Satellite elevation	51.12°	53.83°
Effective Earth radius	8500	
Polarization	LP-H, LP-V, CP	
Percentage % of time of the year	0.001,0.01,0.1,1	

Table 3. Estimated rain fades based on measured rain rate

Downlink Frequency band	Rain Attenuation $A_{0.01\%}$ (dB) at Maximum $R_{0.01\%}$ 51.7 mm/h			Rain Attenuation $A_{0.01\%}$ (dB) at Minimum $R_{0.01\%}$ 4.2 mm/h		
	LP-H	LP-V	CP	LP-H	LP-V	CP
4GHz	0.19	0.12	0.13	0.01	0.01	0.01
12GHz	8.09	7.16	7.61	0.91	0.86	0.88
20GHz	22.62	19.55	21.01	3.14	2.99	2.43
40GHz	63.49	57.98	60.73	12.56	11.71	12.13

Both rain rates are used to predict rain fades at C, Ku, Ka and V-bands with Vertical, Horizontal and Circular polarized waves for two stations in Libya namely, Shahad and Sebha. It has been observed that rain fade is not significant at C and Ku-bands. The maximum rain attenuation corresponding to these frequencies is 0.19dB and 8.09dB respectively. However, it was observed 22.63 dB at Ka-band and 63.5 dB at V-band which can be considered as a challenge to keep link working with 99.99% of reliability. Figures 1 and 2 represent rain attenuation for C, Ku, Ka, V-band down-link frequency at horizontal polarization using measured $R_{0.01}$. For 99.99% of reliability. Figure 3 shows the rain attenuation when the signal is horizontally polarized at 40 GHz for major zones in Libya at different percentage of time and various on rain intensity obtained from long-term measured data. It's clear that rain effect increases with increase of frequency, R and decrease of elevation angle.

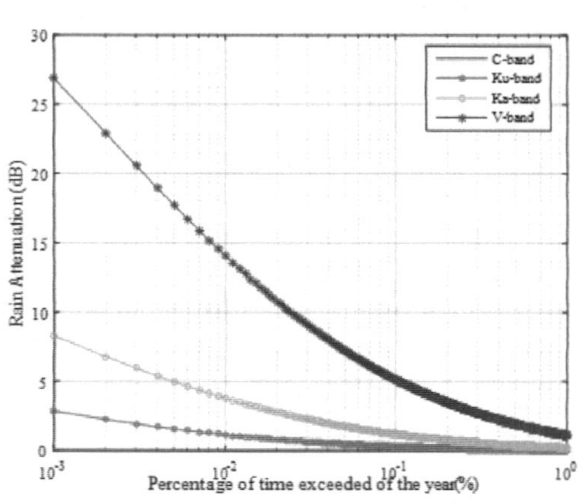

Figure 1. Rain attenuation distribution at Sebha

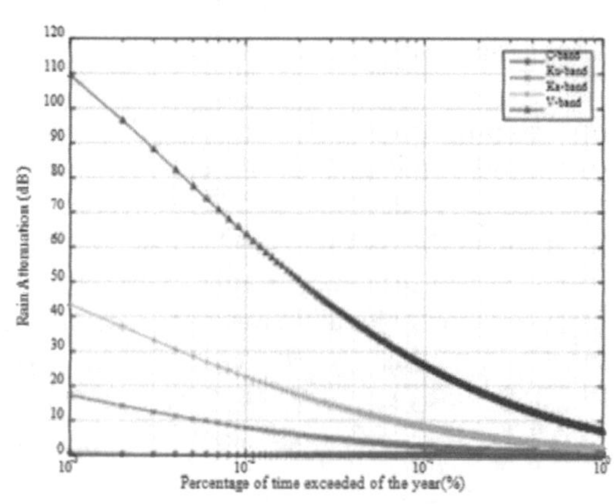

Figure 2. Rain attenuation distribution at Shahat

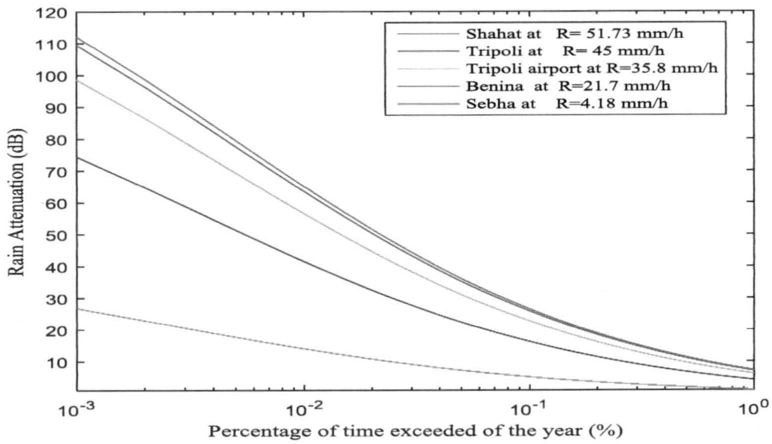

Figure 3. Comparison of Rain Attenuation for Horizontal Polarization in 5 different locations at V- Downlink Frequencies Band

4. Performance Analysis

Performance analysis was performed based on link efficiency in bps/Hz and throughput in Mbps. If the system designed to work at average 7 dB SNR means (BER $=10^{-6}$, 1 dB implementation) at clear air condition at particular reliability. This value of SNR would be decreased by Fade Margin. The spectral efficiency and throughput of downlink satellite- to- earth microwave links changed with transponders bandwidth respectively. The following steps represent the approach of calculation the efficiency and throughput at some vital locations in Libya. By using equation1 and 2 at horizontal polarization. The results of this analysis would assessment the performance of satellite microwave links operating at different frequencies band according to the peculiarity of the calamite condition of the study area of interest during the rainy season.

$$\textit{Spectral Efficiency} = \log(1+SNR/FM)/\log_2 \quad (1)$$

$$\textit{Throughput} = \textit{Spectral efficiency} \times \textit{Transponder BW} \quad (2)$$

In this section, the frequency band utilization efficiency and performance issues of earth-to-satellite communications link would be analytically investigated at Shahat and Sebha. Considering C, Ku, Ka and V- down-link frequency band during rainy events at selected locations, and 4 PSK resource access model. With 7 dB SNR threshold is considered at transponder bandwidths of 36, 54 and 72 MHz respectively. If the link is designed for 99.99% of reliability, a fade margin is introduced to compensate the effects of rain at a given level of availability. The other parameters and values used for the estimation are presented in Tables 5 and 6.

Table 4. Spectral Efficiency and Throughput in Satellite- to- Earth Microwave Links operating at Shahat (Maximum $R_{0.01\%}$)

Parameters	4 GHz	12 GHz	20GHz	40GHz
Average SNR	7 dB	7 dB	7 dB	10 dB
Rain fade margin	0.19 dB	8.09 dB	22.63 dB	63.50 dB
Link Efficiency	2.54 bps/Hz	0.83 bps/Hz	0.04 bps/Hz	0
Transponder BW	Throughput	Throughput	Throughput	Throughput
36MHz	91.32 Mbps	29.88 Mbps	1.40 Mbps	0
54MHz	136.98 Mbps	44.82 Mbps	2.10 Mbps	0
72MHz	182.64 Mbps	59.76 Mbps	2.80 Mbps	0

Table 5. Spectral Efficiency and Throughput in Satellite-to-Earth Microwave Links operating in Sebha (Minimum Measured $R_{0.01\%}$).

Parameters	4 GHz	12 GHz	20GHz	40GHz
Average SNR	6 dB	6 dB	6 dB	6 dB
Rain fade margin dB	0.01 dB	0.91 dB	3.14dB	12.56dB
Link Efficiency bps/Hz	2.31 bps/Hz	2.08 bps/Hz	1.55bps/Hz	0.29 bps/Hz
Transponder Bandwidth	Throughput Mbps	Throughput Mbps	Throughput Mbps	Throughput
36MHz	83.32 Mbps	74.92 Mbps	55.85 Mbps	10.37 Mbps
54MHz	124.98 Mbps	112.37 Mbps	83.77 Mbps	15.56 Mbps
72MHz	166.64 Mbps	149.83 Mbps	111.69 Mbps	20.74 Mbps

The performance of the links is analyzed at two locations, namely Shahat, and Sebha. In (Maximum $R_{0.01}$) the spectral efficiency was 0.04 bps/Hz and 2.08 Mbps throughput. This is the highest efficiency that obtained at Ka –band. When transponder bandwidth is 72 MHz with 7dB SNR. However, at V-band satellite communication link will shut down even if the average value of SNR has been increased to more than 10dB in the northwest and northeast along the Libyan coast. Whereas it can still operate at V-band in the middle of the Libyan coast with an efficiency of 0.25 bps/Hz and 17.48 Mbps throughput at 7dB SNR. In Sebha, V- band satellite communication link may operate even the average of SNR have been reduced to 6 dB with efficiency of 0.29 bps/Hz and 20.74 Mbps throughput with transponder BW is 72 MHz while transponders BW of 72, 54 and 36 MHz are all shut down in Shahat.

5. Conclusions

Rain attenuation is the most significant consideration and challenge to design a reliable earth-to-satellite microwave links for Ku, Ka and V-bands frequencies. Hence, it is essential for satellite link designer to take into account rain fade margin accurately before system implementation. Rainfall statistical data measured and recorded in Libya for the period of 30 years are collected from five locations. Rain attenuation prediction methods require one minute integration time rain intensity. Therefore, collected data were analyzed and processed to convert into one–minute rain rate cumulative distribution in Libya. The model proposed by ITU-R is used to predict and investigate rain fade based on converted 1-minute rain rate data. Rain fade predicted at two locations are used for performance analysis in terms of link spectral efficiency and throughput. V-band downlink shows that 99.99% availability is possible in all the Sothern part stations in Libya at 0.29 bps/Hz spectral efficiency and 20.74 Mbps Throughput when 72 MHz transponder band width. Results of this paper will be a very useful resource to design highly reliable earth-to-satellite communication links in Libya.

Acknowledgement

Authors are thankful to the Libyan National Meteorological Center (LNMC) for their assistance and technical support during data collection. They are also grateful to Research Management Centre, International Islamic University Malaysia to support this research through research grant RIGS16-065-0229.

Reference

[1] G. A. Siles, J. M. Riera, and P. Garc.Del-Pino (2015). Atmospheric attenuation in wireless communication systems at millimeter and THz frequencies [Wireless Corner], IEEE Antennas Propag. Mag., vol. 57, no. 1, pp. 48–61.

[2] M. R. Sujimol, R. Acharya, G. Singh, and R. K. Gupta (2015). Rain attenuation using Ka and Ku band frequency beacons at Delhi Earth Station, Indian J. Radio Sp. Phys., vol. 44, no. March, pp. 45–50.

[3] Badron, K., Ismail, A.F., Islam, M.R., Abdullah, K., Din, J. and Tharek, A.R. (2015). A modified rain attenuation prediction model for tropical V-band satellite earth link, International Journal of Satellite Communications and Networking 33.1 (2015): 57-67.
[4] Kamruzzaman, M., & Islam, M. S. (2014). Rain attenuation prediction for satellite communications link at Ku and Ka bands Over Bangladesh, Proceeding of International Conference on Electrical Engineering and Information & Communication Technology (ICEEICT), pp. 1-6.
[5] Z. Şen and A. Eljadid (1999). Rainfall distribution function for Libya and rainfall prediction, Hydrol. Sci. J., vol. 44, no. 5, pp. 665–680.
[6] M. Tamoiunaite, S. Tamoiunas, V. Daukas, M. Tamoiuniene, and M. linskas (2015). Prediction of Electromagnetic Waves Attenuation due to Rain in the Localities of Lithuania, Elektron. ir Elektrotechnika, no. 9, pp. 9–12.
[7] P. A. Owolawi (2011). Tion and Mapping for the Estimation of Rain," Prog. Electromagn. Res., vol. 112, no. January, pp. 155–181.
[8] Abdulrahman Y, Rahman TA, Islam M. R., Olufeagba BJ, Chebil J. (2015). Comparison of measured rain attenuation in the 10.982-GHz band with predictions and sensitivity analysis. International Journal of Satellite Communications and Networking. 2015 May 1; 33(3):185-95.
[9] Mohammad Mahfujur Rashid, Md. Rafiqul Islam, AHM Zahirul Alam , Othman O. Khalifa, Sheroz Khan and Md. Arafatur Rahman (2009). Rain Fade and It's Effect Analysis for Earth-to-Satellite Microwave Link Based on Measured Rainfall Statistics in Bangladesh, Proceeding in 9th Malaysia International Conference on Communications, Kuala Lumpur, December 15-17, 2009.
[10] ITU-R (2012). Recomendation P.837-6, "Characteristics of precipitation for propagation modelling P Series Radio wave propagation," Radio wave Propag., vol. 6.
[11] Asma Ali Hussein Budalal, Md Rafiqul Islam, Mohamed Hadi Habaebi and Musse Mohamud Ahmed (2016). Rain Rate Distribution for Satellite to Earth Microwave Links Design in Libya", Proceedings of 6th International Conference on Computer and Communication Engineering, DOI 10.1109/ICCCE.2016.99, pp. 443-448.
[12] Md. Rafiqul Islam, Md. Arafatur Rahman, Farhat Anwar and Md. Mahfujur Rashid (2008). Performance Investigation of Earth-to-Satellite Microwave Link Due to Rain Fade in Bangladesh, Proceedings on 11th International Conference on Computer and Information Technology (ICCIT2008), Paper ID 551.
[13] K. I. Elfadli (2009). Precipitation Data of Libya, Libyan National Meteorological Center.

IOP Conf. Series: Materials Science and Engineering **260** (2017) 012042 doi:10.1088/1757-899X/260/1/012042

Speech to Text Translation for Malay Language

Rami Ali Al-khulaidi, Rini Akmeliawati

International Islamic University Malaysia, Department of Mechatronics Engineering Jalan Gombak, 53100 Kuala Lumpur, Selangor, Malaysia

teemomory@gmail.com , rakmelia@iium.edu.my

Abstract

The speech recognition system is a front end and a back-end process that receives an audio signal uttered by a speaker and converts it into a text transcription. The speech system can be used in several fields including: therapeutic technology, education, social robotics and computer entertainments. In most cases in control tasks, which is the purpose of proposing our system, wherein the speed of performance and response concern as the system should integrate with other controlling platforms such as in voiced controlled robots. Therefore, the need for flexible platforms, that can be easily edited to jibe with functionality of the surroundings, came to the scene; unlike other software programs that require recording audios and multiple training for every entry such as MATLAB and Phoenix. In this paper, a speech recognition system for Malay language is implemented using Microsoft Visual Studio C#. 90 (ninety) Malay phrases were tested by 10 (ten) speakers from both genders in different contexts. The result shows that the overall accuracy (calculated from Confusion Matrix) is satisfactory as it is 92.69%.

1. Introduction

Spoken language is a main way of communication between human beings. However, it has become normal in our daily life seeing individuals utter words to devices such as laptop computers or smart phones; it's even a necessity in many cases like when a person with a certain disability must instruct a certain machine or a computer application to perform a certain task for him or her. Speech-to- text analysis attracts more attentions in the last decades. Therefore, such technology can be encountered in many different contexts from the inevitable use of the speech recognition system, along with human robot or animated Avatar, as a mediator between normal and hearing-impaired individuals to the joyful use in computer gaming.

Thus, being a part in the rapid improvement of this technology, we implemented a speech recognition system for Malay Language using C# in Microsoft Visual Studio Platform. It can be easily integrated with other components for controlling tasks. Microsoft visual studio environment has introduced itself as a future promising environment for developers interested in this area. So, it is a momentous advantage the developers can start benefiting thereof. Especially as the Microsoft Visual Studio integrates with controlling platforms such as Arduino and raspberry pi using windows form application and serial communication.

Unlike previously implemented systems that were sophisticated and limited in terms of words, speed and accuracy, our system is adaptable for future addition and edition, and capable to tactfully integrate with other platforms; these features are the major reasons behind implementing the system in Microsoft Visual Studio.

This paper is organised as follows. Section 2 shows some related works. Section 3 presents the review of the system along with some parts of algorithms used. The testing data are presented in Section 4. Section 5 shows the evaluation of the results, and finally, Section 6 presents the conclusion and the future works.

2. Related Work

A huge amount of research has been done in the area of speech processing locally (in Malaysia) and worldwide. [1] developed an algorithm to synthesize Malay spoken language using Hidden Markov Model (HMM) which performed well with a reasonable accuracy. There are also some works worldwide for different languages such as English, Mandarin, Indian, Spanish, French and others.
[2] developed an automatic speech recognition (ASR) system for Malay speaking children. They took advantage of HMM available in HTK Toolkit to build the speech model for Malay language. The developing environment was Linux based. The system is relatively good since it can accurately recognize of up to 76% of test words. However, the ASR focused only on small group of children (six children) to record their voice and include them in the system.

Reference [3] created an application to test word pronunciation for pre-school children aged between 3 and 6 years old. The system focused especially on vowel letters. The interface was in MATLAB using Spectrum Delta (SPD) features and Logistic Regression classification model. Although the accuracy was up to 92.29%, the system only tested vowels letters.

As we can see from both works, [2] and [3], regardless of the accuracy of the systems, they are only focused on a very narrow area which is on the recognition of children pronunciation. Moreover, both systems targeted small groups input, six children voice and three to six-year-old children.

Malay Speech Therapy Assistance Tools (MSTAT) that is used to diagnose children for language disorder and to train children with stuttering problem was proposed in [4]. Voice patterns of the normal and stutter children are used to train the HMM model which is used to evaluate speech problem for stuttering children. The accuracy of the system was reasonable but the system itself is used for diagnostic purposes.

In [5], fusion techniques and adapting filtering is used for Malay language recognition. They used Mel-Frequency Cepstral Coefficients (MFCC) vectors to provide an estimate of the vocal tract filter. Dynamic Time Warping (DTW) and HMM are the two recognition algorithms used DTW is used to detect the nearest recorded voice. Meanwhile, HMM is used to emit a new feature vector for each frame according to an emission probability density function associated with that state. They also use RLS noise cancellation, end point detecting, framing, normalization, filtering. The system is cumbersome since it included many processing and filtering techniques. Therefore, the complexity of the system makes it unsuitable to be applied for additional purposes such as translation because it will be so slow and it may interfere with other functions that improve real time translation or any other kind of performance such as voiced controlled machinery.

All the stated works are about Malay speech recognition and synthesizing. Some of them were focused on small groups of population including children whether in small group [2] and [3] or in general but restricted to patient children only [4]. The others that involved the speech of all age groups for Malay people in general like [5]. Unfortunately, they were quite intricate which make them difficult to be applied in other functions such as controlling other devices. For example, when we want to vocally instruct a robot or at least ordering a certain machine to perform a certain task.

3. System Overview

A speech recognizer is a system that receives an audio stream and converts it into a text transcription. The process of speech recognition is considered as front end and back end. The process of isolating segments of sound, speech, and converting the segments into a series of numeric values that characterize the vocal sounds in the signal is the front-end process. The back end, on the other hand, is a search engine which is specialized to take the output of front end and look out across three databases as in Figure 1:

IOP Conf. Series: Materials Science and Engineering **260** (2017) 012042 doi:10.1088/1757-899X/260/1/012042

- The acoustic model which represents the language acoustic sounds; this can be trained to recognize the characteristics of a certain user's speech patterns and acoustic environments.
- The lexicon includes a large number of the words in the language, and offers information on how to pronounce each word.
- The language model represents the means in which the words of the language are combined.

The quality of the speech recognizer is detected by how good the recognizer at sanitizing its search, removing the poor matches, and choosing the more similar matches. This fully depends on the quality of the language and acoustic models of the recognizer and the algorithms effectiveness, for the processing of input sound and the searching across its models.

3.1 Grammar

The grammar is the most substantial part of the speech recognition system as it limits the number of recognized words and phrases. The speech recognizer often needs to process certain utterances with specific semantic meaning to the intended application rather than using the general language model, such as every day spoken language. This will have several benefits to the system such as, increasing the accuracy of the system, ensuring that all recognized sets are meaningful to the intended application as well as specifying the semantic values inherent in the detected text. This grammar can be authoring programmatically in the Microsoft Speech Platform SDK.

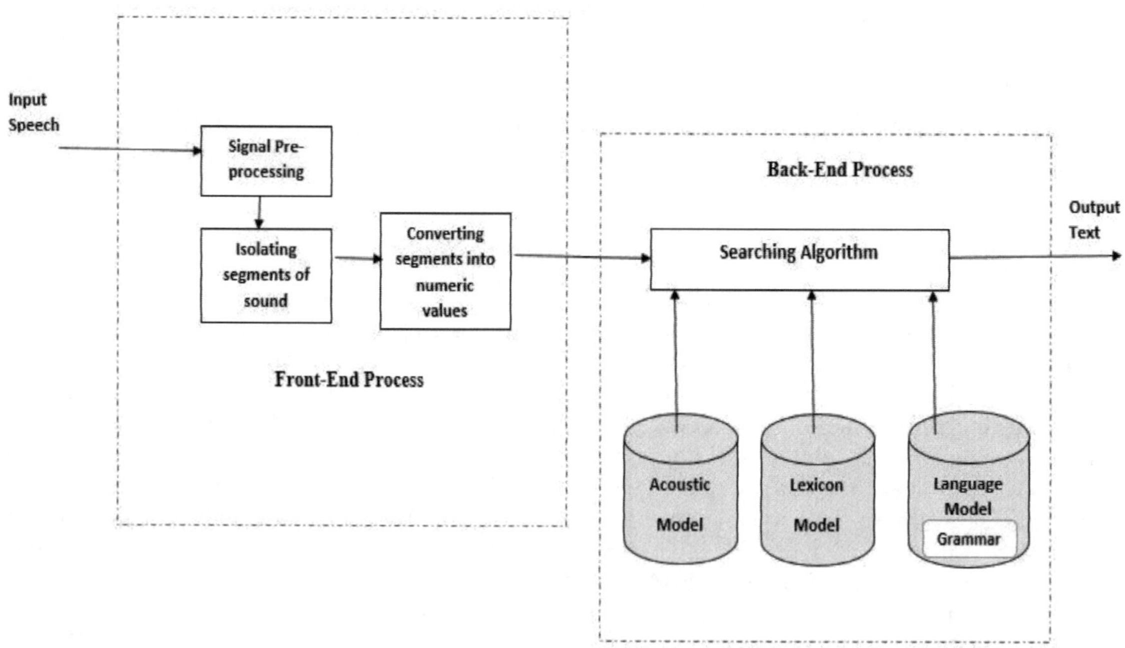

Figure 1. The Flowchart of Speech System

3.2 System Design

The process of implementing speech recognition to a Windows Forms application starts by launching MS Visual Studio and creating a new C# Windows Forms application with the proposed name. Before creating any program with speech recognition system in C#, "System.speech" library has to be added by "using" namespace statement at the beginning of the code, and following by an instance of the "SpeechRecognitionEngine".

The grammar also must be added into the "SpeechRecognitionEngine". Otherwise, the speech recognizer will not be able to recognize the targeted phrases. This could be the counterpart of the pre-recorded database in other software such as, MATLAB. But, here only texts of the exact wording of intended phrases should be inserted instead. The words and phrases included in the Grammar List are

IOP Conf. Series: Materials Science and Engineering **260** (2017) 012042 doi:10.1088/1757-899X/260/1/012042

presented in Table 1. The method "RecognizeAsync" was used, to load grammars asynchronous, as well as event handler which is used to display the recognized word or item. The execution of the program starts by initializing the grammar instance along with grammar builder object. Then, the grammar will be loaded into the speech recognizer following by event handler. Next, the Graphical User Interface will show up. However, the recognizing will not commence unless the "Start" pushbutton is clicked. Once the pushbutton is pressed, the system will be waiting for an utterance to analyse and decide if it's included in the grammar list. Consequently, the method "sRecognize_SpeechRecognized" will be invoked if the speech is recognized. Then, accordingly the result property of event handler will display the recognized phrase if it is included in grammar list. Otherwise, a "Say that again" message will appear telling us to speak again. The program will continue as long as the quit button is not clicked. The flowchart of the main program is shown in Figure 2 below.

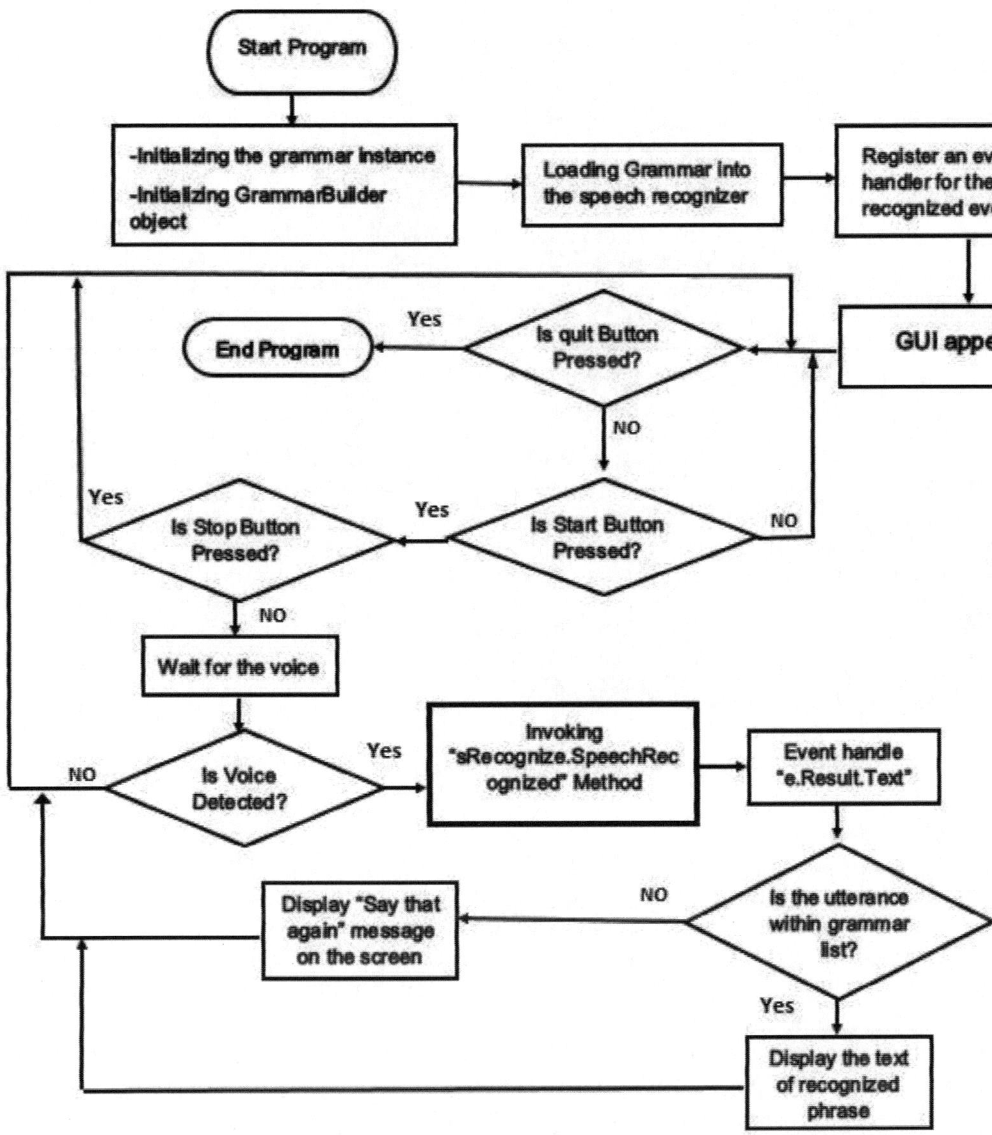

Figure 2. The Flowchart of The Main Program

3.3 User Interface

The design of user interface provides all the information about the recognized items and the control buttons of the system. We can see the data about the words such as the recognized words or phrases as it appears on the screen along with time of recognition and "say that again" message in case the word is not recognized. The control over the user interface includes the "start" button that tells the system to start recognizing the sound and the "stop" button that temporarily forces the system to stop. The user can also quit the interface window by either vocal command, saying "quit program" command, or by clicking on the quit button. Figure 3.

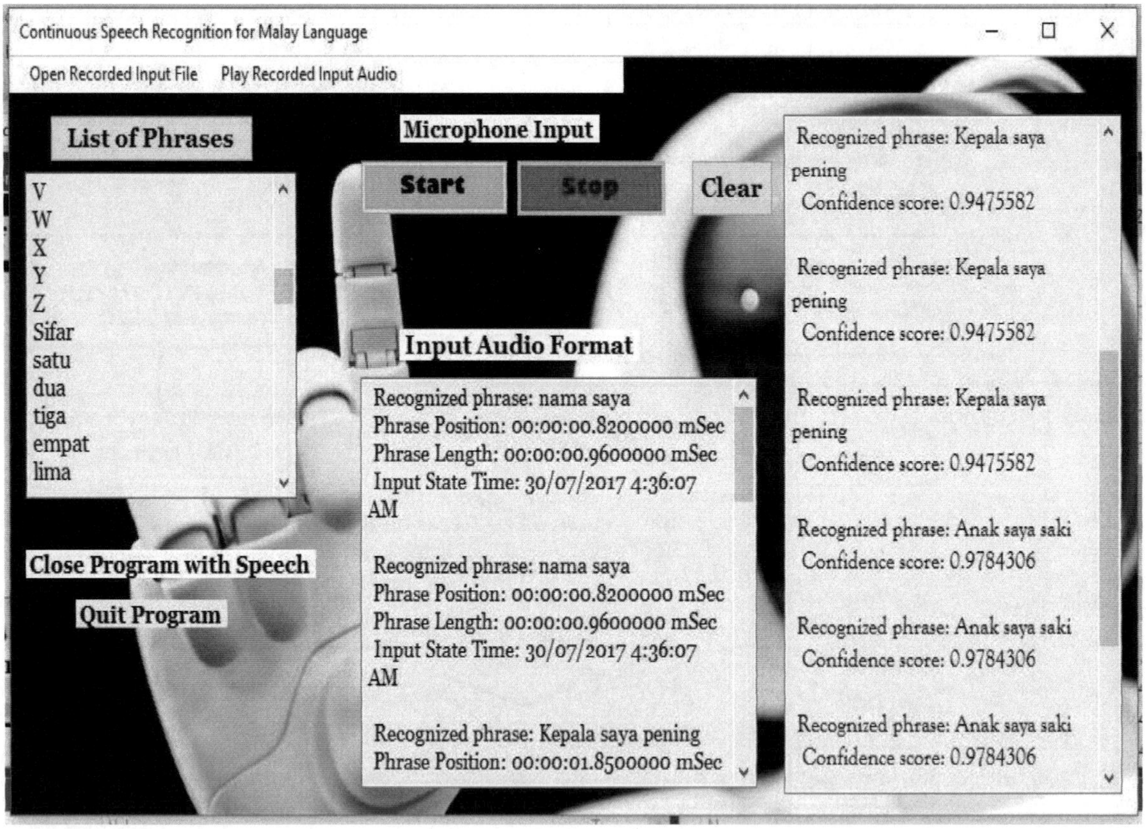

Figure 3. User Interface of Speech Recognition System

4. Testing of the System

Using the standard microphone of a quite old laptop (Dell Inspiron 1545), ninety (90) tested items (alphabets, single words and phrases) were tested by 10 male and female speakers. To ensure the accuracy of the system, we intended to ask each speaker to repeat each tested item five-times. Hence, the total of the data tested is 4500 which will be analysed in section 5. The criteria of choosing such 90 items lies on three crucial reasons: the recurrence of phrases, the diversity of contexts and the urgent situations.

Amongst all the commonly used phrases for Malay language, we carefully picked up the most recurrent phrases; phrases that are used in a daily life context such as salutations, or those used in classroom context such as giving permission for entering or leaving classrooms. In addition, some vital conversational phrases that are used between two or three individuals, considering the phrases that are frequently used as emergency calls or seeking help contexts.

IOP Conf. Series: Materials Science and Engineering **260** (2017) 012042 doi:10.1088/1757-899X/260/1/012042

Table 1 below shows the tested phrases in three different contexts: classroom, normal conversation and some phrases in restaurant. The reason for choosing the restaurant phrases is to test the system in chaotic environments and determine how accurate the system will be under such circumstances.

Table 1: The Tested Phrases

55 Class rooms Phrases	19 Conversational Phrases	6 Restaurant phrases
The 26 Malay alphabets, and 26 numbers (numbers from 0 to 14, and 20, 21, 22, 23, 30, 40, 50, 100, 200, 300, 1000. Three sentences also are included: (Sila Masuk Sila Dudduk and Tolong diam)	Nama saya, jumpa laga, saya datang dari, siapa nama anda, boleh cakap bahasa Melayu, saya sakit, pergi kemana, saya pergi ke hospital, Wang saya dicuri, jalan terus, badan saya ada luka, anak saya sakit, ada pencuri masuk rumah, dokumen saya hilang, kereta saya rusak, tolong saya, saya tak boleh bernafas, saya tak boleh berjalan.	Selamat pagi, selamat tengah hari, selamat petang, selamat malam, selamat tinggal, tumpang lalu.

Ten phrases are also used to check the confusability of the system. In other words, some additional phrases are not included in the algorithm, but added in the test to see if the system will mix them with the 80 included phrases. For example, if the speaker utters a word that is not included in the system and the system confuses it with an included phrase, this will be considered in determining the accuracy of the system.

5. Result and Discussion

The evaluation of data was performed using the confusion matrix which enabled us to get more detailed analysis. The parameters of confusion matrix are as follows:

- True Positive (TP) The correctly recognized words or phrases.
- True Negative (TN) The unconfused words or phrases.
- False Positive (FP) The confused words or phrases.
- False Negative (FN) The unrecognized (missing) words or phrases.

TP means that the words or phrases are predefined in the speech recognition systems and the system recognized them correctly whilst the speaker is saying them. For instance, if the speaker says" Saya sakit", and the recognized phrase is the same ("Saya sakit"), this will be True Positive (TP). However, if the system does not recognize the phrase, this will be False Negative (FN).

TN means that the words the speakers are saying are not predefined in the speech recognition system and the system is not confused with the predefined words or phrases. That is, if the speaker says "apa kabar", for example, the phrase will not be mixed up with other defined phrases. However, if the system mixes it up with a predefined phrase "Saya sakit", for example, this will be a False Positive (FP).

Hence, in order to achieve satisfactory results, the number of TP and TN has to be as high as possible as opposed to FP and FN which are better to be low. Because, the lower the numbers of FP and FN are, the more accurate the system is. Table 2 shows the overall tested data and table 3 shows the Confusion Matrix therefrom the accuracy of the system is calculated.

IOP Conf. Series: Materials Science and Engineering **260** (2017) 012042 doi:10.1088/1757-899X/260/1/012042

Table 2: The Overall Testing Data

Speaker	TP	FN	Sensitivity	TN	FP	Specificity
1	372	28	0.93	43	7	0.86
2	375	25	0.9375	43	7	0.86
3	379	21	0.9475	43	7	0.86
4	371	29	0.9275	39	11	0.78
5	377	23	0.9425	40	10	0.8
6	375	25	0.9375	43	7	0.86
7	365	35	0.9125	46	4	0.92
8	376	24	0.94	41	9	0.82
9	380	20	0.95	42	8	0.84
10	380	20	0.95	41	9	0.82
Total	**3750**	**250**	0.9375	421	79	0.842

Table 3: Confusion Matrix

	Sensitivity		Specificity
TP	**3750**	**FP**	**79**
FN	**250**	**TN**	**421**
	0.9375		0.842

By referring to the Confusion Matrix, the acccuracy of the system is determined as in the next formula.

$$\text{Accuracy} = \frac{\text{TP+TN}}{TP+FP+TN+FN}$$

$$\text{Accuracy} = \frac{3750+421}{3750+79+421+250} = 92.69\ \%$$

The Sensitivity and Specificity curve is shown in figure 4.

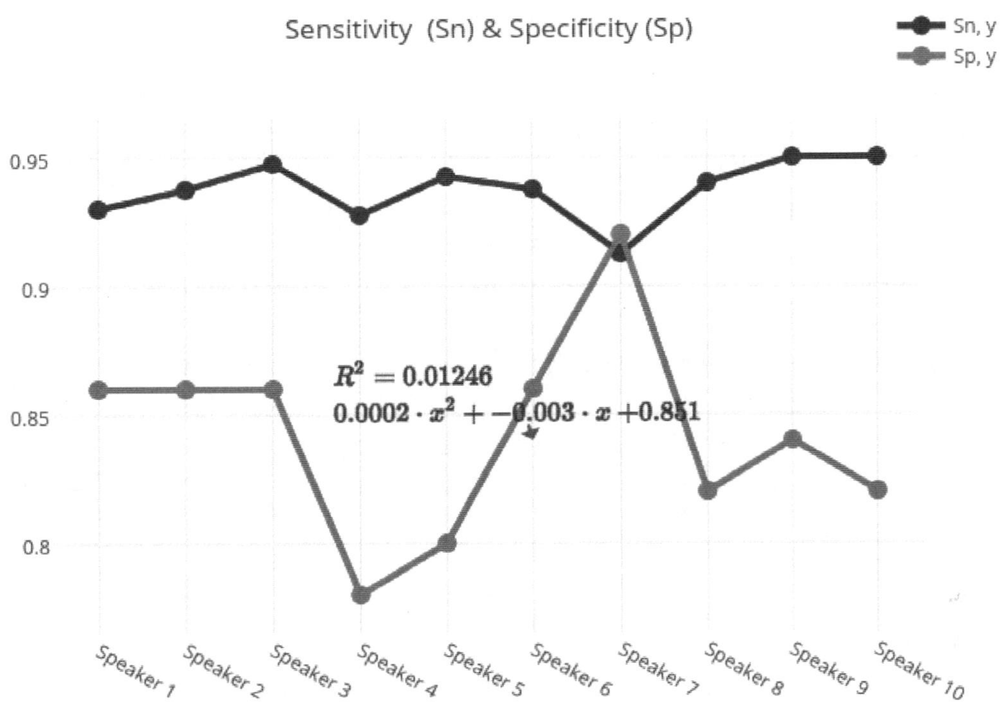

Figure 4. The Sensivity and Specivicity Curve

It can be percieved (from the result) that our system achieved a high accuracy in comparison to previously proposed ones. It can also be noted that the system can easily adapt to future edition as the phrases can be manually added into the algorithm without making recording audios and training them. That is, once a developer, or even a user with basic programming skills, desires to add to some phrases or to replace the existing ones, all he or she can do is writing the phrases literally into the code in the same wording as if he or she is writing in notepad. Therefore, this will facilitate the integration with other platforms such as Arduino and Raspberry Pi for control purposes; for example, when the users wish to control an automobile robot using vocal instructions, they can only add the corresponding word into the code such as, "forward", "backward", "left" or "right". Then by the means of serial communication between Microsoft Visual Studio and Arduino the process will go smoothly. The adaptability of the speech system with control platforms is a significant property that enable us to choose Microsoft Visual Studio to be our platform for speech system.

6. Conclusion

In this work, a speech recognition system that translates the spoken Malay language into texts is presented. We used C# programming language and Microsoft Visual Studio as our platform which performs fast and accurate. The resulted accuracy is very authentic as we use the confusion matrix by adding extra 10 phrases in the test to check the confusability of the system. It is more satisfactory, though, as it is nearly 93%. The accuracy would have been higher than that, if a modern laptop computer, with a high-quality microphone, had been used.

An extension of the work in the future would be in applying the speech system in the control of external device such as a robot or any other automated machine; that will prove the adaptability of the system with other controlling platforms, such as Arduino, so that the features of the proposed speech system will be utilised.

IOP Conf. Series: Materials Science and Engineering **260** (2017) 012042 doi:10.1088/1757-899X/260/1/012042

References

[1] H. A. Maarif, R. Akmeliawati and Z. Z. Htike, "word classification for sign language synthesizer using hidden Markov model," in *The 5th International Conference on Information Information and Communication Technology for The Muslim World (ICT4M), 2014*, 2014.

[2] N. M. M. B. M. S. S. S. Feisal Dani Rahman, "Automatic Speech Recognition System for Malay," in *Third ICT International Student Project Conference (ICT-ISPC2014)* , 2014.

[3] M. Shahrul Azmi, "Malay Word Pronunciation Application for Pre-School Children using Vowel," 2015.

[4] T. T. S. a. S. H. S. Salleh, "Corpus-based Malay Text-to-Speech Synthesis System," in *2008 14th Asia-Pacific Conference on Communications* , Tokyo, 2007.

[5] S. S. A. H. K. I. a. A. N. S.A.R. Al-Haddad, "Robust Speech Recognition Using Fusion Techniques and Adaptive Filtering," *American Journal of Applied Sciences,* vol. 6, no. 2, pp. 290-295, 2009.

IOP Conf. Series: Materials Science and Engineering **260** (2017) 012043 doi:10.1088/1757-899X/260/1/012043

Rapid lard identification with portable electronic nose

Marsad Latief[1], **Aida Khorsidtalab**[1], **Irwan Saputra**[2], **Rini Akmeliawati**[1], **Anis Nurashikin**[3], **Irwandi Jaswir**[2] and **Gunawan Witjaksono**[4]

[1]Department of Mechatronics Engineering
[2]Department of Biotechnology Engineering
[3]Department of Electrical and Computer Engineering
International Islamic University Malaysia, Gombak, Malaysia
[4]Universiti Teknologi Petronas, Tronoh, Perak, Malaysia

rakmelia@iium.edu.my

Abstract. Human sensory systems are limited in many different regards, yet they are great sources of inspiration for development of technologies that help humans to overcome their restraints. This paper signifies the capability of our developed electronic nose in rapid lard identification. The developed device, known as E-Nose, mimics human's olfactory system's technique to identify a particular substance. Lard is a common pig derivative which is often used as a food additive, emulsion or shortening. It's also commonly used as an adulterant or as an alternative for cooking oils, margarine and butter. This substance is prohibited to be consumed by Muslims and Orthodox Jews for religious reasons. A portable reliable device with an ability to identify lard rapidly can be convenient to users concerned about lard adulteration. The prototype was examined using K-Nearest Neighbors algorithm (KNN), Support Vector Machine (SVM), Bagged Trees and Simple Tree, and can identify lard with the highest accuracy of 95.6% among three types of fat (lard, chicken and beef) in liquid form over a certain range of temperature using KNN.

1. Introduction

Lard is pig fat derived from its adipose tissue and is often used in food production as an emulsion, shortening, or as a substitute to butter, margarine or cooking oils. Religions such as Islam and Judaism prohibit the consumption of pig derivatives in any form. Besides this, lard being a common choice of adulterant due to it cheap market price and easy availability is a matter for concern. Hence, detection of such adulteration is of high significance.

Several methods and devices have been developed for lard compound detection in food. Polymerase Chain Reaction (PCR), Gas Chromatography (GC), Gas Chromatography coupled with Mass Spectrometry (GCMS), Differential Scanning Calorimetry (DSC), and Fourier Transform Infrared (FTIR) spectroscopy are among the commonly practiced ones. FTIR is the most commonly used analytical tool in detecting lard adulteration in food and vegetable oils. Some research findings of lard adulterant are reported either by mixing lard with other animal fats or adulterating lard in food [1-3]. The limitation of lard detection using FTIR spectroscopy are identified in Rohman and Che Man [4]. Basically, lard has similar IR spectrum with other animal fats and vegetable oils since they are composed with (Triacylglycerol) TAG, with different length of the fatty acid. Another work on FTIR by [5] reported that the accuracy of the classification of lard adulterated in virgin coconut oil using FTIR Spectroscopy with Partial Least Square and discriminant analysis could be up to 100% accurate. The applied analytical method could detect 1% lard in the mixture. Though FTIR is a common method for the detection of lard mixture with other animal fats as well as adulteration in food products, it however requires laboratory scale equipment's. Hence, it is not a method that offer portability.

Each of the aforementioned methods have drawbacks which include but are not limited to their requirement for high power supplies, specific skilled operator, the requirement for controlled environment such as laboratory for testing, bulky sizes and high cost.

An electronic nose functions like a biological nose in recognizing the type of vapour that touches its "receptors". Receptors in an electronic nose are its sensors which can be selected to detect an odour of interest. In general, an electronic nose is a combination of an odour delivery system, sensor (array), data acquisition and data analysis unit [14]. Electronic noses have various applications in environmental and medical researches. They also play an important role in food industry. Some of the applications of electronic noses include detecting different flavours of milk [6,7], meat [8,9], tea [10], spoiled beef [11] and spoiled fish [12].

As to date, there is no portable E-Nose specifically designed for lard detection. Our developed portable E-Nose has the potential to become a reliable and affordable device for rapid detections of lard. This study reports the developed E-Nose and presents its performance. This paper is organized as follows. Section II describes the features characteristics of the developed E-Nose. Section III explains the experimental procedure. Section IV discusses the results for lard identification and Section V concludes the paper.

2. Device Design

2.1. Features

The version of E-Nose used in this study consists of a single metal oxide gas sensor, an AVR microcontroller, signal conditioning circuits, power supply module, peltier to heat the sample, display module and axial fans to contro the air flow. The general architecture is shown in figure 1. The gas sensor used is selected to be sensitive to volatile organic compounds (VOCs).

2.2. Sensor Selection

Animal fats have several chemical compositions which mostly include TAG. In fact, fats share the same fatty acid compounds but different concentration [13]. According to [4], analysis of fats/oils is possible by focusing on lipids components as fats belong to a group of biological substances called lipid. A triglyceride is a chemical compound that is formed from one molecule of glycerol and three fatty acids [15]. Triglyceride is the main constituent of animal fat including pig fat. In addition, lard is mainly composed of saturated fatty acid [15].

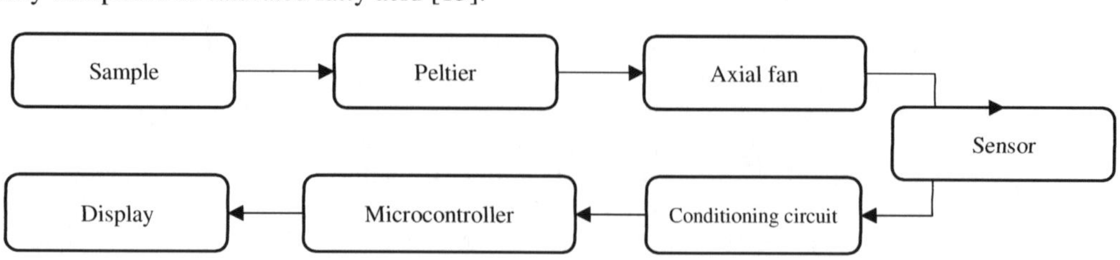

Figure 1: General architecture of the E-Nose for lard detection

IOP Conf. Series: Materials Science and Engineering **260** (2017) 012043 doi:10.1088/1757-899X/260/1/012043

In [15] the composition and classification of pig fat is elaborated. It is reported that lard contains up to 10% linoleic acid and small amount of arachidonic acid, which are in the group of unsaturated fatty acids. Another study on lard adulteration using FTIR spectroscopy resulted in identifying both saturated acyl groups and oleic acyl group in lard [2].

The selected gas sensor for this study is capable of detecting VOCs rapidly. Metal oxide gas sensors are commonly used in electronic noses for various applications including pork adulteration [16] quality control [17,10] and as a formaldehyde sensor [18]. Sensing materials used in metal oxide sensor are Tin dioxide [SnO_2] and tungsten trioxide [WO_3] as both materials are claimed to be highly sensitive to various types of volatile compounds.

Sensor selection was based on the literature, which listed the chemical compounds found in lard [13]. Based on the study, decanal was the chemical compound found abundantly in lard but didn't have significant presense in chicken fat and beef fat. Table 1 lists the decanal content in the fats of interest in terms of Kovats indices. A set of experiments by [19] revealed that both the sensing materials used in metal oxide sensors are sensitive to the presence of aldehydes. The study investigated the effect of aldehydes when exposed to a based metal oxide sensor. It was revealed that decanal, an aldehyde gives a significant signal change when exposed to the sensor [19]. As such the selected sensor is expected to be significantly more sensitive towards lard than other fats.

The sensor used in our case is coated with a SnO_2 layer. Exchange of oxygen molecules due to a reaction of the sample odour molecules and the sening layer takes place in the identification process. This exchange changes the electrical conductivity of the sensor [20] and is recorded.

Table 1 - Decanal profile, measured in Kovats indices [13].

Animal Fat	Decanal (dimensionless)
Lard	17 444
Chicken	586
Beef fat	408.5

The selected sensor also consumes low power and is small in size. These properties of the sensor contribute towards the portability of E-Nose.

3. Experiment

3.1. Sample Preparation

Adipose tissues from various parts of slaughtered pigs (lard), cows (beef) and chickens were obtained from a local market. The tissues were cut into small pieces, melted at 75°C, and strained through a triply folded muslin cloth. The filtered samples were put through a centrifuge to allow the fat to rise. The separated fat was collected and kept in a tightly closed container under nitrogen blanket in refrigerator for later analysis. Sample preparation was carried out intentionally without the addition of any chemical substances. Addition of chemicals will alter the original nature of the samples.

3.2. The effect of Temperature

It is known in analytical chemistry that changes in samples' temperatures causes changes in chemical structure. Organic substances like fat are sensitive to temperature. Temperature change applied on fat samples implies change in their physical and chemical properties.

IOP Conf. Series: Materials Science and Engineering **260** (2017) 012043 doi:10.1088/1757-899X/260/1/012043

The chemical properties will change due to a radical reaction triggered by breaking and forming of new bonds. Radical reaction occurs in three stages namely initiation, propagation and termination, each of which depend on the molecular concentration in the samples. An oil-profile comparison of different fats proved that all four examined fats, including beef, chicken, lamb and lard shared the same fatty acid components but different concentrations [13]. The same study indicates that concentration of fatty acids determines the type of fat and its chemical and physical properties.

In order to understand the effect of temperature and whether there would be any changes in the odor produced, experiments were performed at the end different temperature levels – 40°C, 50 °C and 60°C.

3.3. Scanning Process
A sample fat can be scanned by placing it in the device heating chamber. Odor fumes generated are transferred to the sensing chamber using the axial fan. At a sampling rate of 2 hertz the test in run for 1 minute. The data collected is stored on the device memory.

3.4. Pre-processing
Electronic noses are known to suffer from noise and drift. Hence, besides using software based filtering, the collected data also compensated for noise and drift. Compensation is done using a differential technique using equation (1).

$$Y(t) = X(t) - X(0) \tag{1}$$

where Y(t) is the Settle Point, X(t) is steady state value of a response signal and X(0) is the base or starting value of the response signal.

3.5. Post-processing
The collected labelled data set has two features for each observation, settle point and temperature. In post-processing the dataset formed is put through statistical analysis and various classifiers to achieve the results. All the post-processing was performed offline on a desktop computer.

3.6. Cleaning process
It is believed that nitrogen gas in some experiments [21] is necessary to "wash away" the odour from E-Nose as some of the gases leave very strong and pungent odour behind. However, in this study the principle of "Sniff and Purge" [22] which uses the ambient air as the "cleaner" to clean the sensor after every reading is applied. Ambient air was enough to clean the sensor as well as the chamber and the final result was not affected.

4. Results and Discussion
A scatter plot of the entire dataset is shown in figure 2. The dataset consists of 9 unique classes of 3 types of fat each experimented with 3 different temperatures. Each class consists of 10 observations. Each class is represented in the plot by a unique symbol and an abbreviation where the letter "L" represent a lard sample, "C" represents a chicken fat sample and "B" a beef fat sample. The numbers 40, 50 and 60 after the letters represent the temperature in degree Celsius. Three colours, red for lard, blue for chicken fat and green for beef fat have been used to help visualize the plot. The scatter plot shows the two features – temperature and settle point. Temperature implies the temperature of the peltier while the respective test was conducted. Settle Point signifies the overall response of the E-Nose sensor free of drift and noise. A clear separability can be seen overall in the plot as except classes "L40" and "C40" where there are no overlaps. This clarity in separability proves the potential of an E-Nose to distinguish between the three fats and identify lard. The chemical structure of a

chicken fat is very similar to lard and studies conducted with other techniques have proven that as well.

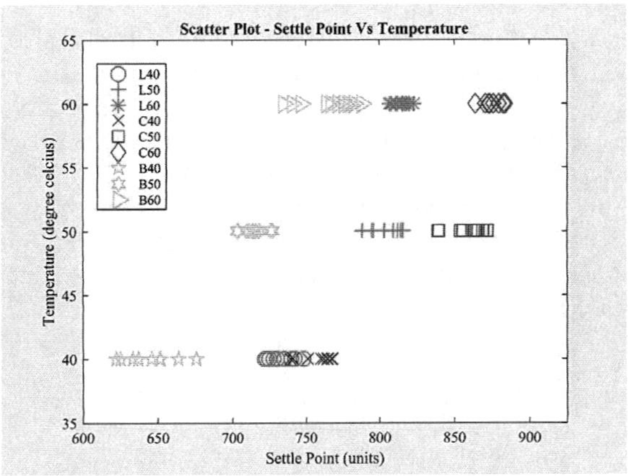

Figure 2 - Scatter plot of the entire dataset

As mentioned before in Section III, temperature was expected to affect the sensor response due to the change in the chemical structure of the fat samples. Figure 3 shows the individual plot of the three classes and their responses at different temperatures. Linear regression lines in the background show an upward trend in sensor response, with lard having the highest gradient out of the three. To claim that this upward trend was due to the chemical changes that must have occurred in the fat sample cannot be explicitly made. The reason being that electronic noses collect only qualitative data and to prove that the chemical changes are responsible to a higher sensor response will require further study. However, due to increase in temperature the density and rate at which the odor fumes are produced must increase, thus giving rise to a higher sensor response. Besides, this lard has the lowest melting point among the three fats and will therefore melt and turn to gaseous state faster. The highest gradient in the upward trend proves this point as well. Chicken fat is supposed to have a lower melting point than beef fat. Even though the sensor response of beef fat has a higher gradient but the settle point values are still significantly lower. In terms of settle point values, chicken fat scored highest above the two as more evident from Figure 2. Melting points of fat vary depending on which part of the animal body it belongs to. This study did not take that into account. However, the higher settle point values of chicken fat can be explained by the fact that chicken fat melting points are very close to that of lard. While as lard fat melting points can vary from 25°C to 45°C, chicken fat melting points range only between 27°C to 35°C. Hence, less variance in the melting point could be a reason for higher settle point values overall.

Table 2 - List of classifiers applied.

Classifier Name	**Accuracy (%)**
KNN	95.6
SVM	95.6
Bagged Trees	93.3
Simple Tree	91.1

A study by Che Man [23] finds a differentiation regression model between palm oil and lard successfully. Another work by Man Che [24] could classify lard from lamb fat. Their work was only capable to classify a mixture of lard and lamb body fat up to 30% adulteration.

In this study, a classification between the three types of fats was also done. Table 2 lists the various classifiers applied with their respective accuracies in descending order.

Higher accuracies of KNN and SVM are expected because the scatter plot in Figure 2 showed clear and separate clusters of each classes overall. Furthermore, Figure 4 shows the confusion matrix of KNN and SVM. Both classifiers have the same confusion matrix results. Interestingly, the classes "L40" and "C40" are confused among each other. However, chicken is being classified as lard 3 out of 4 wrong predictions and lard as chicken as once. This can still be considered in our favour as we are more concerned about lard than chicken.

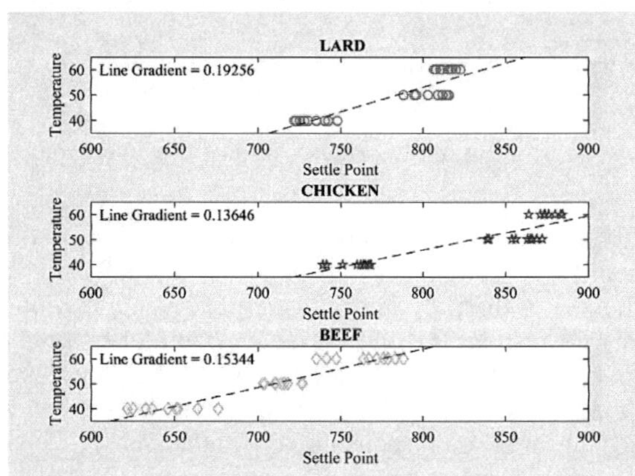

Figure 3 - Response towards change in temperature

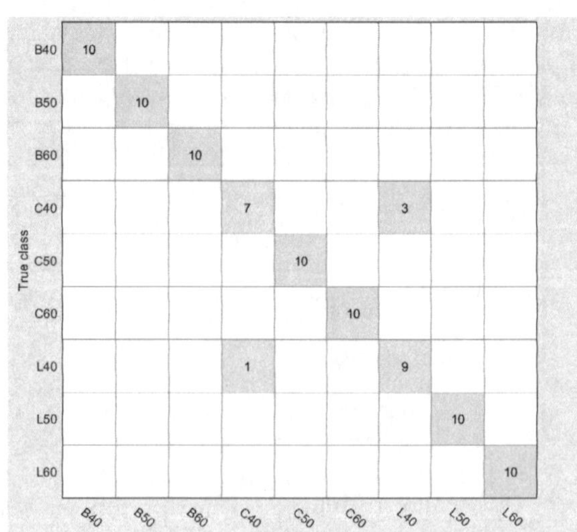

Figure 4 - Confusion matrix of KNN and SVM

Higher accuracies of KNN and SVM are expected because the scatter plot in Figure 2 showed clear and separate clusters of each classes overall. Furthermore, Figure 4 shows the confusion matrix of KNN and SVM. Both classifiers have the same confusion matrix results. Interestingly, the classes "L40" and "C40" are confused among each other. However, chicken is being classified as lard 3 out of 4 wrong predictions and lard as chicken as once. This can still be considered in our favour as we are more concerned about lard than chicken.

5. Conclusion

The findings show that the E-Nose could successfully identify and differentiate lard among three different types of fats. The required sample preparation is simple. There is no need for additional chemical composition for lard detection. The required temperature for the experiment (40°C to 60°C) is easy to be achieved and the total testing time for each reading is 60 seconds only. The experiment, i.e. the detection does not need to be conducted in laboratory. The developed E-Nose is portable, light and cost effective. There were no errors in the conducted experiments which shows the device is reliable.

The direction of future work will be to conduct experiments on fats collected from different parts of the corresponding animal body so as to tune the classifiers further to accommodate that variable. Also, the range of temperature has to be increased, and the focus has to be to reduce the testing time further to around 30 seconds.

Nevertheless, so far, our developed E-Nose has been promising and has shown potential for many applications.

References

[1] Rohman, A., Erwanto, Y., & Man, Y. B. C. (2011). "Analysis of pork adulteration in beef meatball using Fourier transform infrared (FTIR) spectroscopy". Meat Science, 88(1), 91-95.

[2] Syahariza, Z., Che Man, Y. B., Selamat, J., & Bakar, J. (2005). "Detection of lard adulteration in cake formulation by Fourier transform infrared (FTIR) spectroscopy." Food Chemistry, 92(2), 365–371.

[3] Che Man, Y. B., Syahariza, Z.A, Mirghani, M.E.S., Jinap, S. & Bakar., J. (2005). "Analysis of potential lard adulteration in chocolate and chocolate products using Fourier transform infrared spectroscopy." Food Chemistry, 90(4), 815–819.

[4] Rohman, A., & Che Man, Y. B. (2012). "Quantification and Classification of Corn and Sunflower Oils as Adulterants in Olive Oil Using Chemometrics and FTIR Spectra." The Scientific World Journal, 2012-2017.

[5] Mansor, T.S.T & Che Man, Y.B & Rohman, A. (2011). "Application of Fast Gas Chromatography and Fourier Transform Infrared Spectroscopy for Analysis of Lard Adulteration in Virgin Coconut Oil." Food Anal. Methods, 4, 365-372.

[6] Amari, A., El Bari, N., & Bouchikhi, B. (2009). "Conception and development of a portable electronic nose system for classification of raw milk using principal component analysis approach." Sensors & Transducers, 102(3), 33.

[7] Wang, B., Xu, S. & Sun, D-W. (2010). "Application on the Electronic Nose to the Identification on Different Milk Flavorings." Food Research and International, 43(1), 25.

[8] Llobet, E., Bari, N.E., Correig, X., & Bouchikhi, B. (2008). "Electronic Nose based on Metal Oxide Semiconductor as an Alternative Technique for the Spoilage Classification of Red Meat." Sensors, 8(1), 142-156.

[9] Balasubramanian, S., Panigrahi, S., Logue, C. M., Gu, H., & Marchello, M. (2009)." Neural networks-integrated metal oxide-based artificial olfactory system for meat spoilage identification." Journal of Food Engineering, 91(1), 91–98.

[10] Tudu, B., Metla, A., Das, B., Bhattacharyya, N., Jana, A., Ghosh, D. et al. (2009)." Towards Versatile Electronic Nose Pattern Classifier for Black Tea Quality Evaluation: An Incremental Fuzzy Approach." IEEE Transactions on Instrumentation and Measurement, 58(9), 3069-3078.

[11] Panigrahi, S., Balasubramanian, S., Gu, H., Logue, C., & Marchello, M. (2006). "Neural-network-integrated electronic nose system for identification of spoiled beef. LWT-Food" Science and Technology, 39(2), 135-145.

[12] Perera, A., Pardo, A., Barrettino, D., Hierlermann, A. & Marco, S. (2010). "Evaluation of fish Spoilage by means of a single Metal Oxide Sensor under Temperature Modulation." Sensors and Actuators B: Chemical, 146(2), 477-482.

[13] Nurjuliana, M., Che Man, Y. B., & Hashim, D. M. (2011). "Analysis of Lard's Aroma by an Electronic Nose for Rapid Halal Authentication." Journal of the American Oil Chemists' Society, 88(8), 75–82.

[14] Knobloch, H., Tuner, C., Spooner, A. & Chambers, M. (2009). "Methodological variation in headspace analysis of liquid samples using electronic nose." Sensors and Actuators B: Chemical, 139(2), 353-360.

[15] Asif, M. (2011). "General Chemistry, Composition, Identification and Qualitative Tests of Fats or Oils." Journal of Pharmaceutical Research and Opinion, 1(2), 52–64.

[16] Tian, X., Wang, J., & Cui, S. (2013). "Analysis of pork adulteration in minced mutton using electronic nose of metal oxide sensors." Journal of Food Engineering, 119(4), 744–749.

[17] Barbri, N. El, Llobet, E., Bari, N. El, Correig, X., & Bouchikhi, B. (2008). "Electronic nose based on metal oxide semiconductor sensors as an alternative technique for the spoilage classification of red meat." Sensors, 142–156.

[18] Xu, K., Zeng, D., Tian, S., Zhang, S., & Xie, C. (2014). "Hierarchical porous SnO2 micro-rods topologically transferred from tin oxalate for fast response sensors to trace formaldehyde." Sensors and Actuators B: Chemical, 190, 585–592.

[19] Kohl, D., Heinert, L., Bock, J., Hofmann, T., & Schieberle, P. (2000). "Systematic studies on responses of metal-oxide sensor surfaces to straight chain alkanes, alcohols, aldehydes, ketones, acids and esters using the SOMMSA approach." Sensors and Actuators B: Chemical, 70(1-3), 43–50.

[20] Baietto, M., Pozzi, L., Wilson, A. D., & Bassi, D. (2013). "Evaluation of a portable MOS electronic nose to detect root rots in shade tree species." Computers and Electronics in Agriculture, 96, 117–125.

[21] Hao, H. C., Tang, K. T., Ku, P. H., Chao, J. S., Li, C. H., Yang, C. M., & Yao, D. J. (2010). "Development Of A Portable Electronic Nose Based On Chemical Surface Acoustic Wave Array With Multiplexed Oscillator And Readout Electronics." Sensors and Actuators B: Chemical, 146(2), 545–553.

[22] Abdullah, A. H., Adom, A. H., Shakaff, A. Y. M., Ahmad, A., Zakaria, F.S.A., Saad, C.M.N.C., et al. (2012). "Hand-Held Electronic Nose Sensor Selection System for Basal Stamp Rot (BSR) Disease Detection." Proceedings of 2012 Third International Conference on Intelligent Systems Modelling and Simulation. Kota Kinabalu, Malaysia, 8-10 February, 2012.

[23] Che Man, Y. B., Marina, A. M., Rohman, A., Al-Kahtani, H. A., & Norazura, O. (2014). "A Fourier Transform Infrared Spectroscopy Method for Analysis of Palm Oil Adulterated with Lard in Pre-Fried French fries." International Journal of Food Properties, 17(2), 354-362.

[24] Man, Y. C., & Mirghani, M. E. S. (2001). "Detection of lard mixed with body fats of chicken, lamb, and cow by fourier transform infrared spectroscopy." Journal of the American Oil Chemists' Society, 78(7), 753-761.